CONSTRUCTION CONTRACTING

Couch and Howard, P.C.

PUBLIC PROJECT - UNITED STATES GOVERNMENT

Contractor must furnish payment bond.
IF YOU ARE NOT PAID within 90 days after your last labor performed or material supplied you may sue the Bonding Company on the bond.

BUT a supplier to subcontractor must have given written notice to prime contractor within 90 days of last material and labor, by registered mail, stating amount due, before he can sue the Bonding Company.

NO LIENS MAY BE FILED - This is ONLY recourse other than Contractor by whom employed.

PUBLIC PROJECT - NEW YORK STATE OR SUBDIVISION

Contractor must furnish payment bond.
If not paid within 90 days of last labor may sue the Bonding Company on the bond. Must sue within one (1) year.

BUT supplier to subcontractor must have given written notice to prime contractor within 90 days of last material and labor, by registered mail, stating amount due, the name of the party to whom supplied.

LIEN - Within 30 days of completion and acceptance of Project. Expires in six (6) months - must be renewed or foreclosed or lost.

PRIVATE PROJECT - NEW YORK STATE

CONSTRUCTION CONTRACTING

Fourth Edition

Richard H. Clough

DEPARTMENT OF CIVIL ENGINEERING
THE UNIVERSITY OF NEW MEXICO
ALBUQUERQUE, NEW MEXICO

A WILEY-INTERSCIENCE PUBLICATION

JOHN WILEY & SONS

New York   Chichester   Brisbane   Toronto   Singapore

Construction Contracting, fourth edition, has been prepared to present a comprehensive review of the business of construction contracting as conditions exist at the time of publication. The reader is reminded that major changes in the information presented can occur at any time. The material contained herein is designed only to inform the reader and provide background information, and is not intended to serve as business, financial, technical, legal, or accounting advice.

The construction industry is replete with words such as manpower, craftsman, foreman, workmen's compensation, manhours, and others of masculine gender. These have been used throughout this book because they are an established and standard part of the construction vocabulary. Additionally, the English language has no neuter personal pronoun. Where masculine pronouns are used, they are intended to apply equally to men and women.

Library of Congress Cataloging in Publication Data:

Clough, Richard Hudson.
    Construction contracting.

    "A Wiley-Interscience publication."
    Includes index.
    1. Building—Contracts and specifications.
I. Title.

TH425.C55   1981   692'.8   81-7449
ISBN 0-471-08657-6   AACR2

Printed in the United States of America

*To my daughter, Janet, and
my grandson, Steven*

PREFACE

The first three editions of *Construction Contracting* were directed toward providing interested readers with current and authentic information about construction contracting and the construction industry. During these past years, there has been a tremendous upsurge of academic interest in the business and management aspects of construction. Educational programs at many levels have been established to prepare people for careers in this vital industry. Contractors, architects, consulting engineers, public agencies, and other employers associated with the construction industry now recognize the advantages to be gained by broadening the knowledge of their management and supervisory personnel about the industry of which they are a part. Indeed, there now exists a greater demand than ever for a concise, authoritative, and readily available text that is suitable for both study and reference. This fourth edition is a serious attempt to fill this need.

The extensive acceptance and use of the earlier editions of *Construction Contracting* have been most gratifying and indicate that the book serves a continuing need. However, change is perpetual, and any book about the construction industry must be updated at regular intervals if it is to best serve its readers. To this effect the fourth edition has been prepared.

The text of this edition has been largely rewritten and the figures and appendixes have been extensively reworked to reflect current practices and procedures. The fourth edition presents a considerably greater breadth and depth in the coverage of subject matter. Completely rewritten discussions of prehire labor contracts, statutes of limitations, OSHA, equipment accounting, ERISA, architect-engineer liability, withdrawal of bids, subcontractor agreements, and comprehensive general liability insurance are just a few examples. The chapter on project planning and scheduling has been completely reworked, changing the network convention from arrow to precedence notation. A new chapter on project management has been added, discussing in detail those management procedures that are applicable to the field construction process.

I wish to express my deep appreciation to the many people who have provided me with data and information. It is not possible to thank individually all those who contributed in some way to the preparation of this fourth edition. However, it must be reported that many people of assorted callings cheerfully contributed of their time and talents to make this fourth edition as up-to-date and accurate as possible. Special credit must go to my wife, Jean, for her skilled editing and typing of the manuscript.

RICHARD H. CLOUGH

Albuquerque, New Mexico
July 1981

CONTENTS

THE CONSTRUCTION INDUSTRY

1.1 THE CONSTRUCTION PROJECT

Humans are compulsive builders who, throughout the ages, have demonstrated a remarkable and continually improving talent for construction. As knowledge and experience have been gained, humanity's ability to build structures of increasing size and complexity has increased enormously. In this modern world, everyday life is maintained and enhanced by an impressive array of construction, awesome in its diversity of form and function. Buildings, tunnels, highways, pipelines, dams, docks, canals, bridges, airfields, and a myriad of other structures are created to provide us with the goods and services we require. As long as there are people on earth, structures will be built to serve them.

Construction projects are complex and time-consuming undertakings. The structure must be designed in accordance with applicable codes and standards, culminating in working drawings and specifications that describe the work in sufficient detail for its accomplishment in the field. The building of a structure of even modest proportions involves many skills, materials, and literally hundreds of different operations. The assembly process must follow a more or less natural order of events that, in total combination, constitutes a complicated pattern of individual time requirements and restrictive sequential relationships among the various segments of the structure.

To some degree, each construction project is unique, and no two jobs are ever quite alike. In its specifics, each structure is tailored to suit its environment, arranged to perform its own particular function, and designed to reflect personal tastes and preferences. The vagaries of the construction site and the infinite possibilities for creative and utilitarian variation of even the most standardized building product combine to make each construction project a new and different experience. The contractor sets up a "factory" on the site and, to a large extent, custom-builds each job.

The construction process is subject to the influence of highly variable and often unpredictable factors. The construction team, which includes various combinations of contractors, owners, architects, engineers, workers, sureties, lending agencies, governmental bodies, insurance companies, material dealers, and others, changes from one job to the next. All of the complexities inherent to different construction sites, such as subsoil conditions, surface topography, weather, transportation, material supply, utilities and services, local subcontractors, and labor conditions, are an innate part of the construction project.

As a consequence of the circumstances just discussed, construction projects are typified by their complexity and diversity, and by the nonstandardized nature of their production. Despite the use of prefabricated units in certain applications, it seems unlikely that field construction will ever adapt itself to the standardized methods and product uniformity of assembly-line production.

1.2 ECONOMIC IMPORTANCE

For the past several years, construction has been the largest single production industry of the American economy. It is not surprising, therefore, that the construction industry has great influence on the state of this nation's economic health. In fact, construction is commonly regarded to be the bellwether industry of the country. Times of prosperity are accompanied by a high national level of construction expenditure. During periods of recession, construction is depressed and the building of publicly financed projects is one of the first governmental actions taken to stimulate the general economy. A high level of construction activity and periods of national prosperity are simultaneous phenomena; each is a natural result of the other.

Some facts and figures pertaining to construction are useful in gaining an insight into the tremendous dimensions of this vital industry. The total annual volume of construction in this country is measured in terms of hundreds of billions of dollars. Construction expenditures are commonly divided into new construction and into remodeling, maintenance, and repairs. During an average year, new construction accounts for about 80 percent, with remodeling and repairs accounting for the remaining 20 percent of the total.

The annual expenditure for new construction normally accounts for about 10–11 percent of the dollar value of our gross national product with total construction usually amounting to 12–13 percent. Thus about one of every eight dollars spent for goods and services in this country is a construction dollar. With regard to new construction, approximately two-thirds of it is privately financed, and one-third is paid for by various public agencies. About one-third of the public projects is financed by the federal government. The U.S. Department of Labor indicates that construction contractors employ approximately six million different persons at various times during a typical year, this providing approximately four million full-time jobs. If the production, transportation, and distribution of construction materials and equipment are taken into account, construction creates directly or indirectly about 15 percent of the total gainful employment in the United States.

1.3 THE OWNER

The owner, whether public or private, is the instigating party that owns and finances the project, either from the owner's own resources or from some source of external financing. Public owners are public bodies of some kind ranging from agencies of the federal government down through state, county, and municipal entities to a multiplicity of local boards, commissions, and authorities. Public projects are paid for by appropriations, bonds, tax levies, or other forms of financing and are built to meet some defined public need. Public owners must proceed in accordance with applicable statutes and administrative directives pertaining to the advertising for bids, bidding procedures, construction contracts, contract administration, and other matters relating to the design and construction process.

Private owners may be individuals, partnerships, corporations, or various combinations thereof. Most private owners have the structure built for their own use: business, habitation, or otherwise. However, many private owners do not intend to become the end users. The completed structure is to be sold, leased, or rented to others.

1.4 THE ARCHITECT-ENGINEER

The architect-engineer, also known as the design professional, is the party or firm that designs the project. Because such design is architectural or engineering in nature, or often a combination of both, the term "architect-engineer" is used in this book to refer to the design professional, regardless of the applicable specialty or the relationship between the designer and the owner.

The architect-engineer can occupy a variety of positions with respect to the owner for whom the design is done. Many public agencies and large corporate owners maintain their own in-house design capability. In such instances, the architect-engineer is a functional part of the owner's organization. The traditional and most common arrangement is where the architect-engineer is a private and independent design firm that accomplishes the project design under contract with the owner. In other instances, the owner contracts with a single party for both design and construction services, this being referred to as design–construct. In such a case, the architect-engineer is a branch of or is affiliated in some way with the construction contractor.

1.5 THE PRIME CONTRACTOR

The prime contractor, also known as the general contractor, is the business firm that is in contract with the owner for the construction of the project, either in its entirety or for some specialized portion thereof. The prime contractor is that party who brings together all of the diverse elements and inputs of the construction process into a single, coordinated effort.

The essential function of the prime contractor is close management control of the construction process. Ordinarily, the contractor is in complete and sole charge of its field operations, including the procurement and provision of necessary construction materials and equipment. The chief contribution of the contractor to the construction process is the ability to marshal and allocate the resources of manpower, equipment, and materials to the project to achieve completion at maximum efficiency of time and cost. A construction project presents the contractor with many difficult management problems. The skill with which these problems are met determines, in large measure, how favorably the contractor's efforts serve its own interests as well as those of the project owner.

1.6 THE SUBCONTRACTOR

Prime contractors may subcontract portions of their work, the type and amount depending on the nature of the project and their own organization. There are instances where the job is entirely subcontracted, the general contractor providing only supervision, job coordination, and perhaps general site services. At the other end of the spectrum, there are those projects where the general contractor does no subcontracting, choosing to do the entire work with its own forces. In the usual case, however, the prime contractor will perform the basic project operations and will subcontract the remainder to various specialty contractors, although subcontracting is applied much more extensively on building construction than it is on engineering and industrial construction.

When the prime contractor engages a specialty firm to execute a particular portion of the overall construction program, the two parties enter into an agreement called a subcontract. No contractual relationship is established thereby between the owner and the subcontractor. The prime contractor, by the terms of its contract with the owner, assumes complete responsibility for the direction and control of the entire construction program. An important part of this responsibility is coordinating and supervising the work of the subcontractors. When the general contractor subcontracts a portion of the work, this contractor remains completely responsible to the owner for the total project and is liable to the owner for any negligent performance of the subcontractor.

Everyday economic facts have confirmed the subcontract system to be efficient and economical in the use of available resources. The operations of the average general contractor are not always sufficiently extensive to afford full-time employment of skilled journeymen in each of the several trade classifications needed in the field. Hence these contractors are able to keep only a limited nucleus of full-time employees and hire from the local pool of skilled labor as additional needs arise. By subcontracting, the prime contractor can obtain workmen with the requisite skills when they are needed, without the necessity of maintaining an unwieldy and inefficient full-time labor force. By the same token, subcontractors are able to provide substantially full-time employment for their workers, thereby affording an opportunity for the acquisition and retention of the most highly skilled and productive journeymen.

Another common reason for subcontracting is when the project requires construction equipment the prime contractor does not have. In many cases it is economically preferable to subcontract that part of the project rather than attempt to acquire the necessary equipment. Qualified subcontractors are usually able to perform their work specialty more quickly and at lesser cost than can the general contractor. In addition, many construction specialties have specific licensing, bonding, and insurance requirements.

Public construction contracts frequently limit the proportion of the total construction the prime contractor is allowed to subcontract. For example, the Public Buildings Service of the General Services Administration, the U.S. Army Corps of Engineers, and other federal agencies have set such limitations. Also, some states have established statutory restrictions on the subcontracting of public works in those states. Such limitations on construction subcontracting are intended to circumvent problems potentially associated with extensive subcontracting. This condition can seriously complicate the overall scheduling of job operations, lead to a serious division of project authority, fragmentize responsibility, make the coordination of construction activities difficult, weaken communication between management and the field, foster jurisdictional disputes between construction unions, and be generally detrimental to job efficiency. Obviously, the extent to which these difficulties may actually occur is greatly dependent on the experience, organization, and management skill of the prime contractor involved.

1.7 CONSTRUCTION CATEGORIES

The field of construction is as diversified as the uses and forms of the many types of structures it produces. However, construction is commonly divided into four main categories, although there is some overlap among these divisions and certain projects do not fit neatly into any one of them.

In general, contracting firms specialize to some extent, limiting their efforts to a relatively narrow range of construction types. Specialization is usual and necessary because of the radically different equipment requirements, construction methods, trade and supervisory skills, contract provisions, and financial arrangements involved with the different construction categories. The four main divisions—housing, nonresidential building, engineering, and industrial construction—are described in the following paragraphs.

Housing Construction Housing construction includes the building of single-family homes; condominiums; multiunit townhouses; low-rise, garden-type apartments; and high-rise apartments. Design of this construction type is done by owners, architects, or the builders themselves. Construction is performed by the owner, builder-vendors, or independent contractors under contract with the owner. This category of construction is dominated by small building firms and normally accounts for about 30–35 percent of new construction during a typical year. Historically, housing construction has been characterized by instability of market demand and is strongly influenced by governmental regulation and national monetary policy. Housing is an area of construction typified by periodic high rates of contractor business failures. A large proportion of housing construction is privately financed.

Most housing construction is accomplished by the traditional step-by-step assembly of components at the construction site. Recent years, however, have seen some usage of prefabricated housing units. Such homes are built in factories as major components, these units being shipped to the site where they are assembled. In some cases, complete modular homes are factory-produced.

Nonresidential Building Construction Nonresidential building construction includes buildings in the commonly understood sense, other than housing, that are erected for institutional, educational, light industrial, commercial, social, religious, governmental, and recreational purposes. In normal business years, private capital finances most nonresidential building construction, this normally accounting for 40–50 percent of the annual total of new construction. Design of this construction category is predominantly done by architects, with engineering design services being obtained as required. Construction of this kind is generally accomplished by prime contractors or construction managers who subcontract substantial portions of the work to specialty firms.

Engineering Construction Engineering construction is a very broad category covering structures that are planned and designed by engineers. This category includes those structures whose design is concerned more with functional considerations than aesthetics and which involve field materials such as earth, rock, steel, asphalt, concrete, timbers, and piping. Most engineering construction projects are publicly financed and account for approximately 20–25 percent of the new construction market. Engineering construction utilizes large quantities of construction equipment, and projects of this type are characterized by large spreads of power shovels, tractor-scrapers, pile drivers, draglines, large cranes, heavy-duty haulers, paving plants, rock crushers, and associated equipment types. Engineering construction is commonly divided into three subgroups: highway and airfield, heavy, and utility construction.

Highway and airfield construction covers clearing, excavation, fill, aggregate production, subbase and base, paving, drainage structures, bridges, traffic signs, lighting systems, and other such items commonly associated with this type of work. Heavy construction is usually construed to include sewage- and water-treatment plants, dams, levees, pipe and pole lines, ports and harbor structures, tunnels, large bridges, reclamation and irrigation work, flood control structures, and railroads. Utility construction involves mostly municipal work such as sanitary and storm drains, curb and gutter, street paving, water lines, electrical and telephone distribution facilities, drainage structures, and pumping stations.

Industrial Construction Industrial construction includes the erection of projects that are associated with the manufacture or production of a commercial product or service. Such structures are highly technical in nature and are frequently built by large, specialized contracting firms that do both the design and field construction. While this category accounts for only 5–10 percent of the annual volume of new construction, it includes some of the largest projects built. The design responsibility for this type of construction is predominantly that of the engineer. Petroleum refineries, steel mills, chemical plants, smelters, electric power generating stations, heavy manufacturing facilities, and similar installations are examples of industrial construction. In the United States, a very large proportion of this category is privately financed.

1.8 THE CONTRACT SYSTEM

The usual mode of accomplishing construction work is where the prime contractor enters into a contract with the owner. The contract describes in detail the nature of the construction to be accomplished and the services that are to be performed. The contractor is obligated to perform the work in full accordance with the contract documents, and the owner is required to pay the contractor as agreed.

The means by which the prime contractor is selected, the form of contract used, and the scope of duties thereby assumed by the contractor can be highly variable depending on the preferences and requirements of the owner. The prime contractor may be selected on the basis of competitive bidding, the owner may negotiate a contract with a selected contractor, or perhaps a combination of the two may be used. The entire project may be included within a single general contract, or separate prime contracts for specific portions of the job may be used. The contract may include project design as well as construction, or the contractor's responsibility may be primarily managerial. Various ways in which owners can select prime contractors will now be presented. This will be followed by a discussion of different types of contracts commonly used between the two parties.

1.9 COMPETITIVE BIDDING

The owner can select a prime contractor to construct the project on the basis of competitive bidding, negotiation, or some combination of the two. A large proportion of the construction work in this country is done by contractors who obtain their work in bidding competition with other contractors. In such cases, the work is normally awarded to the lowest responsible bidder. The competitive bidding of public construction projects is normally required by law and is standard procedure for public agencies. Essentially, all public works are done by this method.

The contracting firm that obtains its work by competitive bidding occupies an essential position in the construction industry. Equipment and know-how are its stock in trade, and it acts as a broker for the necessary construction materials and labor. It is often said that competitive bidding for construction work is the biggest gamble of the business world. When bidding a project, the contractor must estimate how much the structure will cost while it still exists only on paper. To this cost is added what seems to be a reasonable profit and the contractor guarantees to do the project for the quoted price. If its bid is selected by the owner, the contractor must complete the project for the contracted amount. If the actual costs of construction should exceed this figure, the resulting loss is borne by the contractor.

In the construction industry, competitive bidding is traditional and is still widely used. Competitive bidding is used to encourage efficiency and innovation by contractors, thereby providing the owner with a constructed project of specified quality at the lowest possible price. This mode of contractor selection has served its purpose well but, like other contracting procedures, does have its weaknesses. To illustrate, it must be realized that the bidding process places the prime contractor and the owner in adversary positions, a fact that can lead to undesirable side effects. Evidence can be offered to show that competitive bidding can and sometimes does lead to the selection of incompetent contractors, excessive claims by the contractor against the owner, disputes and litigation between the two parties, bid shopping, and other problems.

There are actually two different types of competitive bidding used in the United States, "open" and "closed." The predominant form is open bidding where all contractors use the same proposal form provided with the bidding documents, with the bids being opened and read publicly. In each case the proposal amount is the contractor's final offer, and there is no subsequent bargaining or negotiation. In closed bidding, sometimes used by private owners, there is no prescribed proposal form, and there is no public opening. The competing contractors are required to submit their qualifications along with their bids and are encouraged to tender suggestions as to how the cost of the work might be reduced. The owner selects one of the proposals submitted or perhaps interviews those contractors whose proposals appear most advantageous and negotiates a contract with one of them.

It is interesting to note that the American preoccupation with the low bidder normally being the successful bidder is not shared by many other parts of the world. Some very good arguments can be advanced to show that the lowest bid is not necessarily the best price for the owner, and there are some interesting variations used in other countries. For example, in one European country the work is awarded to the bidder whose bid is nearest to the average of all bids received. The bid which is more than but nearest to the average of all bids received, but is still below the owner's estimate, wins the contract in an Asian nation. In another European country, the successful bidder is the one nearest to the average after the two highest and the two lowest bids have been rejected.

Competitive bidding can also be used where the successful contractor is determined on a basis other than the estimated cost of the construction. When the contract will involve the payment of a fee to the contractor, the amount of the fee is sometimes used as a basis of competition among contractors. To illustrate, construction management services are sometimes obtained by an owner using the fees proposed by the different bidders as one basis for the contract award.

1.10 THE NEGOTIATED CONTRACT

As has been pointed out previously, competitive bidding can have drawbacks associated with it and is not necessarily the optimum way of selecting a contractor to perform all classes of construction work. There are many instances where it can be advantageous for an owner to negotiate a contract for a project with a preselected contractor or small group of contractors. It is common practice for an owner to forgo the competitive bidding process entirely and to handpick a contractor on the basis of reputation and overall qualifications to do the job. The forms of negotiated contracts are almost limitless because such an agreement can include any provisions mutually agreeable to both parties and that are best suited to the particular work involved. Most negotiated contracts are of the cost-plus-fee type, a subject more fully discussed in Chapter 6. Negotiated contracts are normally limited to privately financed work because competitive bidding is a legal requirement for most public projects except under extraordinary or unusual circumstances.

Extensive use of negotiated contracts has been traditional in some areas of the construction industry such as housing and industrial construction. However, recent years have seen increasing application of negotiated contracts across the board in the private sector. This can only be interpreted as a sign that owners are increasingly finding that such arrangements are in their best interests. A large proportion of the annual work volume of many contractors is now made up of negotiated contracts.

1.11   LINEAR CONSTRUCTION

Traditional in the construction industry, and still widely used, linear construction refers to the procedure of design, bidding or negotiation, and construction following one another in single file order. Also referred to as design-then-construct, this process is one where each step is finished before the next one begins. From this, it can be seen that the linear procedure is not always an efficient one with regard to the total time required for the design and construction of a project. While the single file completion of one step before initiating the next is acceptable to some owners in today's market, it is not acceptable to others. A number of financial considerations dictate the earliest possible completion date for many construction projects. It is possible to reduce the total design-bidding-construction time required for some projects by releasing successive phases of the structure for bidding and construction as the design proceeds.

1.12   PHASED CONSTRUCTION

Phased construction, also known as fast-tracking, refers to the overlapping accomplishment of project design and construction. As the design of successive phases of the work is finalized, these work packages are put under contract, either with a single general contractor or to a series of different prime contractors. In the first case, each successive work package is done by the general contracting firm or awarded by it to a subcontractor. In the second instance, each phase is put out for bids and awarded to a series of different prime contractors. In either instance, early phases of the project are under construction while later stages are still on the drawing boards. This procedure of telescoping design and construction can appreciably reduce the total time required to achieve project completion. For obvious reasons, phased construction can offer some attractive advantages to an owner.

1.13 CONSTRUCTION CONTRACT TYPES

A myriad of contract forms and types are available to owners for accomplishing their contruction needs. All call for defined services to be provided under contract to the owner. The scope and nature of such services can be made to include almost anything the owner wishes. The selection of the proper contract form appropriate to the situation is an important decision for the owner and is deserving of careful consideration and consultation. In this regard, public owners must work within the strictures that applicable law places on them.

The construction contract can be made to include construction, design–construct, or construction management services. These concepts are discussed in the following sections.

1.14 CONSTRUCTION SERVICES ONLY

A large percentage of construction contracts provide that the general contractor has responsibility to the owner only for the accomplishment of the field construction. Under such an arrangement, the contractor is completely removed from the design process and has no input into it. Its obligation to the owner is limited to constructing the project in full accordance with the contract terms.

Where the contractor provides construction services only, the owner may have an in-house design capability. The more common arrangement, however, is for a private architect-engineer firm to perform the design in contract with the owner. Under this latter arrangement, the design professional acts essentially as an independent contractor during the design phase and as an agent of the owner during construction operations. The architect-engineer acts as a professional intermediary between the owner and contractor and often represents the owner in matters of construction contract administration. Under such contractual arrangements, the owner, architect-engineer, and contractor play narrowly defined roles, each performing a particular function semi-independently of the others.

1.15 THE SINGLE PRIME CONTRACT

Under the single contract system, the owner awards construction of the entire project to a single prime contractor. The distinctive function of the prime contractor is to coordinate and direct the activities of the various parties and agencies involved with the construction and to assume full, centralized responsibility to the owner for the delivery of the finished project within the specified time. Customarily, the prime contractor will construct certain portions of the job with its own forces and subcontract the remainder to various specialty contractors. The general contractor accepts complete accountability, not only to build the project according to the contract documents but also to ensure that all costs associated with the construction are paid.

The general contractor is the only party having a contractual relationship with the owner and is fully responsible for the performance of the subcontractors and other third parties to the construction contract. It is worthy of note that when a prime contractor sublets a portion of the work to an independent specialty contractor, the general contractor has a nondelegable duty to the owner for the proper performance of the entire work and remains liable under the contract with the owner for any negligent or faulty performance, including that of the subcontractors. When the work is not done in accordance with the contract, the general contractor is in breach of contract and is liable for damages to the owner, regardless of whether the faulty work was performed by the contractor's own forces or by those of a subcontractor.

1.16 SEPARATE PRIME CONTRACTS

When separate prime contracts are used, the project is not constructed under the terms of a contract between the owner and a single prime contractor who exercises centralized control over the entire construction process. Rather, the project is constructed by several prime contractors, each being in contract with the owner and each being responsible for a designated portion of the project. For example, the owner of a building to be constructed could decide to award separate contracts for general construction; plumbing; heating, ventilating, and air conditioning; and electrical work. Each of these four contractors functions as a prime contractor. Each is in contract with the owner, and each performs independently of the others. Subcontracting by any or all of these four prime contractors is still possible, of course.

Separate contracts are now widely used to satisfy a number of owner needs and requirements. Subdividing the work into stages or specialty areas can give the owner considerable flexibility in letting the work out to contract. The use of separate contracts, by breaking the work up into smaller packages, can also result in contracts of shorter duration, the obtaining of highly skilled specialty contractors, and better prices through increased competition. One of the principal advantages for the owner, stemming from the use of separate contracts, is the saving of the markup on work that would otherwise be subcontracted. If the entire work is awarded as a single contract, the general contractor includes in its price a markup on all subcontracted work as a fee for management and coordination of the subcontractors. Separate contracts can reduce this or eliminate it altogether.

The use of separate contracts can be very troublesome, however, unless carefully prepared contract documents are used and the work is rigorously planned, scheduled, coordinated, and controlled. If strong, centralized management control is not exerted over the separate prime contractors, the several possible advantages of separate contracts will likely turn out to be illusory. The owner, architect-engineer, or a construction manager may assume this responsibility. Another approach has been to have all the prime contracts but one include an assignment clause that establishes the remaining prime contractor as the lead contractor. This party then has project coordination and interface responsibility. In any event, if the party assuming resonsibility for project supervision does not possess the experience and skill required to perform the demanding and specialized management functions needed, separate contracts may become a troublesome procedure indeed. Of particular importance is job delay caused by the inadequate performance of one of the prime contractors. In a separate contract system, allocation of liability for project delay can become extremely complex, and responsibility may devolve to that party assuming overall management control.

A few states have enacted statutes requiring that designated separate contracts be awarded on state-financed public works. Many other public jurisdictions make the use of separate contracts optional under administrative authority. The application of separate contracts has proven to be very troublesome at times for some public agencies, causing contract-letting difficulties, litigation, and extra construction expense. Experience would seem to indicate that public agencies might advantageously be left free to select a contract system appropriate to project circumstances rather than be restricted by law to a specific contracting procedure.

1.17 DESIGN–CONSTRUCT

When the owner contracts with a single firm for both design and construction, this is referred to as a design–construct project. The term "turnkey" is sometimes used to refer to a special case of design–build where the contractor performs a complete construction service for the owner. This would be the case where the contractor obtains project financing, procures the land, designs and constructs the project, and hands it over to the owner ready for occupancy. The design–construct concept has been enjoying increasing acceptance during recent years, and all evidence indicates that it will continue to increase in popularity. Its increasing use by owners is largely due to economies of cost and time that can be realized by melding the two functions of design and construction. Injecting contractor expertise and advice into the design process offers the possibility of achieving both time and cost savings for the owner. Because phased construction can be utilized under a design–construct contract, the owner will likely have beneficial use of his structure well before he would under the more traditional linear construction arrangement.

Design–construct utilizes different types of contracts, depending on the construction involved. Housing and standard or prefabricated buildings usually use a negotiated lump-sum contract. Design–construct is extensively used with industrial construction, usually utilizing negotiated contracts with some form of cost-plus-fee arrangement and a guaranteed maximum cost. There are times when two separate contracts are drawn up between the owner and contractor–designer, with the design contract being on a cost-plus arrangement and the construction contract written on a lump-sum basis. The basic feature of the design–construct method is that it entails single responsibility for the entire project. Actually, it is an expansion of the single contract system to include design as well as construction. The advantages of competition are still realized because, in the usual case, much of the actual construction is subcontracted on the basis of competitive bidding.

A key aspect of design–construct is the team concept. The owner and the designer–builder work closely together in the planning, design, cost control, scheduling, site investigation, and possibly even land acquisition and project financing. Without doubt, this contract mode poses a threat to the traditional position of private architectural and consulting engineer firms. Professional architect and engineer societies maintain that having the designer and builder on the same team can imperil the customary fiduciary relationship between the design professional and the owner. It is also argued that failure to preserve the independence of the designer to operate on the owner's behalf can sacrifice objectivity in the design process. It is a fact, nonetheless, that design-construct is now widely and successfully used.

1.18   CONSTRUCTION MANAGEMENT

There is no universally accepted definition for construction management and, unfortunately, it is a term that means different things to different people. In what might be considered as the typical arrangement, the owner contracts with an architect-engineer for design services and with a construction manager (commonly designated as CM) for specified professional services. By terms of the two contracts, a construction team is created consisting of the owner, the architect-engineer, and the CM. At times, a funding agency may also be a member of this team. The objective of this approach is to treat project planning, design, and construction as integrated tasks within a construction system. The team, by working together from project inception to project completion, attempts to serve the owner's interests in optimum fashion. By striking a balance between construction cost, project quality, and completion schedule, the team strives to produce a project of maximum value to the owner within the most economic time frame. Adherence to the established time schedule and cost budget is a prime responsibility of the CM.

The CM participates in and contributes to all aspects of project planning. During design, this party advises and makes recommendations concerning construction alternatives, furnishing information concerning cost, labor and material availability, and time requirements for each. A budget estimate is prepared by the CM, often resulting in a guaranteed maximum price for the project. The CM advises on and coordinates the procurement of materials and equipment items required.

On most construction management projects, phased construction is applied. As the design proceeds, the CM takes competitive bids on the various divisions of the work. Eventually, several different trade contractors will be engaged to accomplish the work in a collective fashion, with the CM performing the coordinative and management functions. In what may be called a pure construction management relationship, the CM acts as the owner's agent by providing managerial services to the project. Under this arrangement, the CM does not contract with the various trades. On the other hand, the CM may guarantee price and schedule and contract directly with the trades, thus functioning in many respects as a prime contractor. In between these two, the CM may carry out any combination of responsibilities usually performed by the prime contractor or architect-engineer. As a general rule, the CM does not carry out any of the construction operations with its own forces, except for general conditions construction (see Section 1.21) and project supervision. This party selects and oversees the various trade contractors, perhaps provides certain basic items of construction equipment, and coordinates the entire construction process. During field construction, the CM typically is responsible for job inspection, safety, labor relations, payment to trade contractors and material suppliers, shop drawings, expediting, field time and cost control, and general administration of the construction contracts.

Selection of the CM by the owner is sometimes accomplished by competitive bidding using both fee and qualifications as bases for contract award. In the usual instance, however, the construction management contract is considered to be a professional services contract and is negotiated after extensive prequalification procedures. These contracts usually provide for a fixed fee plus full reimbursement of field costs.

At the present writing, construction management is extensively utilized in this country. Although there is no obvious limitation to the procedure, its use up to the present time has been generally restricted to large and/or complex construction projects.

1.19 THE BUILDER-VENDOR

There are many instances in which owners choose to act as their own general contractors for the purpose of constructing projects for subsequent sale or lease to other parties. Single-family dwellings, apartment houses, office buildings, shopping centers, and warehousing facilities are common examples of such construction. In tract housing, the builder-vendor acquires land and builds housing for sale to the general public. This is a form of speculative construction in which the owners act as their own prime contractors, build dwelling units on their own accounts, and employ sales forces to market their products. In much of such construction, the builder-vendor builds for an unknown owner. In commercial applications, however, construction will not normally proceed until suitable sale or lease agreements have been made. This is usually necessary for the developer to arrange the financing as well as to build to the buyer's or lessee's specifications.

Most builder-vendors function more as construction brokers than contractors per se, choosing to subcontract all or most of the actual construction work. The usual construction contract between owner and prime contractor is not present in such cases, the developer occupying both roles. The source of business for the builder-vendor is entirely self-generated, as opposed to the professional contractor who obtains work in the open construction marketplace.

1.20 CONSTRUCTION BY FORCE ACCOUNT

Force account is another instance where the owner acts as the prime contractor, except in this case the project is constructed for the use of the owner and not for sale or lease. Under this method, owner-builders may choose to do the work with their own forces, providing the necessary field supervision, materials, construction equipment, and labor. On the other hand, owners may choose to have the work performed entirely by subcontract, subletting individual segments of the work to specialty contractors. In such a case, owners assume the responsibility of coordinating and directing the work of the subcontractors.

Many studies have been made, mostly by public agencies, to compare the costs of construction accomplished by competitive bidding and by force account. These studies have clearly demonstrated that for all but very small projects, the force-account procedure is more expensive than the competitive-bid method. In addition, the quality of the work is usually better when done by a professional contractor.

There are several good reasons why contract construction is usually cheaper and better than that done by force-account means. A qualified prime contractor is an experienced construction manager who is intimately familiar with materials, construction equipment, and field methods. The contractor maintains a force of competent supervisors and workmen and is adequately equipped to do the job. Construction is a specialized business, and a contractor must be adept at it in order to survive. All these factors lead to efficiencies of cost and time that owners find difficult to match unless their own organizations already include trained construction forces.

Because public projects must ordinarily be contracted out on a competitive-bid basis as provided by law, force-account work by a public agency is generally limited to maintenance, repair, small jobs, and cases of emergency. Some very large and important public works have been built by force-account means, however; the Panama Canal and many projects for the Tennessee Valley Authority are examples.

It must be noted at this point that the term force account is sometimes used in a different context than that just described. While a contractor is proceeding with work under contract, it is not unusual for the owner to find it necessary to modify or add to the work. Additional payment to the contractor in such an instance is often established on the basis of a lump-sum or unit prices. There are times, however, when the contractor is authorized to proceed on some kind of a cost-plus arrangement. This is sometimes referred to as proceeding with the extra work on a force-account basis.

1.21 GENERAL CONDITIONS CONSTRUCTION

In the traditional construction process, general contractors customarily provide certain common job services, not only for their own forces but for their subcontractors as well. These services, called general conditions construction or support construction, include many items normally required and described by the general conditions section of the project specifications. They involve such services as temporary heat, light, and water; job fencing; drinking water; sanitary facilities; job security; job sign; trash disposal; and job parking. When separate contracts are used, the contractor designated as the "general contractor" usually provides these services for the entire project, including work performed by the other prime contractors.

There are instances where general conditions construction is the only part of the construction process actually performed by the general contractor. This would be true, for example, when the contractor, builder-vendor, or owner-builder subcontracts the entire project. Under some construction management arrangements, the construction manager provides support construction services.

1.22 SEASONALITY IN CONSTRUCTION

The volume of construction underway in this country at any one point in time varies with a number of factors, a major one being the season of the year. Although such seasonal fluctuation varies with the geographic location and type of construction activity, on a national basis summer is the peak season and winter is a slack period for almost all segments of the construction industry. The reason construction is such a seasonal business is primarily the effect inclement weather and low temperatures have on certain key construction operations. An undesirable result of this uneven work volume is serious shortages of craftsmen during the warm-weather months and extensive unemployment during the winter. Work does not stop during the winter months, of course, but there are layoffs, reductions in crew sizes, fewer project starts, and weather delays. In total, these add up to significant reductions in the work force.

To illustrate this point, total construction employment in this country during a typical year will vary approximately 25 percent from the summer to the winter months. It must also be realized that, like total employment, average weekly hours worked follow a seasonal pattern in the construction industry. Overtime work is common during the summer months while less than full workweeks are frequent during the winter. Seasonality in construction is a serious national problem, resulting in inefficiency, increased costs of production, and the wasteful misuse of valuable skilled manpower. Peak season bottlenecks and the resulting inflationary pressures have led to public concern. Several government groups have studied the problem and recommended actions to provide fuller utilization of construction labor. Recommendations have included the more effective use of science and technology, improved scheduling of public contracts, and the relation of national manpower policy to the stabilization of construction employment. Private action has also been encouraged to better regularize the employment of construction workers throughout the year.

Many foreign countries with severe winter climates have been forced to find ways to lengthen their construction seasons. This has been accomplished by a variety of public policies involving licensing, resource allocation, loans, and the payment of government subsidies of one type or another to encourage winter work.

1.23 LICENSING

Because construction can affect the public interest, there are special laws pertaining to construction that are designed to protect the public health and safety and to afford protection against improper, unlawful, and fraudulent acts of contractors. These requirements include zoning regulations, building codes, environmental regulations, building permits, field inspections, safety and health regulations, and the licensing of contractors and of journeymen in certain construction crafts.

The licensing of contractors is not universally required but many states and local governments do require that some or all contractors be licensed as such. Some of the licensing statutes or ordinances are solely revenue-raising measures. Under these particular laws, the payment of a license fee confers the right to conduct a construction business, with no further conditions having to be met. Generally, however, the statutes provide that the contractor must not only pay a fee but also meet certain minimum qualifications. These statutes establish a board of registration or other regulatory body that administers the law, accepts applications for licensing, gives examinations where required, issues the licenses, collects the fees, and generally enforces the provisions of the licensing law.

Licensing requirements vary widely among those areas that have such laws. In most cases, however, when the law requires a license, it applies equally to general construction and specialty work, even when subcontracted. Most of these laws require licenses only for contractors whose annual volume of business exceeds certain designated amounts. Almost all require that a license be obtained in advance of any bidding within the state, with certain minor statutory exemptions. Some states exclude from licensing those contractors doing work financed by federal funds or performed for the federal government on government-owned land.

Certain states issue various classes of licenses differentiated in accordance with maximum size of contract or annual volume of business. In addition, licenses may be issued that are valid for and apply only to specific construction types, such as general engineering construction, general building construction, or any of several classifications of specialty work. In a number of cases, contractors must pass an examination before the license is issued.

It is very important that a contractor be properly licensed when this is a legal requirement of the area in which the work is to be performed. Most licensing laws provide that acting or offering to act in the capacity of a contractor without a valid license is a misdemeanor. Also important is the fact that a contractor not licensed at the time it contracts to perform contracting services may not sue to collect for work that has been performed or for breach of contract. In addition, the contractor may have no right to file or claim a mechanic's lien against the owner's property.

With regard to licensing, where it is required, it is important to note that the construction company must be licensed in the form in which it contracts or offers to contract. For example, if a licensed partnership changes to a corporation, most licensing laws provide that the partnership license is not applicable and that the corporation must be licensed as such. Another instance is that the law in some areas makes it unlawful for a joint venture to act as a contractor without being licensed as such, even though each member of the joint venture is individually licensed. It is interesting to note that in many jurisdictions where contractors must be licensed, this requirement does not apply to a construction manager, provided the CM does not perform any of the actual construction work.

Many political jurisdictions require the licensing, registration, or certification of journeymen in certain crafts. Plumbers and pipefitters, electricians, welders, riggers, elevator installers, sprinkler fitters, and hoisting machine operators are examples of these. Required qualifications vary somewhat with the geographical area.

1.24 LICENSE BONDS

A few states and many local governments require that all contractors, including subcontractors, who operate within their jurisdictions post permanent surety bonds with the appropriate government authority. These are in the nature of performance and payment bonds, a subject discussed more completely in Chapter 7, and they serve for the protection of the public. The public authority or other party may bring process against such a bond in the event of unpaid debt or malfeasance on the part of the contracting firm. The bond also serves to guarantee the contractor will make proper payment for required permits and inspections.

It is perhaps a somewhat more common practice to require bonds of only certain contractors whose work is intimately associated with public health and safety. A representative listing of bonded specialty contractors is as follows:

    Boiler installation
    Curb, gutter, and sidewalk (on public property)
    Demolition
    Electrical
    Elevator installation
    Gas fitting
    House moving
    Plumbing
    Sign erection

Firms proposing to do such work must post bond with the designated public authority in the required amount before commencing operations.

1.25 BUILDING CODES

Universally applicable are building codes as adopted by the several states, counties, and municipalities. Passed into law to protect the public health and safety, building codes are rules that control the design, materials, and methods of construction, and compliance with them within their jurisdiction is mandatory. These codes cover the construction, alteration, repair, demolition, and removal of new and existing buildings, including service equipment. Most of these local codes are based, in whole or in part, on various model building codes that are sponsored by different groups. Changes, where made, have been generally to accommodate local needs and conditions and to make the requirements more stringent than provided for by the national code itself. There are, of course, codes that are not based on national codes but which have been specifically written and devised for a particular locality. The codes of some large cities are of this type. In addition, there are special-purpose codes such as those now adopted by several states that pertain to prefabricated housing. These state laws regulate operations of housing manufacturers within the state and set standards for such housing coming into the state from other manufacturers. Various boards, commissions, building departments, and other public bodies administer and enforce the provisions of their codes.

Four prominent model building codes mentioned in the preceding paragraph are listed below.

1. Basic Building Code. Published by Building Officials and Code Administrators (BOCA) International, Chicago, this code is used predominantly in the East and Midwest.

2. Uniform Building Code. Published by the International Conference of Building Officials (ICBO), Whittier, California, this code is used principally on the West Coast and some midwestern areas.

3. Standard Building Code. Published by the Southern Building Code Congress (SBCC) International, Birmingham, Alabama, this code is prominent in the South.

4. National Building Code. Published by the American Insurance Association, this code is adopted in various localities across the country.

These four codes are the basic documents of their respective organizations, and each is accompanied by related construction codes covering such items as electrical work, plumbing, fire prevention and safety, mechanical work, property maintenance, energy conservation, elevators, and others. Three such codes are:

1. National Plumbing Code. Published by the American Public Health Association and the American Society of Mechanical Engineers, this code is widely used in all parts of the country.

2. Uniform Plumbing Code. Published by the International Association of Plumbing and Mechanical Officials (IAPMO), this code is used in the West.

3. National Electrical Code. Published by the American Insurance Association, this code is widely used in all parts of the country.

To ensure that applicable codes are adhered to within their corporate limits, as well as to serve other control purposes, political subdivisions normally require various kinds of permits, field inspections at certain stages of construction, and test reports. A general building permit is a universal requisite requiring the filing of complete drawings and specifications prepared by a registered architect-engineer with a designated public office. These documents are reviewed for design conformance with the applicable codes by the responsible building authority. Permits for plumbing, electrical work, heating equipment, signs, air conditioning, elevators and escalators, and refrigeration systems are also normally required. In addition, occupancy permits are sometimes required after the completion of a building, requiring an inspection to ensure compliance with building code standards.

The form and content of building codes vary widely from one political subdivision to another, the nation as a whole resembling a large patchwork quilt in this regard. Considerable variance exists with reference to code provisions and coverage, job inspection, enforcement, and appeal procedures. Such code diversity has been criticized as inhibiting the use of construction mass-production techniques and the introduction of new materials, systems, and procedures. Difficulties faced by design and construction firms who work over wide areas are cited in conjunction with pleas for a uniform national code. Similar problems are faced by building material manufacturers who market their products on a national scale. Code diversity is defended as being necessary to make adequate provision for local conditions such as climate, winds, and earthquake hazards. One thing is sure, however, as long as building regulations remain police powers to be determined by local government, any appreciable degree of building code uniformity is not apt to take place.

Conformity with the applicable building code is primarily the responsibility of the design professional. In the usual sense, the contractor undertakes to construct in accordance with contract documents and is not directly concerned with building codes or potential violations of them. However, the contractor is under a legal duty to notify the owner or design professional concerning any deviation from the requirements of the applicable building code that come to the contractor's attention. The contracting firm is obviously responsible for its own construction methods.

1.26 CONTRACTOR ORGANIZATIONS

A large number of trade, technical, and professional organizations serve the diverse interests of the construction industry. These associations represent the design professionals, general and specialty contractors, home builders, manufacturers and distributors of building materials and construction equipment, insurance and surety companies, financial interests, and others.

There are many associations of contractors throughout the country. Of the national associations that represent general contractors, the largest is the Associated General Contractors of America (AGC) whose membership includes building, heavy, highway, and utility contractors. The National Constructors Association (NCA) is made up of contractors who specialize in the design and construction of large industrial complexes. The Associated Builders and Contractors (ABC) is made up of member contractors who operate open shop, also referred to as merit shop. The National Association of Home Builders (NAHB) is a national group representing the housing industry.

These associations perform a number of valuable services for their members, such as providing a united front for labor negotiations, monitoring both local and federal legislation, sponsoring safety and apprenticeship programs, providing tax information, holding conferences, promoting public relations, and serving as clearinghouses for construction information. These organizations strive to maintain the business and ethical standards of contracting at a high level and to establish the integrity and responsibility of their members in the public mind.

Both local and national specialty contractor organizations also function to promote the mutual benefit of their members and to bring their combined resources to bear on common problems. These associations usually parallel the craft jurisdictions of the building trades and are represented by aggregations of specialty contractors such as electrical, mechanical, utility, masonry, or roofing contractors. In the usual case, there will be several such groups functioning within a city or other local area. These regional groups are frequently affiliated with national organizations such as the National Electrical Contractors Association, the Mechanical Contractors Association of America, the National Utility Contractors Association, the Mason Contractors Association of America, and the National Roofing Contractors Association. The American Subcontractors Association is a nationwide organization that represents subcontractors in all segments of the building construction industry.

1.27 MANAGEMENT IN CONSTRUCTION

On the whole, construction contractors have been slow in applying proven management methods to the conduct of their businesses. Specialists have characterized management in the construction industry as being "weak," "inefficient," "nebulous," "backward," and "slow to react to changing conditions." This does not mean that all construction companies are poorly managed. On the contrary, some of America's best managed businesses are construction firms, and it may be noted with satisfaction that the list of profitable construction companies is a long one. Nevertheless, in the overall picture, the construction industry is at or near the top in the annual rate of business failures and resulting liabilities.

There are several explanations given for why the construction industry has been slow in applying management procedures that have proven effective in other areas. Construction projects are unique in character and do not lend themselves to standardization. Construction operations involve many skills and are largely nonrepetitive in nature. Projects are constructed under local conditions of weather, location, transportation, and labor that are more or less beyond the contractor's control. The construction business is a volatile one, with many seasonal and cyclical ups and downs. Construction firms, in the main, are small operations, with the management decisions being made by one or two persons.

The conclusion cannot be drawn, however, that management problems in the construction industry are uniquely different from those in other industries. The complexity of the product and the lack of production standardization do indeed lead to difficult management problems, but these are not necessarily more complicated than those in other business fields. Many industries are characterized by a large number of relatively small firms. Many areas of business experience wide seasonal or cyclical variations in the demand for their products and services. Construction is certainly not the only industry experiencing keen competition. Ineffective management in construction is not, therefore, the inevitable result of pressures and demands peculiar to the industry. As a matter of fact, the presence of such pressures and demands emphasizes the need for astute management in the construction industry.

It is a moot question as to how many construction firms are well managed and whether many such enterprises enjoy any conscious management at all. It is an inescapable conclusion, however, that skilled management and business survival go together. That this maxim has not been recognized by all construction firms is amply demonstrated by the high incidence of financial failures in the industry.

1.28 BUSINESS FAILURES IN CONSTRUCTION

For the past several years construction contractors have accounted for a disproportionate number of all business failures in this country. For example, during a recent year in which construction accounted for 12 percent of the gross national product, contractors accounted for 20 percent of all financial failures and 23 percent of the resulting liabilities. Studies by Dun & Bradstreet, Inc., disclose that the underlying weaknesses in descending order of importance are as follows:

1. Incompetence.
2. Unbalanced experience in sales, financing, purchasing, or construction methods.
3. Lack of managerial experience.
4. Lack of experience in the firm's line of work.
5. Fraud, neglect, disaster, or unknown.

Consideration of the above factors makes it clear that the financial success of a construction enterprise depends almost entirely on the quality of its management. Many outward manifestations of these root causes can be given for business failure: inadequate sales, competitive weaknesses, heavy operating expenses, low profit margins, cash flow difficulties resulting from slow payments, and overextension. It is obvious, however, that these business inadequacies are simply indicative of poor business management.

BUSINESS OWNERSHIP

2.1 ALTERNATIVE FORMS

The forms of business ownership used by construction contractors are the individual proprietorship, the partnership, and the corporation. Selection of the proper type depends on many considerations and is a matter deserving careful study. Each form of business organization has its own legal, tax, and financial implications which must be thoroughly investigated and understood. The selection of the ownership type that is most advantageous and appropriate for an owner's particular circumstances is a business decision that too often does not receive the serious study it deserves. Each type of organization has its advantages and disadvantages that must be carefully evaluated in the light of the individual context. The advice of a lawyer and a tax specialist is desirable when setting up a new business or changing the form of a going concern.

2.2 THE INDIVIDUAL PROPRIETORSHIP

The simplest business entity is the individual proprietorship, also called sole ownership. It is the easiest and least expensive procedure for establishing a business and enjoys a maximum degree of freedom from governmental regulation. No legal procedures are needed to go into business as a sole owner except the obtaining of required insurance, registration with appropriate tax authorities, and possible licensing as a contractor. The owner has the choice of operating under his or her own name or under a company name. In some jurisdictions a company name must be registered with a designated public authority.

The proprietor owns and operates the business, provides the capital, and furnishes all the necessary equipment and property. As the sole manager of the enterprise, the owner can make any and all decisions unilaterally. Title to property used in the business may be held in the name of the proprietor or in the name of the firm, if they differ. All business transactions and contracts are made in the owner's name. This mode of doing business has many advantages to offer, such as possible tax savings, simplicity of organization, and freedom of action. Other than the usual obligation to complete outstanding contracts and settle financial obligations, there are no legal formalities barring termination of the business.

As an individual proprietor, however, the owner is personally liable for all debts, obligations, and responsibilities of the business. This unlimited liability extends to personal assets, even though they may not be involved in the business. Although managment is immediate and direct, the owner must shoulder alone all of the burdens and responsibilities that accompany this function. A proprietorship has no continuity in the event of the death of the owner, unless there is a direction in the proprietor's will that the executor continue the business until it can be taken over by the person to whom the business is bequeathed. In the case of the sickness or absence of the proprietor, the business may suffer severely unless there is a competent person available to take over. The size of the business is limited by the amount of the owner's personal funds, unless a backer can be found to provide a source of credit. The owner must pay income taxes at normal individual rates on the full earnings of the business whether or not such profits are actually withdrawn. Such earnings are added to income from other sources, and the tax is computed on the whole. It is to be noted that the sole proprietor, as a self-employed person, is not able to take advantage of some of the tax-favored, fringe-benefit plans available to employees.

2.3 THE GENERAL PARTNERSHIP

A general partnership is an association of two or more persons who, as co-owners, carry on a business for mutual profit. The principal benefits to be gained by such a merger are the concentration of assets, equipment, facilities, and individual talents into a common course of action. The pooling of financial resources results in the increased bonding capacity of the business (see Section 7.12), offering the possibility of a greater scope and volume of construction operations than would be possible for the partners alone. Each general partner customarily makes a contribution of capital and shares in the management of the business. Profits or losses are usually allocated in the same proportion as the distribution of ownership. If no such agreement is made in the articles of partnership, however, each partner receives an equal share of the profits or bears an equal share of the losses, regardless of the amount invested.

For most purposes a partnership is not recognized by the law as being an entity separate from the partners. For example, a partnership pays no income tax, although it must file an information return. However, a partnership can operate under a company name, and for certain purposes, a partnership is recognized as a separate legal entity. In many jurisdictions, for example, a partnership can own property, have employees, and sue or be sued. The Internal Revenue Service has ruled that bona fide members of a partnership are not employees of the partnership. Rather, such a person is deemed to be a self-employed individual whose remuneration is not subject to the Federal Unemployment Tax Act, the Federal Insurance Contributions Act, or to income tax withholding. Partners usually receive drawing accounts against anticipated earnings or annual salaries that are considered to be operating expenses of the partnership. In any event, partners must pay annual income taxes at the normal individual rates on their salaries and on their allocated shares of the partnership profits, whether or not these are actually withdrawn.

A partner's share of profits in a partnership can be assigned. However, an individual partner cannot sell or mortgage partnership assets. Neither can a partner sell, assign, or mortgage an interest in a functioning partnership without the consent of the other partners. Partners act as fiduciaries and, commensurately, have an obligation to act in good faith toward one another.

2.4 ESTABLISHING A PARTNERSHIP

State laws regarding partnerships are not completely uniform, although most states have adopted the Uniform Partnership Act more or less verbatim. It is customary and advisable to draw up written articles of partnership, although a binding partnership can, in certain instances, be formed by oral agreement. It is always advisable to obtain the services of a lawyer to assist in the preparation of articles of partnership. Ruling out statutory requirements, these articles may include almost anything the partners believe to be desirable. It is usual that articles of partnership, signed by the parties concerned, contain complete and explicit statements concerning the rights, responsibilities, and obligations of each partner. Although such articles must be individually tailored for each specific case, the following list indicates the types of provisions typically included in partnership agreements.

1. Names and addresses of the partners.
2. Business name of the partnership.
3. Nature of the business with any limitations thereof.
4. Location of the business.
5. Date on which business operations will commence.
6. Contribution of each partner in the form of capital, equipment, property, contracts, goodwill, services, and the like.
7. Division of responsibilities and duties between partners.
8. Any statement requiring partners' full-time attention to partnership affairs.
9. Division of ownership as it affects allocation of profits and losses.
10. Voting strength of each partner.
11. Drawing accounts or salaries of the partners.
12. Any restriction of management authority of individual partners.
13. Specification of need for majority or unanimous decision on management questions.
14. Payment of expenses incurred by a partner in carrying out partnership duties.

15. Rental or other remuneration for use of a partner's personal property or for personal services.
16. Arbitration of disputes.
17. Record keeping and inventories.
18. Right of each partner to full access to books and audits.
19. Rights and responsibilities of the individual partners upon dissolution.
20. Procedure in event of the incapacitating disability of a partner.
21. Extraordinary powers of surviving partners, such as option to purchase interest of deceased or withdrawing partner.
22. Provision for final accounting of business in event of death, retirement, or incapacity of any partner.
23. Provision that executor or spouse of deceased partner continue as partner.
24. Termination date of the agreement, if any.

Unless there is specific agreement to the contrary, general partners are expected to devote full time to the affairs of the partnership. Partners cannot normally withdraw capital from the partnership unless provision for this is made in the partnership agreement. By mutual consent of all the partners, any provision of a partnership agreement can be modified or deleted or new provisions can be added at any time.

2.5 LIABILITY OF A GENERAL PARTNER

The liability of a general partner for the debts of the partnership is based on two legal principles. First, each general partner is an agent of the partnership and can bind the other partners in the normal course of business without express authority. Second, partners are individually liable to creditors for the debts of the partnership. The implications of these two points will now be discussed.

Agency exists when one party, called the principal, authorizes another, called the agent, to act for the former in certain types of business or commercial transactions. An agency relationship exists by agreement, and an agent can be appointed by the principal to do any act the latter might lawfully do. The principal is liable for all contracts made by the agent while the latter is acting within the scope of his actual or apparent authority. The principal is also responsible for nonwillful torts (civil wrongs) the agent commits in the furtherance of the principal's business. A general agent is one empowered to transact all of the business of his principal. Each full member of a partnership is a general agent of the partnership and has complete authority to make binding commitments, enter into contracts, and otherwise act for his fellows within the scope of the business. This is subject, of course, to any limitations set forth in the partnership agreement.

In most states, the responsibility of a partner for the contractual obligations of the partnership is joint with the other partners, meaning that all partners are necessary parties to any contractual action. Liability of a partner for torts committed by the partners in the ordinary conduct of business affairs is joint and several. In any such action, every member of the firm is individually liable and the injured party can proceed against all parties jointly or against such of the partners as he chooses. Each partner assumes unlimited personal liability to third persons for the claims, contracts, and debts of the partnership. If any one partner is not able to pay his proper share of company liabilities, creditors can force the remaining partners to pay the share of the first. A creditor of a partner can attach the latter's interest in the profits of the business but cannot proceed against partnership property unless the partnership is being dissolved. Under the agency principle, notice to any one partner is considered to be notice to all.

It is apparent from the foregoing that careful judgment should be exercised in the selection of business partners. Each general partner accepts unlimited financial responsibility for the acts of the other partners, and he underwrites the debts of the enterprise to the full extent of his personal fortune. If a partner withdraws from a going partnership, he remains personally liable for the partnership obligations outstanding as of the date of the withdrawal. To protect himself from future partnership debts, the retiring partner should give personal notice to firms doing business with the partnership and publish public notice of his departure. In most states, the withdrawing partner can discharge himself of all liability by an agreement to that effect between himself, the firm's creditors, and the partners continuing the business.

2.6 DISSOLUTION OF A PARTNERSHIP

One of the major weaknesses of the partnership form of business organization is its automatic dissolution on the death of one of the partners. However, this possibility can be circumvented by making provision in the partnership agreement that the business will continue and that the surviving partners will purchase the decedent's interest. Dissolution may also be precipitated by bankruptcy, a duration provision in the original partnership agreement, the decision of a partner to withdraw, the insanity of a partner, a court of equity decree, or the mutual consent of all parties. Dissolution is not a termination of the partnership, but simply a restriction of the authority of the partners to those activities necessary for the conclusion of the business. It has no effect on the debts and obligations of the enterprise. Voluntary dissolution is often accomplished by a written agreement of dissolution between the partners. This agreement provides that the partners will undertake to complete all contracts in progress, settle company affairs, and pay all partnership obligations from the proceeds and assets of the business.

In the settlement of partnership debts, outside creditors enjoy a position of first priority. If partnership assets are insufficient to satisfy the outstanding obligations, the partners must personally make up the difference. After the creditors have been satisfied, partnership assets are used to repay any loans or advances that were made by partners above and beyond their capital contributions. Then follows the return to the partners of their capital investments.

If further assets remain, these are treated as profits and are distributed accordingly. When construction contracts involving considerable time for their completion are involved, the clearing of partnership accounts is often subject to considerable delay pending payment of all debts, receipt of all accounts receivable, and discharge of all contract obligations of the partnership.

2.7 THE LIMITED PARTNERSHIP

A general partner is a recognized member of the firm, is active in its management, and has unlimited liability to its creditors. The general partnership has been the basis of discussion in the preceding sections. A limited partner is one who contributes cash or property to the business and commensurately shares in the profits or losses but provides no services and has no voice or vote in matters of management. Unlike the general partners, limited partners are liable for partnership debts only to the amount of their investments in the partnership. This is true, however, only as long as their names are not used as part of the firm name and they do not participate in management. Limited partners can be held fully liable if the true nature of their participation in the partnership has been withheld from creditors. The interest of a limited partner is assignable.

Under the Uniform Limited Partnership Act, adopted by most states, a limited partnership may be formed by two or more persons, at least one of whom must be a general partner. A contract between the partners is drawn up that clearly defines their duties and financial obligations. Also stipulated may be the length of time the agreement stays in effect and the date by which the contributions of each limited partner are to be returned. In addition, a limited partnership certificate must be signed and filed with a designated public office and published as required by state statute. The operation of a limited partnership, as well as its establishment, must be in strict accordance with the laws of the state in which it is formed. The partnership is not automatically dissolved by the death of a limited partner, as in the case of a general partner.

The limited partnership is used principally as a means of raising capital. It is a vehicle for getting money into a business with immunity from any personal liability for the investor. The limited partner is not a creditor of the business, but an owner of a share of the firm and is entitled to a share of the profits while the business is operational and to a return of capital and undistributed profits on its dissolution. Obtaining new investment by bringing in additional general partners, each of whom shares in the management, can result in serious problems of company control and decision making. The addition of a limited partner may also be preferable to borrowing the desired funds from a bank or other source. In this regard, however, one who lends money to the partnership can be in a better position than a limited partner. A lender to whom a firm is in debt can possibly exert some control over the firm, whereas any show of management interference by a limited partner can cause a loss of the limited partner's immunity from personal liability. Moreover, the lender enjoys a higher priority on dissolution of a partnership than a limited partner.

2.8 SUBPARTNERSHIP

A general partner can enter into an agreement with an outsider, called a sub-partner, whereby the latter will share in some designated way the general partner's profits or losses derived from the partnership's business activities. The subpartner is not a member of the firm. Consequently, he performs no active function, has no voice in the management of the partnership, and has no contractual relationship with the partners other than the one with whom he made the profit- or loss-sharing arrangement. Because the subpartner has contracted to participate in the individual partner's share only, he is not personally liable to creditors of the partnership.

2.9 THE CORPORATION

There are several different classifications of corporations: public and private, profit and nonprofit, quasi-public, and foreign and domestic. The following discussion considers only the private corporation that a contractor would establish for ordinary business and profit motives.

A corporation is an entity, created by law, that is composed of one or more individuals united in one body under a special or corporate name. Corporations have certain privileges and duties, enjoy the right of perpetual succession, and are regarded by the law as being separate and distinct from their owners. A corporation is authorized to do business, own and convey real and personal property, enter into contracts, and incur debts in its own name and on its own responsibility. It sues and is sued in its corporate name. The principal advantages of this form of business organization are the limited liability of its owners (stockholders), the perpetual life of the company, the ease of raising capital, the easy provision for multiple ownership, and the economic benefit in that owners pay taxes only on profits (dividends) actually received. In a corporation, unlike a partnership, the owners (shareholders) can also be employees. Fringe benefits and retirement benefits with tax advantages are possible with corporations that are not available to sole proprietors or partnerships.

A corporation is created by obtaining a corporate charter or articles of incorporation granted by the state in which the corporation is to be domiciled. The powers of a corporation are limited to those enumerated in its charter and to those that are reasonably necessary to implement its declared purpose. The provisions of the corporate charter are usually made very broad so that the new corporation will be authorized to do many things that may become necessary or desirable in future years. These powers include the drawing up of bylaws pertaining to the day-to-day conduct of business. Acts of the corporation are governed and controlled by its articles of incorporation, its bylaws, and by applicable state law.

Corporations are regulated by the various states, each having a corporation code that prescribes the formal process that must be followed by the organizers in obtaining a charter. These legal requirements differ somewhat from state to state, and it is always important to secure competent legal guidance when establishing a new corporation or modifying an existing one. In any case, fees must be paid and substantial quantities of information filed with the proper officials. Certain detailed requirements must be satisfied regarding the incorporators, residence, and financial structure.

A corporation is dissolved by the surrender or expiration of its charter. Dissolution of a corporation merely means that the corporation ceases to exist. State laws govern the means by which the business affairs of a dissolved corporation must be settled.

2.10 THE FOREIGN CORPORATION

A corporation is created under the laws of a specific state and within that state is known as a domestic corporation. In all other states it is considered to be a foreign corporation. The state of incorporation may be chosen not on the basis of location of work done by the company but rather because of favorable corporation laws. The incorporated contractor that wishes to extend its operations beyond the borders of its state of incorporation must apply for and receive certification to do business within each additional state. Usual requirements for the certification of a foreign corporation are the filing of a copy of the corporate charter, designation of a local state resident as an agent for the service of process, and the payment of a filing fee.

The acts of a foreign corporation within another state are considered to be unlawful until the corporation has become certified to do business as a corporation in that state. Hence, the incorporated contractor must use care to comply with the corporation laws of those states in which it does business. Failure to be properly certified as a foreign corporation may render construction contracts unenforceable and eliminate the right of lien. Additionally, an uncertified foreign corporation may not have access to the state courts. A foreign corporation is, of course, subject to the control of the state within which it does business. For example, if franchise or income taxes are levied on businesses of that state, the foreign corporation must pay such taxes in accordance with its total business in the state.

2.11 STOCKHOLDERS

The ownership of a corporation is exercised through shares of stock that entitle the owners thereof to a portion of the business profits and the net assets on liquidation. This does not mean the shareholders own the corporate assets, because the corporation itself holds legal title to its own property. However, each share of stock does represent a share in the ownership of the corporation. Under the laws of most states, all the stock of a corporation can be owned by a single individual. A clear distinction between the individual and his corporation is recognized by the courts.

Ordinarily, shares of stock in a corporation are freely transferable, this being one of the advantages of incorporation. Stock certificates are negotiable instruments that may be transferred by endorsement and delivery in the same manner as promissory notes. However, it is not an uncommon desire that the ownership of stock of an incorporated business be limited to members of a family or to a small circle of associates. In this way control of the company is maintained within a select group. Such a corporation whose stock is not available to the public is referred to as a closed or closely held corporation. Any restriction on the ownership or transfer of stock can be accomplished by stamping the nature of the restriction directly on the face of the stock certificate. Another scheme might be to provide in the bylaws that before stockholders can sell their stock to an outside party, they must first offer their stock for sale back to the corporation.

When a corporation is established, its articles of incorporation authorize the issuance of a certain dollar value of stock (the authorized capital stock). The actual capital stock is the amount actually paid to the corporation for its outstanding stock and does not necessarily equal the authorized capital stock. The stock may be common or preferred or both. Preferred stock is subject to less risk than is common stock, because it has a certain preference as to dividends and distribution of assets on liquidation of the corporation. However, preferred stock usually carries no voting privileges. Common stock generally entitles its owners to one vote per share and presents at least the possibility of greater ultimate profits to its owners.

Stockholders have certain rights and privileges that are based on state statute, the charter of the corporation, and the terms of ownership as stipulated on the stock certificate. These rights include the preemptive right to subscribe new stock issues, to carry out reasonable inspection of the corporation's books and records, to bring suit against the corporation, to share ratably in declared dividends, to receive a share of the net assets on dissolution, and to enact bylaws.

The profits of a corporation are subject to federal and state income taxes at the corporate rate. Such taxes are paid by the corporation before any profits can be distributed as dividends to the stockholders. The stockholders are then taxed individually on any such dividends distributed on a cash basis. Dividends can only be paid out of earned surplus and may be declared only if such action does not impair the position of corporation creditors. Such dividends, or surplus distributions, are declared as a fixed sum per share of outstanding stock. This payment is generally made quarterly but may be declared at such intervals as decided on by the board of directors.

An outstanding advantage of the corporate system is the privilege of an individual shareholder being immune from personal liability for corporate debts. With very few exceptions, stockholders have no liability for obligations of the corporation, and their responsibility is limited to their investment in the corporation's stock. This is unlike the case of sole proprietors, general partners, and most joint venturers (see Section 2.15) who are personally liable for all obligations of the entity. A stockholder is not an agent for the corporation and cannot act in any way for or on behalf of the organization.

2.12 CORPORATE DIRECTORS AND OFFICERS

The voting shareholders elect a board of directors that exercises general control over the business and determines the overall policies of the corporation. The directors must conduct the firm's affairs in accordance with its bylaws and are ultimately responsible to the stockholders. Directors have no authority to act individually for the corporation and are not agents of the corporation or of the stockholders. Their powers may be exercised only through the majority actions of the board. Directors occupy the position of fiduciaries and are required to serve the corporation's interests with prudence and reasonable care.

The articles of incorporation normally provide for annual meetings of the shareholders and for periodic meetings of the board of directors. Minutes of these meetings must be kept and filed. Failure to observe such formalities can lead to serious problems in the event of claims or lawsuits. Chronic disregard of corporate technicalities might result, for legal purposes, in the business being treated as a partnership or sole proprietorship rather than as a corporation.

The president, vice-president, secretary, and treasurer are the corporate officers, normally appointed by the board of directors, who carry out the everyday administrative and management functions. The officers are empowered to act as agents for the corporation. The president is authorized to act for the firm in any proper way, including contract making, whereas the lesser officers generally have more limited authority. Officers also serve in the capacity of fiduciaries and may be held personally liable for coporation losses caused by their neglect or misconduct.

2.13 THE SUBCHAPTER S CORPORATION

If a corporation meets the criteria specified by the Internal Revenue Service, none of which restrict the size of business volume, profits, or capital, it has the option of being taxed as a partnership rather than as a corporation. If such an election is made, the company is referred to as a "subchapter S corporation." To be eligible, the corporation must meet certain tests including being a domestic corporation having no more than 35 stockholders who must be individuals or estates and there being only one class of stock outstanding. An eligible corporation must elect to be treated as a subchapter S corporation, and all stockholders, present and future, must consent by signing prescribed forms that are filed with the IRS. The change in tax status does not affect the limited personal liability of the stockholders.

When such an election is made, shareholders annually report for tax purposes their share of corporate earnings regardless of whether these earnings were actually paid out as dividends. No corporate tax is paid. Even so, the corporation may still deduct for tax purposes payments for fringe benefits and retirement and pension funds, something that a partnership cannot do. Once the taxation election is made, this status must continue until it is properly terminated. This may be done by a unanimous vote of the stockholders to terminate or by a new shareholder who will not accept the arrangement. Once terminated, a new election may not again be made for five years.

2.14 CONSTRUCTION CONTRACTING FIRMS

There are approximately 800,000 business firms in the United States that classify themselves as being construction contractors of one type or another. Of these about 77 percent are individual proprietorships, 6 percent are partnerships, and 17 percent are corporations. From this it is easy to see that a large majority of construction firms are small businesses. This follows from the fact that the business scope of a sole owner is entirely dependent on his own resources and personal credit.

Figures made available by the U. S. Bureau of the Census reveal that more than 65 percent of the construction establishments do less than $250,000 worth of business each year, more than 35 percent do less than $25,000, and more than 25 percent gross less than $10,000 per year. More than 50 percent have no payroll as such, being operated by proprietors, partners, or business associates who do the work themselves. Approximately 10 percent of the contractors account for about 80 percent of the annual dollar volume of construction. There are approximately 2500 construction firms in the United States that have a gross annual business of more than $5 million.

2.15 THE JOINT VENTURE

As a means of spreading risks and pooling resources for certain projects, it is common for two or more contracting firms to unite forces through the medium of a joint venture. The members of a joint venture can be sole proprietorships, partnerships, or corporations, but the joint venture itself is a separate business entity. The usual joint venture may best be described as a special-purpose partnership. The determination of the relationship depends on the particular facts and the agreement on which the joint relationship is created. Properly used, joint ventures are well-proved devices for pooling abilities and facilities and for spreading construction risks.

A joint venture relates only to a single project, even though its completion may take a period of years. Normally the participating contractors submit to the project owner a joint venture proposal in which all the interested parties are named and in which all, jointly and severally, are directly bound to the owner for the performance of the contract. Each coventurer is thereby individually liable for the performance of the entire contract and for the payment of all construction costs.

A joint venture serves to combine the resources, assets, and skills of the participating companies. Such a combination offers the multiple advantages of pooling construction equipment, estimating and office facilities, personnel, and financial means. Each joint venturer participates in the conduct of the work and shares in any profits or losses. Many of America's largest structures have been built by joint ventures. Hoover Dam, constructed by an association of six companies, was an early example. Succeeding years have seen joint ventures develop into standard practice in the construction industry.

The participants in a joint venture enter into a written agreement that defines the aims and objectives of the association and clearly delineates such matters as the advance of working capital; the percentage interest of each participant; the division of profits and losses; the specific responsibilities and contributions of the individual parties; the details of contract administration, project management, and any fidelity bond requirement; the supervision of accounting and purchasing; the procedure in case a coventurer defaults on its commitments; and the termination of the agreement. For a joint venture to exist in a legal sense, there must be (1) a contract, (2) a common purpose, (3) a community of interest, and (4) an equal right of control. Writing a joint venture agreement is an important and technical task requiring the services of a lawyer and expert tax counsel. Prior to bidding, it may be necessary to obtain a contractor's license for the joint venture, prequalify the joint venture for bidding, and file a statement of intent to submit a joint venture proposal.

Liability and tax considerations on very large projects can make the partnership form of a joint venture undesirable. The participating companies may, in such cases, determine that a corporate form of joint venture is preferable. This form of joint venture can become very troublesome, however, because of certain aspects of applicable corporate law. Additionally, the owner of the project concerned may not prequalify a corporate joint venture for bidding or negotiation purposes.

3

COMPANY
ORGANIZATION

3.1 BASIC CONSIDERATIONS

All firms of any size achieve their business objectives through the joint efforts of company principals and employees. An indispensable ingredient for a successful enterprise is an efficient, well-trained, and vigorous company organization.

The establishment of an effective operating organization is one of the principal functions of a firm's management and is the subject of this chapter. Organizing is the process of determining what individual job positions are required, defining the responsibilities of each such position, and establishing the relationship of one position with another insofar as carrying out those responsibilities is concerned. The main task of a contractor's organization is to plan, direct, and control the elements associated with the construction of projects so the best combination of operational economy and time efficiency is obtained. To implement this function, the contracting firm must create an organizational structure that is particularly suited to its mode of operation. The organizational framework must be sufficiently stable to assure action but yet be flexible and adaptive. Within the necessary metes and bounds of company discipline, maximum personal freedom of action must be allowed. It is very important that those who shoulder the decision-making responsibilities of the company should be relieved from excessive detail.

Authority, responsibility, and duty are terms useful to the discussion of organization principles. Authority may be defined as the ability to act or to make a decision without the necessity of obtaining approval from a superior. Authority may be delegated to others. Responsibility implies the accountability of a supervisor for an assigned function or duty. Although responsibility may be assigned to subordinates, the supervisor still remains accountable in full to his own superior. A duty is a specifically assigned task that cannot be relegated to another. When devising a company organization plan, the individual positions of authority and responsibility must be identified and the duties of each participant defined.

Management must protect against both under- and over-organization. A balance must be struck between overhead salaries and the monetary benefit actually realized from those salaries. It is a serious organizational failing to burden too few with too much. It is equally undesirable to have an organization top-heavy with half-productive administrators and supervisors.

3.2 PRINCIPLES OF ORGANIZATION

No one organizational pattern could possibly be appropriate for every construction firm. A plan for a highway contractor will not likely fit the needs of a design–construct industrial contractor or a contractor acting as a construction manager. Each company must devise an organizational plan that best suits its own peculiar situation. There are, however, certain well-recognized principles of organization that can be applied by any contracting firm that wishes to formulate an efficient organizational plan for its business. No business is so small that accepted methods of organization cannot be profitably applied. The mere act of making a formalized analysis of the necessary tasks, determining how they relate to the company as a whole, and specifying who is responsible for each task creates a clear understanding of who, what, when, and how. An organizational plan removes confusion, indecision, buckpassing, duplicated efforts, and neglected duties. The following simple steps are suggested for the development of an effective company organization. The accomplishment of these steps should include extensive discussion and consultation with everyone concerned.

1. List every duty for which someone must be made responsible.
2. Divide the duties listed into individual job positions.
3. Arrange these positions into an integrated functional structure showing lines of supervision.
4. Staff the organization.
5. Establish lines of communication.
6. Prepare a manual of policies and procedures.
7. Implement the plan and adjust as necessary.

3.3 LIST OF DUTIES

The conduct of a contracting business involves certain duties regardless of whether the business is large or small. These duties will be carried out by only a few persons in a small organization, whereas many people will be involved in a larger company. In making up a list of company duties, the question immediately arises as to how detailed such a list should be. The answer to this is that a level of breakdown appropriate for the size of the company and the number of employees involved should be used. It seems reasonable to suggest that the larger the company, the finer should be the subdivision of duties. The more detailed the thinking concerning all of the requirements for sound operation, the less likely it is that something essential will be overlooked. The following list of duties will be used for illustrative purposes and is not intended to be complete or to apply to any given construction company.

EXECUTIVE

Banking
Construction loans
Financial structure
Legal matters
Business organization
Company organization
Auditors and audits
Public relations
Industry associations
Labor negotiations
Contract negotiation and execution
Investment
Personnel relations and policies
Long-range planning
Salaries, bonuses, pensions, and profit sharing
Legislative matters
Capital improvements
Scope of operations
Approval of major expenditures
Operating procedures and policies

ACCOUNTING AND PAYROLL

General books of account
Subsidiary records
Cost records and reports
Financial reports
Tax returns and payments
Payment of invoices
Billing
Collections
Assignments
Bank deposits
Personnel records
Payrolls and records
Wage and personnel reports to public agencies
Office services

PROCUREMENT

Requisitions
Purchase orders
Subcontracts

Change orders
Inventories
Ordering and control of stores
Expediting
Licenses
Insurance; project and company
Subcontractors' insurance
Owners' contract bonds
Bonds from subcontractors
Releases of lien
Guarantees and warranties
Routing and scheduling materials
Building permits
Checking and approval of invoices
Information on prices and sources of supply
Verification of quantity and quality of deliveries

ESTIMATING

Decision to bid
Visiting the site
Obtaining bidding documents
Mailing out bid invitations
Prebid conferences
Quantity takeoff
Subcontract and material quotations
Pricing
Checking estimate
Preparation of proposal
Bid bond
Delivering proposal
Bills of materials and subcontractors

ENGINEERING

Project planning
Construction schedules
Project cost accounting
Project monitoring
Project cost breakdowns for pay purposes
Periodic project pay requests
Shop drawings
Project cost reports
Field and office engineering
Safety policies and procedures
Accident reports to insurance companies
Relations with owners and architect-engineers
Labor relations

CONSTRUCTION

Hiring labor crews
Supervision of construction
Coordination of subcontractors
Timekeeping
Project cost data
Project accident reports
Project safety programs
Project progress reports
Construction methods
Storage of materials on project sites
Scheduling construction equipment

YARD FACILITIES

Receipt, storage, and warehousing of project materials
Maintenance and repair of construction equipment
Storage of construction equipment
Maintenance and issue of stores
Issue, receipt, and repair of hand tools
Transportation
Equipment rental
Prefabrication and subassembly
Spare parts

3.4 DIVISION OF DUTIES

After the duties have been listed, the next step in the development of an organizational plan is to subdivide them into groups so duties in each group can become the assigned responsibilities of a single individual. To illustrate, suppose the business is a small partnership consisting of two partners and an employed bookkeeper. One partner is in charge of the office, and the other supervises the field operations. The three people concerned must, in a collective way, carry out all the duties listed. The executive duties would normally be carried out by both partners acting together. The bookkeeper would perform the accounting and payroll tasks. The office partner could do all duties related to procurement and estimating and be responsible for the bookkeeper. The field partner's duties could include those associated with engineering, construction, and yard facilities. It is obvious the duties could be distributed among the three participants in any way desired. The experience, education, expertise, and talents of individuals will normally be the basis for the allocation of company responsibilities. It is important that every duty be assigned, and, conversely, that every position created include responsibility for a specific list of duties. The list of duties for any given position is known as its "job description."

From the foregoing, it is clear that a member of a small firm will normally be responsible for a considerably broader range and diversity of things to do than would be the case if it were a large company. It is an intrinsic characteristic of a large organization that most of its members have relatively narrow job responsibilities and tend to be considerably more specialized. To illustrate, a small firm may have one person who, single-handedly, does all the takeoff, pricing, bidding, and purchasing for the company. A large contractor may well employ several people to accomplish the same things, each being involved with only a limited aspect of the total process.

3.5 ORGANIZATIONAL STRUCTURE

The organizational procedure followed thus far ensures that, in total combination, the employment positions established will accomplish each and every duty that has been identified as being necessary for proper business operation. It is now desirable to link these positions together into an integrated operational structure. It is common practice in the construction industry to establish company jurisdictions or departments. Each department is roughly equivalent in authority, and although interrelated, each operates semi-independently of the others. This is a functional form of organizational structure, one that is formed by partitioning the work to be done into major functional areas. The functional system of management has the advantage that an individual or group can specialize to some extent in some particular aspect of the work. Semiautonomous departments are created, each of which performs a specialized function. How far this horizontal division is carried depends on the size of the company involved and the wishes of its managment. Each department is assigned a specific area of responsibility (estimating, for example) and is headed by a manager who possesses training, experience, and skill in that particular aspect of the business.

Each department thus created is then divided vertically. Vertical division refers to the establishment of lines of supervision, with each individual along a line being accountable to the person above and acting in a supervisory capacity to those below. The further down a position appears on the organizational ladder, the more limited is the responsibility and authority of the person concerned.

3.6 ORGANIZATION CHARTS

Organization charts present the company's organizational structure in pictorial form, showing every position of responsibility and all lines of supervision and authority. They provide an understanding of the company's structure at a glance. The organizational chart is a particularly efficient way in which to establish clearly in the minds of the employees involved their individual positions within the overall company, the identities of their supervisors, those whom they supervise, and the nature of their duties. It constitutes an established, permanent reminder of job responsibilities. Such a chart also under-

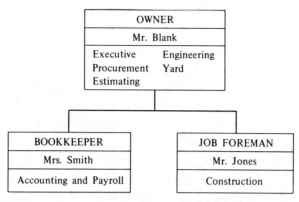

Figure 3.1 Organization chart for small individual proprietorship.

scores to employees the fact that the business is well organized and that top management knows at all times who is responsible for what.

Figure 3.1 shows a typical organizational structure of a small contracting company being operated as an individual proprietorship. The organization chart for a small partnership is presented in Figure 3.2. Figure 3.3 shows a typical organization chart for a moderately large corporate firm. The organizational structure of contractors can be infinitely varied, and the charts shown are intended only to illustrate frequently used schemes. The organizational structure and the assignment of duties shown in these charts are for illustrative purposes only, and the allocation of duties is not intended to be complete. The reader should note that for each position shown, a person's name appears, together with a list of major job responsibilities. In addition, each member of the team reports to only one superior.

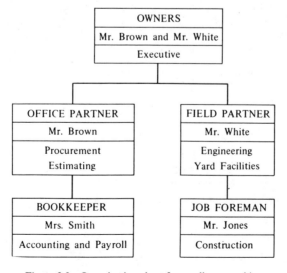

Figure 3.2 Organization chart for small partnership.

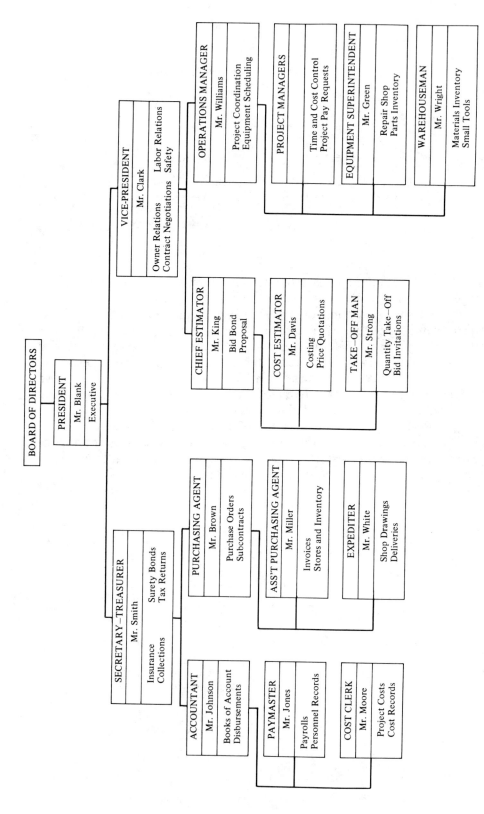

Figure 3.3 Organization chart for moderately large corporation.

3.7 STAFFING

The organizational structure devised must now be staffed; that is to say, a person must be assigned to each of the positions created. A company is what its people make it, and it is impossible to overemphasize the importance of selecting the right person for each position. In the construction industry, supervisory personnel are typically selected on the basis of their job knowledge or technical ability rather than on management training or experience. The reasons for this are obvious, but it should be pointed out that the selection of an individual for a supervisory position only on the basis of construction competence is no guarantee the most effective person will thereby be obtained.

Admittedly, detailed knowledge of the construction process is an important attribute for the construction manager, but other qualifications are also required. Positions on the lower rungs of the organizational ladder are concerned primarily with specific and technical details. Correspondingly, it is entirely appropriate that people be selected for such positions largely on the basis of their job knowledge because their management function is limited. However, in selecting individuals for progressively higher positions, more and more attention must be paid to managerial ability and general experience in the industry. Many examples can be cited to show that technical ability alone does not assure mangerial success.

3.8 AUTHORITY AND RESPONSIBILITY

Persons who are responsible for various company areas do not necessarily personally do everything for which they are accountable. Responsibility implies making sure tasks are done, regardless of who actually performs them. This means that, as shown in Figure 3.3, the president is responsible for every action necessary to the proper operation of the company. The president must be given, by the board of directors, full authority over the operating departments of the business. Otherwise the president cannot be held responsible for results since, without full authority, that officer cannot control what occurs.

According to the organization chart in Figure 3.3, the president has delegated appropriate responsibilities to the secretary-treasurer and the vice-president. In turn, these two officers have delegated parts of their responsibilities to those under their supervision. However, such downward delegation does not relieve the original person of the prime responsibility. To illustrate, the expediter is responsible for the duties specifically assigned to him. The purchasing agent is responsible for his own duties as well as those of the expediter and the assistant purchasing agent. The secretary-treasurer is responsible for his own work plus that of the accountant, paymaster, cost clerk, purchasing agent, assistant purchasing agent, and expediter.

Authority commensurate with responsibility must apply at all operating levels. When the president delegates responsibility for departmental operation to others and they, in turn, delegate parts of their responsibilities to their subordinates, adequate authority must be delegated at the same time. In every instance, authority must be given to the individuals concerned to do their specific jobs in any way they choose as long as the results are satisfactory and the procedures conform to established company policies. Authority and responsibility must be specifically delegated to an individual for each element of the company and necessary resources provided. Top management must be able to hold one person fully responsible for each aspect of company operations.

3.9   COMMUNICATIONS

The proper functioning of a business depends on the exchange of information of many kinds, both within the firm itself and with external agencies. A good system of company communications is essential for smooth and profitable operation. Top management must be kept apprised of job costs and job progress. Procurement personnel must receive purchasing information concerning materials and subcontractors required for a new project. The project superintendents must be kept advised of contract changes such as drawing revisions and change orders. Subcontractors must be consulted about project planning and informed of their part in the construction schedule. Procurement people must keep the yard and the projects aware of the delivery status of materials and be told of the nature and extent of delivery shortages or damage. Information about job accidents must be conveyed from the projects to company management. The payroll department must be informed concerning hirings and layoffs. Information on back charges against material suppliers and subcontractors must get to the accounting department. These are only a few examples of necessary communications within a company.

Communicative needs that are repetitive and routine in nature, such as the examples cited in the preceding paragraph, can be met by the establishment of set procedures. The next section discusses the manual of policies and procedures, which has, as one of its principal purposes, the description of routine communicative processes.

Periodic meetings of various groups within company management are a necessity. These meetings provide an opportunity for them to exchange ideas, resolve misunderstandings, and decide on future courses of action. Also, this mode of communication helps to establish team spirit. Brainstorming sessions of such groups have proven to be extremely effective in producing new ideas and managment innovation.

To disseminate company information of general interest, bulletin boards and a company publication are very effective. Issued periodically, the publication can contain matters such as the firm's safety record, new projects, personnel changes, company policies, and other pertinent information.

3.10   THE MANUAL OF POLICIES AND PROCEDURES

An organization chart is a very useful and effective management device, but it alone cannot completely describe the total workings of the organization. There is an obvious need for further development and description of the organizational plan. Once the organizational structure has been established, written company policies and operating procedures can be prepared that augment the organization chart and work in conjunction with it. These policies and procedures can be set forth in a manual made available to all company personnel concerned. If a loose-leaf manual is used, it is easy to make revisions and add new sheets.

Decision making is the essence of management. Some business situations requiring decisions are unique and must be handled separately and individually. Other situations are recurring. Company policies provide standard decisions for such repetitive cases. Policies serve as guides for action by all levels of company management and provide uniform and consistent guidance in the handling of problems that recur frequently. Obviously, policies are designed to meet specific situations. Policies enhance the effectiveness of the organization and are important to the conduct of everyday business affairs.

Operating procedures establish general rules governing communications, the flow of paperwork, and other routine company operations. Such rules remove the element of decision making from office routine and allow duties to be assigned to the lowest practical management level. The very act of reducing these procedures to writing helps to clarify ambiguities, removes areas of overlap, and reveals discrepancies and other shortcomings in the organizational plan. The written procedures ensure uniformity of action, are valuable in training new personnel, and reduce the need for close supervision.

The first step in writing any such set of company procedures is generally to have the supervisor of each department consult with his personnel and then write down the procedures that apply to his own area of responsibility. These various procedures are then incorporated into a single set of rules through a series of interdepartmental meetings in which the proposed procedures are coordinated and adjustments made.

The manual must be explicit concerning the keeping of records. Government agencies require that certain kinds of records be kept. In addition, company records pertaining to project costs, shop drawings, equipment maintenance, inventory, job progress, estimating, personnel, and other aspects of the business are of real importance. Although too few records can be costly, excessive record keeping can become equally expensive. It makes no sense to keep records, other than those required by law, that are not actually used and whose potential value does not at least equal their cost of preparation. Record forms should be carefully selected to yield a maximum of information with a minimum of effort and clerical time. The manual can include samples of all standard record forms with explanations and illustrations of their use. Printed forms for records and communications are great labor savers. They simplify the task of adhering to company policies and procedures and make it possible to use clerical people with less supervision. Information repeatedly presented in the same format can be quickly understood by all who use it.

Another matter to be covered by the operating manual concerns routine company reports—who prepares them, when they are prepared, and to whom they are to be routed. Effective decision making depends to a large degree on the timely and continuous flow of needed management information. Reports on project costs, accident experience, current financial status, cash projections, and similar matters present vital operating information in a condensed and summary form.

The development of an effective manual of policies and procedures requires time and constant reevaluation. The manual evolves gradually over a period of time, additions and changes being made as needed. For effective standardization, the policies and procedures must be firmly established, and this generally requires a period of adjustment as modifications are made. In a large organization, manuals are usually issued in parts, each section pertaining to the operation of a single department.

3.11 THE OPERATING CHART

A valuable adjunct to the organization chart and the company manual is an operating chart, an example of which is shown in Figure 3.4. This figure is not intended to be complete or to apply necessarily to the operations of any given construction firm, but it does illustrate the type of information that can be conveyed by such a chart. It serves to establish and define the working relationships among company personnel. It shows who participates, and to what degree, when a given activity is performed. Shown on a horizontal line are all persons who are involved in a given function and the extent and nature of their involvement. A vertical line shows the functions for which an individual is responsible and the nature of this responsibility. It serves as an amplified job description.

3.12 MAKING THE ORGANIZATION WORK

There are further considerations on which action must be taken if the company organization is to function and produce up to its full potential. These are briefly discussed in the following.

1. Motivation.   People within the organization must be encouraged, inspired, and impelled to do what has to be done. The key factor in management motivation is the human need for a feeling of recognition and importance. In this regard, money may not be the most compelling element. Working with other people is the primary means of securing the feeling of recognition and importance; people want to feel they are part of a team. The test of a good manager is to build a closely knit, effective organization.

2. Personnel Development.   People must acquire or improve the skills, attitudes, and abilities necessary to do the work assigned to them. Based on an appraisal of individual strengths and weaknesses, a plan for the personal improvement of company personnel is an important management responsibility. Among the many ways in which this can be accomplished is by encouraging continuing education through home study or attending formal courses and seminars. Paying all or a part of the attendant costs and giving time off from work can be powerful stimuli. Committee participation, job rotation, and coaching on the job can also be effective.

Operation	Position	President	Secy-Treasurer	Vice-President	Accountant	Paymaster	Cost Clerk	Purchasing Agent	Asst. Purch. Agent	Expediter	Chief Estimator	Cost Estimator	Take-off Man	Operations Manager	Project Manager	Equip. Superintendent	Warehouseman
Estimating																	
Decision to bid		3		2							1						
Pre-bid conference				4							4	1		4	4	4	
Quantity take-off													1				
Pricing		2		3							3	1					
Markup		4		4							1						
Bid security											1						
Proposal				3							1						
Purchasing																	
Bill of materials				2				3	4		1	1			2		
Subcontracts			3					3	1								
Purchase orders			3					3	1						2		
Expediting									2	1					4		4

1. Originate; does work
2. Must review
3. Must approve
4. Must be advised

Figure 3.4 Company operating chart.

47

3.   Training of Replacements.   Good management encourages the development of young and upcoming talent. People die, retire, or leave the company and must be replaced. Even in cases of sickness or other temporary absence, someone must fill in. The person in a position of responsibility who, because of a dominant personality or for fear of job security, does not develop likely successors from among subordinates is not a good manager and is not serving the best interest of the company.

4.   Decision Making.   A common source of management difficulty is the failure of some member of the organization to make business decisions promptly and positively. Decisions cannot be rushed and must receive due consideration, but undue delay generally only makes the situation worse. Decisions rarely please everyone, but indecision, procrastination, and vacillation serve only to exacerbate the matter further. At each level of the organization, there should be one individual whose function is to decide on operational courses of action. This person should have the benefit of full consultation and recommendations, but the responsibility for decisions rests here, and the individual must act accordingly.

3.13 THE COMPUTER AND COMPANY MANAGEMENT

The extent to which contractors are making computers an integral part of their management procedures is highly variable from one company to another. Larger construction firms are increasingly using computer-based management systems, whereas smaller companies are displaying more reluctance to do so. However, it certainly is not axiomatic that a small construction firm cannot profitably utilize computer procedures. Computers can and do assist contractors of all sizes to conduct their businesses more efficiently. The key is to select the right system for the specific needs of a given company.

For the contractor who wants its business to grow and prosper in the highly competitive marketplace, the computer should be given serious consideration as a means of achieving this goal. The contractor has many options when considering computer application, each of which should be investigated and thoroughly analyzed. A company can utilize a service bureau, use time-sharing services, or acquire in-house hardware. Service bureaus and time-sharing services do not require a large capital outlay and are generally feasible for firms of all sizes. Service bureaus are private firms that rent time on their computer equipment. The contractor need only furnish data in a standard format to the service bureau which then processes it and returns the information to the contractor. Computer service firms that specialize in the area of construction are located in many areas. Time-sharing is an arrangement whereby many computer users in various locations can simultaneously use the capabilities of a large, centrally located computer facility. A typewriter terminal that uses a telephone line is located in the contractor's office and is the device for computer input and output.

In-house computer hardware can be purchased, leased, or rented. The acquisition of an in-house capability is considerably more feasible now than it was a few years ago. Desk size, general purpose, stand-alone machines provide substantial computing capability at a reasonably modest cost. Portable machines can be plugged in anywhere they are needed, including the project field office. With their stored program capability, these small systems have the flexibility to operate independently of a central processor, or in conjunction with a large main-frame computer through a network system. These are fully interactive systems that permit the user to communicate directly with the machine through an intelligent terminal. In obtaining in-house facilities, the contractor should prepare detailed specifications of exactly what it expects from each computer application and place considerable emphasis on software selection. Software, after all, determines if a given system will produce the required information. Criteria for final computer system selection should include machine capability, capacity for expansion, reliability, system support, ease of use, and cost.

The functions listed here are computer applications that contractors are increasingly finding to be beneficial to their operations.

1. Office Administration.   The computer is applied to general accounting, record keeping, financial statements, cash flow control, materials and supplies purchasing, delivery scheduling, cash forecasting, and inventory control.

2. Project Cost Estimating.   Using stored information on productivity and cost, the computer generates a summary schedule of estimated quantities and costs.

3. Payroll and Labor Cost.   Using time card input, the computer prepares payroll checks, periodic and special payroll reports, the payroll register, and updates the employee master files and the project cost files. Also produced are labor production rates and labor unit costs.

4. Equipment Accounting.   The computer maintains records of equipment depreciation, ownership and operating costs, hours of operation, maintenance, production rates, and unit costs. The computer charges equipment time and cost to individual projects and records the location and maintenance schedules for equipment items.

5. Projects.   The computer is applied to project budgeting and scheduling, manpower schedules, progress reports, schedule updating, labor and equipment cost reports, job status reports, project cost forecasts, and progress payments.

4

DRAWINGS AND SPECIFICATIONS

4.1 THE ARCHITECT-ENGINEER

The design of construction projects is performed by architects and/or engineers, the division of responsibility between the two depending on the nature of the construction involved. As a general rule, building construction is primarily of an architectural nature, with structural, electrical, mechanical, and other engineers providing supportive services as may be required. Highway, heavy, utility, and industrial projects are predominantly engineering projects, with architectural services obtained if needed. There are many firms that designate themselves as architect-engineers and that perform planning and design work of both classifications. Within this text, "architect-engineer" is used to designate the organization, person, firm, or team that performs the project design, whether it be architectural, engineering, or a combination of both in makeup.

The architect-engineer can occupy a variety of positions relative to the owner and the contractor. Typical among these are the following:

In-House Capability. Owners sometimes have their own in-house design capability. Some large industrial firms and many public agencies maintain their own design departments. In such a case, the architect-engineer is an integral part of the owner's organization. The construction is usually accomplished by a contractor or contractors under contract with the owner.

Owner-Client. The traditional pattern is where the architect-engineer is a professional design firm that performs its function under contract with the owner. Normally the construction is performed by a contractor or contractors under contract with the owner. There is no contractual relationship between the contractor and the architect-engineer.

Construction Management. Contractually, this is much the same arrangement as owner-client. However, the contracts between owner and architect-engineer and between owner and construction manager provide for extensive cooperation between the architect-engineer and the construction manager from project inception.

Design–Construct. In this instance, the owner contracts with a single party for both design and construction. Although the architect-engineer and the contractor may be related in a variety of ways, three are mentioned here as being commonly used: the contractor may have its own design capability, with architects and engineers on its payroll; the architect-engineer can be a corporate affiliate or subsidiary of the contractor; or the contractor and architect-engineer, both independent firms, can form a joint venture for a given project or contract.

4.2 SERVICES PROVIDED BY THE ARCHITECT-ENGINEER

When the owner has an in-house design capability, there is no need to choose an architect-engineer to design the project. If not, some selection process must be followed. The usual procedure is for the owner to notify various design firms, selected on the basis of personal acquaintance, reputation, or recommendation, about the forthcoming project. Each architect-engineer firm that is interested then presents the owner with information concerning its previous projects, names of ex-clients, qualifications of its staff, special abilities, and similar data. After evaluation of these submissions, discussions are held with those firms who rank highest. The owner then selects the architect-engineer that seems to have the best ability to perform the required services within the stated time frame and in accordance with the owner's budget requirements. A contract is then negotiated which includes mutually agreeable compensation. A private owner may not choose to follow this total procedure, preferring to negotiate immediately a design contract with a selected firm. A standard form of design contract published by The American Institute of Architects (AIA) and widely used for building construction is reproduced in Appendix A. This and other standard forms and documents reproduced in this text are for illustrative purposes only. Because AIA documents are revised from time to time, users should obtain from the AIA the current editions of the documents reproduced herein.

The scope of services required of the architect-engineer is subject to considerable variation, depending on the needs and wishes of the owner. Basic to such services, however, is ascertaining the needs and desires of the owner, developing the design, preparing the several documents required for bidding or negotiation as well as for contract purposes, making an estimate of construction cost, and aiding in the selection of a contractor. In the case of competitive bidding, the documents prepared by the architect-engineer are referred to as the "bidding documents" during the bidding phase. After the contract is signed, these same documents form the contract.

The scope of services provided to the owner by the architect-engineer during field construction depends on the needs and preferences of the owner. Responsibility of the architect-engineer to the owner may cease when the contract documents are finalized and delivered. On the other hand, the owner may require full construction-phase services, including project inspection, the checking of shop drawings, the approval of periodic payments to the contractor, the issuance of certificate of completion, and the processing of change orders. Although the architect-engineer is not a party to the construction contract between the owner and contractor, this contract often conveys certain powers to the architect-engineer such as the authority to decide contract interpretation questions, judge performance, condemn defective work, and stop field operations.

4.3   CONTRACTOR INPUT INTO DESIGN

Although the function of the contracting firm is not project design as such, its experience and expertise can make it a valuable member of a planning and design team. In the traditional and still predominant linear construction process, input from the contractor into the design process does not normally occur. There are occasional instances where the owner or architect-engineer obtains the consultation services of a contractor during the planning and design phases, but this is much more the exception than the rule. However, extensive contractor input into design is a normal part of design–construct and construction management where the team concept prevails.

Where the input of experienced construction people into planning and design is a part of the procedure, the contractor normally enters the picture at an early date. This party does not do the design but provides continued advice concerning general site planning, local work practices and union jurisdictions, labor availability and costs, material availability, delivery times, and alternative work methods and procedures. The contractor prepares cost estimates and construction and procurement schedules and participates in the value-engineering program. The contractor's knowledge of prices and of the availability of materials and services is a valuable source of information to those making design decisions. As one who is familiar with performance, maintenance, and installation costs, the contractor can help assess life-cycle costs and benefits. Under the right circumstances, contractor input into the design process can result in substantial benefits to the owner.

4.4   FEE FOR DESIGN SERVICES

When an architect-engineer acting as a private practitioner performs a design service for a client, the fee may be determined in any one of several possible ways, the most commonly used being the following:

1. Percentage of construction cost.
2. Multiple of salary cost.
3. Multiple of salary cost plus nonsalary expense.

4. Fixed fee or lump-sum fee.
5. Total expense plus professional fee.

Fees for the design professional on publicly financed projects are often subject to statutory or administrative limitations. There are instances in which the design contract between the owner and architect-engineer provides that payment of the fee is contingent in some way on the project cost being within the budget established by the owner. For example, such contracts may provide that the architect-engineer shall not be entitled to any increase in fee when project redesign is required to keep construction costs within an agreed-on amount. It is common for design contracts to provide that the architect-engineer's fee be paid in installments as designated phases of the designer's services are completed.

Traditionally, professional societies of architects and engineers have opposed competitive bidding and price competition as a means of selecting firms to perform professional design services and have included in their codes of ethics prohibitions against competitive bidding by their members. These organizations advocate that design professionals be selected by the owner based on their qualifications and competence with the fee to be determined through negotiation. Objection to competitive bidding for planning and design services is voiced on the grounds that it is not in the client's or public's best interests.

However, in 1978 the U.S. Supreme Court ruled that the National Society of Professional Engineer's ethical ban on competitive bidding for engineering services was not permissible under the Sherman Antitrust Act. The result of this decision is that the professional societies may not bar competitive bidding procedures if those in practice wish to engage in such methods. As a result, it is not now considered unethical to submit or invite such bids. Professional societies, however, may encourage owners to continue using the traditional selection and fee-setting procedures. Professional designers are not required to submit bids and owners may, if they wish, shun competitive procedures in their procurement of professional design services. For instance, most agencies of the federal government are barred from using competitive bidding in the procurement of design services by the Brooks Act (1972). This act mandates the negotiation of design contracts on the basis of demonstrated competence and qualifications for the type of professional service required. Many states have passed mini-Brooks Laws that set forth professional selection and negotiation procedures for state-financed projects.

Following the 1978 Supreme Court decision on competitive bidding, there has been some movement by public agencies to the use of price as one criterion in the selection of architect-engineer firms. The competitive element has been used on a trial basis by the federal government and several states now have such requirements in their procurement statutes.

4.5 RESPONSIBILITY TO THE OWNER

Although the owner and architect-engineer are usually joined together by contract, their exact relationship depends somewhat on the duties being performed by the architect-engineer. In the preparation of construction documents, the architect-engineer firm functions primarily as an independent contractor, but its role is more that of an agent of the owner during the construction phase. The architect-engineer has a fiduciary obligation to its client requiring fairness, trust, and loyalty. The design professional must avoid any conflict of interest which could work to the disadvantage of the owner.

As a professional, the architect-engineer is required to exercise ordinary care and diligence in carrying out its responsibilities. Learning, skill, and experience are expected to the degree customarily regarded as being necessary and sufficient for the ordinary practice of that profession. By the contract of employment, the architect-engineer implies that it possesses ordinary skill and ability and that it will carry out the design with promptness and a reasonable exactness of performance. The designer is responsible for the adequacy of the materials, components, and equipment that are selected and specified. The architect-engineer also bears the responsibility for preparing design documents that are in conformance with applicable building codes, setback requirements, zoning regulations, and environmental requirements. It is not expected that the architect-engineer produce a perfect design or even satisfactory end results. However, the law expects the design professional to carry out the tasks it has undertaken with reasonable standards of care, skill, and performance. If the owner suffers loss or injury because this standard is not met, the architect-engineer is liable.

Liability of the designer to the owner may stem from an express or implied warranty in the design contract, or from negligence in the performance of the architect-engineer's duties under the contract. If the design professional is found to have given an express or implied warranty of the sufficiency of the design or that the structure would be reasonably suitable for the purpose intended, then the architect-engineer is strictly liable for damages caused by a breach of the warranty, and it is not necessary that the owner prove any specific negligence.

Although the architect-engineer firm can be held responsible for lack of care, diligence, or skill, it is not normally considered to be negligent because of errors in judgment. However, there is a definite trend in the courts toward expecting a greater degree of perfection and foresight on the part of the architect-engineers and holding these parties responsible for their negligence. In addition, if the architect-engineer represents itself to be a specialist in a certain type of work, it will likely be held responsible for a greater degree of competence and care than would a general practitioner.

If the architect-engineer selects and employs consultants such as electrical engineers, mechanical engineers, acoustical engineers, or landscape architects to design or advise concerning specialized portions of the project, the architect-engineer remains responsible to the owner for the overall adequacy of the completed design. In this regard, the design professional has a nondelegable duty to the owner and impliedly promises that all services will be properly performed. If a consultant's work should prove to be faulty, the architect-engineer is liable to the owner for any resulting damages. The consultant is, of course, accountable to the architect-engineer for which the special work was done.

If the architect-engineer has an inspection responsibility during construction, it has the duty to ensure that the contractor materially follows the drawings and specifications. It has the responsibility to see that the owner gets substantially the structure called for by the construction contract.

Architect-engineers are liable for damages resulting from their design errors, and the rule regarding the absence of responsibility of the contractor for design defects in the completed structure has become well settled in most jurisdictions. Where the contractor has followed the plans and specifications prepared by the architect-engineer and these documents prove to be defective or insufficient, the contractor will not be responsible for any loss or damage resulting solely from the design defect. This rule is subject to the absence of any negligence on the part of the contractor, or any express warranty by the contractor that the drawings and specifications were sufficient or free from defects. In addition, the contractor has a duty to notify the owner and/or architect-engineer if a design defect should be detected during the construction of the project. The liability of architect-engineers for design errors is not usually affected by contract clauses requiring the contractor to visit the site, check the drawings and specifications, and to be informed concerning the requirements of the work.

Architect-engineers sometimes include exculpatory clauses in their design contracts in an attempt to limit their professional liability. Some of these clauses provide that the architect-engineer will not be liable to the owner for damages resulting from negligence of the designer. Other clauses limit the potential liability of the architect-engineer to a specific amount. However, the extent to which such exculpatory provisions are enforceable by the courts is often uncertain.

4.6 LIABILITY TO THIRD PERSONS

Although it was not always so, the law now generally recognizes that an architect-engineer can be held liable to a third party if the designer caused bodily injury or property damage to that party by reason of negligence or failure in duty of the designer. In addition, the architect-engineer may in some instances be held liable to third parties for economic loss suffered as a result of its negligence. In this context, "third party" refers to any party who is a stranger to the design contract between the architect-engineer and owner.

Third-party liability can arise from a project either during its construction or after its acceptance and occupancy by the owner. Under this concept of liability, architect-engineers now find themselves subject to damage claims by contractors, subcontractors, construction workers, sureties, suppliers, lenders, and outsiders lawfully on the project premises where the negligent performance of duty by the architect-engineer allegedly caused or contributed to harm or injury.

In actions commonly referred to as third-party suits, recent years have seen architect-engineers held liable for a variety of injuries suffered by workers during construction and members of the general public after project completion. There have been several cases in which an architect-engineer firm, being responsible in contract with the owner for job inspection, has been judged responsible for the safety of the work and has been made liable for damages when it failed to take corrective measures although it knew or should have known of a dangerous job condition. Negligence suits stemming from injuries suffered on completed projects have been filed by parties whose injuries were caused by alleged improper or inadequate design. An architect-engineer's continuing liability for completed projects often is limited by special statutes of limitations, a subject discussed in the following section.

In any discussion of tort liability arising out of negligence, the matter of strict liability arises. There is an increasing trend in this country toward imposing strict liability for injuries caused to the user or consumer of mass-produced products on the basis of implied warranty. Product liability, or strict liability, refers to liability without proof of fault; that is, liability for damages is not based on a demonstration of negligence on the part of the producer of the goods. Under this theory, the person suffering injury or damages can receive compensation if it can be proved the product was defective and this defect caused the injury or other loss. The courts have not yet applied product liability to architect-engineers although it has been applied to builder-vendors who produced and sold tract houses that turned out to be defective. These findings were based on implied warranties of workmanship and habitability.

4.7 STATUTES OF LIMITATIONS

Most states now have special statutes of limitations that apply to the accountability for damages that arise out of a defective and unsafe condition created as the result of an improvement to real property. These statutes apply to architect-engineers and construction contractors and establish a time beyond which these parties are no longer liable for damages arising out of completed construction projects. Most of these statutes provide that the time during which the architect-engineer and contractor remain liable starts with the substantial completion of the work. These statutory periods vary from four to twenty years in the various states, with the average being about seven years.

In those states without such statutes, the architect-engineer and contractor must rely on the states' general statutes of limitations. The time period within which a given action must be brought under these statutes varies, but a typical statute provides for a three-year period for torts (negligence) and a six-year period for breach of contract. However, a serious question concerning these general statutes of limitations is when the statutory time begins. The usual provision for negligence suits against architect-engineers and contractors is that time starts when the plaintiff is injured. The right of the owner to sue for breach of contract ordinarily starts when the work is accepted. However, there can be exceptions to this when the construction defect is concealed. The result of these conditions is that in those states without special statutes of limitations, the architect-engineer and contractor are indefinitely vulnerable to suits charging negligence or breach of contract. In this regard, it must be noted that some of these special statutes of limitations have been declared unconstitutional during recent years.

4.8 PROJECT DESCRIPTION

The nature and extent of the construction to be done, the materials to be provided, and the quality of workmanship required are described by the drawings and specifications. Complementing each other very closely, the drawings and specifications present a complete description of the work. The drawings portray pictorially the extent and arrangement of the components of the structure. The specifications describe verbally the materials and workmanship required. These documents serve three important functions. First, they serve as a basis for competitive bidding or contract negotiation. Second, they serve as contract administration documents during the construction phase, describing the work to be accomplished and defining the rights and duties of the participants. Third, they are contract documents that constitute the basis for the settlement of claims, disputes, and breaches of contract. It is important, therefore, that these documents be carefully examined by the contractor during the bidding or negotiation phase.

When lump-sum, unit-price, or guaranteed maximum-price contracts are involved, the drawings and specifications must be in a complete and detailed form before the contractors are brought in for bidding or contract negotiation. Contractors can scarcely be expected to obligate themselves to the terms of a fixed-sum contract without being able to establish with some precision the nature and extent of the work involved. When a cost-plus type of contract is under consideration, the documents may be more rudimentary. In such cases, detailed drawings and specifications still must be provided, but they can follow as needed during the construction period. However, it is always preferable to have the drawings and specifications in as complete and detailed condition as possible before the owner discusses costs, completion dates, and other such matters with the contractor. When incomplete and preliminary drawings and specifications form the basis for such discussions, subsequent development of the design nearly always creates changes and complicating features that alter the original agreements and understandings.

4.9 OWNERSHIP OF THE DESIGN

Ownership of the drawings and specifications and their possible unauthorized use can be a matter of considerable importance to the architect-engineer and the owner. Where the owner is a public agency, these documents often become the sole property of the owner as a matter of law. Otherwise, most design contracts between the owner and architect-engineer explicitly designate which of the two parties has legal title to the design documents that are the subject of the agreement. The usual provision is that the design documents are and remain the property of the architect-engineer. However, many owners include a specific provision in the design contract that conveys all rights to and ownership of the documents to the owner. Agreements of this type, however, do not bind third parties to the contract. Accordingly, such a provision does not protect the interests of the owner or architect-engineer by preventing third parties from making unauthorized use of the drawings and specifications, such as building a duplicate structure.

To protect drawings and specifications against unauthorized usage, a statutory copyright can be obtained in accordance with the U.S. Copyright Law of 1976. This federal statute gives copyright protection to the architect-engineer unless there is an express agreement otherwise. The owner for whom the documents are prepared can obtain the copyright if they were "made for hire." This condition is generally met if the drawings and specifications are prepared by an employee as a part of his regular employment or where the preparation is specially ordered or commissioned by the owner. Even if the design is made for hire, however, the copyright law provides that the drawings and specifications belong to the architect-engineer unless there is a written agreement otherwise. Although a copyright does not protect ideas or processes contained in the documents, it does prohibit any display, sale, or physical copying of the design. Expiration dates for the copyright of works is 50 years after the creator's death or, if the work was made for hire, 100 years after its creation.

4.10 THE DRAWINGS

The drawings, or plans, are instrumental in the communication of the architect-engineer's intentions concerning the structure it has conceived and designed. They portray the physical aspects of the structure, showing the arrangement, dimensions, construction details, materials, and other information necessary for estimating and building the project. Drawings are individually prepared for almost every project. A job covered by drawings that are complete, intelligible, accurate, detailed, and well correlated can be priced much more realistically and be better constructed than one described by sketchy, poorly drawn, ambiguous, and incomplete documents. When well prepared documents are provided, disputes and claims for extra payment during construction are minimized, and the owner is likely to get a much better finished product at a lesser cost.

From original drawings prepared by the architect-engineer, booklets or rolls of drawings are reproduced. Custom and usage have evolved more or less standard classifications of drawings and a prescribed order in which they appear in a set. To illustrate, the drawings for a building typically include subgroups such as plot plan, structural, architectural, plumbing, mechanical, and electrical, appearing in that order. The sheets of each category are designated by an identifying capital-letter prefix and are numbered separately and consecutively. The drawing format varies with the type of construction, however. The makeup of the drawings differs considerably among building, industrial, heavy, highway, and utility construction.

Interactive computer systems are now being used to produce finished drawings from rough, handmade sketches. Thus far, most of the application of computerized drawing has been with schematic drawings such as flow diagrams and piping and electrical layouts that are not drawn to scale.

4.11 STANDARDIZED DRAWINGS

The use of standardized drawings is quite widespread in certain areas of the construction industry. Such drawings are usually standard design details, showing materials, dimensions, arrangement, configuration, and other information. Designs for sewer and street work and other municipal construction are common applications. Street and alley paving, manholes, curb and gutter, drop inlets, catch basins, sewer connections, valve chambers, hydrant and motor settings, and gate wells are examples of municipal work readily susceptible to the standardization of design. Many engineering aspects of bridge and highway construction are also illustrated by standard design details, which may be physically incorporated into the project drawings or included by reference only. The development of such details is of considerable convenience to the construction process and is conducive to economies of fabrication and installation.

4.12 THE SPECIFICATIONS

Specifications are written instructions concerning project requirements. The drawings show what is to be built, and the specifications describe how the project is to be constructed and what results are to be achieved. Historically, the word "specifications" has referred to specific statements concerning technical requirements of the project, such as materials, workmanship, and operating characteristics. However, it has become customary to include the bidding and contract documents together with the technical specifications, the entire aggregation being variously referred to as the project manual, the construction documents book, or most commonly, simply as the specifications or "specs."

Large projects or those involving an unusual number of legal or contractual formalities may have a separate booklet of nontechnical forms and instructions. These document forms are required for bidding and to inform the contractor concerning the contract terms and provisions. A contractor cannot be expected to enter into a contract with the owner without first having seen the several contract documents. Such nontechnical forms vary considerably with the project and can include such things as invitation to bid, instructions to bidders, general conditions, supplementary conditions, proposal form, bid bond form, contract bond forms, list of prevailing wage rates, noncollusion affidavit, agreement form, and list of subcontractors. All of these items are discussed in this chapter or in the chapters following.

The preparation of the specifications is an important part of an architect-engineer's responsibility and is an arduous chore requiring considerable knowledge and writing ability. Every specification writer must rely on material and equipment manufacturers to furnish technical data and accurate descriptions of their products. The architect-engineer combines this information with its own experience and judgment to specify the right product or material. The volume of manufacturers' literature has reached such enormous proportions that sophisticated systems of storing and retrieving product information have been developed. There are a number of commercial information retrieval systems currently available. Many architect-engineers have established computerized systems of specification writing using word processors and stored master specifications.

4.13 SPECIFICATION DIVISIONS

Specifications are issued by the architect-engineer in duplicated form with stiff paper covers. For large projects, this document can assume the dimensions of a good-sized catalog and may be issued in more than one volume. It is usual practice for the specifications to be segmented into standardized divisions generally used by the construction industry. As mentioned previously, the initial divisions of the specifications normally contain the nontechnical provisions of the contract. The succeeding divisions, which consitute the bulk of the specifications, contain the technical provisions that describe the workmanship and materials for each of the individual construction segments, such as site work, concrete, masonry, carpentry, mechanical, and electrical.

The technical divisions that appear within a set of specifications depend on the nature of the project and the general category of construction. Specifications covering the construction of a building, for example, will have only a general resemblance to those for an engineering work such as a highway, tunnel, or marine structure. Appendix B presents two specification outlines that illustrate this point. The first part of this appendix presents the Broadscope Section Titles of MASTERFORMAT as promulgated by the Construction Specifications Institute (CSI) and Construction Specifications Canada (CSC). This format is used primarily for building construction. The second part of Appendix B is a typical specification outline for engineering construction.

The CSI 16-division format has been widely adopted by many trade, technical, professional, and public groups associated with design and construction. It is broadly used by architect-engineers in the preparation of specifications for building construction projects. Most building materials are now identified by their CSI classification numbers for filing purposes. However, there are other established specification formats in the construction industry. For example, many state highway departments follow the standard format for highway construction specifications sponsored by the American Association of State Highway and Transportation Officials (AASHTO). Many projects, of course, have job specifications whose formats are nonstandard.

4.14 THE GENERAL CONDITIONS

The general conditions, sometimes called general provisions, set forth the manner and procedures whereby the provisions of the contract are to be implemented according to accepted practices in the construction industry. These conditions are intended to govern and regulate the obligations of the formal contract; they exert no effect on any remedy at law that either party to the contract may possess. They are not intended to regulate the internal workings of either party to the agreement, except insofar as the activities of one may affect the contractual rights of the other party or the proper execution of the work.

Although the headings and topics included within different sets of general conditions vary, there is a certain similarity of subject matter. The general conditions will typically include the following items:

1. Definitions.
2. Contract documents.
3. Rights and responsibilities of owner.
4. Duties and authorities of architect-engineer.
5. Rights and responsibilities of contractor.
6. Subcontractors.
7. Separate contracts.
8. Time.
9. Payments and completion.
10. Changes in the work.
11. Protection of persons and property.
12. Insurance and bonds.
13. Disputes.
14. Termination of the contract.
15. Miscellaneous provisions.

4.15 STANDARDIZED GENERAL CONDITIONS

Standardized sets of general conditions have been developed by various segments of the construction industry. The American Institute of Architects; the National Society of Professional Engineers; the Associated General Contractors of America; various branches of the federal, state, and municipal governments; and others have compiled standardized sets of general conditions. The general conditions compiled and published by The American Institute of Architects have found wide acceptance by the building construction industry and constitute a part of the specifications of most private and many public building projects. These conditions are reproduced in full as Appendix C.

Standard forms of general conditions have the advantage that their record of use has proven them to be workable, and many of the provisions have been tested in the courts. In addition, they have evolved into a form that has stood the test of time and experience and have become familiar to contractors who clearly understand their meanings and implications.

4.16 SUPPLEMENTARY CONDITIONS

Any standard set of general conditions is intended to apply to a relatively broad range of construction and must be amended and/or supplemented at times to conform to the idiosyncracies of a given project. This is accomplished by a section of the specifications, called the supplementary conditions, which immediately follows the general conditions. Supplementary conditions are occasionally also referred to as special conditions. Common examples of necessary amendments to the general conditions are the number of sets of contract documents to be furnished the contractor, limitations on surveys to be provided by the owner, special instructions to the contractor when requesting material substitutions, changes in insurance requirements, and special documentations required by the owner as a condition of final payment.

Additional articles must frequently be included that supplement those of a standard set of general conditions. Conditions of project location, order of procedure, times during which the work must proceed, owner-provided equipment, other contracts, unusual contract administration requirements, early occupancy by owner, time of project completion, and liquidated damages are examples of such distinctly individual contract requirements. Appendix D, which contains a representative set of supplementary conditions, illustrates the specialized nature of this division of the specifications.

4.17 THE TECHNICAL SPECIFICATIONS

The technical provisions of the specifications are normally presented in approximately the same general sequence as the corresponding construction operations actually proceed in the field and are subdivided in accordance with the usual construction craft jurisdictions. It is customary that a separate division of the specifications be devoted to each major type of construction operation that will be involved, such as excavation, concrete, structural steel, piping, insulation, and electrical work.

Many projects, buildings for example, include elements of such a nature that it is impractical or impossible to verify their adequacy or quality by field tests after completion of the construction. Under such circumstances, the technical specifications prescribe the materials and the workmanship standards required.

It is possible with other construction elements to measure, test, or otherwise prove the service performance of the finished product. In such cases, the specifications frequently specify only the desired end result. Such performance or end-result specifications are now widely used.

4.18 PERFORMANCE SPECIFICATIONS

A performance specification is one that describes the required performance or service characteristics of the finished product or system without specifying in detail the methods to be used in obtaining the desired end result. This type of specification makes the contractor responsible for obtaining the results expected, and an end product is required that will meet the acceptance tests and standards specified. The selection of construction methods and procedures is left up to the contractor.

When service requirements can be established and measured by means of some practical test procedure, the use of the end-result type of specification can be advantageous. The contracting firm is made responsible for obtaining a satisfactory result. By leaving it free to exercise its ingenuity, skill, and experience to the fullest extent in achieving the desired result, the cost of construction may well be reduced.

4.19 MATERIAL AND WORKMANSHIP SPECIFICATIONS

A material and workmanship specification, also referred to as a design specification, describes the kinds and types of materials to be provided, their physical and performance properties, their sizes and dimensions, the standards of installation and workmanship, and inspection and tests required for verification of quality. It is common practice that a specific brand name or names be listed with model numbers and other data to establish the standard of quality desired. The use of substitutes by the contractor is discussed in Section 4.22.

In some material and workmanship specifications, the architect-engineer need not stipulate detailed construction methods except with reference to the quality of workmanship. For example, matters of finish, appearance, tolerances, clearances, noise, and like properties of the finished product are indicators of workmanship standards and are made subject to final approval by the architect-engineer. Many building specialties, such as painting, are examples of this. However, in other areas, such as structural work, it may be necessary that the architect-engineer specify exact dimensions, methods, materials, weights, quantities, and procedures. In such a case, the contractor must perform and construct in accordance with the contract and has little or no discretion. Here is a case where the architect-engineer normally assumes the responsibility for the adequacy of the end product.

4.20 MATERIAL STANDARDS

There are many standard material specifications that are sponsored by the federal government and a variety of technical societies. These specifications have been devised by specialists in their fields and are accepted as authoritative by the industry. Federal specifications prescribe technical requirements for materials, products, and services procured by the federal government. The American Society for Testing and Materials (ASTM) promulgates large numbers of specifications pertaining to the physical, chemical, electrical, thermal, performance, and acoustical properties of a wide variety of materials, including practically all those used in construction. ASTM standards are widely incorporated into construction specifications by reference as a means of controlling the quality of portland cement, reinforcing steel, asphalt, insulation, and a host of other construction products. The American National Standards Institute (ANSI) publishes widely used material and product standards of great variety including those used in the construction industry. The American Association of State Highway and Transportation Officials (AASHTO) publishes test standards pertinent to highway and airfield construction.

In addition to these standards, which cover broad ranges of materials, there are many other standard material specifications of a more or less specialized nature. The American Institute of Steel Construction, the American Water Works Association, the American Institute of Timber Construction, the American Concrete Institute, the American Welding Society, the Structural Clay Products Institute, and many others issue standards that apply to the quality and construction application of their product specialties. These standards establish reliable quality criteria for particular work classifications.

4.21 CLOSED SPECIFICATIONS

If a material or equipment specification is worded such that only one or a few proprietary products will be acceptable and if no provision is made for substitutions, it is known as a closed specification. A closed specification is achieved either by listing an exclusive brand-identified product or products or by wording a performance requirement such that only one or a few trade-name products can satisfy it. The underlying purpose of a closed specification is to ensure that only construction products of a desired type or level of quality will be furnished by the contractor.

On private work it is possible to use a closed material specification in which only one manufacturer is named as acceptable. This procedure, however, is not generally considered to be in the best interests of the owner because it eliminates the several advantages of competition. It is more common practice on private projects to name two or three manufacturers, the product of any one of which will be acceptable. This method of specifying is sometimes used for public construction although, more commonly, public projects must be bid under completely open conditions.

4.22 OPEN SPECIFICATIONS

On public works, the architect-engineer cannot ordinarily write a closed material specification because of laws requiring the use of nonrestrictive specifications. Consequently, the architect-engineer firm must word its specifications so that the products of various manufacturers are acceptable, whether or not they are mentioned by name, providing they can meet the prescribed operational and quality standards.

In order to establish a standard of quality, the architect-engineer will sometimes specify a brand name or names, followed by the term "or approved equal." In this context, a substitute need not be identical in every respect to the product specified as a standard of quality. A substitute may be equal but be different in appearance, size, configuration, or design. The equality of an alternate product is established on the basis of the quality and performance of the substitute when compared to the brand-name product specified. The equality of substitutes proposed by the general contractor or a subcontractor is decided by the architect-engineer or owner. For each proposed substitution, samples, descriptive and technical data, test reports, and other information must be submitted by the contractor as a means of demonstrating equality. Even though the or-equal clause is useful in allowing substitutions when cost savings are possible or when the availability of materials specified is uncertain, this provision has been and continues to be troublesome for both the architect-engineer and the contractor.

Because it is often not possible to obtain approval for a material substitution before bids are submitted on a competitively bid job, contractors are frequently faced with difficult decisions as to what material prices to use. It is not uncommon that the lowest price received by the contractor for a certain item applies to a brand not listed in the specifications. The contractor is faced with the dilemma of whether or not to use this low price and gamble that the architect-engineer will subsequently approve the substitution. Alternatively, if the bidding firm states in its proposal that the bid amount is based on stipulated substitutions, its proposal may be rejected by the bidding authority. Although the contractor has the obligation of providing proof that a substitution meets the standards originally specified, the architect-engineer must bear the responsibility for its decisions of material equality. Architect-engineers can be' held liable for the inadequacy of approved substitutions and must make such decisions with care. In spite of the troubles encountered, however, the or-equal approach to specification policy is commonly used.

Materials for federal government projects are designated by federal specification numbers. Agencies of the federal government use a standardized system of specifying desired materials, each specification bearing a distinctive identification number and being listed in the federal catalog of standard specifications. When no federal specification exists, it is usual practice to specify a standard brand or brands with provision for the substitution of alternate products of equal quality and performance.

4.23 OTHER SPECIFICATION TYPES

As discussed in the preceding sections, a closed specification has the advantage of ensuring the desired quality of material but the disadvantage of limiting or eliminating competition among suppliers. An open specification will provide competition but, by permitting substitutions, introduces the possibility of materials being used that are inferior to those desired. There have been several specification schemes devised to obtain the advantages and minimize the disadvantages of the open and closed concepts. Unfortunately, a completely satisfactory procedure has yet to be devised.

One of these combination type specifications is called a base-bid material specification or substitute-bid specification. When this procedure is followed, items of equipment and materials are identified in the technical specifications by a manufacturer's name, model, catalog number, or other specific data. The words "or equal" do not appear. The intent is that only those items listed are to be used in preparing the base bid. However, the bidding contractors may offer alternate items, either on the proposal form or as an attachment to the proposal. These alternate proposals are accompanied by full descriptions and technical data, together with a statement of the cost additional to or deductive from the base bid if the substitution is approved. The architect-engineer or owner does not approve or disapprove such alternates before the bid opening, and the submission of material or equipment alternates is voluntary with the contractor. The low bidder is determined on the basis of the base bid. After the bids have been submitted, decisions about whether to accept or reject any or all of the alternate proposals suggested by the bidding contractors are made. If decisions concerning approved substitutions are made prior to execution of the contract, they can be suitably incorporated into the contract documents. Otherwise, approved substitutions can subsequently be incorporated into the contract by change order.

Another scheme sometimes used is for a closed specification to be written, but with a provision that proposals for substitutions can be submitted by a bidder up to a stipulated number of days prior to bid opening. Notice of any approved product substitution is circulated to all bidding contractors by addendum (see Section 4.25). After the deadline for substitute requests has passed, no further substitutions are permitted.

4.24 STANDARD SPECIFICATIONS

Standardized specifications, including both the technical and nontechnical provisions, are used by some segments of the construction industry. These have found considerable application in highway, bridge, and utility construction. These preprinted standard specifications are issued by the contracting agency and may be obtained by any interested party. Although they may not form a physical part of the specification booklet for a specific project, these specifications are made a part thereof by reference. The issued specifications for the individual projects consist merely of the proposal, bond, and agreement forms, together with any necessary modifications and special provisions to the standard specifications. This practice can save considerable time and effort in the preparation of project specifications and is conducive to bidding and construction uniformity.

Similar standard specifications in other fields of construction have been prepared, covering the work of the various trades. Many government, state, and city agencies utilize these standard trade specifications, augmenting them with modifications in order to make them conform to the peculiarities of the project under consideration. However, to date, the use of standard specifications by the construction industry has met with only limited acceptance, with the notable exception of the highway, bridge, and utility categories. Building construction in particular continues to favor the customized specification approach.

4.25 ADDENDA

When a project is being competitively bid, it is occasionally necessary during the bidding period to make changes, modifications, corrections, or additions to the bidding documents. Notice of such revisions is made by means of an addendum issued by the owner or architect-engineer and sent to all bidders of record. On public projects, an addendum is sometimes referred to as an amendment. These addenda serve to notify the bidders of changes in the owner's requirements, corrections of detected errors or oversights, clarifications or interpretations of the various provisions of the bidding documents, changes in the design, and similar matters. Changes in the drawings are accomplished by the submittal of revised sheets.

Such corrections, changes, interpretations, and clarifying answers must be in writing. It is the responsibility of the architect-engineer or owner to see that copies of all addenda promptly reach the parties who hold plans and specifications. Each addendum is made a part of the contract documents and must receive the full attention of all parties who are preparing bids covering any or all portions of the project.

4.26 ON-SITE INSPECTION SERVICES

To ensure that the construction work is carried out in accordance with the contract documents and that the owner receives the structure contracted for, some form of on-site inspection is customary. These services are normally provided by either the architect-engineer or the owner. The on-site person is referred to by a wide variety of titles, there being no generally accepted nomenclature in this regard. Many public owners refer to their representatives as project engineers, field engineers, or resident engineers. Architect-engineers commonly use the terms of observer, field representative, contract administrator, or inspector. Small to medium projects do not ordinarily require the full-time presence of a job inspector. In such cases, the job is visited frequently enough to ascertain that the work is being properly performed. Ordinarily, large projects deserve and require continuous inspection and observation.

The purpose of field inspection is to see that the intent of the drawings and specifications is carried out and to help the contractor avoid making errors. The inspector has no authority to give directions, render interpretations, or change the contract requirements. The inspector is not present to manage the job, direct the work, or to relieve the contractor from any of its obligations. He cannot tell the contractor what to do or how to do it, nor interfere in field operations unless it is to prevent something from being done improperly. The function of the inspector is to observe the construction process and to ensure compliance with contract requirements by the contractor.

The providing of field inspection services significantly increases the architect-engineer's or owner's vulnerability to a wide range of potential liabilities. To illustrate, the responsibility for field inspection has led to the liability of architect-engineers for construction defects caused by a contractor's failure to follow requirements of the contract documents. Design professionals have been held responsible for contractors' safety procedures and construction methods and have been held accountable for damages to workmen and members of the public injured by construction operations.

As a result of these liabilities, architect-engineers have made extensive revisions of the contract language that defines their field inspection responsibility. The word "supervision" has been especially troublesome for architect-engineers because that word suggests that the designer has some degree of control and thus responsibility for the contractor's day-to-day operations. Subparagraph 1.5.4 of Appendix A illustrates the current wording as used by The American Institute of Architects. It is to be noted that this particular provision calls for only periodic job visits. More extensive full-time inspection requires a further agreement between the owner and architect-engineer.

5

COST ESTIMATING
AND BIDDING

5.1 GENERAL

Construction estimating is the compilation and analysis of the many items that
influence and contribute to the cost of a project. Estimating, which is done
before the physical performance of the work, requires detailed study of the
bidding documents. It also involves a careful analysis of the results of the
study in order to arrive at the most accurate estimate possible of the probable
cost, consistent with the bidding time available and the accuracy and completeness of the information submitted.

Construction costs are estimated to serve a variety of purposes, and much
of the credit for the success or failure of a contracting enterprise can be ascribed to the skill and astuteness, or lack thereof, of its estimating staff. If the
contracting firm obtains its work by competitive bidding, it must be the low
bidder on a sufficient number of the projects it bids if it wishes to stay in business. However, the jobs it obtains must not be priced so low that it is impossible to realize a reasonable profit from them. In an atmosphere of intense
competition, the preparation of realistic and balanced bids requires the utmost
in good judgment and estimating skill.

Although negotiated contracts frequently lack the competitive element, the
accurate estimating of construction costs constitutes an important aspect of
such contracts. The contractor is expected to provide the owner with reliable
advance cost information, and its ability to do so determines in large measure
its continuing ability to attract owner-clients. In design–construct and construction management contracts, the contractor is called on to provide expert
cost assistance and advice as the design develops. The estimation of costs is
a necessary part of any construction operation.

It must be understood that construction estimating bears little resemblance
to the compilation of industrial "standard costs." By virtue of standardized

conditions and close plant control, a manufacturing enterprise can arrive at the total cost of a unit of production in an almost exact way. Construction estimating, by comparison, is a relatively crude process. The absence of any appreciable standardization of conditions from one job to the next, coupled with the inherently complicating factors of weather, materials, labor, transportation, locale, and a myriad of others, makes the advance computation of exact construction costs a matter more of accident than of design. Nevertheless, on the whole, construction estimators do a remarkably good job considering the project imponderables involved.

There are probably as many different estimating procedures as there are contractors. In any process involving such a large number of intricate manipulations, innovations and variations naturally result. The form of the worksheets, the order of procedure, the mode of applying costs—all are subject to considerable diversity, procedures being developed and molded by the individual construction company to suit its own needs. Rather than attempt any detailed discussion of estimating procedures, this chapter presents only the general aspects of construction estimating.

5.2 LUMP-SUM ESTIMATES

Cost estimates in the field of building construction are customarily prepared on a lump-sum basis. Under this procedure, a fixed sum is compiled for which the contractor agrees to perform a prescribed package of work in full accordance with the drawings and specifications. Lump-sum estimates are applicable only when the nature of the work and the quantities involved are well defined by the bidding documents.

Lump-sum estimating requires that a detailed quantity survey, or "quantity takeoff," be made. This is a complete listing of all the materials and items of work that will be required. Using these work quantities as a base, the contractor computes the costs of the materials, labor, equipment, subcontracts, taxes, insurance, overhead, and contract bond. The sum total of these individual items of cost constitutes the anticipated overall cost of the construction. Addition of a markup yields the lump-sum estimate that the contractor submits to the owner as its bid for doing the work.

5.3 UNIT-PRICE ESTIMATES

Engineering construction projects are generally bid not on a lump-sum basis but as a series of unit prices. Unit-price estimates can be compiled when quantities of work items may not be precisely determinable but the nature of the work is well defined. The bidding schedule of Appendix E shows a typical list of bid items. It should be noted that an estimated quantity is shown for each item. These quantity estimates are those of the architect-engineer and are not guaranteed to be accurate. When unit-price proposals are involved, a somewhat different estimating procedure from that described in the previous section must be followed.

A detailed quantity survey is made, much as for lump-sum estimates, but a separate survey is needed for each bid item. This survey not only serves as a basis for computing costs but also checks the accuracy of the architect-engineer's estimated quantities. A total project cost, including labor, equipment, materials, subcontracts, overhead, taxes, insurance, markup, and contract bond, is compiled just as in the case of a lump-sum estimate, but all costs are kept segregated according to the individual bid item to which they apply. More information on this process is presented in Section 5.38.

When computing unit prices, the contractor must keep several important factors in mind. The quantities as listed in the schedule of bid items are estimates only. The contractor will be required to complete the work specified in accordance with the contract and at the quoted unit prices, whether quantities greater or less than the estimated amounts are involved. This requirement is often modified, however, by contract provisions for the equitable adjustment of the contract price when the actual quantity of a pay item varies more than a stipulated percentage above or below the quantity estimated by the architect-engineer. Values of 15 to 25 percent are used in this regard. All items of material, labor, supplies, or equipment that are not specifically enumerated for payment as separate items, but which are reasonably required to complete the work as shown on the drawings and as described in the specifications, are considered as subsidiary obligations of the contractor. No separate measurement or payment is made for them.

5.4 APPROXIMATE ESTIMATES

The fixed-sum costing procedures referred to in Sections 5.2 and 5.3, which are based on a detailed quantity survey, furnish the most accurate and reliable estimates possible of what future construction costs will be. The results of these exhaustive and very detailed methods are often referred to as detailed estimates. A detailed estimate of project cost is what the contractor will normally compile for bidding or negotiation purposes when the design documents are finalized and a final working estimate is required.

For a variety of reasons, however, the contractor may wish to determine an approximate or conceptual cost estimate by means of some shortcut method. It may be desirable to make a quick and independent check of a detailed cost estimate. The general contractor may wish to compute an approximate cost of work normally subcontracted, either to serve as a preliminary cost in its bid or to check quotations already received from subcontractors. Negotiated contracts with owners are sometimes consummated while the drawings and specifications are still in a rudimentary stage. The contractor, in such a case, must compute a target estimate for the owner by some approximate method. Approximate estimates are necessarily involved when the contractor is called on to provide cost information as the design develops and progresses.

The making of preliminary estimates is an art quite different from the making of the final, detailed estimate of construction cost. Fundamentally, all approximate price estimates are based on some system of gross unit costs obtained from previous construction work. These unit costs are extrapolated forward in time to reflect current prices, market conditions, and the peculiar character of the job now under consideration. Some methods commonly used to prepare preliminary estimates are listed here.

Cost-per-Function Estimate.   An analysis based on the estimated cost per item of use, such as cost per patient, student, seat, car space, or unit of production.

Square-foot Cost Estimate.   An approximate cost obtained by using an estimated price for each square foot of gross floor area.

Cubic-foot Cost Estimate.   An estimate based on an approximated cost for each cubic foot of the total volume enclosed.

Modular Takeoff Estimate.   An analysis based on the estimated cost of a representative module, this cost being extrapolated to the entire structure, plus the estimator's assessment of common central systems.

Partial Takeoff Estimate.   An analysis using quantities of composite work items that are priced using estimated unit costs. Preliminary costs of projects can be computed on the basis of making estimates of the probable costs of concrete in place, per cubic yard; structural steel erected, per ton; excavation, per bank cubic yard; hot-mix paving in place, per ton; and the like.

Panel Unit Cost Estimate.   An analysis based on assumed unit costs per square foot of floors, perimeter walls, partition walls, and roof.

Parameter Cost Estimate.   An estimate involving unit costs, called parameter costs, for each of several different building components or systems. The costs of site work, foundations, floors, exterior walls, interior walls, structure, roof, doors, glazed openings, plumbing, heating and ventilating, electrical work, and other items are determined separately by the use of estimated parameter costs. These unit costs can be based on dimensions or quantities of the components themselves or on the common measure of building square footage.

The unit prices used in conjunction with the foregoing approximate cost methods can be extremely variable, depending on specific contract requirements, geographical location, weather, labor productivity, season, transportation, site conditions, and other factors. There are many sources of such cost information in books, journals, magazines, and the general trade literature. Unit costs are also available commercially from a variety of proprietary sources as well as from the contractor's own past experience. In addition, there are many forms of national price indexes which are useful in updating cost information of past construction projects. When using such costs or cost indexes, care must be taken that the information is adjusted as accurately as possible to conform to local and current project conditions.

5.5 ADVERTISEMENT FOR BIDS (PUBLIC CONTRACTS)

In all jurisdictions laws regulate and control the award of public construction projects. These legal requirements start with the first step in the construction process; that is, notice must be given to interested and qualified members of the construction industry in advance of the bidding on any project financed by public funds. In addition, all bidders must be treated alike and afforded an opportunity to bid under the same terms and conditions.

Public agencies must conform with applicable regulations relating to the dissemination of information pertaining to the bidding of construction projects. The contracting agency may be required to give notice by placing advertisements for bids in newspapers, magazines, trade publications, or other public media. How often and over what period of time such notices must appear vary from jurisdiction to jurisdiction. Weekly notices for two, three, or four consecutive weeks are common. An advertisement of this type is commonly referred to as a "Notice to Bidders" or an "Invitation to Bid."

The advertisement describes the nature, extent, and location of the work and the authority under which it originates, together with the time, manner, and place in which bids are to be received. The place where bidding documents are available and the deposit required are designated, and information is listed concerning the type of contract, bond requirements, dates when the work is to be started and completed, terms of payment, estimate of cost, and the owner's right to reject any or all bids. Statements concerning minimum wage rates and applicable labor statutes are also commonly included. Figure 5.1 is a typical example of such an advertisement.

Rather than advertising as such, agencies of the United States government commonly utilize a standard form, "Invitation for Bids." This form is posted in public places and is generally distributed to the local construction community as well as to an agency bid list consisting of contractors who have indicated an interest in bidding on work within a given geographical area. It conveys essentially the same information as does the advertisement. Figure 5.2 illustrates a typical invitation covering a construction project for the U.S. Army Corps of Engineers. In addition to the invitation for bids, a short synopsis of the proposed project is prepared and sent to trade journals, magazines, and the *Commerce Business Daily,* which is published by the U.S. Department of Commerce and which lists U.S. government procurement invitations including construction. Generally, a federal project can be advertised by any means as long as it results in no cost to the government. Where a public agency requires its bidding contractors to be prequalified (see Section 5.9). a bid invitation may be sent only to those contractors who will be permitted to submit proposals.

<u>ADVERTISEMENT</u> <u>FOR</u> <u>BIDS</u>

Sealed bids for the construction of a Municipal Airport Terminal Building at Portland, Ohio, will be received by the City of Portland at the City Manager's Office, Portland, Ohio, until 2:30 P.M. (E.S.T.), Wednesday, May 19, 19—, and then publicly opened and read aloud. Bids submitted after closing time will be returned unopened. No oral or telephoned proposals or modifications will be considered.

Plans, specifications, and contract documents will be available April 16, 19—, and may be examined without charge in the City Manager's Office, in the office of Jones and Smith, Architect-Engineers, 142 Welsh Street, Portland, Ohio, in Plan Services in Cleveland, Akron, Toledo, and Youngstown, Ohio; Pittsburgh, Pennsylvania; Detroit, Michigan; Chicago, Illinois; and Buffalo, New York. General Contractors may procure five sets from the Architect-Engineer upon a deposit of $100.00 per set as a guarantee for the safe return of the plans and specifications within 10 days after receipt of bids. Others may procure sets for the cost of production.

A cashier's check, certified check, or acceptable bidder's bond payable to the City of Portland in an amount not less than 5% of the largest possible total for the bid submitted including the consideration of additive alternates must accompany each bid as a guarantee that, if awarded the contract, the bidder will promptly enter into a contract and execute such bonds as may be required.

The Architect-Engineer's estimate of cost is $3,800,000.00.

Full compliance with applicable Federal, State, and Municipal Wage Laws is required and not less than the rates of wages legally prescribed or set forth in the Contract, whichever is higher, shall be paid.

Proposals shall be submitted on the forms prescribed and the Owner reserves the right, as its interest may require, to reject any and all proposals; waive any formalities or technicalities. No bidder may withdraw his proposal after the hour set for the opening thereof, or before award of contract, unless said award is delayed for a period exceeding thirty (30) days.

CITY OF PORTLAND, OHIO
By John Doe, City Manager

April 15, 19 — .

Figure 5.1 Advertisement for bids.

5.6 ADVERTISEMENT FOR BIDS (PRIVATE CONTRACTS)

Private owners may proceed in any manner they choose to select a contractor. Public advertising is sometimes used on private projects to obtain the advantages of open and free competition. In general, however, public advertising for bids is not commonly used by private owners. Negotiated contracts, as opposed to competitive bidding, are now widely used in the private sector. Where competitive bidding is used, private owners frequently use a procedure known as "invitational bidding." This is a scheme where the owner selects a few prime contractors, each of which he considers to be reputable and qualified, and invites this limited group to bid on the project. This procedure offers the owner the advantages of competition while restricting the bidders to a preselected number.

When private owners wish to put their projects out for open competitive

	REFERENCE
	DACW54-80-002

INVITATION FOR BIDS
(CONSTRUCTION CONTRACT)

DATE
3 January 19—

NAME AND LOCATION OF PROJECT

SONORA DAM, El Paso County,
El Paso, Texas (Estimated
Construction Cost: Between
$3,000,000 and $5,000,000)
(DAR 14-116)

DEPARTMENT OR AGENCY

DEPARTMENT OF THE ARMY
CORPS OF ENGINEERS

BY (*Issuing office*)
DISTRICT ENGINEER, U.S. ARMY ENGINEER DISTRICT, ALBUQUERQUE CORPS
OF ENGINEERS, P.O. BOX 1580, ALBUQUERQUE, NEW MEXICO 87103

Sealed bids in   ONE COPY   for the work described herein will be received until

1:30 P.M., 7 February 19— (LOCAL TIME AT THE PLACE OF BID OPENING),

at the E. R. Lockhart Room, El Paso Chamber of Commerce Building,
10 Civic Center Plaza, Main and Santa Fe Streets, El Paso, Texas,

and at that time publicly opened.

Information regarding bidding material, bid guarantee, and bonds

Bidding documents may be obtained from the "Issuing Office"
above. Sets of drawings, reduced to half size, with specifications
will be furnished upon receipt of payment of $3.00 per set.
Bidding documents will be on file and may be reviewed at the
following address:

> Corps of Engineers
> 517 Gold Avenue S.W.
> Albuquerque, New Mexico.

A bid guarantee equal to 20% of the bid price must be
submitted with any bid in excess of $25,000. If the contract
exceeds $25,000, a performance bond equal to one hundred percent
(100%) of the contract price and a payment bond in amount
prescribed by the Miller Act will be required.

Description of work

The work consists of furnishing all labor, equipment, and
materials, except as may otherwise be provided herein, to
perform the work described in the specifications, schedules
and drawings entitled:

"SONORA DAM, El Paso County, El Paso, Texas."

20-104   U.S. GOVERNMENT PRINTING OFFICE:1960—O-572728

Figure 5.2   Invitation for bids.

bidding, they need only advise the local construction reporting services (see Section 5.7) and give them the necessary bidding data. These services will immediately pass this information on to their members and subscribers.

5.7 REPORTING SERVICES

Another source of bidding information is the plan service centers that have been established in larger cities about the country. These centers publish and distribute on a regular basis bulletins that describe all projects to be bid in the near future within that locality. In addition, these services keep copies of the bidding documents on file for the use of subscribing general contractors, subcontractors, material dealers and other interested parties.

These plan centers provide a valuable service in making bidding documents available to a wide circle of potential bidders. Prime contractors use the services as a central source of bidding information, to make a quick check of the bidding documents to determine whether to bid, and as an indication of which subcontractors and material dealers will be bidding a given project. The use of plan centers for actual estimating purposes is generally limited to those subcontractors and material vendors whose takeoff is not extensive. Otherwise, prime contractors and others normally obtain sets of bidding documents for their own use from the owner or architect-engineer.

On the national level, *Dodge Reports,* published by the Dodge Division of McGraw-Hill Information Services Company, provide complete coverage of bidding and construction activity within different specified localities. Subscribers receive daily reports concerning jobs to be bid in that area, including all known general contractor bidders. This is valuable in that it informs the general contractor about its competition and tells subcontractors and material vendors who the bidding prime contractors are. After each bidding, the subscribers are informed of the low bidders and are given information concerning the contract award. During construction, reports are issued periodically listing the general contractor and the subcontractors. Other services are also available and subscriptions can be tailored to suit the needs of the contractor.

5.8 THE DECISION TO BID

After learning that proposals are to be taken on a given construction project, the contractors must ascertain whether they are interested in bidding. In a general sense, this depends on the "bidding climate" prevailing at the time. The decision to bid involves a study of many interrelated factors, such as bonding capacity considerations (see Section 7.12), location of the project, the owner and his financial status, the architect-engineer, the nature and size of the project as it relates to company experience and equipment, the amount of work presently on hand, the probable competition, labor conditions and supply, and the completion date. In short, contractors do not always bid every job that comes along. Rather, they do a considerable amount of picking and choosing.

Because it involves considerable expense to bid a job, it goes without saying that when an affirmative decision to bid is made, the contractor will exert every effort to be the successful bidder. Limited data available indicate that estimating expense may be of the order of 0.2 to 0.3 percent of the total bid amount.

It is possible, of course, that the contractor may not be eligible to bid on a particular project. This particularly applies to various kinds of "set-asides" on publicly financed projects. For example, under regulations of the Small Business Administration, certain federal construction projects can be set aside for small contractors. In such an instance, a contractor must satisfy certain business measures to be eligible to bid on such a project. Another instance is afforded by public construction work that is set aside for minority-owned business enterprises.

5.9 QUALIFICATION

Many states have enacted statutes that require a general contracting firm wishing to bid on public work in those states to be adjudged qualified before it can be issued bidding documents or before it can submit a proposal. Other states require that the contractor's qualifications be judged after it has submitted a proposal. The first method is called prequalification, and the second is called postqualification. Many other public bodies, including agencies of the federal government, require some form of qualification for contractors bidding on their construction projects. The relative merits and drawbacks of the process have been debated for many years, but qualification in some form has become almost standard practice in the field of public construction. The obvious purpose to be served by qualification is to eliminate the incompetent, overextended, underfinanced, and inexperienced contractors from consideration.

Prequalification requirements apply almost universally to highway construction, and in certain jurisdictions all construction projects financed with public money require contractor prequalification. In those states in which licenses are required, contractors must be licensed before applying for prequalification. To prequalify, they must submit detailed information concerning their equipment, experience, finances, current jobs in progress, references, and personnel. Evaluation of these data results in a determination of whether the contractor will be allowed to submit a proposal. Highway contractors usually submit qualification questionnaires at specified intervals and are rated as to their maximum contract capacity. Their construction activities are reflected in their current ratings, with proposal forms being issued only to those qualified to bid on each project. The prequalification certificate may also limit the contractor to certain types of work, such as grading, concrete paving, or bridge construction.

In jurisdictions having postqualification, the contractor is called on to furnish certain information along with its bid. The information is much the same as that required for prequalification, but it serves the qualification purpose only for the particular project being bid.

The Associated General Contractors of America publish the following standard forms for use in qualification procedures.

1. Standard Prequalification Questionnaires and Financial Statements for Prospective Bidders, Engineering Construction (for qualifying before bidding).
2. Prequalification Statement for Building Construction and Standard Form of Contractor's Financial Statement (required in advance of consideration of application to bid).
3. Standard Questionnaires and Financial Statement for Bidders, Engineering Construction (for qualifying after bidding).
4. Contractor's Experience Questionnaire and Financial Statement (precontract).

These forms have been approved by The American Institute of Architects and other organizations.

Although the foregoing description of qualification is based entirely on publicly financed projects, the same procedures can also be utilized on private jobs. A somewhat different procedure often used in closed biddings is for the contractor to submit an individualized qualification summary with its bid. This is essentially a sales document and contains information designed to enhance the contractor in the eyes of the owner.

5.10 DEPOSITS

After a general contracting firm has decided it wishes to bid a given project, it will normally obtain the necessary sets of bidding documents from the architect-engineer or the owner. The number of sets needed depends on the size and complexity of the project, the time available for the preparation of the bid, and the distribution the architect-engineer has made to subcontractors, material dealers, and plan services. In general, more sets will be required for shorter times of bidding and for more complex projects. It is customary for architect-engineers to require a deposit for each set of documents as a guarantee for its safe return. This deposit, which may range from fifty to several hundred dollars per set, is usually refundable. There are instances, however, where contractors must pay a fee or nonrefundable "deposit" for each set of bidding documents obtained. This is sometimes applied on public projects. The objective of what amounts to contractor purchase of the documents is to help the owner or architect-engineer recoup some of the printing costs involved.

5.11 AVAILABILITY OF PLANS AND SPECIFICATIONS

Undue limitation by the architect-engineer or owner on the number of sets of plans and specifications made available to the various bidders and plan services is shortsighted. Any aspect of the bidding process that inhibits competition can only result in a higher cost for the owner. Admittedly, the reproduction of these documents is expensive, and certainly no more sets should be provided than are really necessary. However, there can be no argument that sufficient numbers of bidding documents to achieve a reasonably thorough coverage of the total bidding community produce the best competitive prices, particularly for large and complex projects. Plan services perform a valuable function in this regard by making drawings and specifications readily available to interested bidders.

For takeoff and pricing purposes, subcontractors and material dealers can obtain their own bidding documents from the architect-engineer or owner, or they can use the facilities of local plan services. Despite this, however, the prime contractor, in order to obtain certain prices, often finds it necessary to make its own copies of the drawings and specifications available to parties wishing to bid on various aspects of the project. A word of caution to contractors must be injected regarding the lending of their drawings and specifications. Care must be exercised that all bidding documents in their entirety, including addenda, are made available. This foresight will eliminate possible later difficulties with subcontractors or material suppliers claiming that they bid on the basis of incomplete information and hence did not tender a complete proposal. The general contractor, if possible, should set aside a well lighted plan room on its own premises for the use of the estimators of subcontractors and material dealers. When such facilities are available, the drawings and specifications need never leave the office and can be made to serve a wider range of bidders in a considerably more efficient manner.

5.12 THE BIDDING PERIOD

A sufficient allocation of time for the bidding process can be an important consideration for the owner. The estimating process takes considerable time and must be dovetailed into contractors' current operations. Careful study and analysis of the plans and specifications usually result in lower bid prices and substantial savings for the owner. The more complicated and extensive the project, the bigger the dividends a reasonable bidding period will pay to the owner. Unless the work is truly of a rush or emergency nature, it is sound economics for the owner to allow sufficient time for the bidding contractors to make a thorough study and examination of the proposed work.

When an insufficient bidding period is allotted, many of the best prices will not have been received by the contractor in time to pass the savings on to the owner. In addition, when an unreasonable bidding deadline is imposed, a certain amount of guesswork is inevitably injected into the estimating process. The contractors understandably compensate for uncertainty by bidding on the high side, resulting in inflated proposals.

From the contractor's point of view, the owner should avoid setting any day following or preceding a weekend or holiday as the date for receiving bids. Morning hours are also undesirable. Such times impose undue hardships on the contractors and other bidders and greatly restrict the availability of prices and other information. As far as is practicable, a bidding date should not conflict with other construction bid openings.

5.13   INSTRUCTIONS TO BIDDERS

For competitive bidding to be a valid procedure, all competitors must bid under exactly the same conditions for an identical package of work. This requires that a set bidding procedure must be established and all bidders be required to conform. Consequently, when competitive proposals are requested, a considerable amount of information concerning the technicalities of the bidding process must be communicated to the contractors involved. This information usually comprises one of the first divisions of the specifications and is designated as "Instructions to Bidders." These instructions review the requirements the owner or contracting authority has set up for the form and content of the bids and also prescribe certain procedures with which the bidding contractors are required to conform. This document, for example, states conditions pertaining to the form of the bid, where and when it must be delivered, proposal security required (see Section 5.46), and information concerning late bids and bids submitted by mail or telegram. Failure to comply with such instructions can result in a contractor's bid not being accepted.

Clauses are commonly included that give the owner the right to reject any or all bids, to postpone the date of bid opening, and to exercise many prerogatives in the selection of the successful bidder. A typical list of instructions to bidders is contained in Appendix F. Standard forms of instructions to bidders are used by many public agencies. *Document A 701,* "Instructions to Bidders," of The American Institute of Architects is used for the bidding of private building projects. Standard sets of instructions are designed for general use and do not normally provide all of the information peculiar to a specific project that would be required by prospective bidders. Consequently, supplementary instructions or modifications are customary or reference is made to certain specific information of interest to bidders already contained elsewhere, such as in the advertisement or invitation to bid.

5.14   PRELIMINARY CONSIDERATIONS

Before starting the detailed estimating procedure, the estimator must become familiar with the instructions to bidders, proposal form, alternates (see Section 5.45), general and supplementary conditions, drawings, specifications, addenda, and form of contract. Such familiarity is equally as important for the quantity takeoff people as it is for the cost estimator. The form and arrangement of the quantity surveys can be dictated by the requirements of the foregoing documents. For example, alternate proposals often require that the quantity surveys of certain portions of the project be compiled separately. When bidding large projects, it is desirable to hold a prebid meeting between the estimators and the persons who will occupy the principal supervisory positions on the project if the bid is successful. The purpose of this meeting is to explore the alternative construction procedures that might be followed and to make tentative decisions regarding methods, equipment, personnel, and time schedules.

Another type of prebid meeting is sometimes held, this one being initiated and conducted by the owner. This is a case where the owner, early in the bidding period, meets with all of the prime contractors and perhaps the major subcontractors who will be submitting proposals. The purpose of this meeting is to give the owner an opportunity to review the project requirements, emphasize certain aspects of the proposed work, clarify or explain difficult points, and answer questions. This kind of meeting can be very useful when the work will be complex or involve out of the ordinary procedures.

5.15 JOBSITE VISIT

After preliminary examination of the drawings and specifications, the construction site must be visited. Information is needed concerning a wide variety of site and local conditions. Some examples are as follows:

1. Project location.
2. Probable weather conditions.
3. Availability of electricity, water, telephone, and other services.
4. Access to the site.
5. Local ordinances and regulations.
6. Conditions pertaining to the protection or underpinning of adjacent property.
7. Storage and construction operation facilities.
8. Surface topography and drainage.
9. Subsurface soil, rock, and water conditions.
10. Underground obstructions and services.
11. Transportation and freight facilities.
12. Conditions affecting the hiring, housing, and feeding of workers.
13. Material prices and delivery information from local material dealers.
14. Rental of construction equipment.
15. Local subcontractors.
16. Wrecking and site clearing.

It is important that an experienced person perform this site inspection, particularly in locations that are relatively new or strange to the contractor's organization. The visitor should become familiar with the project requirements in advance of the visit and take a set of drawings and specifications along for reference. It is desirable that the information developed during the site visit be recorded in a signed report that is made a part of the project estimate. It is usually helpful to have available a camera, tape recorder, measuring tape, and other aids appropriate to the occasion such as an earth auger or hand level. It is not uncommon that a follow-up visit must be made on important projects. When access to the site is restricted for some reason, the site visit may have to be coordinated with other bidding contractors, a representative of the owner and/or the architect-engineer, and the primary subbidders. Site conditions where remodeling or renovation work is at issue can be especially demanding, involving such things as use of existing elevators, control of noise and dust, matching existing materials and finishes, and maintenance of existing services and operations. *Outline for Reports of Site Investigations,* published by the Associated General Contractors of America, is an excellent guide to follow in making site studies for heavy and utility construction. It offers a good checklist of significant items that might otherwise be overlooked.

5.16 BID INVITATIONS

Early in the bidding period the contractor will usually mail out bid invitations to all material dealers and subcontractors that are believed to be interested and whose bids would be desirable. This mailing, by postcard or form letter, advises of the project under consideration, the item for which a quotation is requested, the deadline for receipt of proposals by the prime contractor, the place where bids will be accepted, the name of the person to whom a proposal should be directed, the place where bidding documents are available, and any special instructions that may be necessary. Specific reminders concerning the status of addenda, alternates, taxes, and bond are desirable. Contractors customarily maintain a card file that lists addresses and other pertinent information about material suppliers and subcontractors. These cards are filed both by geographical area and by specialty.

5.17 QUANTITY SURVEYS

A quantity survey, or takeoff, is the detailed compilation of the quantity of each elementary work item that is called for on the project. The quantity survey is the only accurate and dependable procedure for compiling either a lump-sum or a unit-price estimate. Although the bid documents for a unit-price project customarily provide contractors with estimated quantities of each bid item, these are approximate only, and the architect-engineer assumes no responsibility for their accuracy or completeness. Architect-engineers' quantities are sometimes computed only within so-called "paylines" and may not be representative of work quantities that must actually be done to accomplish the bid item. In addition, more detail is usually required for accurate pricing of the work. Consequently, contractors customarily make their own quantity takeoffs, even on projects for which estimated quantities are given.

The quantity survey provides the contractor with valuable information during the construction period as well as for the initial estimation of project cost. For example, it can provide the contractor with data concerning times to be required in the field for work accomplishment, crew sizes, and equipment needs. The quantity survey provides purchasing information and serves as a reference for various aspects of the field work as it progresses. It is useful in providing the owner with a variety of costs pertaining to the work.

The takeoff is made using pads of standard forms that provide suitable spaces for the entry of dimensions, numbers of units, and extensions. Allowances for the waste of materials such as lumber, concrete, brick, and mortar are made as the takeoff proceeds. In large construction organizations in which there are several takeoff people, it is usual practice for each person to specialize, more or less, along certain lines. For instance, one person may take off only electrical work, whereas another will concentrate on excavation, concrete, and forms. Although this system has many obvious advantages, it also has the disadvantage that there is always the possibility of overlooking items or duplicating them. This possibility can be minimized, however, by careful coordination and checking.

Prime contractors do not ordinarily concern themselves with making quantity surveys for all work categories but limit their takeoff to items that they may reasonably expect to carry out with their own forces. The specialized categories that are done by subcontractors are not usually investigated in detail by prime contractors. However, at times they may analyze any or all these categories as a check against the possibility that they may be able to do the work for less money than the best available subcontract bid. In addition, they may have to price a given trade specialty either to check subbids received for reasonableness or when no subcontractor quotations have been submitted. Contractors may also not compile the quantities of certain materials, relying on the vendors to do this. Reinforcing steel and structural steel are common examples of this practice.

5.18 PROFESSIONAL QUANTITY SURVEYORS

There are a number of proprietary firms whose principal business is making quantity surveys and cost estimates of selected construction projects. This may be done after the design of the project has been completed, or the concern may act as a cost adviser during the design stage. Services of such companies are engaged mostly by owners and architect-engineers as a means of keeping project costs within budgets and to get the most for the construction dollar. Companies engaged in this form of business are commonly called quantity surveyors, a possibly misleading appellation, but one firmly entrenched by usage.

It is standard practice in the United States for bidding contractors to make their own quantity takeoffs. Thus there is considerable duplication of effort. Nevertheless, contractors seldom utilize the services of professional quantity surveyors for purposes of preparing proposals. Contractors prefer to prepare their own quantity surveys, in which they have a sense of confidence. They maintain a staff of experienced estimators whose reliability and accuracy have been well established. During the takeoff process, the estimator must make many decisions concerning procedure, equipment, sequence of operations, and other such matters. All these decisions must be considered when costs are being applied to the quantity surveys. In addition, work classifications used in quantity surveys must conform with the contractor's established cost accounting system.

It is interesting to note that in Great Britain the quantity surveyor is an established part of the entire construction process. This party is hired by the owner at the inception of the project and makes cost feasibility studies, establishes a construction budget, makes cost checks at all stages of the design process, and prepares a final quantity takeoff. This takeoff is submitted to the contractors bidding the project who use the information as the basis for their pricing of the work. The quantity surveyor advises the owner on contractual arrangements and compiles certificates of interim and final payment to the contractor doing the work. In some ways, the duties of the British quantity surveyor and the American construction manager are quite similar.

SUMMARY SHEET

Project Municipal Airport Terminal Building   Work Items Concrete and Forms

Cost Account Number	Work Item	Unit	Total Quantity	Material Cost		Labor Cost		Total Cost
				Each	Total	Each	Total	
240	CONCRETE							
.01	Footings	c.y.	1,040	$39.50	$ 41,080	$4.50	$ 4,680	$ 45,760
.05	Grade beams	c.y.	920	42.50	39,100	6.30	5,796	44,896
.07	Slab	c.y.	2,772	39.50	109,494	6.00	16,632	126,126
.08	Beams	c.y.	508	42.50	21,590	6.30	3,200	24,790
.09	Bond beams	c.y.	62	42.50	2,635	9.00	558	3,193
.11	Columns	c.y.	102	42.50	4,335	8.70	887	5,222
.12	Walls	c.y.	466	42.50	19,805	6.75	3,146	22,951
.16	Stairs	c.y.	317	42.50	13,473	7.80	2,473	15,946
.19	Sidewalks	c.y.	320	39.50	12,640	6.00	1,920	14,560
.20	Expansion joint	l.f.	41,060	0.25	10,265	0.15	6,159	16,424
.40	Screeds	s.f.	148,000	—	—	0.12	17,760	17,760
.50	Float finish	s.f.	22,000	—	—	0.15	3,300	3,300
.51	Trowel finish	s.f.	128,000	—	—	0.25	32,000	32,000
.52	Stair finish	l.f.	2,060	—	—	1.05	2,163	2,163
.60	Rubbing	s.f.	2,530	0.06	152	0.55	1,392	1,544
.91	Curing	s.f.	180,000	0.04	7,200	0.03	5,400	12,600
					$281,769		$107,466	$389,235
	Indirect Labor Cost					39%	41,912	41,912
	Total Concrete				$281,769		$149,378	$431,147
260	FORMS							
.01	Footings	s.f.	1,220	$0.70	$ 854	$1.00	$ 1,220	$ 2,074
.05	Grade beams	s.f.	4,840	0.80	3,872	1.00	4,840	8,712
.08	Beams	s.f.	1,965	0.88	1,729	1.08	2,122	3,851
.11	Columns	s.f.	2,178	0.82	1,786	1.50	3,267	5,053
.12	Walls	s.f.	3,240	0.88	2,852	1.00	3,240	6,092
.17	Risers	l.f.	910	0.68	619	1.28	1,165	1,784
.45	Chamfer	l.f.	510	0.08	41	0.08	41	82
.60	Oil, ties, nails	s.f.	13,443	0.06	807	—	—	807
.63	Anchor slot	l.f.	2,800	0.30	840	0.06	168	1,008
					$13,400		$16,063	$29,463
	Indirect Labor Cost					43%	6,908	6,908
	Total Forms				$13,400		$22,971	$36,371

Figure 5.3 Summary sheet, lump-sum bid.

5.19 SUMMARY SHEETS

On completion of the quantity surveys, the totals are then transferred to summary sheets which list and classify the several items of work. Figure 5.3 presents a form of summary sheet commonly used for a lump-sum project. For this type of contract, a summary sheet is prepared for each major work classification. Figure 5.3 contains final quantities of concrete and form work on the Municipal Airport Terminal Building, a hypothetical project that will be used in this text to illustrate selected procedures. Similar summaries are prepared for all of the other principal work types. When the total quantities of work for each classification have been entered, the summary sheet is then used for the pricing of materials and labor. Figure 5.3 does not contain an equipment cost column, although it could. It is common practice on projects that are bid lump-sum, such as buildings, to compute equipment costs for the entire project on a separate sheet rather than on the individual summary sheets.

Figure 5.4 shows a summary sheet for the unit-price project referred to in Appendix E. For this type of bidding, a summary sheet is prepared for each bid item. As illustrated by Figure 5.4, all work types associated with the accomplishment of that bid item are listed on the summary sheet, which often includes several different work categories. Using the work quantities obtained from the takeoff, the contractor uses the summary sheet to obtain the total direct costs of that bid item. The summary sheets for unit-price projects customarily serve for the computation and recording of material, labor, equipment, and subcontract costs. Figure 5.4 illustrates this point.

It should be noted that each work classification entered on a summary sheet is identified by a cost account number. These are the company's standard cost account numbers which are basic to its cost accounting system. Each estimate is broken down in accordance with the established cost system. Chapter 12 discusses the essentials of a contractor's cost accounting system.

The next several sections are devoted to describing how the many different categories of project cost are evaluated and priced.

5.20 MATERIAL COSTS

Although manufacturer's price lists may occasionally be used, it is customary for the contractor to solicit and receive special quotations for all materials required for the job being bid. Exceptions to this generality are stock items such as plyform, nails, and lumber that the contractor purchases occasionally in large quantities and of which a running inventory is maintained. Written quotations from the various material dealers are desirable so that such important considerations as prices, freight charges, taxes, delivery schedules, and guarantees are explicitly covered. Most suppliers tender their quotations on printed forms that include stipulations pertaining to terms of payment and other considerations.

Material costs as entered on the summary sheets must all be on a common basis: for example, delivered to the jobsite and without tax. These costs will ordinarily include freight, transportation, storage, and inspection. It is com-

SUMMARY SHEET

Project Holloman Taxiways and Aprons Bid Item No. 5 Concrete Pavement, 9-inch

Cost Account Number	Work Item	Quantity	Unit	Material		Labor		Equipment		Subcontract
				Unit Cost	Total	Unit Cost	Total	Unit Cost	Total	
250.03	Concrete, production	23,625	c.y.	$20.60	$486,675	$15.27	$360,754	$3.44	$ 81,270	—
254.01	Concrete, hauling	23,625	c.y.	—	—	0.30	7,088	0.35	8,269	—
258.02	Concrete, lay-down	90,000	s.y.	—	—	1.06	95,400	0.49	44,100	—
270.08	Reinforcing	50	ton	550.00	27,500	350.00	17,500	—	—	—
248.20	Grooves	105,000	l.f.	0.04	4,200	0.03	3,150	0.01	1,050	—
240.95	Curing	90,000	s.y.	0.10	9,000	0.04	3,600	0.005	450	—
	Totals				$527,375		$487,492		$135,139	—
	Labor Indirect Cost, 37%						$180,372			
	Labor Total						$667,864			

Figure 5.4 Summary sheet, unit-price bid.

mon practice to enter material prices without tax, adding this expense as a lump-sum amount on the final "recap" sheet (see Sections 5.36 and 5.38).

It is not unusual for the bidding documents to provide that certain materials will be supplied by the owner. In such cases, the owner will provide the designated materials to the contractor, and no material cost need be added into the estimate. However, all other costs, such as handling and installation expenses, must be included.

5.21 DIRECT LABOR COSTS

Direct labor cost is determined from basic wage rates, that is, the hourly rates used for payroll purposes. Direct labor costs are difficult to evaluate and cannot be determined with the same degree of precision as can most other elements of construction cost. The pricing of labor represents one of the two largest areas of uncertainty in the estimating process. The costs associated with construction equipment are the other, a subject to be discussed later in this chapter. To do the best job possible of establishing these two costs, the estimator must make an accurate and detailed quantity takeoff, analyze job conditions carefully, obtain advance decisions concerning the construction methods to be used, and maintain a comprehensive library of costs and production rates from past projects.

Contractors differ widely in how they estimate labor costs. Some choose to include all elements of labor expense into a single, hourly rate for each craft. Others evaluate direct labor cost separately from indirect labor cost (see Section 5.22). Some estimators compute regular and overtime labor costs separately, while others combine scheduled overtime with straight time into an average hourly rate. Some evaluate labor costs using production rates directly; others use labor unit costs. There are usually good reasons why a given contractor evaluates labor costs the way it does, and there certainly is no single correct method that must be followed. The procedures described in this chapter compute direct labor costs separately from indirect labor costs, a common practice.

Basic to the determination of the labor cost associated with any work category is the production rate. Labor production rates can be expressed in two ways, one of which is the labor-hours or crew-hours required to accomplish a unit or prescribed number of units of a given work type. Also used is the number of units of work accomplished per man-hour or per crew-hour. As an example, six hours of carpenter time and six hours of carpenter-tender time may be required to assemble, oil, erect, plumb, brace, strip, and clean 100 square feet of rectangular concrete column forms using adjustable steel form clamps. To obtain the estimated labor cost for a quantity of such work, the labor hours required per unit of work can be multiplied by the appropriate wage scales per hour and again multiplied by the total number of units of work of this category as given by the quantity survey. By following the same procedure for all such categories, the total labor costs associated with the overall project can be determined.

Many, probably most, cost estimators prefer to work with a labor cost per unit of work. For example, if carpenters earn $14.70 per hour and a carpenter-tender's wage scale is $10.35 per hour, the rectangular column forms cited previously will cost $1.50 per square foot. These "labor unit costs" are easy to apply, as illustrated by Figure 5.3. However, the cost estimator must make sure that his labor unit costs are kept up-to-date and that they reflect the actual levels of work production and current wage rates.

The most reliable labor productivity information is that which the contractor obtains from its own completed projects. Information on labor productivity and costs is also available from a wide variety of commercial sources. However, while information of this type can be very useful at times, it must be emphasized that labor productivity differs from one geographical location to another and is variable with season and many other factors. When labor unit costs are being used, care must be exercised that they are based on the appropriate levels of work productivity and the proper wage rates. Also, estimators must be very circumspect when using labor unit costs that they have not developed themselves. For the same work items, different estimators will include different expense items in their labor unit costs. It is never advisable to use a labor unit cost derived from another source without knowing exactly what it does and does not include.

5.22 INDIRECT LABOR COSTS

Indirect labor costs are those expenses that are additions to the basic hourly rates and that are paid by the employer. Indirect labor expense involves various forms of payroll taxes, insurance, and employee fringe benefits of wide variety. Employer contribution to social security, unemployment insurance, workmen's compensation insurance, and contractor's public liability and property damage insurance are all based on payrolls. Employers in the construction industry typically provide for various kinds of fringe benefits such as pension plans, health and welfare funds, employee insurance, paid vacations, and apprenticeship programs. The cost of these benefits is customarily based on direct payroll costs. Premiums for workmen's compensation insurance and most fringe benefits differ considerably with craft and geographical location.

Indirect labor costs are substantial in amount, usually constituting a 35 to 50 percent addition to the direct payroll costs. Indirect labor costs can be included in the cost estimate in different ways. For example, labor unit costs or hourly labor rates can be used that include both the direct and indirect costs of labor. However, this procedure does not always interrelate well with a contractor's cost accounting methods. For this reason direct and indirect labor costs are often computed separately when job costs are being estimated. One commonly used scheme is to add the proper percentage allowance for indirect costs to the total direct labor cost, either for the entire project or for each major work category. Because of the appreciable variation of indirect costs from one labor classification to another, it may be preferable to compute indirect labor cost at the same time that direct labor expense is obtained on the summary sheets. The 39 and 43 percent additions to the labor column of Figure 5.3 represent the total indirect labor costs associated with the concrete placing and forming, respectively.

5.23 EQUIPMENT COST ESTIMATING

Unfortunately, the term "equipment" does not have a consistent meaning in the construction industry. A common usage refers to tractors, power shovels, pile drivers, and other such items used by contractors to accomplish the work. However, equipment is also used with reference to various kinds of electrical and mechanical furnishings that become a part of the completed project, such as boilers, escalators, transformers, and blowers. As used herein, equipment refers to a contractor's equipage used for the physical accomplishment of the job. The term "materials" is construed to include all items that become a part of the finished structure, including electrical and mechanical machinery.

Equipment costs, like those of labor, are difficult to evaluate with any exactness. Equipment accounts for a substantial proportion of the total cost of most engineering projects, but is less significant for buildings. When the nature of the work requires major items of equipment such as earth-moving machines, concrete plants, and draglines, detailed studies of the associated costs must be made. Costs associated with minor equipment items such as power tools, concrete vibrators, and power buggies are not normally subjected to detailed study. A standard cost allowance for each such item required is included in the estimate, usually based on the total time it will be required on the job. The cost of hand tools, wheelbarrows, water hose, extension cords, and the like can be covered by a lump-sum allowance calculated as a small percentage of the total direct labor cost of the project and included on the job overhead sheet (see Section 5.31).

To estimate the expense of major equipment items as realistically as possible, early management decisions must be made concerning the equipment sizes and types to be required and the manner in which the necessary units will be provided to the project. A scheme sometimes used when the duration of the construction period will be about equal to the service life of the equipment is to purchase all new or renovated equipment for the project and sell it at the end of construction activities. The difference between the purchase price and the estimated salvage value is entered into the job estimate as a lump-sum equipment expense.

Equipment is frequently rented or leased. Rental can be especially advantageous when the jobsite is far removed geographically from the contractor's other operations, for satisfying temporary peak demand, or for providing specialized or seldom used equipment. Leasing is a common and widely used means of acquiring construction equipment and may be a desirable alternative to equipment ownership. Leasing can improve a contractor's working capital position by avoiding having funds tied up in fixed assets. Under certain circumstances, lease payments compare favorably with ownership costs. Many leases provide that at the expiration of the lease period the contractor has a purchase option if there is a continuing need for the machine and if it is worth the additional payment. Lease agreements for construction equipment normally extend for periods of two years or more, whereas renting is for shorter terms. The equipment expense associated with rental or lease costs is figured into the job by applying the lease or rental rates to the time periods that the equipment will be needed on the project.

The most common manner in which equipment is provided to construction projects is where the contractor owns the equipment and uses it on a succession of projects during its economic lifetime. Rather than a lump-sum equipment purchase expense or cost of rental or lease, ownership expense is involved in this instance. Ownership expense is of a fixed nature and includes depreciation, interest on investment or financing cost, taxes, insurance, and storage.

Regardless of the mode of supplying equipment to a project, equipment operating costs must also be computed and included in the project estimate. Operating costs include expenses such as fuel, oil, grease, filters, repairs and parts, tire replacement and repairs, and maintenance labor and supplies. There is some difference of opinion about whether the wages of equipment operators should be included in equipment operating cost. Some contractors prefer to regard the labor associated with equipment operation as a labor expense rather than an equipment cost. Others include the labor cost as a part of the equipment operating expense. Logically, it would seem preferable to treat equipment-operating labor as any other labor cost rather than include it in with the equipment cost. For purposes of discussion herein, equipment operators' wages are treated as a labor cost and are not included as a part of equipment expense.

5.24 EQUIPMENT EXPENSE

As described in Section 5.21, direct labor costs are computed by combining a labor production rate with the applicable hourly wage scales. Most equipment costs are calculated in much the same fashion except, of course, that equipment production rates and hourly equipment costs must be used. The hourly wage rates of various labor categories are immediately determinable, usually from applicable labor contracts, prescribed prevailing wage rates, or established area practice. This is not true for equipment. Contractors must establish their own hourly equipment rates as well as their equipment production rates. For most items of operating equipment, ownership, lease, or rental expense is combined with operating costs into an estimated total cost per operating hour. Power shovels, tractor scrapers, and ditchers are examples of equipment whose costs are usually expressed in terms of hourly rates.

There are some classes of construction equipment, however, for which it is more appropriate to express costs in terms of time units other than operating hours. The cost of commercial prefabricated concrete forms might be better spread over an estimated number of reuses. Items such as towers and scaffolding are required at the jobsite on a continuous basis during particular phases of the work, and operating hours have no significance in such cases. Costs in terms of other time units such as calendar months are more appropriate for such equipment items. The costs of some classes of production equipment are frequently expressed in terms of expense per unit of material produced. Portland cement concrete mixing plants, asphalt paving plants, and aggregate plants are familiar instances of this.

Move-in, erection, dismantling, and move-out expenses, also called mobilization and demobilization costs, are entirely independent of equipment operating time and production and are not, therefore, included in equipment hourly rates. These equipment expenses are separately computed for inclusion in the estimate, often as an item of field overhead.

5.25 EQUIPMENT COST RATES

The determination of the expense rates for the equipment items to be used on a project being estimated is an important matter and requires considerable time and effort. It must be remembered that estimators are always working in the future tense and that the equipment rates they use in their estimates are their best approximations of what such expenses will turn out to be when the project is actually being built. Equipment expense rates are approximations at best and must be regarded as such.

The standard way in which future equipment expense is computed is through the use of historical equipment cost records as contained in a contractor's equipment accounting system. The source of ownership and operating cost data for a specific piece of equipment is its ledger account. An important part of a contractor's general accounting system is the equipment accounts where actual ownership and operating expenses are recorded as they are incurred. A separate account is set up for each major piece of equipment. This account serves to maintain a detailed and cumulative record of the use of and all expenses chargeable to that equipment item. All expenses associated with that piece of equipment, regardless of their nature or the project involved, are charged to that account. These expenses include depreciation, investment costs, operating expenses, tire replacement, overhauls, and painting. Also maintained is a cumulative record of that equipment item's use in the field. These data constitute a basic resource for reducing ownership and operating expense to a total equipment cost rate. The sum of ownership and operating expense expressed as a cost per unit of time (operating hour, week, month) is often referred to as the "use rate," "budget rate," or "internal rental rate" of an equipment item. This latter term refers to the usual construction accounting practice where equipment time on the project during construction is charged against the job at that rate.

The preceding discussion assumes that equipment accounting is done on an individual machine basis. However, contractors vary somewhat in how they maintain their equipment accounts, some preferring to keep equipment costs by categories of equipment rather than by individual unit. These firms use a single account for all equipment items of a given size and type and compute an average budget rate based on the composite experience with all of the units included. Thus the same expense rate is applied for any unit of a given equipment type regardless of differences in age or condition. When estimating equipment cost on a future project, it would seem logical to use an average hourly cost rate for any given equipment type and capacity because it is often not possible to predict exactly which equipment units will be placed on a particular project.

There are many external sources of information concerning ownership and operating costs for a wide range of construction equipment. Manufacturers, equipment dealers, and a large assortment of publications offer such data. It must be realized, however, that these are typical or average figures and that they must be adjusted to reflect contractor experience and methods and to accommodate the specific circumstances of the project being estimated. Climate, altitude, weather, job location and conditions, operator skill, field supervision, and other factors can and do have a profound influence on equipment costs. When a new piece of equipment is procured for which there is no cost history, equipment expense must be estimated using available sources of information considered to be reliable.

5.26 OWNERSHIP AND OPERATING COSTS

Hourly expense rates are computed and periodically updated for each major piece of equipment. Figure 5.5 illustrates a commonly used procedure for this, using as an example a bottom-dump hauler. The average annual usage of this hauler is approximately 1200 hours and it is assumed that the useful life of the machine will be a total of 12,000 operating hours or 10 years.

In Figure 5.5, the original price of the hauler, less tires, is depreciated out uniformly over 12,000 hours. The cost of the tires is deducted from the original acquisition price because the tires have a shorter service life than the mechanical part of the equipment. Depreciation is equipment expense caused by wear and obsolescence and allows the recovery of the invested capital over the useful life of the equipment. The method by which the depreciation expense is spread uniformly over the total service life is referred to as straight-line depreciation. Regardless of the mode of equipment depreciation a contractor may use for tax and other accounting purposes, the use of straight-line depreciation for estimating and job cost accounting is usual practice. Depreciation methods are discussed in Chapter 9.

Taxes, insurance, storage, and interest on investment, referred to together as investment cost, are customarily based on the average annual value of the equipment. This example assumes taxes as 1 percent, insurance and storage as 3 percent, and interest as 12 percent, for a total annual cost of 16 percent of the average yearly value. A standard way of computing the average annual value, based on straight-line depreciation is by the equation

$$A = \frac{C(n + 1) + S(n - 1)}{2n}$$

where A = average annual value
C = delivered cost
n = number of years of useful life
S = salvage value

In Figure 5.5, with a salvage value of zero,

$$A = \frac{\$77{,}794\ (11)}{20} = \$42{,}787$$

Operating costs are computed on the basis of the contractor's own experience or using available national averages. Equipment manufacturers and dealers provide typical ownership and operating costs. It must be realized, however, that these are average figures and that they may not reflect the conditions of the project being estimated. Job estimates must be based on individual job conditions, and the contractor's past experience is the best guide in this regard. Nevertheless, there are times when national averages can be very useful. In the computation of tire replacement cost in Figure 5.5, a new set of tires is assumed to have a life of 3500 operating hours.

Figure 5.5 discloses that for the bottom-dump hauler, the ownership cost is $11.31 per operating hour and the estimated operating costs are $12.40 per

---

ESTIMATE
Hourly Ownership and Operating Cost for
DP-12 Bottom-Dump Hauler

Ownership Costs

1. Depreciation   Purchase price   = $76,356
                   Freight   =   1,438
                   Delivered price   = $77,794
                   Less tires   = 10,510
                   Depreciable value = $67,284

                   Hourly depreciation $= \dfrac{\$67{,}284}{12{,}000}$ = $ 5.61

2. Interest, taxes, insurance, and storage

$$\frac{\$42{,}787\ (0.16)}{1{,}200}$$ =   5.70

   Total hourly ownership cost   = $11.31

Operating Costs

3. Fuel, 4 gallons per hour @ $1.31   = $ 5.24
4. Oil, lubricants, filters ($\frac{1}{3}$ of fuel cost) = $\frac{1}{3}$ ($5.24)   =   1.75
5. Repairs, parts, and labor (35% of depreciation) = 0.35 ($5.61) =   1.96
6. Tire Replacement $= \dfrac{\$10{,}510}{3{,}500}$ =   3.00
7. Tire repairs (15% of tire replacement cost) = 0.15 ($3.00)   =   0.45

   Total hourly operating cost   = $12.40

Total estimated hourly ownership and operating cost   = $23.71

Figure 5.5   Estimate of equipment ownership and operating costs.

operating hour. This piece of equipment operating in the field will cost the contractor a total of $23.71 per working hour, not including operating labor.

It is important to note that ownership expense goes on whether or not the equipment is working. If the bottom-dump hauler actually works 1200 hours or more per year, the owner recovers this cost. If the hauler works less, the contractor does not recover all the annual ownership costs, and the deficit comes out of general company profits. Operating costs are, of course, incurred only when the equipment is actually being used. When equipment is leased or rented rather than owned, there is no ownership cost involved. Rather, the lease or rental cost, expressed on an operating hour basis, is used instead. Operating costs apply as before with the sum of the two again giving the rental rate of the item.

The costing of minor equipment items is done in a somewhat similar although less exact manner. A common example might be the pricing of the expense of a concrete vibrator whose initial cost is $840. Suppose the service life of this vibrator is approximately four years and about $240 per year must be spent on servicing, repairs, and new parts. With an annual investment cost of approximately $67, this translates into a total cost of $2068 spread over 48 months, or $43 per month. If concrete pouring on a project will extend over four months, the equipment expense to be added into this job for the concrete vibrator will be $172 plus the estimated fuel cost if applicable.

5.27 EQUIPMENT PRODUCTION RATES

In addition to equipment cost, equipment production rates are also needed for the computation of equipment expense in a construction estimate. Paralleling the case of labor, applying equipment hourly expenses and production rates to job quantities enables the estimator to compute the total equipment expense for the project. Equipment unit costs, which are equipment costs per unit of production, can also be determined. Equipment production rates, like those of labor, are subject to considerable variation and are influenced by a host of job-site conditions. In addition, some equipment production rates must be computed using specific job conditions such as haul distances, grades, and rolling resistance. Estimators must consider and evaluate these factors when they are pricing a new project.

There are several sources of equipment production information. The most reliable are production records from past projects. Advice from the equipment operators themselves can be very useful at times. If a new piece of equipment is involved with which there has been no prior experience, production information provided by the equipment manufacturer or dealer can be of assistance. There are many rules of thumb and published sources of information concerning average equipment production rates. Stopwatch "spot checks," made to obtain the productivity of specific equipment items, can be of value. In this regard, however, it should be mentioned that production rates of labor and equipment used for estimating purposes should be average figures taken over a period of time. Daily job production tends to be variable, and this is a hazard of using spot checks. Production records from past projects produce good time average values while a spot check may unknowingly catch production at a high or low point.

Production rates may be applicable to a single piece of equipment or to an equipment spread. A spread of equipment is an aggregation of equipment items working together as a group to accomplish a given aspect of the job. Figure 5.6 illustrates how a production rate and equipment unit cost are obtained for earth hauled by our bottom-dump haulers. As is true for all such production equipment, the manufacturer of the bottom-dump hauler provides the contractor with much valuable information concerning its capacity and performance characteristics. As illustrated by Figure 5.6, the contractor can use this information, together with operating characteristics of the loading power shovel and the length, grade, and condition of the haul road, to determine the hauling production rate of the bottom-dump unit. Applying the cost figures derived in Figure 5.5 to the production rate in Figure 5.6 discloses that the equipment unit hauling cost is $0.293 per bank cubic yard for a fleet of four such units. If an equipment unit cost that includes both loading and hauling is desired, the shovel cost per bank cubic yard must be obtained and added to the hauling cost. Such equipment unit costs are utilized in the same fashion as labor unit costs. "Bank" measure refers to soil in its original, undisturbed

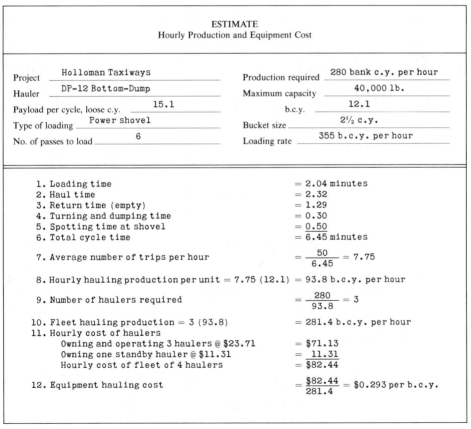

Figure 5.6 Estimate of equipment production rate and unit cost.

condition, as compared to "loose" measure, which applies to the soil after excavation. Bank soil generally is more dense and occupies less space than the same soil after it is excavated.

Equipment production rates given by the manufacturer or observed on the job are usually those that occur when the equipment is turning out its maximum possible output. Obviously, such rates are very seldom continuously maintained and allowances must be made for time losses and delays. The "50-minute hour" is commonly used with regard to excavating and other forms of operating construction equipment. This term means that, on the average, the equipment is producing a net time of 50 minutes per hour. The remaining time is consumed by nonproductive actions such as greasing, fueling, minor repairs, maintenance, shifting position, operator delays, and waiting. The application of the 50-minute hour is illustrated in Item 7 of Figure 5.6.

5.28 SUBCONTRACTOR BIDS

Quotations from subcontractors can be an important element in the compilation of a project estimate, especially on building construction where subcontracting is utilized much more extensively, on the average, than it is on engineering or industrial construction. Unfortunately, many of these bids may not be submitted to the prime contractor until shortly before closing time, a factor that makes it difficult to study and analyze them. Submission by telephone of bids by subcontractors, together with the practice of making last-minute deductions, can make the final bid compilation a difficult matter during that hectic period immediately before bid opening. The submission of written quotations by subcontractors, with sufficient time to allow the general contractor to analyze the bids, is to be preferred. Stipulations regarding the use of hoisting facilities; the supply of electricity, water, and heat; conditions to be met before starting work; the delivery and storage of materials; and others are commonly attached to and made a part of subcontract proposals. Because subbid stipulations usually require some commitment on the part of the prime contractor, they must receive careful attention during the bidding period so that all costs pertaining to them can be included in the prime contractor's estimate.

The estimator must be sure what each subbid includes and what it does not, a matter that can require considerable checking when many subbids are involved. The inclusion of taxes and surety bond and the acknowledgment of addenda must either be stipulated by the quotation or determined by the general contractor. On building construction in particular, many items of work must be checked with the bidding subcontractors to be sure of their inclusion. For example, the prime contractor must check with the plumbing and electrical subcontractors to see whether their bids include an allowance for connecting up temporary water, sanitary, and electrical services. Another consideration is the need to make a background check on those subcontractors hitherto unknown to the general contractor.

To obtain time for subbid evaluation and study, it may be necessary for the general contractor to establish some deadline beyond which subbids or revisions thereof will not be accepted. Two hours before bid time is used in this regard by many contractors. Such an arrangement works best when all of the general contractors in a given area agree to abide by it. If a contractor establishes this practice unilaterally, it may well miss out on some favorable last-minute figures.

In general, the prime contractor has complete freedom in selecting its subcontractors. On some public projects, however, the prime contractor must award some stipulated percentage of its total contract to minority business enterprises. In such cases, the contractor must seek out and use the bids of such companies. On private projects, there may occasionally be listed in the instructions to bidders the names of subcontractors acceptable to the owner for major divisions of the work such as electrical, plumbing, heating and air conditioning, and elevators. In such cases, the general contractor can subcontract these items of work only to one of those subcontractors listed. Otherwise, it is a common requirement that all subcontractors be approved by the owner or architect-engineer (see Section 6.36). At this point, it suffices to say that the low subbid is not necessarily the *best* subbid. The general contractor must consider, in addition to immediate cost, the subcontractor's integrity, organization, financial strength, equipment, and reputation for getting the job done.

5.29 BID DEPOSITORIES

The problems associated with the submission of subcontract bids (and certain material prices) were mentioned in the preceding section. The last-minute placing and changing of price quotations can result in a mad scramble as the general contractor attempts to summarize the cost figures and compute its final bid. The process is particularly difficult when there are several alternates or bid items. Serious errors can be made as a result of the haste with which the final figures must be compiled. The use of bid depositories is one means of circumventing some of the problems associated with the bids of subcontractors and material suppliers. Bid depositories can also reduce bid shopping, a topic discussed in Section 5.49.

Although there are about as many types of bid depositories as there are groups using them, all depositories are designed to accomplish a central purpose. A bid depository is a facility created by a group of contractors or a trade association and operated by an independent agency, such as a bank, to collect bids from certain subcontractors and material suppliers and convey them to the general contractors. These quotations must be submitted to the depository before an announced deadline, four hours before the time of bid opening being a common requirement. After the deadline, these price quotations are transferred by the depository to the bidding general contractors for their use in compiling their final cost figures. Although practice varies, many depositories tabulate the subcontractors' quotations after the bid opening and disseminate this information to the bidders.

Bid depositories are not a new idea and have had a long and sometimes stormy history. Few depositories have experienced unqualified success. It is axiomatic that to be workable the depository must have the wholehearted support of a large majority of the members of the industry within the area served. Some depositories have been found to be in illegal restraint of trade and in violation of antitrust laws. If depositories engage only in the unrestricted collection and distribution of bids, without restraining competition, it seems improbable that such difficulties will be encountered. Rules prohibiting subcontractors that are members of the depository from submitting bids to general contractors that are not members, and forbidding general contractor members from accepting bids from nonmember subcontractors have led to serious legal difficulties.

5.30 ALLOWANCES

On occasion the architect-engineer will designate in the specifications a fixed sum of money the contracting firm is directed to include in its estimate to cover the cost of some designated item of work. These are called "allowances" and are often used with regard to the costing of designated materials such as finish hardware, face brick, or electric light fixtures. Such an allowance establishes an approximate material cost for bidding purposes in those instances where the designer does not prepare a detailed material specification. In such cases, the contractor does not solicit material prices for these items but just includes the allowance amount in its bid. After the contract is let, a final material choice is made by the owner or architect-engineer.

The architect-engineer may also state that a given sum is to be included in the estimate to act as a contingency for extras that might prove to be necessary as the work progresses. Such contingency provisions are sometimes utilized for remodeling and other types of work that have many uncertainties associated with them.

Occasionally, allowances are used for subcontracted work. For example, the owner may solicit bids from specialty contractors for specific portions of the project in advance of the date for the opening of the prime proposals. After the low bidding subcontractors have been determined, the bidding prime contractors are advised of the amounts and nature of the subbids that are to be included within their estimates. As an alternative to this, the specifications may merely indicate a sum of money to serve as an allowance for certain subcontracted work that will be let after the award of the prime contract. Landscaping is a common example of this. In either case, however, the prime contractor must be cognizant of all conditions pertaining to the award of the subcontracts and the exact nature of the work covered by them.

Serious operational problems can develop when the general contractor does not select its own subcontractors, as may be the case with subcontract allowances. If the two parties should be incompatible or have had previous difficulties, job relationships can deteriorate badly. This situation can be especially miserable when one is a union contractor and the other is not.

The inclusion of an allowance in the proposal does not necessarily mean the contractor is entitled to receive payment in the same amount. The contract documents provide that, after the completion of construction, an adjustment in the contract amount will be made equal to any difference between the specified allowance and the actual cost eventually associated with it.

5.31 PROJECT OVERHEAD

Overhead expenses are costs that do not pertain directly to any given construction work item but are nevertheless necessary for ultimate job completion. Project overhead refers to costs of this type that are incurred on the project itself. Office overhead is discussed in the next section. Job overhead costs can be estimated with reasonable accuracy and are compiled on a separate overhead sheet. This overhead is a significant item of expense and will generally run from 5 to 15 percent of the total project cost, depending somewhat on where certain project costs are included in the cost estimate.

Job overhead expense should be computed by listing and costing each item of overhead individually rather than by using an arbitrary percentage of project cost. This is true because different projects can have widely varying job overhead requirements. The only way to arrive at an accurate estimate of job overhead is to analyze the particular and peculiar needs of each project.

Typical items of job overhead are listed below. This list is not represented as being complete, nor would all the items necessarily be applicable to any one project.

Job mobilization	Timekeepers
Project manager	Watchmen and guards
General superintendent	Engineering services
Nonworking foremen	Job sign
Heat	Temporary lighting
Electricity	Drinking water facilities
Water	Badges
Storage buildings	Equipment move-in and assembly
Sanitary facilities	Equipment dismantling and move-out
Field office supplies	Worker transportation
Job telephone	Worker housing
Small tools	Legal expenses
Temporary enclosures	Surveys
Temporary stairs	Field office
Permits and fees	Parking areas
Special insurance	Security clearances
Builder's risk insurance	Load tests
Photographs	Temporary roads
Barricades	Storage area rental
Winter operation	Travel expenses
Concrete and other tests	Protection of adjoining property
Cutting and patching	First aid
Drayage	Storage and protection
Cleanup	Temporary partitions

5.32 OFFICE OVERHEAD

Office overhead includes general business expenses such as office rent, office insurance, heat, lights, office supplies, furniture, telephone and telegraph, legal expenses, donations, advertising, travel, association dues, and the salaries of executives and office employees. The total cost of this overhead expense generally ranges from 3 to 10 percent of a contractor's annual business volume. This percentage represents inescapable costs of doing business, and the contractor must include in the cost estimate of each project an allowance for office overhead expense.

From the preceding it is easy to see that office overhead includes costs that are incurred in support of the overall company construction program and that generally cannot be charged to any specific project. For this reason, office overhead is normally included in an estimate as a percentage of the total estimated job cost. Details of how the contractor does this vary, but the net effect is the same. A percentage of the total estimated cost of the job can be added as a separate line item, or a suitable markup percentage can be applied that will include office overhead as well as an allowance for profit.

5.33 PROJECT TIME SCHEDULE

When the cost of a project is being estimated, it is important that a general timetable of construction operations be devised. Small projects usually require little investigation in this regard, but larger projects need and deserve attention to this matter because construction time is of first-rate importance. One reason for this is that the bidding documents normally stipulate the length of time in calendar days within which the construction must be completed. In some cases, however, the contractor must indicate on the proposal form its own time requirement for completing the project. In either case, construction time becomes a contract provision, and failure to complete the project on time is a breach of contract that can subject the contractor to damage claims by the owner. Because of the importance of time, the contractor cannot accept at face value the construction period stipulated in the bidding documents but must make its own independent assessment of the construction time that will actually be required. When the contractor's estimate of construction time is called for on the proposal form, it must be as accurate as possible.

An approximate construction schedule is also needed for pricing purposes. For one reason, most items of job overhead expense are almost directly related to the duration of the construction period. In addition, a general calendar schedule of the major job segments provides the estimator with valuable information concerning times available and weather to be expected. This provides a basis for decisions regarding equipment and labor productivity, cold weather operations, need for multiple shifts or overtime, and other such considerations. In cases where the contracting firm believes the stipulated contract time to be inadequate for project completion using normal construction procedures, it may request the owner and architect-engineer to reconsider and allow a longer construction period. Otherwise, the feasibility of overtime or multiple-shift work must be investigated. If liquidated damages will be assessed against the contractor for failure to complete the project within the designated period, the contractor may decide simply to add the anticipated amount of such damages into its estimate of job cost.

A standard way of devising a general job plan for estimating purposes is to divide the project into major segments, estimate the time necessary to accomplish each of these segments, and establish the sequence in which they will proceed. The result is a bar chart (see Section 11.24) that shows the start and finish dates for the overall project and the approximate calendar times during which the various parts of the job will proceed. A bar chart is a series of bars plotted to a horizontal time scale. Each bar represents the beginning, duration, and completion of some designated segment of the total project. In combination, the bars make up a calendar time schedule for the entire job.

It must be noted that the bar chart just discussed is by no means the equivalent of a detailed network analysis of the project as discussed subsequently in Chapter 11. A complete and detailed time analysis of the project requires considerable time and effort. On a competitively bid job it is not usually possible for contractors to go through the extensive mechanics of a project network time study during the bidding period. When prepared with care, however, the much more approximate bar chart can suffice for estimating purposes.

5.34 MARKUP

On competitively bid projects, the markup or margin is added at the close of the estimating process and is an allowance for profit plus other items such as office overhead and contingency. Markup, which may vary from 5 to more than 20 percent of the job cost, reflects the contracting firm's considered appraisal of a whole series of imponderables that can influence the probability of it being the low bidder and its chances of making a reasonable profit if it is. Many factors must be considered in deciding a markup figure, and each can have an influence on the value chosen. The size of the project and its complexity, its location, provisions of the contract documents, the contractor's evaluation of the risks and difficulties inherent in the work, the identity of the owner and/or the architect-engineer, and other intangibles can have a bearing on how a contractor marks up a particular job. Many contractors use different markup figures for different parts of the project. For example, it is common practice for a contractor to apply a given markup to the work that it will accomplish with its own forces and a smaller markup (perhaps half of the former) for work that is subcontracted.

The contractor is required to bid under a form of contract that has been designed to protect the owner and architect-engineer. In their attempts to afford these parties protection against liabilities and claims arising out of the construction process, the writers of contract documents usually incorporate a great deal of "boiler plate" including disclaimers of one sort or another. A common disclaimer provides that neither the owner nor architect-engineer assumes any responsibility for the accuracy or completeness of boring logs or other subsurface information submitted with the bidding documents and that the contractor must assume full responsibility for any and all unknown physical conditions, including subterranean, that may be found at the construction site. Most of the court cases involving contract disclaimers pertaining to subterranean conditions have upheld them in the absence of fraud, warranty of the information by the owner, or deliberate failure to disclose all available information. As a general rule, the courts have placed the risk of unknown subsurface conditions on the contractor unless there are contractual provisions that give the contractor a right to additional costs (see Section 6.23).

Other disclaimer clauses sometimes force the contractor to assume liability for every conceivable contingency, some of which involve eventualities over which it would have no control whatsoever. For example, provisions are found in some contract documents that absolve the owner from all claims for damages caused by project delay, even if caused by the owner's own fault or negligence. Provisions are sometimes included which indicate that almost nothing will provide a valid basis for an extension of time. When such severe language is used in private contracts, the contractor can sometimes protect itself by inserting protective language on the proposal form or by submitting a qualifying letter with its bid. This is not usually possible on public contracts because of the risk of bid disqualification. When the general tone of the bidding and contract documents places the risk of the "unexpected" entirely on the contractor, the markup figure must obviously include some contingency allowance. It is debatable whether such one-sided and burdensome contract language truly serves the owner's best interests. For one thing, such contract terms can result in owners paying for contingencies that never happen.

The contracting firm's objective when it selects a markup figure is to include the maximum possible profit, at the same time keeping its bid at a competitive level. Stated another way, it wants to be the low bidder but with a minimum spread between its low bid and that of the second-low bidder. Procedures have been developed to assist the contractor in selecting markup figures that will maximize its profits over the long term. The two best-known and most widely accepted of these "bidding strategies" are known as the Friedman Model and the Gates Model. The Friedman Model is the simpler of the two and assumes that all bidders are acting independently of one another. This means that the probability of underbidding a group of competitors equals the product of the probabilities of underbidding each competitor separately. The Gates Model assumes that bids from competing contractors are not totally independent of one another and procedures are applied to take this into account. Appendix G discusses a simple bidding strategy that is based on the Friedman Model using concepts of mathematical expectation and elementary probability.

5.35   THE CONTRACTOR'S FEE

Normally associated with cost-plus and management-type contracts, a contractor's fee can be determined in a number of ways. This matter is discussed in Chapter 6. Decisions regarding fees are a management responsibility worthy of careful study. All too often the only job factor given consideration by the contractor is the estimated project cost. However, the nature and complexity of the construction operations, geographical location, equipment and manpower requirements, and estimated construction time must also be considered. A project must not be allowed to immobilize company funds, manpower, equipment, and supervision without the contractor realizing a proper return on its investment. The implication of this is that the fee designated be based not only on the size and nature of the project but also on the contractor resources required and the time period they will be needed. Additionally, the contractor's fee must often include an allowance for costs not considered to be reimbursable, such as office overhead (see Section 6.9).

5.36 THE LUMP-SUM BID

The preparation of a detailed cost estimate results in the determination of a great many separate prices that are scattered throughout a number of work sheets of various kinds. All these costs are finally brought together and entered on a recapitulation or "recap" sheet. Figure 5.7 illustrates a typical recap sheet for a lump-sum bid. The costs associated with work items the contractor proposes to carry out with its own forces are posted from the summary sheets and entered in the material and labor columns. All subcontract bids are shown under the subcontract heading. Equipment costs are brought forward from either an equipment sheet or the individual summary sheets. The equipment

RECAP SHEET

Project _____ Municipal Airport Terminal Building _____

Work Item	Material	Labor	Subcontract	Total
1. Clearing and Grubbing	$ 1,150	$ 8,620	$ —	$ 9,770
2. Excavation and Fill	14,060	15,780	—	29,840
3. Concrete	281,769	149,378	—	431,147
4. Forms	13,400	22,971	—	36,371
5. Masonry	—	—	422,000	422,000
6. Carpentry	19,120	13,640	—	32,760
7. Millwork	54,060	20,211	—	74,271
8. Steel and Misc. Iron	251,890	73,340	—	325,230
9. Kitchen Equipment	—	—	59,981	59,981
10. Insulation	—	—	22,280	22,280
11. Caulk and Weatherstrip	—	—	4,980	4,980
12. Lath, Plaster, and Stucco	—	—	170,200	170,200
13. Ceramic Tile	—	—	19,333	19,333
14. Roofing and Sheet Metal	—	—	182,770	182,770
15. Resilient Flooring	—	—	22,624	22,624
16. Acoustical Tile	—	—	28,019	28,019
17. Painting	—	—	83,340	83,340
18. Glass and Glazing	—	—	62,319	62,319
19. Terrazzo	—	—	84,108	84,108
20. Miscellaneous Metals	61,228	11,315	—	72,543
21. Finish Hardware	43,600	4,430	—	48,030
22. Plumbing, Heating, Air-Cond.	—	—	542,120	542,120
23. Electrical	—	—	338,738	338,738
24. Clean Glass	—	—	2,800	2,800
25. Paving, Curb, and Gutter	—	—	79,553	79,553
26. Construction Equipment	56,159	—	—	56,159
27. Job Overhead	39,550	88,271	—	127,821
	$835,986	$407,956	$2,125,165	$3,369,107
			Changes (−)	95,452
				$3,273,655
			Markup, 8%	261,892
				$3,535,547
			Bond	26,454
				$3,562,001
			Tax, 3%	106,860
				$3,668,861

Figure 5.7   Recap sheet for lump-sum bid.

cost shown in Figure 5.7 was compiled on a separate equipment sheet, as is commonly done with lump-sum bids, and is entered in the material column. Project overhead costs are entered from an overhead sheet, with wages placed under the labor heading and the total of all other expenses in the material column. The total of the material, labor, and subcontract columns of the recap sheet must equal the sum of the total column, which is a useful check on the accuracy of the additions. The bid price is arrived at by the final inclusion of changes, markup, bond, and tax.

Taxes applicable to construction projects vary with locality and can differ between public and private projects. A common tax provision requires the contractor to pay a gross receipts tax or sales tax on its total business volume. Figure 5.7 includes such a tax in the amount of three percent.

5.37 CHANGES

The recap sheet must be totaled and checked at a time when subbids and material prices are still being received and changed. The contractor must have some method for incorporating last-minute price revisions into the final bid amount. Figure 5.8 shows a change sheet which is commonly used by estimators to list and summarize such changes and as a means for adding in items previously forgotten or overlooked.

To illustrate the procedure, let us suppose that somewhat in advance of bid time, the recap sheet in Figure 5.7 was totaled to obtain the direct cost of $3,369,107. This was based on the best plumbing and heating bid of $542,120. Shortly thereafter, a telephone call cut this bid by $46,300. This is entered into Figure 5.8 in the deduct column. In the brief period before the bid must be finalized and the proposal form filled in, other changes are entered, some additive and some deductive. The final algebraic total of the changes is shown in Figure 5.8 as a deduction of $95,452.

CHANGE SHEET

Project _____ Municipal Airport Terminal Building _____

Item	Increase	Deduct
1. Plumbing, Heating, Air-Cond.	$ —	$46,300
2. Lath, Plaster, and Stucco	—	9,400
3. Roofing and Sheet Metal	—	16,480
4. Electrical	—	22,000
5. Ready-mix Concrete	—	5,250
6. Metal Doors	1,815	—
7. Chain Link Fence	2,163	—
8.		
9.		
10.		
Total Changes	$3,978	$99,430
		$95,452

Figure 5.8 Bid change sheet.

5.38 THE UNIT-PRICE BID

In a manner similar to that for a lump-sum bid, all costs associated with a project being bid as unit prices are accumulated and summarized on a final recap sheet. Figure 5.9 illustrates a typical recap sheet for a unit-price bid. To produce a separate bid value for each item designated in the proposal form, this recap sheet must maintain all costs of each bid item separately. The costs associated with each bid item are entered from its summary sheet onto the recap sheet. In Figure 5.9 the total direct cost of the entire quantity of each bid item is obtained by adding its labor, equipment, material, and subcontract costs. The sum of all such bid-item direct costs gives the estimated total direct cost of the entire project ($4,796,527). To this are added the job overhead, markup, bond, and tax giving the total bid price ($5,950,695). Dividing the total project bid price by the total direct project cost gives a factor of 1.241. By multiplying the total direct cost of each bid item by this factor, the total bid amount for that work item is obtained. Dividing the total bid amount of each work item by the quantity gives the unit price for that bid item.

There are instances in which projects are bid as a combination of lump-sum and unit prices. An example of this is a powerhouse, built under a single contract, whose foundation was bid on the basis of unit prices and the structure itself was bid lump-sum. Another instance is when the proposal form of a building being bid lump-sum asks for certain unit prices as well. This might be done when some changes to the work during the construction phase appear likely and bid prices are requested on a unit basis for the possible later addition or deletion of work items such as concrete or excavation.

5.39 COST OF CONTRACT BOND

A large proportion of construction contracts require that a contract bond be provided by the prime contractor to the owner as a guarantee against any default or failure on the part of the contractor during the construction period. The workings and characteristics of contract bonds are discussed in Chapter 7. For the present we are concerned more with the cost of their premiums than with their nature.

The contract documents may require a performance bond only or both a performance and a payment bond. Many different combinations are used. For example, the contract documents may require a performance bond in the amount of 100 percent of the contract price and no payment bond. On the other hand, a performance bond of 50 percent and a payment bond of 50 percent are sometimes required. Many of the professional societies recommend a 100 percent performance bond and a 100 percent payment bond. Federal projects require a 100 percent performance bond and a sliding-scale payment bond (see Section 7.6). Generally, however, the contract amount, not the total face value of the bonds, determines the premium the contractor must pay. The explanation for this is that the risk depends on the nature and size of the contract rather than on the bond penalty. Consequently, the bond premium is the same for any of the preceding combinations of performance or performance and

RECAP SHEET

Project   Holloman Taxiways and Aprons   Bid Date _____   August 9, 19—

Bid Item	Unit	Quantity	Total Material Cost	Total Labor Cost	Total Equip. Cost	Total Sub. Cost	Total Direct Cost	Bid Unit	Bid Total
1. Clearing	l.s.	Job	$ —	$ 3,426	$ 7,398	$ —	$ 10,824	$13,433.00	$ 13,433
2. Demolition	l.s.	Job	—	3,817	4,922	—	8,739	10,845.00	10,845
3. Excavation	c.y.	127,000	—	50,424	116,796	—	167,220	1.63	207,010
4. Base course	ton	79,500	106,319	235,113	434,663	—	776,095	12.11	962,745
5. Concrete pavement, 9 in.	s.y.	90,000	527,375	667,864	135,139	—	1,330,378	18.34	1,650,600
6. Concrete pavement, 11 in.	s.y.	115,400	827,978	1,048,546	212,168	—	2,088,692	22.46	2,591,884
7. Asphalt concrete surface	ton	150	1,472	690	869	—	3,031	25.08	3,762
8. Concrete pipe, 12 in.	l.f.	1,000	10,318	3,455	3,626	—	17,399	21.60	21,600
9. Concrete pipe, 36 in.	l.f.	300	4,539	1,958	2,504	—	9,001	37.23	11,169
10. Inlet	ea.	2	534	74	109	—	717	444.50	889
11. Fiber duct, 4-way	l.f.	600	3,417	2,185	6,911	—	12,513	25.89	15,534
12. Fiber duct, 8-way	l.f.	1,200	10,730	6,023	13,717	—	30,470	31.51	37,812
13. Electrical manhole	ea.	6	—	—	—	6,210	6,210	1,284.00	7,704
14. Underground cable	l.f.	34,000	—	—	—	170,638	170,638	6.22	211,480
15. Taxiway lights	ea.	120	—	—	—	30,473	30,473	315.00	37,800
16. Apron lights	ea.	70	—	—	—	17,542	17,542	311.00	21,770
17. Taxiway marking	l.s.	Job	3,122	2,373	450	—	5,945	7,378.00	7,378
18. Fence	l.f.	26,000	86,318	19,004	5,318	—	110,640	5.28	137,280
Totals			$1,582,122	$2,044,952	$944,590	$224,863	$4,796,527		$5,950,695

	$4,796,527
Job overhead	201,394
	$4,997,921
Markup, 15%	749,688
	$5,747,609
Bond	29,765
	$5,777,374
Tax, 3%	173,321
Total project cost	$5,950,695

Factor = $\dfrac{\$5,950,695}{\$4,796,527}$ = 1.241

Figure 5.9   Recap sheet for unit-price bid.

payment bonds. Special rules apply when low-percentage performance bonds or payment bonds only are required.

The premium the contractor must pay for the required contract bonds is based on the project completion time, the class of construction, the total contract amount, and any deviations from regular bureau rates that may apply. These matters are discussed in Section 7.8. In the following example it is assumed that the contractor must pay the regular, undeviated manual rate. Figure 5.7 illustrates the addition of bond premium as the next-to-last step in arriving at the lump-sum bid. The amount of $26,454 is obtained by applying the rates shown in Figure 7.4 for Class B construction (buildings) to the sum of $3,535,547 obtained from the recap sheet in Figure 5.7. The first $500,000 is assessed at $12 per $1000 or at a rate of 1.2 percent. The next $2,000,000 is charged at the rate of 0.725 percent, and everything over $2,500,000 up to $5,000,000 has 0.575 percent applied to it. The sum of these is $6000 + $14,500 + $5954 = $26,454. The bond premium rates given in Figure 7.4 apply for a construction time of up to 24 months or 731 days. For longer periods, the premium is increased by 1 percent per whole month. The estimated time for the Metropolitan Airport Terminal Building is given in Section 5 of Appendix D as 380 calendar days.

5.40 ESTIMATING BY COMPUTER

The discussions in this chapter have been based on manual estimating methods. However, it is important to note that computers are being increasingly utilized in conjunction with construction estimating. This does not mean that drawings and specifications are fed into one end of a machine and the finished proposal is printed out at the other. The estimator must still tabulate quantities and dimensions from the drawings and make basic decisions concerning construction procedures, equipment types, sequence of operations, and prices to be used. In addition, it is not generally feasible to estimate the entire project using computerized procedures. Ideal for the computer are situations where there are large quantities of similar work items. Portions of the project involving small quantities or general requirements are probably best handled manually.

Computer estimating starts with the estimator making the usual quantity takeoff from the drawings. Quantity takeoff can be assisted by an assortment of special mechanical devices. Special probes enable estimators to obtain item counts, linear footage, and areas directly from the drawings and to store these values in the computer memory bank. Volumes can also be found by inputting cross-sectional data directly into the computer.

A computerized estimating system can accept quantity data such as count, length, area, and cross sections directly from the estimator or as it is received from a quantity takeoff device. The computer stores this information; recalls it when needed; applies unit costs of materials, labor, and equipment; and computes summary costs. Computerized procedures systematize the accumulation and analysis of detailed cost information from past projects and make it easier for the estimator to apply this historical experience to the project being estimated. One of the most important aspects of computer application is that it relieves the estimator of much of the drudgery associated with routine, repetitive, and time-consuming extensions and calculations. Thus the estimator is left with more time to study and analyze the project.

The computer has become especially valuable in estimating equipment costs. Computer application to equipment accounting makes it possible to record and summarize equipment ownership and operating costs with an accuracy and completeness hitherto not possible. The computer assists materially in obtaining reliable hourly equipment rates for use in estimating the cost of future work. Computer reports show where equipment is located and its availability. Computers have become especially useful in applying simulation techniques to the evaluation of construction methods, vehicle characteristics, spread balancing, and fleet size. Computer services provided by major equipment manufacturers can be especially helpful in this regard.

Because estimating requires almost immediate results, computerized estimating procedures normally require that contractors purchase or lease their own computer equipment or have direct access to computers on a time-share basis. In addition, contractors must have the use of computer programs that will carry out the necessary computations and print out the information in the desired format. Computer programs for estimating are available from most of the computer manufacturing companies. However, some of the larger contractors have developed their own programs. In addition, there are proprietary programs available for sale. In either case, the contractor's investment in computer software can become substantial. For use in conjunction with its estimating program, each contracting firm must develop its own library of production rates and unit costs that the machine can use during the pricing stage.

Computerized estimating is expensive to initiate and requires considerable time and many check runs before it can be used with confidence. Nevertheless, once such a system has been established and is operational, it can offer real advantages to the contractor. The computer can speed up the estimating process considerably and minimize computational errors. Of real benefit is the computer's ability to make rapid cost comparisons between alternate methods of construction and different equipment choices.

5.41   RANGE ESTIMATING

The estimation of future construction costs always involves many subjective judgments and a considerable degree of uncertainty. Every project estimate is in error. The amount of the error and whether it is plus or minus is unknown until completion of the project. There are certain probabilistic procedures that can be introduced into the estimating process that enable the estimator to establish the mathematical probability that the actual cost of the project will not overrun any particular figure. This technique of risk analysis, known as "range estimating," is not for every project but can be of considerable value to owners, architect-engineers, contractors, and construction managers on large and complex projects.

In an attempt to measure uncertainty, the estimator fixes an upper and lower cost limit, as well as a target value, for each critical element of the project. Included with this is the confidence level or chance that the actual cost of the project element will be at least as favorable as the target estimate. In effect, this range of possible costs for each critical part of the job brackets the traditional single-value estimate or target cost. In this context, a critical element is a project segment whose cost can be subject to substantial uncertainty. The results of these price distributions are fed into a computer that runs the project through thousands of simulations ranging from conservative to optimistic perspectives of job experience. The computer reports its results in a profile showing different total project costs versus the probability of overrunning each such total cost. Thus, individual uncertainties are combined in such a way that the uncertainty associated with the project's total cost is determined and presented for decision-making purposes. Range estimating can provide management with a means of coping with the problems that major project uncertainties present.

5.42 THE PROPOSAL

A proposal or bid is a written offer, tendered by the contracting firm to the owner, which stipulates the price for which the contractor agrees to perform the work described by the bidding documents. A proposal is also a promise that, on its acceptance by the owner, the bidder will enter into a contract with the owner for the amount of the proposal. Thus timely acceptance of a proposal by the owner is automatically binding on the bidder.

When open bidding is being used, a prepared proposal form is included with the contract documents and must be used by the contractor to present its bid. Failure to do so will normally result in disqualification. The prepared proposal form is both desirable and necessary so that all bids will be presented and evaluated on the same basis. It facilitates the detection of omissions and other irregularities and makes the comparison and analysis of the figures an easier matter for the owner and the architect-engineer.

A typical example of a lump-sum proposal is shown in Appendix H. The usual basis of contract award, assuming that all bidders are considered to be qualified, is the base bid plus or minus any alternates (see Section 5.45) accepted by the owner.

Appendix E illustrates the usual nature of a unit-price bid form. It to be noted that each bid item in Appendix E is accompanied by an estimate made by the architect-engineer of the quantity of that work item which is to be done. These estimated quantities are multiplied by the respective quoted unit prices, and the results are added. This sum represents an approximate total project cost and is the basis for the contract award. In cases of error in multiplication or addition by the contractor in preparing the proposal, it is usual for the unit prices to control and the corrected total sum to govern.

Contractors sometimes find that they would very much like to include certain information concerning their bid that is not provided for on the proposal form. This is often in the nature of some form of limitation on the bid, either placing certain restrictions on owner acceptance or establishing special conditions pertaining to the conduct of the work. Under such circumstances the contractor may be tempted to "qualify" its bid. In general, this is not permissible on public bidding, because any bid qualification will make the bid subject to rejection. A private owner generally has more latitude regarding proposal qualifications.

When closed bidding is at issue, the form of the quotation is frequently left up to the individual bidder. In such biddings, the job is not necessarily awarded to the lowest bidder but may be negotiated by the owner from among the two or three lowest bidders. Consequently, the wording and content of the proposal are designed by the contractor to meet circumstances and to put its bid in the best possible light with the owner.

5.43 LIST OF SUBCONTRACTORS

It is a common requirement, especially on public projects, that the prime contractor must submit, as a part of its bid, a listing of the subcontractors whose bids were used in the preparation of the prime contractor's proposal. In some cases, the dollar amounts of the subcontractors' bids must also be included. Details of this listing are variable but will typically require a listing of the subcontractors doing major categories of work and other categories comprising more than a given percentage of the total bid price. This listing requirement can serve the purpose of approval of subcontractors by the owner, but it is designed primarily to minimize bid shopping. Under some bidding directives of this type, subsequent changes in the named subcontractors can be made only under certain designated circumstances and with permission of the owner. In others, the general contractor has a prescribed length of time after the bidding within which it can make changes to its subcontractor list. In general, subcontractor listing is favored by subcontractors and is opposed by prime contractors.

The listing of subcontractors can be troublesome on the bidding of projects with alternates. The problem arises because the identity of the low-bidding subcontractor for a given work specialty can depend on which alternates the owner may choose to accept or reject. In such an instance, the prime contractor will normally list its subcontractors on the basis of the base bid excluding the alternates.

5.44 UNBALANCED BIDS

On a unit-price project, a balanced bid is one in which each bid item includes its own direct cost plus its pro rata share of the project overhead, markup, bond, and tax. For a variety of reasons, a contractor may occasionally raise the prices on certain bid items and decrease the prices on others proportionately so that the bid for the total job remains unaffected. This is called an unbalanced bid. It is common practice, for example, for a contractor to increase certain unit prices for items of work that are accomplished early in the course of construction operations and reduce proportionately prices for certain bid items that follow later. This process serves the purpose of making the early progress payments to the contractor of such disproportionate size that a minimum of its own capital is required to finance its initial operations. Such early overpayment is of considerable assistance in helping the contractor to recover heavy move-in and other start-up costs that are involved before the actual start of construction. Unbalancing is unnecessary when a pay item covering job mobilization is provided.

Unbalanced unit-price quotations may be submitted for other reasons. When the contractor detects what it believes to be a substantial error in the quantities listed in the proposal form, some unbalancing of unit costs is likely to be necessary in order that fixed costs such as equipment and overhead will be properly distributed over the true quantities of work. Also, for profit motives, the contractor may increase unit prices on items that it believes will substantially exceed the estimated quantities and lower unit costs commensurately on other items. The contractor may simply juggle certain unit costs to disguise the makeup of its prices. When submitting an unbalanced bid, the contractor must be willing to assume the risk of having its proposal declared unacceptable by the contracting authority. Bid rejection because of unbalancing is rare, however, because it is difficult to detect and to substantiate. Some unbalancing is standard practice in most unit-price biddings.

To illustrate some of the elements of unbalanced bidding, let us consider a numerical example. To understand the rationale behind unbalancing, two characteristics of a unit-price contract must be kept in mind. First, the low bid is determined on the basis of total cost, using the engineer's quantity estimates and the unit prices bid by the contractor. Consequently, the raising of some prices and the corresponding reduction of others is normally done in such a way that the total amount of the bid remains unchanged. Second, the contractor is paid on the basis of its quoted unit prices and the quantities of work actually done.

Suppose, for illustration, we consider two bid items, one of which the contractor believes to be substantially in error. The engineer's estimate for ordinary excavation is 150,000 cubic yards, but the contractor's takeoff indicates the actual amount will be about 200,000 cubic yards. The contractor decides to unbalance its bid as follows:

Bid Item	Engineer's Estimates	Straight Bid		Unbalanced Bid	
		Unit	Total	Unit	Total
Ordinary excavation	150,000 c.y.	$1.00	$150,000	$1.50	$225,000
Selected excavation	100,000 c.y.	$3.10	$310,000	$2.35	$235,000
			$460,000		$460,000

Let us assume that ordinary excavation actually turns out to be 200,000 cubic yards and selected excavation to be 100,000 cubic yards. Payment made on the basis of actual quantities of work done could vary as follows between a straight bid and the unbalanced bid the contractor decided to use:

Bid Item	Actual Quantities	Straight Bid		Unbalanced Bid	
		Unit	Total	Unit	Total
Ordinary excavation	200,000 c.y.	$1.00	$200,000	$1.50	$300,000
Selected excavation	100,000 c.y.	$3.10	$310,000	$2.35	$235,000
			$510,000		$535,000

As can be seen, the contractor's use of unbalanced bidding results in its receiving an increased payment of $25,000.

5.45 ALTERNATES

Proposals on lump-sum biddings are sometimes solicited for two or more alternative ways of accomplishing the same job of work. A more common procedure involves a request for the costs of additions or deductions, called alternates, to a defined basic structure. Proposal forms frequently request price quotations for alternative methods, materials, or scope of construction. These are called alternate proposals and may be additive or deductive to the base bid. Appendix H illustrates the use of alternate proposals. Such quotations must be complete within themselves and include all direct costs, job overhead, markup, bond, and tax. Alternates can be of special importance to the owner as a means of ensuring he receives a bid within his limited financing or providing him with an opportunity to make the most judicious selection of a material or process. However, there is no arguing the fact that alternates complicate the bidding process, and a multiplicity of them can work against accurate bidding. Many architect-engineers maintain that the most effective way to set up alternates is to make them additive to the base bid rather than deductive. This point is debatable but the practice of it is widespread.

The award of lump-sum contracts is based on the total of the base bid and any alternates accepted by the owner. When there are several alternates, it may be possible for the owner to juggle acceptance of them so that a preferred contractor receives the contract. To combat this possibility, it is usual for the bidding documents to state the order of acceptance of the alternates.

On private projects a bidding contractor may include with its proposal unsolicited alternate prices that it considers potentially attractive to the owner. If a substantial savings is possible for an acceptable substitution, it may win the job for the contractor. Unsolicited alternates normally may not be accepted by a public owner.

5.46 BID SECURITY

With few exceptions, the proposal must be accompanied by some form of security as a guarantee that the contractor, on being declared the successful bidder, will enter into a contract with the owner for the amount of its bid and will provide contract bonds as required. Bid bonds, also called proposal bonds, are widely used for the purpose of bid security, although some owners require that each bidding contractor submit a certified check, cashier's check, or other form of negotiable security. When a contractor becomes the successful bidder, the owner retains the security until such time as the contract is signed and satisfactory contract bonds are provided. Ordinarily, the owner returns the bid security of the unsuccessful bidders shortly after the opening of proposals, possibly retaining that of the next lowest bidder or two until after contract signing.

Bid bonds, which are provided by the contractor's surety for a small annual service charge, have the advantage of not immobilizing appreciable sums of the contractor's money. Bid bonds may be executed on the standard form of the surety, or the contract documents may include a specific form of bid bond that the contractor is required to use. On public works the bid bond must be of such form as to satisfy any statutory requirements, and many public agencies require the use of their own standard bid bond forms. A form used by The American Institute of Architects, reproduced in Figure 5.10, illustrates the usual style and content of a bid bond.

The minimum bid security required by the instructions to bidders may be stated as a given percentage of the maximum possible contract amount, including alternates, or as a designated lump sum. Bid security in the amount of 5 or 10 percent of the maximum bid price is a common requirement, although larger percentages are also used. For example, many federal projects require 20 percent bid security. Because bid bonds are prepared in advance of the bidding and before the proposal amount is accurately known, the contractor must arrive at some advance rough estimate of project cost for the use of its bonding company. If a fixed-sum bid bond is to be written, the amount of which is to be at least a given percentage of the maximum possible contract amount, the preliminary estimate of cost must be made conservatively high to ensure that sufficient bid security is available.

Should the contractor refuse or be unable to enter into a contract for the amount of its bid or to provide contract bonds as required, the bid security can be proceeded against by the owner as reimbursement for the resulting damages. In this regard, there are two forms of bid bonds in general use. One of these is a "liquidated damages" bond in which the surety agrees to pay the owner the entire bond amount as damages for default. The second type is a "difference-in-price" bond that provides the surety must pay the owner the difference between the defaulted low bid and the price the owner must pay to the next lowest responsible bidder, up to the face amount of the bid bond. This is the provision of the bid bond shown in Figure 5.10. However, statutes pertaining to some public works provide that, if the amount of the bid security is not sufficient to pay the owner the additional cost of procuring the work, the contractor is also liable for the deficit. In any event, the contractor is liable to its surety for any such damages the surety may have to pay to an owner. When the formal application for bond service is signed, the contractor agrees to indemnify the surety company from all claims that may be made against it under the bond.

On deciding to bid a project, the contractor should immediately so inform its surety company, particularly if the project is unusually large or different, or if the contractor is already burdened with a near-capacity volume of work.

THE AMERICAN INSTITUTE OF ARCHITECTS

AIA Document A310

Bid Bond

KNOW ALL MEN BY THESE PRESENTS, that we

(Here insert full name and address or legal title of Contractor)

as Principal, hereinafter called the Principal, and

(Here insert full name and address or legal title of Surety)

a corporation duly organized under the laws of the State of
as Surety, hereinafter called the Surety, are held and firmly bound unto

(Here insert full name and address or legal title of Owner)

as Obligee, hereinafter called the Obligee, in the sum of

Dollars ($),
for the payment of which sum well and truly to be made, the said Principal and the said Surety, bind ourselves, our heirs, executors, administrators, successors and assigns, jointly and severally, firmly by these presents.

WHEREAS, the Principal has submitted a bid for

(Here insert full name, address and description of project)

NOW, THEREFORE, if the Obligee shall accept the bid of the Principal and the Principal shall enter into a Contract with the Obligee in accordance with the terms of such bid, and give such bond or bonds as may be specified in the bidding or Contract Documents with good and sufficient surety for the faithful performance of such Contract and for the prompt payment of labor and material furnished in the prosecution thereof, or in the event of the failure of the Principal to enter such Contract and give such bond or bonds, if the Principal shall pay to the Obligee the difference not to exceed the penalty hereof between the amount specified in said bid and such larger amount for which the Obligee may in good faith contract with another party to perform the Work covered by said bid, then this obligation shall be null and void, otherwise to remain in full force and effect.

Signed and sealed this   day of   19

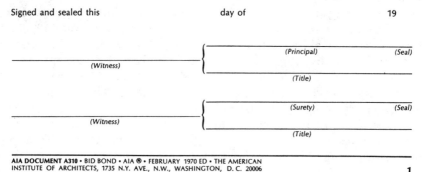

(Witness)	(Principal)	(Seal)
	(Title)	
(Witness)	(Surety)	(Seal)
	(Title)	

AIA DOCUMENT A310 • BID BOND • AIA ® • FEBRUARY 1970 ED • THE AMERICAN
INSTITUTE OF ARCHITECTS, 1735 N.Y. AVE., N.W., WASHINGTON, D. C. 20006

1

Figure 5.10 Bid bond. AIA copyrighted material has been reproduced with permission of The American Institute of Architects under application number 80063. Further reproduction is prohibited.

When the surety writes the bid bond, it is ordinarily obligated to provide the necessary contract bonds should the contractor become the successful bidder or, alternatively, pay the owner in accordance with the terms of the bid bond. Bidding documents often require certification from the surety providing the bid bond that required contract bonds will be provided to the owner if the contractor becomes the successful bidder. Consequently, considerable investigation by the surety may be required before the bid bond can be written, and such investigation can be time-consuming. This matter is discussed more fully in Section 7.11.

With regard to bid bonds, it is to be noted that general contractors, on occasion, may require bid bonds of certain of their bidding subcontractors. This is sometimes done when the general contractor is new to the area and has no knowledge of the local subcontractors or their records of performance. Bid bonds may be required of the subcontractors bidding major portions of the work as a check of sorts on their financial stability.

5.47 SUBMISSION OF PROPOSALS

It is the responsibility of the contractor to deliver its bid to the proper place prior to the deadline time designated. The completed proposal form, together with the bid security and other necessary supplementary information, is sealed in an envelope that is addressed as directed by the instructions to bidders and clearly labeled as a proposal for the project being bid. Bids may be submitted at any time prior to the deadline scheduled for their acceptance. If feasible, it is usual practice for the contractor to deliver the sealed bid shortly before opening time. However, proposals may also be dispatched by letter, telegraph, or messenger service.

Public contract policy dictates that all proposals shall be opened publicly and read aloud, a process of open bidding that is also commonly used on private projects. Such bid openings are usually well attended by the bidding general contractors, subcontractors, material vendors, and other interested parties. The ceremony consists of the owner or architect-engineer opening each bidder's sealed envelope, noting the type and amount of bid security, verifying receipt of addenda, and reading the amount of each bid item. Other bid formalities may also be noted for the record. At many bid openings, the estimate of cost prepared by the architect-engineer is also read or distributed to those attending. Bid tabulation forms are made available to those attending so they can make a record of the proceedings. It is customary that the opening adjourns without an official announcement concerning the identity of the successful bidder. Before the contract award can be made, the bids must be carefully studied and evaluated by the owner and architect-engineer, a process sometimes referred to as "canvassing."

After the bids have been opened and read, the contractor communicates the results of the bidding to its surety company. This information is incorporated into the surety's permanent file on the contractor and constitutes an important part of the contractor's record of performance.

In closed bidding, the amounts of the bids are not necessarily disclosed. After delivery of the proposals, the owner uses the bids in any way he sees fit to serve his own best interests. He can select any of the bidders he chooses, or reject all bids. The owner often makes a final selection of the successful contractor only after extensive negotiations.

5.48   COMPLIMENTARY BIDDING

Complimentary bidding refers to the submission by a contractor of a proposal that the contractor did not prepare itself but obtained from another contractor. Complimentary bids, sometimes called courtesy bids, may be used for a variety of reasons. For example, there are times when every contractor finds it impossible or undesirable to bid a project when asked to do so by an architect-engineer. To keep the goodwill of this architect-engineer, the contractor may obtain a complimentary figure from a fellow contractor who is bidding the job. Complimentary proposals may also be submitted to obtain the refund of plan deposits or to please an owner-client, as well as for other legitimate reasons.

The contractor furnishing the complimentary figure to another contractor makes it safely larger than its own bid but sufficiently close to appear to be a bona fide proposal. It is obvious that the use of complimentary bids smacks of collusion, even when none is intended or involved. Contractors who provide or use complimentary bids must be very circumspect in this regard, especially on publicly financed projects where complimentary bidding is considered to be an unlawful form of price-fixing. In addition, many biddings on public projects require that each bidder must certify that its bid is noncollusive, that it has compiled its bid independently, that it has not disclosed its bid to any other bidder, and that it has made no attempt to have any other contractor bid or not bid.

5.49   BID ETHICS

How the general contractor handles its subcontract bids and translates them into subcontracts involves ethical considerations that can be very troublesome. Bargaining is the essence of competition but there are occasions where the propriety of certain procedures is questionable, to say the least. For instance, in an attempt to improve its bidding position, a prime contractor may attempt, during the bidding period, to "chisel" down certain of the subcontract bids that have been received. To do this, the contractor might lead subcontractors to believe that lesser bids than theirs have been received. Or, the prime contractor may actually reveal low subbids in an attempt to get better prices. The fear that their bids may be disclosed during the bidding period at least partially explains the propensity of many subcontractors not to submit their final bids until the last minute.

On being designated the successful bidder, the general contractor is then faced with making contract awards to its subcontractors. The conventional concept of this process would normally envisage the awarding of subcontracts to those firms whose bids were used in making up the general contractor's estimate. Thus in most instances the low-bidding subcontractor for a given portion of the project would automatically get the work except in those relatively unusual cases in which the low bidder was not considered to be capable or desirable and its bid was not used. In any case, the subcontract would be for the amount of the subcontractor's bid.

On the other hand, a completely different approach is possible, one in which the contractor regards the procurement of subcontractor services as being basically no different than bargaining for any other commodity on the open market. Under this philosophy the general contractor would feel no obligation to those subcontractors whose prices were incorporated into its own bid. Rather, after being declared the successful bidder, the general contractor would feel free to seek lower subcontract prices than those originally used. This practice is often referred as "bid shopping." At this point, the low-bidding general contractor has a very strong bargaining position, and any lower subcontract prices that can be obtained represent windfall profits. The prime contractor may attempt to persuade an original low-bidding subcontractor to reduce its quotation or to get another subcontractor to underbid the original low bidder. Subcontractors may take the initiative and attempt to better the lowest subcontract prices, an activity sometimes referred to as "bid peddling." These practices are obviously distasteful to subcontractors and often lead them not to submit their bids to those prime contractors who resort to such methods. Another reaction of subcontractors is to inflate their original bids when it appears there is going to be hard bargaining at a later date.

Bid shopping on publicly financed projects has come in for a great deal of criticism, and considerable effort has been expended to minimize or eliminate this practice. Several of the states have passed legislation that imposes regulations on the bidding of state-financed projects. For example, heating, ventilating and air conditioning, plumbing, and electrical work on state-financed projects must be awarded as separate contracts and not as a portion of the overall prime contract in some states. Another procedure requires general contractors bidding on a public project to name their subcontractors at the time they submit their bids to the owner. A system of bid filing is used on public contracts in a few states. Under this system, subcontractors file their bids with the awarding authority. The authority then furnishes the prime contractors with a list of the subbidders and the amounts of their bids. The general contractors then use these subbids as they choose in the preparation of their own proposals.

5.50 RESPONSIVE BID AND TECHNICALITIES

On public projects, an acceptable bid must be "responsive" to the invitation for bids and the instructions to bidders. Responsiveness is determined by whether the bid as submitted is an offer to perform, without exception, the exact work as called for by the invitation, and upon acceptance will require the contractor to perform in accordance with all the terms and conditions thereof. A bid is nonresponsive if it contains qualifications or conditions not in the invitation or if it offers performance which varies from the invitation. Additionally, a bid may be nonresponsive if it does not conform to the technical bidding requirements established in the instructions to bidders.

The instructions to bidders set forth specific requirements for the submission of the proposal. The following checklist can assist in avoiding most of the commonly occurring bidding irregularities.

1. File notice of intention to bid, if required.
2. Date the proposal with the day of the bid opening.
3. Fill in all blanks on the proposal form.
4. Make no interlineations or qualifications if prohibited by the instructions to bidders.
5. Properly countersign all erasures and corrections.
6. Make sure that signatures agree exactly with names as typed.
7. Check that the bid is signed by the proprietor, partners, or duly authorized corporate agent.
8. Be sure the individual signing for a corporation is so authorized by proper action of the corporation, this action having been entered into the minutes as a matter of formal record.
9. Attach corporate seal, if applicable.
10. Get signatures of witnesses, if required.
11. Have proposal notarized, if required.
12. Enter contractor's license number, if required.
13. Submit noncollusion affidavit, if required.
14. Acknowledge receipt of addenda.
15. Submit required number of copies of proposal.
16. Check multiplications and additions on unit-price proposal.
17. Check to see that unit prices are on the basis of the correct unit, for example, per square yard rather than per square foot.
18. Quote a price for all items indicated, including all bid items and alternates.
19. Enter prices both in writing and in figures, where required.
20. Do not include any bid stipulation (special conditions or limitations on owner acceptance).*
21. Submit representations and certifications form, if required.
22. Submit all supplementary information required, such as experience record, list of equipment, proposed plan of operations, list of subcontractors, and list of values of major trades.
23. Check face value of bid security against that required by the bidding documents.
24. Post mailed bids in ample time to ensure their receipt before the time of opening.
25. Follow up a telegraphed bid or modification with a written confirmation.
26. Submit minority hiring plan, if required.

---

*A common exception to this is on large projects bid as several separate sections where the contractor stipulates that it will accept any section and no more, or that it will accept all sections or none.

Any deviation from the stipulated requirements constitutes a bidding informality that can result in rejection of the proposal. A saving clause is usually present that allows the owner to waive any informalities, irregularities, or minor defects in bids received. A private owner has considerable latitude in this regard. Public owners must be much more circumspect in such waivers and will generally insist on strict compliance with all bidding technicalities.

5.51 THE ACCEPTANCE PERIOD

The bidding documents contain a provision that gives the owner a stated period of time after the opening of bids to make acceptance. Thirty days is common, although other periods of time are specified. During this waiting period, the subcontractors and material dealers are presumably obligated to the general contractor to stand by their price quotations just as the prime contractor is obligated to the owner.

Contractors are sometimes requested by the owner to grant an extension of the acceptance time. The contractor is generally willing to oblige, but sometimes in its eagerness to get the job it will agree to such an extension without giving the matter sufficient consideration. Such action means that the completion date of the project will be set back by a length of time equal to the extension of the acceptance period. Therefore, raises in labor wages that may occur during the construction period will be in force for the additional time. Also, the extension means that the placing of material orders will be delayed with the ever-present possibility of price advances. Another important aspect of this matter is that a subcontractor or supplier, having submitted what it considers to be an excessively low price, may not be willing to stand by its price quotation past the original acceptance period.

All these factors represent potential elements of additional cost that the contractor may have to assume should it too hastily grant an owner additional acceptance time. When increased costs are anticipated and the contractor does not wish to absorb them, it should quote the required additional amount to the owner in exchange for extending the acceptance period.

5.52 WITHDRAWAL OF BID BY PRIME CONTRACTOR

Under ordinary contract law, a bidder can withdraw or revoke its offer anytime prior to its acceptance. The situation is considerably different in construction, however. Despite the common-law doctrine of revocability, a prime contractor's bid proposal is normally considered to be irrevocable after the bid opening and during the acceptance period prescribed by the bidding documents. This results, on public projects, from many and varied statutes and local ordinances. Regarding bid withdrawal on private projects, courts often hold that bids are irrevocable, basing this on various legal grounds, although findings on this matter are not entirely consistent. In addition, the bidding documents themselves generally disallow bid withdrawal. It is only realistic, therefore, for a general contractor to regard as normally impossible the withdrawal of its bid after the deadline for the receipt of proposals without forfeiture of the bid security. The courts have long held that a bidder cannot unreasonably refuse to comply with its bid without penalty.

There is, however, a generally accepted legal basis for relieving the prime contractor from its bid, this being the doctrine of mistake. Prohibiting the withdrawal of bids, either by provisions created by the owner or by rules of law, is almost uniformly construed as inapplicable to prevent relief from the submission of a bid containing a gross mistake, provided the following conditions are met.

1. The mistake is of such grave consequences that to enforce the contract as offered would be unconscionable.
2. The mistake relates to a material feature of the contract.
3. The mistake has not come about because of the violation of a positive legal duty or from culpable negligence.
4. The owner is put in a status quo position to the extent that he suffers no serious prejudice except the loss of his bargain.

When the preceding conditions apply and when the mistake is excusable and one of fact (as opposed to a mistake of judgment), when the error is of a mechanical or clerical nature, and when the contractor acts promptly to notify the owner of the mistake and to rescind the bid, the courts almost unanimously permit withdrawal of the bid and the return of the contractor's bid security. There are state statutes and provisions in federal procurement regulations that specifically provide for relief from a unilateral mistake under certain circumstances.

When a bidding error is made, prompt action by the contractor can be important. The granting of relief as described in this section is the rule where the owner is advised of the error and the request to withdraw the bid is made before contract award. If notice is not given until after contract award, the contractor may have to show that the award was invalid on the basis that the owner should have known an error was made because of the disparity in bid amounts and did not ask the contractor to verify its bid. Relief for the contractor will likely be denied where it proceeds with the work without first seeking recision of the bid.

In granting relief to the contractor, it must be noted that the courts rarely will allow the erring contractor to correct its bid, only to withdraw it without penalty. However there have been instances where a public agency directed that the bid be corrected and did not allow it to be withdrawn. Such action has been limited to those cases where the bid as corrected remained the lowest bid.

5.53 WITHDRAWAL OF BID BY SUBCONTRACTOR

Occasionally the tables are turned and the general contractor is faced with a request from a subcontractor to withdraw its subbid. The legal question of revocation of a bid offer by a subcontractor is a tangled one and is very much a matter of the specific facts of the case. Nevertheless, there are some generally applicable concepts that can be discussed. In many jurisdictions the subcontractor can rescind its bid anytime prior to its formal acceptance by the prime contractor. There are some exceptions to this such as where informal understandings were made between the prime contractor and subcontractor prior to the award of the general contract. Oral acceptance of the subbid by the general contractor may not be enforceable because of the applicable state statute of frauds which requires certain contracts to be in writing before they are enforceable. Additionally, the courts have sometimes ruled that the general contractor and subcontractor did not intend to bind themselves contractually until a written subcontract was executed. This entire matter is further complicated by the fact that the general contractor is understandably loathe to accept a subbid until the owner has made a formal award of the general contract. In light of these complications, it is not surprising that attempts by general contractors to hold low-bidding subcontractors to their bids on the basis of acceptance prior to withdrawal have met with variable results.

When the low-bidding prime contractor bases its proposal on a subcontract bid that is substantially lower than any other submitted, the withdrawal of this low subbid can cause the contractor substantial loss. As a result of this condition, the courts have sought rationales by which subcontractors would be required to hold their prices firm for some reasonable time such as the acceptance period specified in the bidding documents. Probably the most common approach to the matter of holding a subcontractor to its bid is now based on the doctrine of promissory estoppel. This doctrine avoids entirely the question of whether the subcontractor's bid was formally accepted by the prime contractor. The elements of promissory estoppel can bind a subcontractor to its bid price if the prime contractor can prove (1) that it received a clear and definite offer from the subcontractor, (2) that the subcontractor could expect that the general contractor would rely on the offer, (3) that the contractor actually did rely on the offer, and (4) that this reliance worked to the general contractor's detriment. Where the general contractor can establish these points, a subcontractor's low bid is generally binding if the detriment can be avoided only by enforcement of the bid.

For the contractor to have the defense of relying on the subbid, it must notify the subcontractor when the bid is so low that it suggests a mistake might have been made and request the subcontractor to verify the bid's correctness. In addition, the contractor must accept the subcontractor's bid with reasonable promptness after its own bid is accepted by the owner. In accepting the subbid, the general contractor may not change the terms on which the initial bid was solicited.

With regard to "freezing" subcontract bids once they have been tendered to the general contractor, it is interesting to note that, in general, it does not necessarily work the other way, although this is a matter of applicable state law. There have been recent court decisions to the effect that a general contractor's use of a subcontractor's bid or the listing of the subcontractor's name in the prime contractor's proposal to the owner does not by itself create a binding contract between the two contractors, and the prime contractor is not bound to award the subcontractor the work involved. This is in the absence of any prior agreement or understanding. The subcontractor's bid is an offer and does not ripen into a contract until it is voluntarily and expressly accepted by the prime contractor. There are some state statutes, however, that impose penalties on the prime contractor if work awards are not made to the subcontractors listed on its proposal.

5.54 REJECTION OF PROPOSALS

The contract documents reserve the right for the owner to reject any or all bids. An individual bidder may be rejected because of insufficient finances to handle the project, lack of experience, unsatisfactory reputation indicating irresponsibility or unreliability, inadequate personnel or equipment, or failure to submit a responsive bid. Private owners are free to ignore technical inadequacies if it is in their interest to do so, but public owners must be very careful to observe applicable laws in this regard. Rejection of bids on public projects because of technical bidding errors is relatively common. Rejection may be also based on irregularities in the bidding procedure, suspected collusion, flagrant unbalancing of bid unit costs, not enough bona fide bidders, or unexpectedly high proposals. The lowest proposal sometimes far exceeds the available funds, and the project must then be abandoned or revamped. Public contracts are subject to statutory regulations in this regard.

In the event the owner wishes to reduce the project cost by a minor amount, he is likely to negotiate a final contract with the low-bidding contractor. If major savings are required, it is usual that all proposals are rejected. After redesign or other corrective procedures, the project is readvertised and new bids are taken.

5.55 STATE PREFERENCE STATUTES

Several states have enacted statutes pertaining to the award of state-financed construction projects that give preference to bidding contractors who are domiciled in that state or who have previously satisfactorily performed public contracts and have paid specified state taxes for a statutory period. These statutes typically provide that construction contracts for state public works shall be awarded to the lowest responsible resident bidder if its bid is not more than five percent above that of the lowest responsible nonresident bidder.

6

CONSTRUCTION
CONTRACTS

6.1 AWARD OF THE CONTRACT

After competitive bid proposals have been submitted and found to be responsive, the owner, after careful study and evaluation of the bids received, must identify the contractor to whom the project will be awarded. Open biddings, involving both private and public owners, customarily award the job to the "lowest responsible bidder." This is mandatory on publicly financed projects.

If the bidding contractors have been prequalified, as is common on public projects, or if the bidding has been limited to an invited list of contractors, a practice occasionally used by private owners, the matter of responsibility has already received attention. In either case, the owner has restricted the bidders to those judged qualified and capable, using whatever criteria that are believed to be suitable for the purpose. In the absence of bidding discrepancies, the successful bidder is determined on the basis of lowest total project cost.

Where prequalification or invitational bidding is not applied, the owner must make the determination of responsibility after the bids have been opened. The term "lowest responsible bidder" has been held to mean the lowest bidder whose offer best responds in quality, fitness, and capacity to the particular requirements of the proposed work. In the case of public contract-awarding bodies, the law gives them discretionary power as to which contractor is the lowest responsible bidder, such discretion not to be interfered with by the courts in the absence of fraud, collusion, or bad faith.

It is a difficult matter to establish practicable criteria by means of which qualification can be measured and judged. Basically, the owner must evaluate by some means the bidder's capacity, or incapacity, to perform. The owner may request qualification information to be submitted with the bids. Default by the contractor on a previous contract, proof of dishonesty, past difficulties in completion of projects on time, or a reputation for uncooperativeness and cutting corners may be taken as direct evidence of contractor irresponsibility. In any event, disqualification of a bidder on the grounds of irresponsibility is a ticklish matter and may lead to litigation.

With regard to the judgment of contractor responsibility, the owner will often rely on the contractor's surety company to do this for him. Before a surety will provide a contractor with a bid bond and contract bonds (Chapter 7), it will investigate the contractor's finances, experience record, and other qualifications. By requiring such bonds, the owner may simply assume that if a contractor can "bond" the job, it is a responsible firm.

On selection of the contractor, the owner advises it in writing that its proposal has been accepted. This acceptance is conveyed by the issuance of a "notice of award," which is forwarded to the contractor, together with information concerning arrangements for the signing of the contract. This notice, usually in letter form, sets forth the conditions pertaining to the award.

6.2 LETTER OF INTENT

Occasionally the owner may want the contractor to start construction operations before the formalities associated with the signing of the contract can be completed. However, the contractor must proceed with caution in placing material orders, issuing subcontracts, or otherwise obligating itself before it has an executed and signed contract in its possession. In such urgency, it is common practice for the owner to authorize the start of work by a "letter of intent." This letter is prepared for the signatures of both parties and states their intent of entering into a suitable construction contract at a later date. When signed, the letter is binding on both parties and furnishes the contractor with sufficient authority to proceed with construction in the interim before the contract is formally executed. This interim authorization contains explicit information about settlement costs in the event the formal contract is never executed and will often limit the contractor to certain procurement and construction activities. The contractor should have its lawyer examine this document before it is signed.

6.3 THE CONTRACT DOCUMENTS

Under ordinary circumstances, a contract arises from the proposal, which is an offer, and the notice of award, which is an acceptance. Although this sometimes suffices for small jobs, it is universal practice in construction for the contract to be formalized by a written document. The basic purpose of a written contract is to define exactly and explicitly the rights and obligations of each party to it. The complex nature of construction dictates a form of contract that is relatively lengthy, sacrificing brevity in order to describe precisely the legal, financial, and technical provisions. Construction contracts are substantially different from the usual commerical variety. The commodity concerned is not a standard one but a structure that is unique in its nature and whose realization involves considerable time, cost, and hazard. Construction contracts involving public bodies are regulated by statute as to content and procedure.

Actually, the usual construction contract consists of a number of different documents. Exactly which documents constitute the construction contract is variable. The following is a listing of those documents that are essential to the bidding, negotiation, and construction process.

1. Invitation to bid.
2. Instructions to bidders.
3. General conditions.
4. Supplementary conditions.
5. Technical specifications.
6. Drawings.
7. Addenda.
8. Proposal.
9. Bid bond.
10. Agreement.
11. Performance bond.
12. Labor and material payment bond.

All of the contract documents are construed together for purposes of contract interpretation, giving meaning and effect to each part because it is presumed that everything in the contract was inserted deliberately and for a purpose. In general, the intention of the contracting parties is determined from the final contract executed by them, rather than from preliminary negotiations and agreements. Contracts are interpreted strictly and to the letter where possible and feasible under the law. It is presumed that those who enter into contracts know what they want, say what they mean, and understand what they have said. It is an accepted rule of law that a person has a duty to read and understand a contract before executing it, and failure to do so is no excuse for not rendering proper performance.

When a conflict exists in a contract, the specific provision prevails over the general provision, the handwritten provision prevails over the typed provision, and the typed provision prevails over the printed provision. In the event there is an inconsistency where numbers are expressed in words and figures, the words govern. When there are ambiguous terms they are interpreted against the party who drafted the document. If a conflict exists between the drawings and the specifications, it is usual that the specifications control, although the general conditions may provide that the architect-engineer or owner shall determine which is to prevail. Should a construction requirement appear only in the specifications and not on the drawings, or vice versa, the contractor must provide the requirement just as though it were included in both places (see Subparagraph 1.2.3 in Appendix C).

In general it may be said that contractors prefer contract documents whose arrangement, form, and content are familiar to them. Consequently, standard forms of contract documents are in wide use throughout the construction industry. Standardization of contract documents has done much to eliminate areas of disagreement among owners, architect-engineers, and contractors.

6.4 THE AGREEMENT

The agreement is a document specifically designed to formalize the construction contract. It acts as a single instrument that brings together all of the contract segments by reference and it functions for the formal execution of the contract. It serves the purpose of presenting a condensation of the contract elements, stating the work to be done and the price to be paid for it, and provides suitable spaces for the signatures of the parties. The agreement usually contains a few clauses that are closely akin to the supplementary conditions and serve to amplify them. To illustrate, it is common for the agreement to contain clauses that designate the completion time of the project, liquidated damages, and particulars concerning payments to the contractor, and that list the contract documents. However, practice varies in this regard, with such clauses appearing sometimes in the supplementary conditions, other times in the agreement, and perhaps in both.

The agreement can be a standard form, or it may be specially prepared for the purpose. Appendix I illustrates the use of *Document A101* of The American Institute of Architects, which is a popular choice for building construction contracts. *Document A101* is designed for use with the AIA "General Conditions of the Contract for Construction," *Document A201,* as contained in Appendix C. Appendix J reproduces *Document A111,* a form of agreement between the contractor and owner that is widely used for cost-plus-fee contracts. A standard agreement form recommended for engineering construction projects has been prepared by the National Society of Professional Engineers. Municipal construction contracts can use a standard form devised by the American Public Works Association together with the Associated General Contractors of America. Construction contracts with the U.S. government use Standard Form 23 of the General Services Administration. Each of these forms must be used in conjunction with its own particular set of general conditions.

On public contracts, statutory requirements must have been observed by the responsible public officials for a valid contract to exist. In most jurisdictions the omission of any required procedural step can leave the contractor without remedy to obtain payment for work performed. If a government body enters into a contract without complying with the statutory requirements pertaining to the bidding and awarding of public construction contracts, the contract is considered beyond the power of the public agency and may be declared void. There is a growing body of legal precedent, however, that when an imperfect public contract was entered into in good faith and is devoid of fraud or collusion, the contractor is entitled to relief based on the equitable doctrine of unjust enrichment. Notwithstanding this, however, the public may become the unintended beneficiary of the work without paying for it. One who does business with a public entity must be aware of the laws governing its administration and the limitations on the powers and authorities of the public officers involved.

In a similar vein, before affixing its signature to a contract involving private financing, the contractor should make some investigation of the financial integrity of the owner or the source of the financing for the project. This applies particularly to projects of a speculative or promotional nature. There are many instances where private owners have defaulted on their payment obligation to the contractor because their financing went awry. Subparagraph 3.2.1 of Appendix C specifically provides that the contractor can request and receive reasonable evidence from the owner that suitable financial arrangements have been made.

6.5 TYPES OF CONSTRUCTION CONTRACTS

Although there are many different types of construction contracts, they may be grouped together into two large divisions. One division includes those contracts for which the contractor is selected on the basis of competitive bidding. Almost all public construction contracts, as well as much private work, are in this category. Competitive-bid contracts are customarily prepared on a fixed-price basis and consist of two types. The unit-price contract is drawn on the basis of estimated quantities of specified work items and a unit price for each item. In the lump-sum contract, the second of the two types, the contract amount is a fixed sum that covers all aspects of the work described by the contract documents.

The second major division of construction contracts consists of those that result from direct owner-contractor negotiation. Negotiated contracts can be on any mutually agreeable basis: lump-sum, unit-price, or cost-plus-fee. Most negotiated contracts are on a cost-plus-fee basis, in which the owner reimburses the contracting firm for all construction costs and compensates it for services rendered. The provisions regarding compensation of the contractor are the major difference between the various types of negotiated contracts. The contractor's fee can be designated as a fixed percentage of the cost of the work, a sliding-scale percentage of the cost of the work, a fixed fee, a fixed fee with a guaranteed top price, a fixed fee with bonus, or a fixed fee with an arrangement for sharing any cost saving.

6.6 THE LUMP-SUM CONTRACT

The lump-sum contract is one in which the contractor agrees to carry out a stipulated job of work in exchange for a fixed sum of money. The satisfactory completion of the work for the stated number of dollars remains the obligation of the contractor, regardless of the difficulties and troubles that may be experienced in the course of construction activities, even though the total cost of the work may turn out to be greater than the contract price. However, the contractor can be relieved of this contractual responsibility because of impossibility of performance, where there is contract provision for price adjustment in the event of changed conditions (see Section 6.23), and possibly because of other contingencies.

This type of contract is popular from the owner's viewpoint for the obvious reason that the total cost of the project is known in advance. Its use is limited, however, to construction programs that can be accurately and completely described at the time of bidding or negotiation. If the work is of such a type that its nature and quantity cannot be accurately determined in advance of field operations, the lump-sum type of contract is not suitable.

6.7   THE   UNIT-PRICE   CONTRACT

This type of contract is based on estimated quantities of certain well-defined items of work and costs per unit amount of each of these work items. The estimated quantities are compiled by the architect-engineer, and the unit costs are those bid by the contractor for carrying out the stipulated work in accordance with the contract documents. The total sum of money paid to the contractor for each work item remains an indeterminable factor until completion of the project, however, because payment to the contractor is made on the basis of units of work actually done and measured in the field. Therefore the exact ultimate cost of the construction is not known to the owner until completion of the project. In addition, the owner often must support, either directly or through the architect-engineer, a field force for the measurement and determination of the true quantities of work.

The contractor is obligated to perform the quantities of work actually required in the field at the quoted unit prices, whether they are greater or less than the architect-engineer's estimates. This is subject to any contract provision for redetermination of unit prices when substantial quantity deviations occur (see Section 5.3). The same requirement for contractor performance described in the previous section for lump-sum contracts also applies to unit-price contracts.

Unit-price contracts offer the advantages of open competition on projects involving quantities of work that cannot be accurately forecast at the time of bidding or negotiation. An example might be the driving of piles or the excavating of foundations. A price per linear foot of pile or per cubic yard of excavation allows a reasonable variation in the driven length of the individual piles or the actual quantity of excavation as a result of job conditions that cannot be determined precisely before actual construction operations. However, drawings and specifications must be available for bidding that are complete enough for the contractor to assess the overall magnitude of the project and the general nature and complexity of the work.

6.8   COST-PLUS   CONTRACTS

Contracts of the cost-plus variety are used where, in the judgment of the owner, a fixed-sum contract is undesirable or inappropriate. Cost-plus contracts are normally negotiated between the owner and the contractor. Most cost-plus contracts are open-ended in the sense that the total construction cost to the owner cannot be known until completion of the project. When the drawings and specifications are not complete at the time of contract negotiation, the owner and contractor negotiate what is commonly called a "scope contract."

Based on preliminary drawings and outline specifications, the contractor arrives at a project "target estimate." The contract provides that the original contract documents shall be subsequently amplified "within the original intent of the preliminary drawings and specifications."

When negotiating contracts of the cost-plus type, the contractor and the owner must pay particular attention to four important considerations.

1. A definite and mutually agreeable subcontract-letting procedure should be arranged. Competitively bid subcontracts are generally preferred by both parties when they are feasible. If the nature of the work is such that competitive subbids cannot be compiled or are not desirable, a mutually agreeable negotiation procedure will have to be devised.

2. There must be a clearly understood agreement concerning the determination and payment of the contractor's fee. Fees may be determined in many different ways, the most important of these being discussed in subsequent sections. Involved here is not only the amount of the fee but also the method by which it will be paid to the contractor during the life of the contract. A statement concerning any variation of fee with major changes in the work should be included.

3. A common understanding regarding the accounting methods to be followed is essential. Many problems and controversies can be avoided by working out in advance the details of record keeping and purchasing, and the reimbursement procedure. Some owner-clients have need of accurate and detailed cost information for tax, insurance, and depreciation purposes. Owner requirements of this type made known at the beginning of the contract enable the contractor to better serve the owner.

4. A list of job costs to be reimbursable to the contractor should be set forth. Articles 8 and 9 of Appendix J present typical lists of reimbursable and nonreimbursable items.

6.9 SPECIAL REIMBURSABLE COSTS

With respect to reimbursable costs, two categories of expense can be particularly troublesome and merit special attention. One of these is the contractor's office overhead. Involved are the costs associated with the preparation of payrolls, purchasing, record keeping, engineering, preparation of working drawings, and similar office functions that are necessary elements of a construction project. If the job is of reasonably large extent, the contractor can establish a project field office that performs the necessary office functions directly on the site. All expenses, including salaries, incurred by the project office are directly assignable to the job and are considered to be legitimate costs for reimbursement. When the project is smaller, office functions usually are carried out in the contractor's main office, and items of office overhead directly assignable to a specific project are difficult to establish. It is common practice under such circumstances to eliminate office overhead altogether as a reimbursed cost and to increase the contractor's fee by a reasonable amount to provide for it.

The second troublesome category of reimbursed expense is that associated with construction equipment. When the equipment is owned or leased by the contractor, it is usual that rental rates for the various equipment types be established which the owner will pay to the contractor for the time the equipment is required on the job. It is important that specific rates be established on an hourly, weekly, or monthly basis, and that it be clearly understood whether these rates include the costs of overhauls, operating labor, and operating expense. The costs of move-in, erection, dismantling, and move-out of equipment would ordinarily be separate reimbursable items.

In the event equipment is to be required that the contractor does not own or lease, provision must be made for its purchase by the owner or for rental from a third party. When the project is large or when the equipment has a limited useful life, it is common practice for the contractor to purchase the equipment for the owner on a reimbursement basis. At the end of the project, whatever salvage or resale value the equipment might have reverts to the owner. When rental of equipment from third parties is involved, it is best to be as explicit as possible concerning item requirements and rental rates to be paid.

6.10 COST-PLUS-PERCENTAGE-OF-COST CONTRACTS

From the contractor's point of view, one of the most advantageous ways of determining the fee in a cost-plus contract is as a percentage of the cost of construction. This percentage may be a fixed amount or may vary in accordance with a prescribed sliding-scale arrangement. This type of contract is particularly well-fitted to cover work whose scope and nature are poorly defined at the outset of operations. Work may be involved of such an emergency nature that time is not available for the advance preparation of contract documents and for the usual bidding routine. Wars and other periods of extreme urgency afford instances of this type. On the other hand, the work entailed may be such that no one can ascertain what difficulties will be encountered or even of what order of magnitude the eventual cost may be. Cleanup and repair of damage such as that resulting from fire, storm, and flood afford many examples of situations where the cost-plus-percentage contract is suitable. Remodeling, expansion of facilities where services must be maintained, underpinning, and certain classes of demolition work also are occasionally done under this form of construction contract.

The cost-plus-percentage-of-cost contract does not provide any direct incentive for the contractor to minimize construction costs. Rather, it would seem to work the other way. As a result, this mode of carrying out construction is generally confined to the exceptional and unpredictable described in the previous paragraph. Public owners are prohibited from negotiating such contracts except under extraordinary circumstances. There has been a modest revival recently of this contract form on private construction where the contract includes a maximum or "upset" price. Contractors must practice strict economy in the interest of their owner-client and be satisfied with a reasonable profit if they are to enjoy continuing success with this type of contract.

6.11 COST-PLUS-FIXED-FEE CONTRACTS

A popular type of cost-plus contract is one in which the contractor's fee is established as a fixed sum of money. When this scheme is utilized, the work must be of such a nature that it can be fairly well defined and a reasonably good estimate of cost can be approximated at the time of the negotiations. The contractor computes the amount of the fee on the basis of the size of the project, estimated time of construction, nature and complexity of the work, hazards involved, location of the project, equipment and manpower requirements, and similar considerations.

Under this arrangement, the contractor's fee is fixed and does not fluctuate with the actual cost of the project. Hence it is to the advantage of the contractor to prosecute the work in as diligent a manner as possible. Failure to do so will cause additional office overhead expense to be incurred for which the contractor is not reimbursed. Expeditious handling of construction operations will minimize time and cost, two factors that contribute directly to overhead. Speed is also desirable in order to free workers and equipment for other contracts.

A variant of the fixed-fee arrangement is what is often referred to as cost-plus-award-fee, sometimes used on public projects. Under this arrangement the contractor is paid a fixed base fee. However, the contractor is afforded the opportunity, through superior performance, to earn an additional award fee.

6.12 INCENTIVE CONTRACTS

To provide the contractor with an incentive to keep the cost of the work and/or the time of construction to a minimum, various bonus and penalty provisions can be applied to the determination of the fee. Under a cost-incentive contract, the contractor and owner agree to a target estimate of construction. Bonus or penalty arrangements are tied to this target figure. Hence this type of cost-plus contract must of necessity be applied to work of a fairly definite nature for which drawings and specifications are sufficiently developed to enable a reasonably accurate target cost to be determined.

As an incentive for the contracting firm to minimize costs, a bonus clause can be written according to which it shall receive, in addition to a base fee, a stated percentage of the amount by which the total actual cost is less than the target estimate. A figure of 50 percent is common for the contractor's share of the savings. There may also be a provision whereby the contractor's fee is reduced if the construction cost exceeds the target estimate.

When time of completion is of great importance to the owner, the contract can be made to provide that the contractor shall receive, in addition to the base fee, a fixed sum of money for each day of beneficial occupancy realized by the owner before the originally agreed-on completion date. This can be extended to provide that the contractor's fee will be reduced by the same amount for each day completion is delayed. When such a bonus-penalty arrangement is stipulated, the penalty need not be considered as liquidated damages (see Section 6.20) but can be assessed strictly as a penalty.

6.13 GUARANTEED MAXIMUM COST

An objection of many owners to the cost-plus type of contract is that the cost of the project is not accurately known until after completion of the construction. One answer to this problem has been to provide for a guaranteed maximum cost to the owner. Under this form of contract, the contractor guarantees the project will be constructed in full accordance with the drawings and specifications and the cost to the owner will not exceed some total upset price. In return for its services, the contractor receives a prescribed fee. If the cost of the work exceeds the assured maximum, the contractor pays for the excess. In this way, a ceiling price is established, and the owner is assured it will not be exceeded.

The determination by the contractor of the upset cost must be based on careful estimates made from complete drawings and specifications. Such a contract establishes an ironclad maximum cost above which the contractor will not be reimbursed. Any overage is its own direct responsibility. An incentive for the contractor to keep costs below the guaranteed maximum is sometimes provided by a bonus clause stating that the contractor and the owner will share any savings.

Contracts providing for a fixed fee with a guaranteed maximum price are often competitively bid in a manner similar to that for lump-sum contracts. The successful bidder is determined on the combined basis of the quoted maximum price and fixed fee. The contractor's share of any savings below the guaranteed maximum may also be a criterion in the owner's determination of the successful bid.

6.14 CONTRACT CLAUSES

Construction contracts contain many nontechnical provisions that pertain to the conduct of the work. These general clauses, covenants, and agreements constitute the general conditions (Appendix C), supplementary conditions (Appendix D), and provisions of the agreement (Appendixes I and J).

These clauses are devised predominantly for the protection of the owner and must be carefully examined and studied by the contractor so the obligations to be assumed are thoroughly understood. When standard document forms are involved, such as those of The American Institute of Architects, the federal government, the American Society of Civil Engineers, or the Associated General Contractors of America, there is seldom need for the contractor to concern itself with ferreting out "weasel" clauses and "loaded" provisions. These standard contract clauses have well-established records of service and have become familiar to the industry. Many of them have received legal interpretation by the courts.

The contract clauses of many construction agreements are not of the standard variety, however, but have been specially prepared by the architect-engineer or owner. These sometimes impose burdensome, one-sided obligations on the contractor, who must single out such documents for more than the usual hurried scanning. Obviously, the time for a careful reading of all such contract articles is before rather than after the contract is signed. After execution of the contract, the contractor is bound by all its provisions.

Contractors must recognize that they are not lawyers and hence are not competent to appraise the legal implications of contract clauses. Should they have reason to question any legal aspect of a contract document, they should seek the assistance of their attorneys. Failure to do so can result in serious complications that could have been avoided had competent legal advice been obtained.

During the bidding period, the contractor must evaluate each clause with regard to its possible or probable contribution to the cost of construction. On becoming the successful bidder, the contract must again be examined, but with a different purpose in mind. Many provisions require specific actions on the part of the contractor during the life of the contract.

It is beyond the scope of this book to attempt a complete discussion of the meanings, implications, and legal record of all contract clauses. Nevertheless, the most important aspects of the principal contract provisions are discussed under appropriate topic headings. The following sections consider contract clauses of special significance that are not treated elsewhere herein.

6.15 PROGRESS PAYMENTS

It is customary that projects of more than very limited duration require the owner or construction lender to make periodic payments or cost reimbursements to the contractor during the construction period. It is normally not practicable or desirable for the contractor to finance the construction from its own resources. Because the contractor often operates on borrowed funds, the terms of payment provided for in the contract are especially important. In general, the contractor must make application for progress payment a prescribed number of days before it is due, or on completion of designated phases of the work. In the former case, each payment is based on the value of the work put into place, including that performed by subcontractors, during a prescribed period of time. In the latter case, a fixed amount of money or designated percentage of the total contract becomes due as each prescribed construction stage is finished.

Competitive contracts commonly provide that the contractor shall submit applications for payment at the end of each month and that the owner will make payment on or about the tenth day of the month following (see Subparagraph 9.3.1 of Appendix C and Article 5 of Appendix I). Depending on contract terms, the request may be submitted to the architect-engineer or directly to the owner. When it is submitted to the architect-engineer, this party checks the pay request and issues a "Certificate of Payment" which is sent on to the owner or lending institution.

Under lump-sum contracts the degree of completion of each major work category is usually expressed as a percentage (see Section 9.24). The quantities of work done on unit-price contracts are determined by the field measurement of work put into place. Materials suitably stored on the site or other agreed-on locations are customarily taken into account, as well as prefabrication or preassembly work that the contractor may have done at some location other than the jobsite. It is to be noted that partial payments, limited occupancy by the owner, inspection of the work in the field, or acceptance of the work by the architect-engineer do not constitute an acceptance of the partially completed work by the owner and do not serve as a waiver of any claim the owner may subsequently have against the contractor for defective work or failure to construct in accordance with the contract.

Cost-plus contracts usually provide for the contractor to submit payment vouchers to the owner at specified intervals during the life of the contract. A common contract provision is for weekly or biweekly reimbursement of payrolls and monthly reimbursement of all other costs, including a pro rata share of the contractor's fee.

6.16   RETAINAGE

Many construction contracts, especially those that involve competitive bidding, provide that a certain percentage of the progress payments will be retained by the owner. In the usual instance, the accumulated retainage remains in the possession of the owner until the project is completed and final payment is made, with the owner paying no interest on these funds. A retainage of 10 percent for the entire project has been typical, although reduced percentages and other retainage arrangements are now the rule. In any event, retainage on larger projects results in the owner having custody of large sums of the contractor's funds for extensive periods of time.

There has been considerable discussion and study during recent years about the need for retainage when the contractor has already provided the owner with performance and payment bonds. However, it must be recognized that contract bonds come into play only on breach of contract by the contractor. Owners look on retainage as further protection against possible eventualities such as contractor failure to remedy defective work, settlement of liens or other claims against the project, collection of damages from the contractor for late completion, payment of damages to others caused by the contractor's performance, and similar claims that the owner may be called on to settle.

Despite these considerations, however, retainage does have some undesirable aspects for owners, general contractors, and subcontractors alike. Subcontractors are involved because general contractors apply retainage to their subcontractors in the same percentage as the owner applies it to the general contractor. Retainage can and does produce real cash-flow problems for contractors, resulting in substantial borrowing at hefty interest rates. This results in higher construction costs for owners. In addition, it discourages contractors from bidding some projects, reducing competition. Withholding retainage from a subcontractor until completion of a project, even though its work may have been satisfactorily completed long ago, is particularly unfair. Yet the general contractor cannot be expected to remedy this situation from its own funds.

To mitigate the undesirable effects of retainage, a number of changes and innovations have been introduced during recent years. All of these changes have tended to ease the withholding requirement and its impact on the contractors involved. For example, it is now common that when job progress is satisfactory, and with consent of the surety, 10 percent retainage is withheld only during the first half of the project with subsequent progress payments being made in full. Article 5 of Appendix I illustrates this point. An alternate to this is to apply five percent retainage to the entire project. A more recent development is where 10 percent is retained on each work category of the project up until the time of 50 percent completion of that category, after which full payment is made if the work is proceeding satisfactorily. On some public projects, 10 percent is withheld for the entire project but the contracting officer may authorize full payment when satisfactory progress is being achieved. Another common provision on public contracts is that on substantial completion of the project, and at the discretion of the contracting officer, a portion of the retainage can be returned to the contractor. Certain federal agencies have even eliminated retainage altogether on a temporary, trial basis. A number of state highway departments have approved plans for the contractor to place bonds or other securities in escrow as an alternative to retainage. The contractor is allowed to withdraw its retainage after depositing with an escrow agent approved securities having a market value equal to the amount withdrawn. Interest and other income derived from the deposited securities accrue to the contractor.

It is common with cost-plus contracts for the owner to pay all contractor vouchers in full without deducting any percentage as retainage. There is a procedure sometimes used on such contracts that is analogous to retainage. The owner makes full reimbursement to the contractor up until some designated percentage (80 percent may be used) of project completion. Further payment by the owner is then withheld until some specified amount of money has been retained. This reserve is kept by the owner until final payment is made.

6.17   ACCEPTANCE AND FINAL PAYMENT

The making of periodic payments by the owner as the work progresses does not constitute acceptance of the work. Construction contracts generally provide that acceptance of the contractor's performance is subject to the approval of the architect-engineer, the owner, or both. Most contracts provide that mere occupancy and use by the owner do not constitute an acceptance of the work or a waiver of claims against defects. Acceptance of the project and final payment by the owner must proceed in accordance with the terms of the contract. Procedure in this regard is somewhat variable, although inspection and correction of deficiencies are usual practice.

On building construction, the contractor will normally advise the owner or architect-engineer when substantial completion has been achieved. An inspection is held, and a list of items, called a "punch list," requiring completion or correction is compiled. The architect-engineer issues a "Certificate of Substantial Completion" at that time. This action by the architect-engineer certifies that the owner can now occupy the project for its intended use. A provision appearing in many contracts is that, upon substantial completion, the contractor is entitled to be paid up to a specified percentage of the total contract sum (95 percent is common), less such amounts as the owner may require for uncompleted work and unsettled claims. The Certificate of Substantial Completion also serves other purposes including shifting the responsibility for maintenance, heat, utilities, and insurance on the structure from the contractor to the owner. After the contractor has attended to the deficiency list, a final inspection is held and a final certificate for payment is issued. A common contract provision is that final payment is due the contractor 30 days after substantial completion. This is applicable, however, only with consent of the surety and provided all work has been completed satisfactorily and the contractor has provided the owner with all required documentation. Final payment includes all retainage still held by the owner.

A legal question that frequently arises concerning final acceptance and payment involves the degree of contract performance required of the contractor. The modern tendency is to look to the spirit and not the letter of the contract. The vital question is not whether the contractor has complied in an exact and literal way with the precise terms of the contract but whether it has done so substantially. Substantial performance may be defined as accomplishment by the contractor of all things essential to fulfillment of the purpose of the contract, although there may be inconsequential deviations from certain terms. This is a recent principle, peculiar to construction contracts generally, and has been evolved by the courts to mitigate the severity of the rule of exact performance.

A contractor who has in good faith endeavored to perform all that is required by the terms of the contract and has, in fact, substantially done so is ordinarily entitled to recover the contract price less proper deductions for minor omissions, deviations, and defects. If defects are easily correctable, the owner is entitled to deduct the cost of correction. Otherwise, the difference in value between the structure specified and that actually built is the usual measure of recovery by the owner. The burden of proving defective performance and the amount of the setoff rests on the party asserting the deficiency, the owner in this instance. In evaluating substantial performance, consideration must be given to the contractor's intentions, the amount of work performed, and the benefit received by the owner.

Certainly, the substantial performance concept does not confer on the contractor any right to deviate freely from the contractual undertaking or to substitute materials or procedures that it may consider equivalent to those actually called for by the contract. Only if defects are purely unintentional and not so extensive as to prevent the owner from receiving essentially what was bargained for does the principle come into play.

6.18 THE WARRANTY PERIOD

Acceptance of the work by the owner and his payment for it normally constitutes a waiver of his rights for damages on account of defects in the structure if no claim is made within a "reasonable time." Many construction contracts obligate the contractor to make good all defects brought to its attention by the owner during some warranty period after either the time of substantial completion or final acceptance, the point at which the period begins being variable with the form of contract. One year is a commonly specified warranty period, although periods of up to five years are sometimes required on certain categories of work, such as utility construction. The contractor is required, on notice, to make good at its own expense defects detected during this period. In most cases, the prescribed warranty period fixes the "reasonable time" and releases the contractor from further responsibility after its expiration. Warranty periods required by the contract are normally covered by the performance bond, and they do not operate to defer final payment.

An exception to the reasonable-time concept occurs when a defect caused by the contractor's inadequate performance is latent in nature and could not have been detected by the owner during ordinary use and maintenance of the structure. In such a case, the owner has a right of action against the contractor beyond the warranty period for such longer period of time as may be provided for by the contract or as prescribed by law. For example, the owner would ordinarily have a right of action against the contractor for faulty construction up until the end of the period designated by the applicable statute of limitations for breach of contract.

A contractor's warranty does not imply, however, that the contracting firm is liable for the sufficiency of the plans and specifications unless it prepared them or unless it guaranteed their adequacy. A contractor is required to construct in accordance with the contract documents, and when it does so it cannot ordinarily be held to guarantee the design will be free from defects or the completed job will accomplish the purpose intended. The contractor is responsible only for improper workmanship, inferior materials, or other faults resulting from its failure to perform in accordance with the contract.

6.19 CONTRACT TIME

Most construction contracts are explicit regarding construction time, designating either a completion date or a specific number of calendar days within which the work must be finished. The term "calendar days" includes Saturdays, Sundays, and holidays and is used rather than "working days" to eliminate possible controversy concerning weather, overtime, multiple shifts, weekends, and holidays. When the contract declares that time is "of the essence," this signifies that the stipulated completion date or time is considered to be essential to the contract and is an important part of the contractor's obligation. Failure to complete the project within the time specified is considered to be a material breach of contract and could possibly afford the owner the right to terminate the contract. In addition, any delay in project completion can make the contractor liable to the owner for damages. There have also been instances where late completion has made the contractor liable for business losses suffered by a tenant when the contractor negligently failed to prosecute the job with due diligence. In such cases, some courts have ruled that the contractor has a duty of care to the tenant where the injury was reasonably foreseeable.

When the contract time is stated to be a given number of calendar days, the date on which the time begins is an important matter. Construction contracts usually state that the time shall begin on the date the contract is signed or on the date the contractor receives a formal "notice to proceed" (see Section 6.35) from the owner. The contract may establish the finish date as that of substantial completion, final completion, or just "completion" of the project. In general, completion is related to the structure's capability of being utilized for its intended purpose despite small defects to be corrected or minor items not yet accomplished; that is, substantial completion. When the contract stipulates a completion date, the contractor must protect against any delays in getting construction operations under way. If the contract is not promptly executed, a letter of intent may be needed or the contractor may have to inform the owner that an extension of contract time (see Section 6.21) will be claimed.

If the construction contract is silent regarding contract time, the contractor is expected to perform the work within a reasonable time. This also applies when the contract requires project completion "at the earliest possible date," "as soon as possible," or "without delay." As one might expect, such imprecise statements regarding contract time can be troublesome. What constitutes a reasonable time is subject to wide differences of opinion and may have to be decided by the courts in case of dispute.

6.20 LIQUIDATED DAMAGES

Many projects are of such a nature that the owner will incur hardship, expense, or loss of revenue should the contractor fail to complete the work within the time specified by the contract. Where the contract makes time an essential part of the contract, either by the phrase "time is of the essence" or by explicit reference to the stated time requirement, failure to complete the project on time is a breach of contract and can make the contractor liable to the owner for damages. The amount of such damages may be determined by agreement or by litigation. Compensatory damages of this sort are difficult to determine exactly, and in construction contracts it is common practice to provide that the contractor shall pay to the owner a fixed sum of money for each calendar day of delay in completion.

This assessment against the contractor, known as "liquidated damages," is used in lieu of a determination of the actual damages suffered. The word "liquidated" in this instance merely signifies that the precise amount of the daily damages to the owner has been established by agreement. Liquidated damages, when provided for by contract, are enforceable at law provided they are a reasonable measure of the actual damages suffered. The effect of a clause for liquidated damages is to use the amount stipulated in lieu of the actual damages that would be caused by late completion and thereby preventing a controversy between the parties over the amount of damages. The owner deducts any liquidated damages from the sum due the contractor at the time of final payment. Typical values of liquidated damages appearing in construction contracts vary from $50 to $1000 per calendar day, although much higher values are occasionally stipulated.

It must be emphasized that the courts enforce liquidated damage provisions in construction contracts only when they represent a reasonable forecast of actual damages the owner would be expected to suffer upon breach and when it is impossible or very difficult to compute or make an accurate estimate of actual damages. When it has been established that the amount was excessive and unreasonable, the courts have ruled that such payment by the contractor to the owner constituted a penalty and was not enforceable. Punitive damages (those intended to punish) are not ordinarily recoverable for breach of contract. It may be noted that architect-engineers are careful when writing such provisions to state that the sum named is in the nature of liquidated damages and not a penalty. Article 5b of Appendix D illustrates this point.

6.21 EXTENSIONS OF TIME

During the life of a contract, there are often occurrences that cause delay or add to the period of time necessary to construct the project. Just what sort of delays will justify an extension of time for the contractor depends on the provisions of the contract. In the complete absence of any clause on excusable delay, the contractor can normally expect relief only from delays caused by the law, the owner, the architect-engineer, or by an act of God. For this reason construction contracts contain an extension-of-time provision that defines which delays are excusable. The provisions of such clauses are quite variable. Because of the importance of time in construction contracts, it is usual for such extensions of time to be formalized by a signed instrument, called a change order (see Section 6.25), that constitutes a binding change to the contract.

The addition of extra work to the contract is a common justification for an extension of time. In this case the additional contract time made necessary by the broadened scope of the work is negotiated at the same time the cost of the extra work is determined. Delay of the project caused by the owner or his representative is another frequent cause of contract-time extensions.

Many other circumstances can contribute to late project completion. Contracts often list specific causes of delay deemed to be excusable and beyond the contractor's control. Certain causes are essentially of an undisputed nature, provided the contractor is not directly at fault and has exercised due care and reasonable foresight. These causes include flood, earthquake, fire, epidemic, war, riots, hurricanes, tornadoes, and similar disasters. Other legitimate but less spectacular causes may or may not be included; some contracts are unduly severe in this regard. Strikes, freight embargoes, acts of the government, project accidents, unusual delays in receiving ordered materials or equipment, and owner-caused delays usually constitute acceptable reasons for an extension of contract time. Subparagraph 8.3.1 of Appendix C may be considered a typical contract provision in this regard. Unreasonable language relating to extensions of time must be given proper consideration when bids are being prepared or the contract is being negotiated.

As a general rule, claims for extra time are not considered when based on delays caused by conditions that existed at the time of bidding and about which the contractor might be reasonably expected to have had full knowledge. Also, delays caused by failure of the contractor to anticipate the requirements of the work in regard to materials, labor, or equipment do not constitute reasons for a claim for extra time. Normally, adverse weather does not justify time extensions unless it can be established that such weather was unusually severe at the time of year in which it occurred and in the location of its occurrence.

Previously discussed is the fact that on unit-price contracts actual quantities of work can vary from the architect-engineer's original estimates. Minor variations do not affect the time of completion stated in the contract. However, if substantial variations are encountered, an extension of contract time may very well be justified.

Whenever a delay in construction is encountered over which the contractor has no control, the contractor must bring the condition to the attention of the owner and/or architect-engineer in writing within a designated period of time after the start of the delay. Ten to twenty calendar days is the time period commonly specified in construction contracts. This communication should present specific facts concerning times, dates, and places and include necessary supporting data about the cause of delay. Failure by the contractor to submit a timely written request with justification may seriously jeopardize the chances of obtaining an extension of time.

6.22 CHANGES

It is standard practice that a construction contract gives the owner the right to make changes in the work within the general scope of the contract during the construction period. Depending on the contract and its terms, such changes might involve additions to or deletions from the contract, modifications of the work, changes in the methods or manner of work performance, changes in owner-provided materials or facilities, or even changes in contract time requirements. Changes may have to be made to correct errors in the drawings or specifications. Owner requirements and circumstances sometimes change after the contract award, and changes must be made to meet such conditions. Changes are even occasionally made as the result of suggestions by the contractor. How changes are handled depends on the contract provisions as normally contained in a "changes" clause. Typical language in this regard provides that the owner may make changes in the work and that an equitable adjustment of price and time shall be made by a change order to the contract (see Section 6.25).

If the contractor detects a job condition that it believes will require a contract change, the contractor should so advise the owner or architect-engineer immediately in writing. If the contractor believes the work will require an increase in contract price or an extension of time, it should so state. Many contract documents require that such a notice must be given before work constituting an extra to the contract is performed.

When a change is to be made, a suitable description is prepared by the owner or architect-engineer that includes any necessary supplementary drawings and specifications. This information is then presented to the contractor for its action. On a unit-price contract, changes are automatically provided for unless the changed work involves items that were not included in the original contract, unless the changes are so extensive the contractor or owner is authorized by the contract to request adjustments in the unit prices affected, or unless an extension of time will be required. When a change to a publicly financed project is involved, the contractor may be well advised to determine if the change is within the scope of the changes clause of the contract and that the public agency has followed the procedural requirements of the applicable statute (see Section 6.4). Where such is not the case, the resulting change order may be declared void, thereby jeopardizing payment to the contractor for the work involved.

When the contract is lump-sum, the contractor will determine the cost and time consequences of the change and advise the owner or architect-engineer of them. The adjustment in the contract amount occasioned by the change may be determined as a lump-sum, by unit prices, or as the additional cost plus a fee. In this regard contracts often limit the contractor's markup or direct cost of the change to some stipulated figure, such as 15 percent for overhead and profit or 10 percent for overhead and 10 percent for profit. Cost-plus contracts present few difficulties in accommodating changes to the work. The contractor's fee might have to be adjusted in accordance with contract terms to reflect the change being made, and target or upset prices must be adjusted by the cost of the change.

Construction contracts universally provide that the contractor is not to proceed with a change until it has been authorized in writing by the owner or his representative. To proceed without written authorization may make it impossible for the contractor to obtain payment for the additional work. Notwithstanding this requirement, however, the courts, under certain circumstances, have ruled that the lack of written authorization does not automatically invalidate a contractor's claim for extra work, even when authorization in writing is expressly required by the contract. When the owner or his agent approved orally the additional work and promised to pay extra for it, had knowledge that it was being performed without written authorization and did not protest, the courts have ruled either that the work was done under an oral contract separate and distinct from the written one or that the clause requiring written authorization was waived. In either event, the owner was judged to have made an implied promise to pay and the contractor was allowed to recover the cost of the additional work. The courts have sometimes disregarded the requirement for a written order on the grounds that it is unjust enrichment for the owner to enjoy the benefit without having to pay for it.

6.23 CHANGED CONDITIONS

A "changed condition" refers to some physical aspect of the project or its site that differs materially from that indicated by the contract documents or that is of an unusual nature and differs materially from the conditions ordinarily encountered. The most common instances are subsurface soil and water conditions. Unforeseen conditions resulting from hurricanes, floods, abnormal rainfall, or nonphysical conditions are not considered to be changed conditions.

The construction cost and time of many projects depend heavily on local conditions at the site, conditions that cannot be determined exactly in advance. This applies particularly to underground work, where the nature of the subsoil and groundwater has great influence on costs. At the time of bidding, the owner and his architect-engineer must make full disclosure of all information available concerning the proposed project and its site. The contractor uses the information as it sees fit and is responsible for its interpretation of it. The contractor is expected to perform its own inspection of the site and to make a reasonable investigation of local conditions.

Who pays for the "unexpected" depends on the contract provisions. No implied contractual right exists for the contractor to collect for unforeseen conditions. Any ability to collect extra payment for changed conditions must be provided by contract. This is subject to the exception, however, of where the drawings and specifications contain incorrect and misleading information. The fact that a contractor is not liable for the consequences of defects in the contract documents has been previously discussed.

Many contracts provide for an adjustment of the contract amount and time if actual subsurface or other physical conditions at the site are found to differ materially from those indicated on the drawings and specifications or from those inherent in the type of work involved. This proviso is a "changed-conditions clause," the objective of which is to provide for unexpected physical conditions that come to light during the course of construction. When such a contract clause is present, the financial responsibility for unexpected conditions is placed on the owner. Documents of The American Institute of Architects (see Subparagraph 12.2.1 of Appendix C) and the federal government contain such provisions.

It is the purpose of a changed-conditions clause to reduce the contractor's liability for the unexpected and to mitigate the need for including large contingency sums in the bid to allow for possible serious variations in site conditions from those described and intimated by the contract documents. It must be clearly understood at this point that for a claim to be sustained under this clause, the unknown physical conditions must truly be of an "unusual nature" and must differ "materially" from those ordinarily encountered and generally regarded as inherent in work of the character provided for in the contract. The essential question is whether the conditions are different from those the contractor should have reasonably expected and whether these conditions caused a significant increase in the project time or cost.

Various disclaimers and exculpatory clauses denying liability and responsibility for actual field conditions frequently appear in construction contracts. A common clause provides that neither the owner nor the architect-engineer accepts responsibility for the accuracy or completeness of subsurface data provided, and that all bidders are expected to satisfy themselves as to the character, quantity, and quality of subsurface materials to be encountered. Where a changed-conditions clause has been in the contract, such disclaimers have not acted as a bar to contractor relief under the clause.

When the contract has not included a changed-conditions clause and the owner disclaimed all responsibility for the subsurface information that was given, some courts have held that the contractor could not recover extra costs. Enforcement by the courts of such provisions in construction contracts is subject to many exceptions, however. Instances of such exceptions are when the site conditions were deliberately or negligently misrepresented on the contract documents, when the owner did not reveal all of the relevant information he possessed or of which he had knowledge, or when there was a breach of warranty of the correctness of the bidding information. Disclaimer clauses have also been denied force or effect when they were too strict, when they were in direct contradiction with specific representations, and on the basis that it was unreasonable to expect the contractor to make an extensive investigation of underground conditions during the bidding period.

It must be noted that, regardless of the contract wording, if changed conditions are found at the site and if the contractor intends to claim additional cost or time, it must promptly notify the owner in writing, and must also keep detailed and separate cost records of the additional work involved. Failure to do so may make it impossible for the contractor to prevail.

6.24 OWNER-CAUSED DELAY

There are many instances in which the work is delayed by some act of omission or commission on the part of the owner or by someone for whom the owner is responsible such as the architect-engineer or other contractors. Examples are delays in making the site available to the contractor, failure to deliver owner-provided materials on time, unreasonable delays in approval of shop drawings, delays caused by another contractor, delays in issuance of change orders, and suspension of the work because of financial or legal difficulties. Additionally, work that is unchanged can be prolonged or disrupted because of changes or changed conditions associated with other portions of the project.

Owner-caused delay is of concern to the contractor for two reasons. First, when the overall completion date of the project is affected, a suitable extension of time must be obtained. As a rule, contracts provide for extensions of time on account of owner-caused delay, and this matter is seldom troublesome. The second and more difficult problem for contractors is the "ripple" effect of such delays, that is, the "consequential damages" or "impact costs" associated with unchanged work that can result from project delay. Examples of impact costs are standby costs of nonproductive workmen, supervisors, and equipment; expenses caused by disrupted construction and material delivery schedules; start-up and stopping costs such as those incurred in moving workers and equipment on and off the job; and additional overhead costs. It sometimes happens that work is delayed until cold weather, the rainy season, or spring runoff, resulting in greatly increased operating costs. Project delay can defer work until higher wage rates have gone into effect, until material prices have gone up, or until bargains or discounts have been lost.

A contractor has the right to recover damages for its increased costs of performance caused by delays attributable to the owner. Such claims are based on the owner's implied warranty that the drawings and specifications provided by him are adequate for construction purposes and his implied promise that he will not disrupt or impede the performance of the construction process. In order to recover damages, the contractor must prove the delays were caused by the owner and that the contractor was damaged as a direct result of the delays.

Construction contracts contain a variety of provisions concerning unexcused owner-caused delay. On federal projects the government has assumed responsibility for payment of the impact cost of changes and changed conditions as they affect unchanged work. Federal contracts and many others contain suspension-of-work clauses that provide for the extra cost occasioned by owner-caused delay or suspension. Some contracts contain provisions that expressly limit contractor relief in cases of job delay to time extensions only. Others indicate clearly that the provision for time extension does not preclude delay damages. Many contracts contain "no damage" clauses whereby the owner is excused from all responsibility for damages resulting from owner-caused delay. Where such exculpatory clauses are present, they are generally given effect by the courts, although they are strictly interpreted. Two exceptions that allow the contractor to collect increased costs are where the owner actively interferes with the contractor's work and when the delay is for an unreasonable period of time. Even when consequential damages are awarded, they often are made to reimburse the contractor for extra costs only with no allowance for profit.

The contractor frequently is unable to recover delay damages because of an inability to prove the exact extent of the loss suffered. For this reason, when such a delay occurs, the contractor must keep careful and detailed records if it expects to recover extra costs from the owner. An additional point is that many contracts require the contractor to notify the owner within a designated time after any occurrence that may lead to a claim for additional contract time or extra cost.

6.25 CHANGE ORDERS

When additions, deductions, or changes in the work are made by the owner, a supplement to the contract between the owner and the prime contractor is prepared that can be on the basis of a lump-sum, unit-price, or a cost-plus arrangement. This supplement, called a change order, is consummated by a written document that describes the modification to be made, the change in the contract amount, and any authorized extension of contract time. The change-order form identifies the change being made as a modification to the original construction contract and normally bears the acceptance signatures of the owner and the prime contractor. If it is not the owner who signs, the contractor must be sure the party executing the change order has the authority to obligate the owner and to make a binding change to the contract. For example, by virtue of its position alone, the architect-engineer has no authority to order changes and must be authorized in some way by the owner to act on the owner's behalf. Many architect-engineer firms have developed their own individual change-order forms, an example of which is shown in Figure 6.1. The American Institute of Architects has prepared a form that is in wide use

JONES AND SMITH
ARCHITECT–ENGINEERS
PORTLAND, OHIO

CONTRACT CHANGE ORDER

Project: Municipal Airport Terminal Building   Change Order No. __1__
For: City of Portland, Ohio   Date: July 30, 19_____
To: The Blank Construction Co., Inc.
      1938 Cranbrook Lane
      Portland, Ohio

                                    Revised Contract Amount
                              Previous contract amount $3,602,138.00
                              Amount of this order
                                  (decrease) (increase)   5,240.00
                              Revised Contract Amount $3,607,378.00

An (increase) (decrease) (no change) of _____ days in the contract time is hereby authorized.

This order covers the contract modification hereunder described:
     Providing and installing folding partitions and ornamental
     screens as shown and described by Supplemental Drawing X-1
     attached hereto. This change includes all grounds, nailer
     blocks, and other provisions required for the satisfactory
     installation of said partitions and screens.

The work covered by this order shall be performed under the same terms and conditions as included in the original construction contract.

Changes Approved   Jones and Smith, Architect–Engineers

_____   by_____
         (Owner)

by_____

_____
        (Contractor)

by_____

Figure 6.1   Contract change order.

for this purpose on building construction. The previous contract amount shown in Figure 6.1 can be verified by reference to Appendix I.

A change order, as a modification of the contract, means that the parties have entered into a new contract. It is presumed that such a contract modification has taken into account all prior negotiations and understandings leading up to its signing and also that the terms of each change order reflect proper consideration of these negotiations. Mention has been made previously in this text of the usual contract requirement that the contractor is not to proceed with any contract change before it has been authorized in writing by the owner or his agent. There are often instances where it is desirable to proceed with changed work before the formal change order has been executed. Many construction companies utilize "field orders" for this purpose, these being forms that authorize the contractor to proceed until the formal change order can be processed. Caution is needed in relying on such field orders, however, especially when substantial changes to the contract are involved.

6.26 CLAIMS AND DISPUTES

The nature of construction is such that there are quite frequently disputes between the owner and the prime contractor concerning claims by the latter for increased costs and/or extensions of contract time. It certainly is not an unusual circumstance for the contractor to believe it has a valid claim against the owner based on breach of contract by the owner or by someone for whom the owner is responsible, or where the work went beyond that required by the contract. Disputes and claims can and do stem from a wide variety of job circumstances. Disagreements between the two contracting parties can arise from such considerations as interpretation of the contract, changes made by the owner, owner-caused delays, differing site conditions, changed conditions, acceleration, suspension of the work, and faulty drawings and specifications. Construction contracts routinely include procedures to be followed in the settlement of claims and disputes.

Although provisions vary, construction contracts typically require the contractor to formally advise the owner within a specified time period when a situation arises that could lead to a claim. In the event of a dispute concerning a potential claim, the contractor usually cannot refuse to proceed with the work without committing breach of contract. Although the disputed work may be performed under protest, the contractor must continue field operations with diligence, relying on remedies in the contract to settle the questions of compensation and extension of time. The requirement that the contractor must proceed with the disputed work does not apply, however, if the change or modification involved is beyond the scope of the contract. The distinction between what is and what is not within the scope of the contract is normally difficult to establish, however.

Contracts normally stipulate that claims and disputes be first submitted to the owner or his representative. If the claim is denied at this first level, the resulting dispute can then be submitted to various levels of appeal, such as appeals boards, arbitration, or the courts. The remedies available depend considerably on whether the dispute involves questions of fact or law. Most public owners have statutory or administrative procedures established for the settlement of contract disputes. Settlement of such disputes involving the federal government is in accordance with The Contract Disputes Act of 1978. Contractors will generally avoid litigation if possible. There can be many advantages to the negotiation of an early settlement to a contract dispute. However, it must be recognized that there are times when it is not possible to compromise or negotiate differences. In any event, there are steps the contractor can take to strengthen its case when the possibility of a claim arises. This matter is discussed in detail in Chapter 10.

Construction contracts frequently contain a provision that acceptance of final payment by the contractor shall be considered as a general release in full of all claims against the owner arising out of or in consequence of the work. Other contracts provide for a conditional release, the contractor being permitted to maintain other causes of action. Subparagraph 9.9.5 of Appendix C provides that the acceptance of final payment shall constitute a waiver of all claims by the contractor except those previously made in writing and still unsettled. This can be an important point because, for one reason or another, contractors often do not or cannot file claims until after the work has been completed. Where such contract provisions appear, they are given effect by the courts.

A common action by the owner is to make a notation on the check used for final payment to the effect that, "By endorsement, this check is accepted in full payment of the account indicated." Ordinarily, if the contractor cashes the check, it is barred from suing to collect an additional sum. This follows from a generally recognized rule of law that when there is a dispute between debtor and creditor over the amount due, acceptance of such a check amounts to an agreement to accept it as final payment. In most states, however, if there is no reasonable basis for a debtor's denial that he owes the sum claimed, acceptance of a check for a smaller sum does not constitute a binding agreement for final payment, even if there is a notation on the check. The contractor should consult with its attorney when such matters are at issue.

6.27 VALUE ENGINEERING

Many construction contracts, principally those with public agencies, include what are called "value-engineering" incentive clauses. In this context value engineering applies after the contract is awarded and is concerned with the elimination or modification of any contract provision that adds cost to a project but is not necessary to the structure's required performance, safety, appearance, or maintenance. The basic objective of value engineering is often stated to be the elimination of "gold plating." A clause of this type provides an opportunity for the contractor to suggest changes in the plans or specifications and to share in the resulting savings. These changes may involve substitutions of materials, modifications of design, reductions in quantities, or procedures other than those set forth and required by the contract documents. Value engineering is designed, not to enable the contractor to second guess the designer, but to take advantage of the contractor's special knowledge and to cut the cost of a project to the lowest practicable level without compromising its function or sacrificing quality or reliability. In short, the contractor is encouraged to develop and submit to the owner cost-reducing proposals leading to changes in the plans or specifications. If the owner accepts them, a change order is processed, and the savings are usually shared about equally by the owner and the contractor.

It is to be noted that value engineering refers to making a change in the drawings, specifications, or other contract provision and requires the approval of the owner. Savings that can be realized without a contract change properly belong to the contractor.

Value engineering is applied to the planning and design of construction projects as well as to their construction. For example, value engineering during the planning and design phases has been an in-house activity of the U.S. Army Corps of Engineers for several years. Some public agencies are now including a requirement for value engineering in many of their professional service contracts for architect-engineer and construction management services. This is a contract requirement above and beyond the normal design considerations of cost and function. It requires a systematized, formal effort directed at isolation and elimination of unnecessary costs. It might be described as a "second look" by a qualified team of specialists whose purpose is to challenge the basic system and the choice of materials in the design with the objective of eliminating excess costs without diluting the quality or function of the various segments of the project. Value engineering at the design stage is geared to long-term cost rather than just to original construction expense. The Society of American Value Engineers (SAVE) is a professional group concerned with value engineering and its application.

6.28 RIGHTS AND RESPONSIBILITIES OF THE OWNER

Construction contracts typically reserve several rights to the owner. Depending on the type of contract and its specific wording, the owner may be authorized to award other contracts in connection with the work, to require contract bonds from the contractor, to approve the surety proposed, to retain a specified portion of the contractor's periodic payments, to make changes in the work, to carry out portions of the work himself in case of contractor default or neglect, to withhold payments from the contractor for adequate reason, and to terminate the contract for cause. The right of the owner to inspect the work as it proceeds, to direct the contractor to expedite the work, to use completed portions of the project before contract termination, and to make payment deductions for uncompleted or faulty work are common construction contract provisions.

By the same token, the contract between owner and contractor imposes certain responsibilities on the owner. For example, construction contracts make the owner responsible for furnishing property surveys that describe and locate the project site, securing and paying for necessary easements, providing certain insurance, and making periodic payments to the contractor. The owner is required to make extra payments and grant extensions of time in the event

Net after taxes 105,900
Gross 720,000
Annual Volume 4,600,000

PROBLEM SET 3

Bold Construction, Inc.
Comparative Balance Sheet on a Certain Date

Line No.		Before Assumptions	After Assumptions
	ASSETS		
	Current Assets		
1	Cash	$ 60,000	
2	Accounts Receivable	500,000	
3	Notes Receivable	20,000	
4	Inventory	160,000	
5	Cost in Excess of Billings	150,000	
6	Prepaid Expenses	10,000	
7	Total Current Assets	$ 900,000	
	Fixed Assets		
8	Land	$ 35,000	
9	Buildings	45,000	
10	Equipment	140,000	
11	Office Equipment	10,000	
12	Total Fixed Assets, at Cost	$ 230,000	
13	Less: Accumulated Depreciation	(120,000)	
14	Net Book Value, Fixed Assets	$ 110,000	
15	TOTAL ASSETS	$1,010,000	
	LIABILITIES & NET WORTH		
	Current Liabilities		
16	Accounts Payable	360,000	
17	Accrued Expenses (Taxes, etc.)	20,000	
18	Notes Payable (Current Portion)	30,000	
19	Billings in Excess of Cost	90,000	
20	Total Current Liabilities	$ 500,000	
	Long-Term Liabilities		
21	Mortgage Payable	$ 80,000	
22	Total Liabilities	$ 580,000	
	Net Worth		

of certain eventualities provided for in the contract. When there are two or more prime contractors on a project, the owner may have a duty to coordinate them and synchronize their field operations.

It is important to note that the owner cannot intrude on the direction and control of the work. By the terms of the usual construction contract, the contractor is known at law as an "independent contractor." Even though the owner enjoys certain rights with respect to the conduct of the work, he cannot issue direct instructions as to method or procedure, 'unreasonably interfere with construction operations, or otherwise unduly assume the functions of directing and controlling the work. By so doing, the owner can relieve the contractor from many of the latter's rightful legal and contractual responsibilities. If the owner oversteps his rights, he may not only assume responsibility for the accomplished work but may also become liable for negligent acts committed by the contractor in the course of construction operations. Under the laws of most states, the owner is responsible to the contractor for the adequacy of the design. If the drawings and specifications are defective or insufficient, the contractor can recover the resulting damages from the owner.

6.29   DUTIES AND AUTHORITIES OF THE ARCHITECT-ENGINEER

Other than cases in which both design and construction are performed by the same contracting party or in which the owner has an in-house design capability, the architect-engineer firm is not a party to the construction contract, and no contractual relationship exists between it and the contractor. It is a third party who derives its rights and authorities over the construction process from the general contract between the owner and the prime contractor. When private design professionals are used by the owner, the effect on the construction contract is to substitute the architect-engineer for the owner in many important respects. However, the jurisdiction of the architect-engineer to make determinations and render decisions is limited to and circumscribed by the terms of the contruction contract. The architect-engineer represents the owner in the administration of the contract and acts for him during the day-to-day construction operations. The architect-engineer advises and consults with the owner, and communications between owner and contractor are made through the architect-engineer. Paragraph 2.2 of Appendix C contains typical provisions regarding the architect-engineer's role in contract administration.

Construction contracts often impose many duties and bestow considerable authority on the design firm. All construction operations are conducted under its surveillance, and it generally oversees the progress of the work. It has a direct responsibility to see that the workmanship and materials fulfill the requirements of the drawings and specifications. To ensure this fulfillment, the architect-engineer exercises the right of job inspection and approval of materials. Also involved may be the privilege of approving the contractor's general program of field procedure and even the construction equipment the contractor proposes to use. Should the work be lagging behind schedule, the architect-engineer may reasonably instruct the contractor to speed up its activities. Many contracts bestow upon the architect-engineer the right to stop the project or any part thereof to correct unsatisfactory work or conditions. The specific language in this regard is widely variable, ranging from a broad to a limited right.

The foregoing paragraph does not mean that the architect-engineer assumes responsibility for the contractor's methods merely because it retains the privilege of approval. The rights of the architect-engineer are essentially concerned with verifying that the contractor is proceeding in accordance with the contract documents. It should be pointed out, however, that the architect-engineer cannot unreasonably interfere with the conduct of the work or dictate the contractor's procedures. Here again, if the direction and control of the construction are taken out of the hands of the contractor, the construction firm is effectively relieved of many of its legal and contractual obligations.

The contract documents often authorize the architect-engineer to interpret the requirements of the contract and to judge the acceptability of the work performed by the contractor. In addition, the architect-engineer is commonly made the arbiter of disputes between the owner and contractor. Some contracts provide that the architect-engineer's decisions are not final and that the owner and contractor can exercise their rights to appeal, arbitration, or the courts, providing the architect-engineer has rendered a first-level decision. Other contracts state that the architect-engineer's decisions are final and binding on both parties with regard to artistic effect only. Still other contracts give the architect-engineer broad authority to make final decisions concerning the quality and fitness of the work and to interpret the contract documents.

Where the architect-engineer is given final and binding authority, this is necessarily restricted to questions of fact. In the absence of fraud, bad faith, and gross mistake, the decision of the architect-engineer can be considered as final, provided the subject matter falls within the proper scope of authority given to the design firm by the construction contract. With respect to disputed questions of law, however, the architect-engineer has no jurisdiction. It cannot deny the right of a citizen to due process of law, and the contractor has the right to submit a dispute concerning a legal aspect of the contract to arbitration or to the courts, as may be provided by the contract. Matters pertaining to time of completion, extensions of time, liquidated damages, and claims for extra payment usually involve points of law.

6.30 RIGHTS AND RESPONSIBILITIES OF THE CONTRACTOR

As one might expect from a document prepared especially for the owner, the contractor has few rights and many obligations under the contract. Its major responsibility, of course, is to construct the project in conformance with the contract documents. Despite all the troubles, delays, adversities, accidents, and mischances that may occur, the contractor is expected to "deliver the goods" and finish the work in the prescribed manner. Although some difficulties justify allowing more construction time, only severe contingencies, such as impossibility of performance, can serve to relieve the contractor from its obligations under the contract. In an era of rapidly rising costs of fuel, materials, and labor, the contract may provide a degree of financial protection by the inclusion of an escalation clause. Such a clause transfers some of the risk of inflation from the contractor to the owner by providing that the contract amount shall be adjusted according to how designated key cost factors change during the contract period.

The contractor is expected to give its personal attention to the conduct of the work, and a responsible company representative must be on the jobsite at all times during working hours. The contractor is required to conform with laws and ordinances concerning job safety, licensing, employment of labor, sanitation, insurance, zoning, building codes, and other aspects of the work. Many contracts now include tough rules designed to decrease air and noise pollution on construction projects, these rules imposing regulations and restrictions concerning trash disposal, pile driving, riveting, demolition, fences, and housekeeping.

The contractor is required to follow the drawings and specifications and cannot be held to guarantee that the completed project will be free of defects in design or will accomplish the purpose intended. However, the contractor is responsible for and warrants all materials and workmanship, whether put into place by its own forces or those of its subcontractors. Contracts typically provide that the contractor shall be responsible for the preservation of the work until its final acceptance, although engineering projects commonly include an "act of God" exclusion (see Section 8.8). Even though the contractor has no direct responsibility for the adequacy of the plans and specifications, it can incur a contingent liability for proceeding with faulty work whose defects are obvious. Should an instance occur in which the contractor is directed to do something it feels is not proper and is not in accordance with good construction practice, it should protect itself by writing a letter of protest to the owner and the architect-engineer, stating its position before proceeding with the matter in dispute.

Insurance coverage is an important contractual responsibility of the contractor, both as to the type of insurance and the policy limits. The contractor is required to provide insurance not only for its own direct and contingent liability, but frequently also for the owner's protection. The contractor is expected to exercise every reasonable safeguard for the protection of persons and property in, on, and adjacent to the construction site.

The most important contractor rights concern progress payments, recourse should the owner fail to make payments, termination of the contract for a sufficient cause, right to extra payment and extensions of time as provided, and appeals from decisions of the owner or architect-engineer. Subject to contractual requirements and limitations therein, the contractor is free to subcontract portions of the contract, purchase materials where it chooses, and process the work in any way and in any order that it pleases.

6.31 INDEMNIFICATION

The vulnerability of architect-engineers to third-party suits has been previously discussed (see Section 4.6). In like manner, owners are subject to claims by third parties to the construction contract for damages arising out of construction operations. The rule that one who employs an independent contractor is not liable for the negligence or misconduct of the contractor or its employees is subject to many exceptions. To protect the owner and architect-engineer from third-person liability, most construction contracts now include indemnification or "hold-harmless" clauses. Indemnification is where one party compensates a second party for a loss that the second party would otherwise bear.

Hold-harmless clauses typically require that the contractor indemnify and hold harmless the owner and the architect-engineer, and their agents and employees, from all loss or expense by reason of liability imposed by law on them for damages because of bodily injury or damage to property arising out of or in consequence of the work. The present tendency is for parties injured by construction operations to bring suit against practically everyone associated with the construction. Additionally, the courts show a growing inclination to ascribe liability, in whole or in part, for such damages to the owner or architect-engineer. For example, owners and architect-engineers have been made responsible for damages caused by construction accidents where adequate safety measures were not taken on the job.

Hold-harmless clauses are not uniform in their wording. However, these clauses can be grouped into three main categories:

1. Limited-form indemnification. The limited form holds the owner and architect-engineer harmless against claims caused by the negligence of the prime contractor or a subcontractor.

2. Intermediate-form indemnification. The intermediate form includes not only claims caused by the contractor or its subcontractors but also those in which the owner and/or architect-engineer may be jointly responsible. This is the most prevalent type of indemnification clause, an example of which can be seen in Paragraph 4.18 of Appendix C. It is to be noted in Subparagraph 4.18.3 that the obligation assumed by the contractor does not extend to liability arising out of certain acts of the architect-engineer.

3. Broad-form indemnification. The broad form indemnifies the owner and/or architect-engineer even when the party indemnified is solely responsible for the loss.

When broad-form hold-harmless clauses are used, contractual liability coverage (see Section 8.25) is expensive and some provisions may not be insurable. The effect is compounded when the general contractor logically requires the same indemnification from its subcontractors. Some courts frown upon the use of broad-form indemnification, but such clauses have usually stood up legally. Most states have enacted laws that make it impossible for a contractor to indemnify an architect-engineer from liability arising out of defects in plans and specifications, and many states have passed laws that make void and unenforceable certain forms of broad-form indemnification clauses in construction contracts.

Indemnification clauses are also routinely included in subcontracts, under which the subcontractor agrees to hold harmless the general contractor, the owner, and perhaps the architect-engineer. In the presence of such language these indemnified parties are protected against liability that may devolve to them out of or in consequence of the performance of the subcontract. Just as the general contract can relieve the owner and/or architect-engineer from liability even when they are at fault, a subcontract indemnity clause can be worded so that it acts to relieve the general contractor, owner, and architect-engineer from liability even when they may be at fault.

It is a principle of long standing that a party may protect itself from losses resulting from liability for negligence by means of an agreement to indemnify. However, the rule is restricted to the extent that contractual indemnity provisions will not be construed to indemnify a party against its own negligence unless such intention is expressed in unequivocal terms. When such a clause is to be included in subcontracts, it must explicitly provide that the subcontractor indemnifies the designated parties whether or not the liability may be caused solely by them.

6.32 ARBITRATION

Customarily, when a contractual dispute arises between the contractor and the owner, the matter is referred first to the owner or architect-engineer. However, as has been pointed out previously, almost all contracts provide for certain appeals from such first-level decisions. In the construction industry, arbitration is frequently the next and last step in the settlement of such controversies between the owner and prime contractor. Arbitration is the reference of a dispute to one or more impartial persons for final and binding determination.

Court action can impose delay, expense, and inconvenience on both the contractor and the owner. For this reason most construction contracts provide for the arbitration of disputes. Actually, no contract clause is necessary for arbitration, because any dispute can be arbitrated at any time by mutual consent of the parties. Arbitration implies a common consent by the disputants to have their differences settled. It offers the advantages of a settlement that is prompt, private, convenient, and economical. Arbitration is not a replacement for the law but rather an adjunct to it.

Arbitration makes it possible for a construction dispute to be judged by professionals experienced in the construction industry. Although arbitration is an orderly proceeding governed by rules of procedure and standards of conduct, it is informal and need not conform to the adversary rules of conduct that the courts require. Finality is an important reason for the use of arbitration. Court decisions are open to lengthy appeals, resulting in long and costly delays in settling many cases. The award in an arbitration hearing cannot be changed without both parties agreeing to reopen the case.

Contract clauses that provide for arbitration are essentially phrased so that the parties to the contract agree to submit to arbitration any future disputes that may arise during the course of construction operations. Subparagraph 7.9.1 of Appendix C illustrates this point. However, while all arbitration statutes make agreements to arbitrate existing disputes irrevocable and enforceable, agreements to arbitrate unknown future disputes are not enforceable in all states. Where certain state arbitration statutes apply, either party can refuse to submit to arbitration even though the party may have promised to do so by the terms of a contract. However, most construction involves interstate elements and thereby is covered by the Federal Arbitration Act which applies to interstate commerce. The federal statute makes agreements to arbitrate future disputes binding.

Arbitration is not usually provided for in public contracts. Responsible for this are legal problems concerning who has the authority to commit a public body to binding arbitration. Public corporations lack such authority unless specifically authorized by statute. In the absence of a definitive resolution of this question, an arbitration award against a public body may be held void and unenforceable. In lieu of arbitration, contract appeals to various boards and commissions are normally possible with public agencies.

Most construction contracts that provide for arbitration stipulate that it shall be conducted under the *Construction Industry Arbitration Rules* as administered by the American Arbitration Association (AAA). These rules are reproduced in full in Appendix K. The AAA neither gives legal advice nor arbitrates disputes, but does provide assistance in obtaining arbitrators, furnishing rules of procedure, and giving other help. In exchange for its assistance, the AAA charges a nominal fee. Where contract bonds or insurance may be involved in the arbitration, the contractor should give its surety or insurance company advance notification. To illustrate this point, an arbitration award is not binding upon a party who did not agree to submit its rights to arbitration. An insurer, for example, who has not specifically agreed to be bound by an arbitration award or to indemnify the contractor for any liability imposed by arbitration, might claim to be a nonconsenting third party and thus would not be bound by the arbitration award. Some insurance policies specifically provide that the insurer will defend arbitration proceedings as well as lawsuits and will pay arbitration awards as well as court judgments. The contractor should check this matter before signing an agreement containing an arbitration clause.

General contractors commonly include arbitration clauses in their subcontracts. Such clauses provide that disputes between the subcontractor and general contractor that cannot be settled by mutual agreement shall be referred to arbitration. When a contract with the owner provides for arbitration, the prime contractor may assume the arbitration requirement extends on to the subcontractors. Experience indicates, however, that when the general contractor wishes arbitration to be used, the inclusion of a specific arbitration provision in the subcontracts will help ensure the primacy of arbitration with respect to disputes between the general contractor and its subcontractors.

6.33 ARBITRATION PROCEDURE

Whether or not conducted strictly under the rules of the American Arbitration Association, the general arbitration procedure is well established. The party wishing to initiate arbitration makes a written demand on the other side, stating the nature of the dispute, the amount involved, and the remedy sought, and requests that the matter be submitted to arbitration. A board of arbitration consisting of one or three persons is then selected. Arbitrators are picked not only for their impartiality and disinterest in the subject at arbitration but also for their experience in and knowledge of the construction field. No arbitrator should have any family, business, or financial relationship with either party to the controversy. A point to be stressed is that the arbitrator's authority to hear and decide exists only by virtue of the agreement of the parties. He is endowed with only such authority as they may confer upon him.

After the board has been selected, a hearing is conducted during which each side is free to call witnesses and to present such evidence as it wishes and the arbitrators consider admissible. Each party can be represented by counsel and is entitled to question the other party and its witnesses. Arbitration is a less formal process than litigation in a court of law. The parties, having elected to resolve their controversy by arbitration rather than in a lawsuit, have themselves agreed not to be bound by strict rules of evidence. The principal legal requirement is that a fair and full hearing for both sides be held.

After the hearing has been completed, an award is made within a reasonable period of time. A written copy of the findings and award signed by the arbitrators is sent to each of the parties. The arbitrators are not required to explain the rationale of their findings, only to decide all of the questions that were submitted by the disputants. It is usual practice that the arbitrators stipulate how the fees and costs shall be apportioned between the parties. Once the disputants have submitted to arbitration and an award is handed down, the parties are bound to it, and the award is enforceable by law. No appeal can be taken against the arbitrators' findings, although the award can be challenged if a party thinks the award was not within the submission or that there was bias, misconduct, or prejudice on the part of the board. Reversal of awards on such grounds is seldom requested and almost never found justified. Courts give every reasonable presumption in favor of the award and of the arbitrators' proceedings. The burden rests on the party attacking the award to produce evidence sufficient to invalidate it.

In recent years, the American Arbitration Association has introduced a new method of dealing with disputes in the construction industry, that of mediation. Mediation is a less formal procedure than arbitration and is an alternative course of action in the early stages of a dispute. Mediation involves a mutually acceptable and impartial third party who attempts to assist the parties in reaching a mutually agreeable settlement, while lacking any power to impose a decision. Mediation has long been associated with labor contract negotiations. It is now being applied in construction contract situations where the disputants need outside assistance in settling their differences. Mediation is a completely voluntary process. The mediator cannot impose a settlement but can only seek to assist the parties to make a direct settlement between themselves.

6.34 TERMINATION OF THE CONTRACT

Construction contracts may be ended in a variety of ways, full and satisfactory performance by both parties being the usual manner of contract termination. However, there are other means of bringing a contract to an end that are of interest and importance to the construction industry.

Breach of contract by either party is occasionally a cause for contract termination. Failing to make prescribed payments to the contractor or causing unreasonable delay of the project are probably the most common breaches by the owner. In such circumstances the contractor is entitled to damages caused by the owner's failure to carry out his responsibilities under the contract.

Default or failure to perform under the contract are the usual breaches committed by contractors. Nonperformance, faulty performance, failure to show reasonable progress, failure to meet financial obligations, persistent disregard of applicable laws or instructions of the architect-engineer—these are examples of material default by contractors that convey to the owner a right to terminate the contract. When the owner terminates the contract because of breach by the contractor, the owner is entitled to take possession of all job materials and to make reasonable arrangements for completion of the work. It is worthy of note that when failure to complete a project within the contract time is the breach involved, the owner probably will not be awarded liquidated damages if he terminates the contract and does not allow the contractor to finish the job late when the latter is making a genuine effort to complete the work. The record is not completely clear on this point, however.

A third way in which a contract can be terminated is by mutual agreement of both parties. This is not common in the construction industry, although there have been instances. For example, it sometimes happens that the contractor faces unanticipated contingencies such as financial reverses, labor troubles, or loss of key personnel that make proper performance under the contract a matter of considerable doubt. Under such circumstances both the owner and the contractor may agree to terminate the contract and to engage another contractor. When little or no work has been done, termination of the contract by mutual consent can sometimes be attractive to both parties.

Construction contracts, particularly those publicly financed, normally provide that the owner can terminate the contract at any time it may be in his best interests to do so by giving the contractor written notice to this effect. When such a provision is present, the contractor has agreed to termination at the prerogative of the owner. However, in such an event the contractor is entitled to payment for all work done up to that time, including a reasonable profit, plus such expenses as may be incurred in canceling subcontracts and material orders and in demobilizing the work. If the owner terminates the contract capriciously, he may become liable for the full amount of the contractor's anticipated profit plus costs.

A contract may be rescinded because of impossibility of performance under circumstances beyond the control of either party. Impossibility of performance is not the case when one party finds it an economic burden to continue. To be grounds for termination of a contract, it must indeed be impossible or impracticable to proceed. For instance, unexpected site conditions may be found that make it impossible to carry out the construction described by the contract. Operation of law may render the contract impossible to fulfill. However, the doctrine of legal impossibility does not demand a showing of actual or literal impossibility. Impossibility of performance has been applied to cases in which, even if performance were technically possible, the costs of performance would have been so disproportionate to that cost reasonably contemplated by the contracting parties as to make the contract totally impracticable in a commercial sense. The deciding factor was that an unanticipated circumstance made performance of the contract vitally different from what was reasonably to be expected.

6.35 THE NOTICE TO PROCEED

The beginning of contract time is often established by a written notice to proceed, which the owner dispatches to the contractor. The date of its receipt is normally considered to mark the formal start of operations. This notice, in the form of a letter advising the contractor that it may enter the site immediately, directs the contractor to start work. If by some chance the owner does not yet have clear title to the property, the notice will relieve the contractor of liability caused by its construction operations. Contracts usually require that the contractor shall commence operations within some specified period, such as ten days, after receipt of the notice to proceed.

6.36 SUBCONTRACTS

A subcontract is an agreement between a prime contractor and a subcontractor under which the subcontractor agrees to perform a certain specialized part of the work. A subcontract does not establish any contractual relationship between the subcontractor and owner, and neither is liable to the other in contract. A subcontract binds only the parties to the agreement: the prime contractor and the subcontractor. Nevertheless, construction contracts frequently stipulate that all subcontractors shall be approved by the owner or architect-engineer. The bidding documents may require that a list of subcontractors be submitted with the proposal or that a similar list be submitted for approval by the low-bidding prime contractor after the owner has accepted his proposal. (See Subparagraph 5.2.1 of Appendix C.) When the owner is given the right of approval, the general contractor is not relieved of any of its responsibilities by the owner's exercise of his prerogative.

If approval of subcontractors is required, such approval must be obtained before the general contractor enters into agreements with its subcontractors. There have been cases in which the owner refused to accept a subcontractor with which the prime contractor had already signed an agreement. It is possible in such a predicament that the general contractor must abrogate the existing subcontract and thereby render itself liable for damages to the subcontractor. Actually, the disapproval of a subcontractor is not a common occurrence. This is particularly true on public projects where disapproval of a subcontractor may result in litigation and can be difficult to sustain.

Although informal letters of proposal and acceptance may suffice for subcontracted work of small consequence, it is preferable that the prime contractor formalize all of its subcontracts with a written document that sets forth in detail the rights and responsibilities of each party to the contract. A well-prepared subcontract document can eliminate many potential disputes concerning the conduct of subcontracted work. The prime contractor may use a standard subcontract form, or it may develop its own special form to suit its particular requirements. The Associated General Contractors of America has prepared a subcontract form that is presented in Appendix L. Subcontract forms must be prepared with extreme care and with the advice of an experienced lawyer.

Change orders to construction contracts often involve modifications to subcontracted work. When this is the case, the general contractor and the subcontractor must execute a suitable change order to the subcontract affected. The prime contractor normally uses a standard form for this purpose similar to that shown in Figure 6.1.

6.37 SUBCONTRACT PROVISIONS

Subcontracts, in many respects, are similar both in form and content to the prime construction contract. Two parties contract for a specific job of work that is to be performed in accordance with a prescribed set of contract documents. On some projects, the owner reserves the right to approve the form of subcontract to be used by the general contractor. Frequently, the general contractor is required to include in its subcontracts express provisions that the subcontractor is bound to the prime contractor in the same manner as the prime contractor is bound to the owner. Additionally, the subcontract must provide that the subcontractor assumes all obligations to the prime contractor that the prime contractor assumes to the owner. An illustration of this is afforded by Subparagraph 5.3.1 of Appendix C. With such wording, provisions of the general contract such as changes, changed conditions, prevailing wages, warranty period, compliance with applicable laws, and approval of shop drawings extend to the subcontractors. Other owners require the general contractor to include in its subcontracts certain provisions or designated clauses that appear in the general construction contract.

The U.S. Army Corps of Engineers and other federal agencies stipulate that general contract provisions pertaining to apprentices and trainees, payrolls and basic records, applicable federal laws, withholding of funds, contract termination-disbarment, and others be included in subcontracts that apply to their projects.

In addition to the provisions of the general construction contract, there are many relationships applicable to the conduct of the subcontracted work itself for which provision must be made. Many of these are in the nature of general conditions to the subcontract, these being made an integral part of the printed subcontract form. Clauses pertaining to temporary site facilities to be furnished, insurance and surety bonds to be provided by the subcontractor, arbitration of disputes, and indemnification of the general contractor by the subcontractor are typical instances. Others are peculiar to the project and must be entered in spaces provided. For example, a detailed description of the work to be accomplished by the subcontractor must be included, this normally including reference to page numbers of drawings and sections of the specifications. A statement concerning the starting time and schedule of the subcontracted work, including any liquidated damages, is especially important. Special conditions must often be included such as security clearances for workers, job site storage, work to be accomplished after hours, and other requirements that arise from the peculiarities of a given project.

Terms of payment to the subcontractor and retainage by the general contractor are, of course, especially significant. Payment to the subcontractor can be established as a lump-sum or unit-prices or on a cost-plus basis. A matter of continuing concern to all parties is payment to the subcontractor by the general contractor when, for some reason, the general contractor has not been paid by the owner. Different subcontracts make quite different provisions in this regard. All subcontracts stipulate that the general contractor shall pay its subcontractors promptly after payment is actually made by the owner, although some obligate the contractor to pay the subcontractors *only* after owner payment. Paragraph 2.1 of Appendix L provides that payment to the subcontractor is conditioned on receipt by the prime contractor of its payment from the owner.

It should be noted that the courts do not always enforce such a contingent payment provision, especially when the owner's failure to pay was the general contractor's own fault. Recently, the courts have begun to rule that subcontractors must receive progress payments and final payments within a reasonable time from prime contractors, whether or not the prime contractors have been paid by the owners, and even when the lack of payment is not the fault of the prime contractor. Some forms of subcontracts require the unpaid general contractor to pay a subcontractor unless payment has been withheld by the owner because of the fault of the subcontractor.

Progress payments to subcontractors by the prime contractor are normally subject to the same retainage provisions as apply to payments made by the owner to the prime contractor. Such retainage is provided for by the subcontract instrument.

6.38 PURCHASE ORDERS

A purchase order is a written document that defines and prescribes the conditions pertaining to a purchase of materials. When a purchase order is signed by both the buyer (contractor) and the seller (material dealer), it becomes a purchase contract between the two parties. Although there can be oral purchase agreements, such contracts are not enforceable in most states if the price of the goods is $500 or more. This is in accordance with the statute of frauds section of the Uniform Commercial Code. Oral supply contracts, just as oral subcontracts, often lead to disagreements concerning the terms of the agreement and are to be avoided. Contracts for the sale of materials may, like other contracts, contain conditions. Purchase order forms used in the construction industry normally include characteristic clauses or terms pertaining to the materials being ordered and their delivery.

Normally, a purchase order is prepared by the contractor using its own standard printed form. Any terms and conditions included are, therefore, directed primarily toward protecting the interests of the contractor and ensuring that the materials and their delivery conform to the requirements of the construction contract and the contractor. It is not unusual, however, that vendors will sell materials only on the basis of their own terms and conditions. Under such circumstances the contractor must sign a purchase order form prepared by the seller. By so doing, the contractor accepts all terms and conditions of the resulting purchase contract.

Purchase orders must be prepared with care and attention to detail. It is of great importance that they contain all of the agreements, promises, and provisions pertaining to the procurement at hand. For instance, purchase orders not only describe and specify the materials and quantities to be provided but also define the conditions pertaining to taxes, freight, shipping and packaging, insurance, time of delivery, discounts, credit terms, guarantees, payment terms, and other such important matters. A properly prepared purchase order can be of great importance to the contractor in the event of improper shipment, loss, damage, late delivery, shortage, or other procurement problem.

7

CONTRACT BONDS

7.1 SURETY BONDS IN CONSTRUCTION

The use of contract bonds in the construction industry is but one of many applications of surety bonds in business and commerce. In law, a surety is a party that assumes liability for the debt, default, or failure in duty of another. A surety bond is the contract that describes the conditions and obligations pertaining to such an agreement. Surety bonds act as an extension of credit by the surety, not in the sense of a financial loan but as an endorsement.

By the terms of a contract bond the surety agrees to indemnify the owner, called the obligee, against any default or failure in duty of the prime contractor, called the principal. Contract bonds are three-party agreements that guarantee the work will be completed in accordance with the contract documents and that all construction costs will be paid. When the contractor properly discharges his obligations and after any warranty period covered by the bond expires, the bond agreement is discharged and is no longer of any force or effect. Regardless of the reason, if the prime contractor fails to fulfill its contractual obligations, the surety must complete the contract and pay all costs up to the face amount of the bond. It is to be noted that dual obligee bonds are sometimes used on both public and private projects. Such bonds protect both the owner and the lending institution that advances construction funds to the owner.

A contract bond form is a simple document that makes no attempt to describe in detail the specific liabilities of the surety. The bond guarantees the construction contract in all its provisions, and the obligations of the bond are identical with the provisions of the contract. The bond does not impose on the surety any obligations that are separate from or in addition to those assumed by the general contractor. By the same token, by being bonded, the contractor assumes no additional obligation to the owner that it has not already assumed by contract or by operation of law. The bond can be invoked by the owner only if the contractor is in breach of contract. The statutes of frauds of the various states universally require that contracts of suretyship must be in writing to be enforceable. For this reason, contract bonds are always written documents.

Contract bonds are not required of the contractor on all construction projects. Although such bonds are required by law on public jobs, a substantial proportion of private construction is not bonded. This is especially true when invitational bidding or negotiation is used. With regard to contract bonds, the owner must keep one important fact in mind. Contract bonds cannot be viewed as a satisfactory substitute for the able, honest, adequately financed contractor. Unfortunately, owners must sometimes look to the surety for job completion and make the best of a bad situation.

7.2 FORMS OF CONTRACT BONDS

By the terms of the construction contract with the owner, the prime contractor accepts two principal responsibilities: to perform the objective of the contract and to pay all costs associated with the work. Both of these obligations can be included within a single bond instrument, and combined performance and payment bonds are written on a few projects, almost all of these being privately financed. However, it is usual practice for construction contracts to require two separate contract bonds, one bond covering performance of the contract and the other guaranteeing payment for labor and materials. The separate forms bear the endorsement of The American Institute of Architects, and virtually all statutory bonds on public work are in separate forms.

Under the single type of bond, there is a potential conflict of interest between the owner and persons furnishing labor and materials. Because the owner has priority, the face value of the bond can be used up in satisfying his claims. Thus in many instances the single bond form has afforded little or no protection for material dealers, workmen, and subcontractors. In addition, there have been serious problems with the priority of rights of the persons covered. The double form of bond covers separately the interest of the owner and of subcontractors, material suppliers, and workmen. The premium cost of the bond protection is not increased by furnishing two separate bonds rather than one.

7.3 PERFORMANCE BONDS

The owner is entitled to receive what he contracted for or its equivalent. A performance bond acts primarily for the protection of the owner. It guarantees that the contract will be performed and that the owner will receive his structure built in substantial accordance with the terms of the contract. This bond customarily covers any warranty period that may be required by the contract, the usual bond premium including one year of such coverage. All performance bonds, as well as payment bonds discussed in the section following, have a face value which acts as an upper limit of expense the surety will incur in finishing the contract should that action become necessary. This face value is expressed as a fixed sum of money, the amount of which is usually derived from some percentage, 100 percent for instance, of the total contract price.

Figure 7.1 reproduces AIA *Document A311,* "Performance Bond," of The American Institute of Architects and illustrates the provisions of a typical performance bond used on private projects. The standard form used by the federal government provides that the bond shall apply to all contract modifications and that the life of the bond must include all extensions of time and any guarantee period required.

7.4 PAYMENT BONDS

A payment bond acts primarily for the protection of third parties to the contract and guarantees payment for labor and materials used or supplied in the performance of the construction. The private owner is thereby protected against liens (see Section 9.32) that can be filed on his project by unpaid parties to the work. Although a private owner ordinarily decides for himself whether to require contract bonds from the contractor, it is worthy of note that there are a few state statutes that require payment bonds on privately financed work. Figure 7.2 presents the labor and material payment bond of The American Institute of Architects. This bond, typical of common-law payment bonds employed by corporate sureties for privately financed projects, provides the following:

1. The claimant must have had a direct contract with either the general contractor or a subcontractor.
2. Labor and materials include water, gas, power, light, heat, oil, gasoline, telephone, and rental of equipment directly applicable to the contract.
3. Written notice must be given by the claimant, other than one having a direct contract with the general contractor, to any two of these: general contractor, owner, or surety, within 90 days after claimant performed its last work or furnished the last of the materials.
4. The owner is exempted from any liabilities in connection with such claims.
5. Claims must be filed in the appropriate court.
6. No claims shall be commenced after the expiration of one year following the date on which the general contractor stopped work, barring a statute to the contrary.

Payment bonds exclude from their coverage parties who are remote from the general contractor. It is to be noted that the bond form in Figure 7.2 includes only those claimants who have a direct contract with the prime contractor or one of the subcontractors. Whether specific instances of labor, material suppliers, or sub-subcontractors on public projects are protected by statutory payment bonds depends on the language of the related statute. Because liens cannot be filed against public property, the payment bond may well be the only protection that vendors, workmen, and subcontractors have for payment on public projects.

Standard forms of payment bonds do not require unpaid parties who deal directly with the prime contractor to give it written notice of the outstanding

THE AMERICAN INSTITUTE OF ARCHITECTS

AIA Document A311

Performance Bond

KNOW ALL MEN BY THESE PRESENTS: that

(Here insert full name and address or legal title of Contractor)

as Principal, hereinafter called Contractor, and,

(Here insert full name and address or legal title of Surety)

as Surety, hereinafter called Surety, are held and firmly bound unto

(Here insert full name and address or legal title of Owner)

as Obligee, hereinafter called Owner, in the amount of

Dollars ($

for the payment whereof Contractor and Surety bind themselves, their heirs, executors, administrators, successors and assigns, jointly and severally, firmly by these presents.

WHEREAS,

Contractor has by written agreement dated   19   , entered into a contract with Owner for
(Here insert full name, address and description of project)

in accordance with Drawings and Specifications prepared by

(Here insert full name and address or legal title of Architect)

which contract is by reference made a part hereof, and is hereinafter referred to as the Contract.

AIA DOCUMENT A311 • PERFORMANCE BOND AND LABOR AND MATERIAL PAYMENT BOND • AIA ®
FEBRUARY 1970 ED. • THE AMERICAN INSTITUTE OF ARCHITECTS, 1735 N.Y. AVE., N.W., WASHINGTON, D. C. 20006   **1**

Figure 7.1   Performance bond. AIA copyrighted material has been reproduced with permission of The American Institute of Architects under application number 80063. Further reproduction is prohibited.

debt. It is rightfully assumed that this contractor is well aware of its financial obligations. After payment by the general contractor is more than 90 days in arrears, a suitable claim can be filed with its surety. When the first-line suppliers and subcontractors don't pay *their* debts, the second-line providers of materials and labor are required to give written notice to any two of the general contractor, the owner, or the surety company. This must be done within 90 days after the unpaid claimant has performed its last work or furnished the last of the materials.

PERFORMANCE BOND

NOW, THEREFORE, THE CONDITION OF THIS OBLIGATION is such that, if Contractor shall promptly and faithfully perform said Contract, then this obligation shall be null and void; otherwise it shall remain in full force and effect.

The Surety hereby waives notice of any alteration or extension of time made by the Owner.

Whenever Contractor shall be, and declared by Owner to be in default under the Contract, the Owner having performed Owner's obligations thereunder, the Surety may promptly remedy the default, or shall promptly

1) Complete the Contract in accordance with its terms and conditions, or

2) Obtain a bid or bids for completing the Contract in accordance with its terms and conditions, and upon determination by Surety of the lowest responsible bidder, or, if the Owner elects, upon determination by the Owner and the Surety jointly of the lowest responsible bidder, arrange for a contract between such bidder and Owner, and make available as Work progresses (even though there should be a default or a succession of

defaults under the contract or contracts of completion arranged under this paragraph) sufficient funds to pay the cost of completion less the balance of the contract price; but not exceeding, including other costs and damages for which the Surety may be liable hereunder, the amount set forth in the first paragraph hereof. The term "balance of the contract price," as used in this paragraph, shall mean the total amount payable by Owner to Contractor under the Contract and any amendments thereto, less the amount properly paid by Owner to Contractor.

Any suit under this bond must be instituted before the expiration of two (2) years from the date on which final payment under the Contract falls due.

No right of action shall accrue on this bond to or for the use of any person or corporation other than the Owner named herein or the heirs, executors, administrators or successors of the Owner.

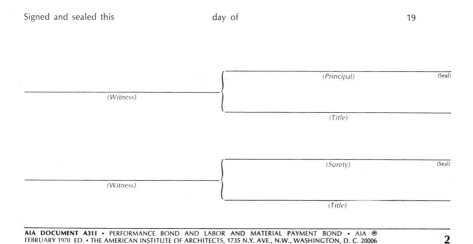

Signed and sealed this   day of   19

_____   (Principal)   (Seal)
    (Witness)

                                   (Title)

_____   (Surety)   (Seal)
    (Witness)

                                   (Title)

AIA DOCUMENT A311 • PERFORMANCE BOND AND LABOR AND MATERIAL PAYMENT BOND • AIA ®
FEBRUARY 1970 ED. • THE AMERICAN INSTITUTE OF ARCHITECTS, 1735 N.Y. AVE., N.W., WASHINGTON, D. C. 20006   2

Figure 7.1 Performance bond. (*continued*).

7.5 STATUTORY AND COMMON-LAW BONDS

Payment bonds are either statutory or common-law, there being important differences between the two. The bonding requirements on public projects are prescribed by law, and a statutory bond, at least by implication, contains the provisions of the statute that makes the bond a requirement. Private projects use common-law bonds whose coverage and functionings stand entirely on the provisions contained in the bond instrument itself.

THE AMERICAN INSTITUTE OF ARCHITECTS

AIA Document A311

Labor and Material Payment Bond

THIS BOND IS ISSUED SIMULTANEOUSLY WITH PERFORMANCE BOND IN FAVOR OF THE
OWNER CONDITIONED ON THE FULL AND FAITHFUL PERFORMANCE OF THE CONTRACT

KNOW ALL MEN BY THESE PRESENTS: that

(Here insert full name and address or legal title of Contractor)

as Principal, hereinafter called Principal, and,

(Here insert full name and address or legal title of Surety)

as Surety, hereinafter called Surety, are held and firmly bound unto

(Here insert full name and address or legal title of Owner)

as Obligee, hereinafter called Owner, for the use and benefit of claimants as hereinbelow defined, in the

amount of
(Here insert a sum equal to at least one-half of the contract price)   Dollars ($),
for the payment whereof Principal and Surety bind themselves, their heirs, executors, administrators,
successors and assigns, jointly and severally, firmly by these presents.

WHEREAS,

Principal has by written agreement dated   19   , entered into a contract with Owner for
(Here insert full name, address and description of project)

in accordance with Drawings and Specifications prepared by

(Here insert full name and address or legal title of Architect)

which contract is by reference made a part hereof, and is hereinafter referred to as the Contract.

AIA DOCUMENT A311 • PERFORMANCE BOND AND LABOR AND MATERIAL PAYMENT BOND • AIA ®
FEBRUARY 1970 ED. • THE AMERICAN INSTITUTE OF ARCHITECTS, 1735 N.Y. AVE., N.W., WASHINGTON, D. C. 20006   **3**

Figure 7.2 Labor and material payment bond. AIA copyrighted material has been reproduced with permission of The American Institute of Architects under application number 80063. Further reproduction is prohibited.

The distinction between statutory and common-law bonds is an important matter to the parties for whose protection the payment bond is written. On public projects the action of claimants to obtain protection under the bond must be in accordance with the applicable statute. This applies whether or not the statutory requirements are contained in the language of the bond itself. When payment bonds are required by statute on public projects, the right to recover on them is limited by the conditions of the statute to the same extent as though the provisions of the statute were fully incorporated into the bond

LABOR AND MATERIAL PAYMENT BOND

NOW, THEREFORE, THE CONDITION OF THIS OBLIGATION is such that, if Principal shall promptly make payment to all claimants as hereinafter defined, for all labor and material used or reasonably required for use in the performance of the Contract, then this obligation shall be void; otherwise it shall remain in full force and effect, subject, however, to the following conditions:

1. A claimant is defined as one having a direct contract with the Principal or with a Subcontractor of the Principal for labor, material, or both, used or reasonably required for use in the performance of the Contract, labor and material being construed to include that part of water, gas, power, light, heat, oil, gasoline, telephone service or rental of equipment directly applicable to the Contract.

2. The above named Principal and Surety hereby jointly and severally agree with the Owner that every claimant as herein defined, who has not been paid in full before the expiration of a period of ninety (90) days after the date on which the last of such claimant's work or labor was done or performed, or materials were furnished by such claimant, may sue on this bond for the use of such claimant, prosecute the suit to final judgment for such sum or sums as may be justly due claimant, and have execution thereon. The Owner shall not be liable for the payment of any costs or expenses of any such suit.

3. No suit or action shall be commenced hereunder by any claimant:

a) Unless claimant, other than one having a direct contract with the Principal, shall have given written notice to any two of the following: the Principal, the Owner, or the Surety above named, within ninety (90) days after such claimant did or performed the last of the work or labor, or furnished the last of the materials for which said claim is made, stating with substantial

accuracy the amount claimed and the name of the party to whom the materials were furnished, or for whom the work or labor was done or performed. Such notice shall be served by mailing the same by registered mail or certified mail, postage prepaid, in an envelope addressed to the Principal, Owner or Surety, at any place where an office is regularly maintained for the transaction of business, or served in any manner in which legal process may be served in the state in which the aforesaid project is located, save that such service need not be made by a public officer.

b) After the expiration of one (1) year following the date on which Principal ceased Work on said Contract, it being understood, however, that if any limitation embodied in this bond is prohibited by any law controlling the construction hereof such limitation shall be deemed to be amended so as to be equal to the minimum period of limitation permitted by such law.

c) Other than in a state court of competent jurisdiction in and for the county or other political subdivision of the state in which the Project, or any part thereof, is situated, or in the United States District Court for the district in which the Project, or any part thereof, is situated, and not elsewhere.

4. The amount of this bond shall be reduced by and to the extent of any payment or payments made in good faith hereunder, inclusive of the payment by Surety of mechanics' liens which may be filed of record against said improvement, whether or not claim for the amount of such lien be presented under and against this bond.

Signed and sealed this   day of   19

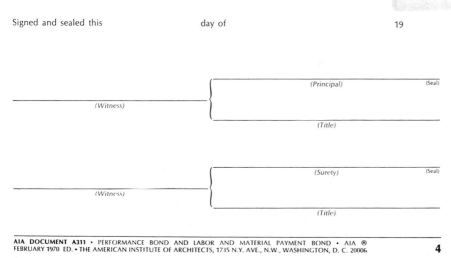

AIA DOCUMENT A311 · PERFORMANCE BOND AND LABOR AND MATERIAL PAYMENT BOND · AIA ®
FEBRUARY 1970 ED. · THE AMERICAN INSTITUTE OF ARCHITECTS, 1735 N.Y. AVE., N.W., WASHINGTON, D. C. 20006   **4**

Figure 7.2 Labor and material payment bond. (*continued*).

instrument. If a claimant fails to comply with the statutory requirements applying to enforcement of rights under the bond, he will not be permitted to recover.

A common-law bond is used when there are no statutory requirements. It is a contract that stands by its own language and that is enforced in the usual manner for contracts. In this case a claimant must proceed as described on the face of the bond.

On private projects the use of the standard common-law payment bond used by the various surety companies is standard practice. This form is standardized nationally and is approved by professional groups such as The American Institute of Architects. This form is also used on public construction, where bonding statutes do not apply. When statutory bonds are required, most public agencies that have substantial building programs have developed standard bond forms that conform with the applicable statute. Because the laws pertaining to bonding requirements differ somewhat from one jurisdiction to another, bond forms for public contracts are not standardized nationally. The federal government and many states and municipalities use their own standardized bond forms.

The standard payment bond used by the federal government is written to comply with the provisions of the Miller Act (see Section 7.6). This statutory bond protects laborers, material vendors, and subcontractors who perform work or supply materials for the project, although the extent of this protection depends on how far removed the unpaid party is from the general contractor.

7.6 THE MILLER ACT

The Miller Act prescribes the requirements of performance and of payment bonds used in conjunction with federal construction projects. Enacted in 1935 and subsequently amended, this statute provides that on all federal construction contracts of more than $25,000, the contractor shall furnish a performance bond for the protection of the United States and a payment bond for the protection of persons supplying labor and materials in the prosecution of the work.

The act provides that the performance bond be written in such amount that, in the opinion of the contracting officer, the interests of the United States are adequately protected. Under issued regulations of the comptroller general, federal agencies customarily require a performance bond in the amount of 100 percent of the contract amount. Payment bond amounts are established in accordance with the following sliding scale: 50 percent if the contract if $1,000,000 or less; 40 percent if the contract is more than $1,000,000 and not in excess of $5,000,000; and a fixed sum of $2,500,000 if the contract price is above $5,000,000.

The Miller Act gives workers, subcontractors, and material vendors who deal directly with the prime contractor the right to sue on the prime contractor's payment bond if payment is not received in full within 90 days after the date on which the last of the labor was done or the last of the materials was furnished. The law further provides that any person having a direct contractual relationship with a subcontractor but no contractual relationship with the prime contractor shall have a right of action on the prime contractor's payment bond provided the claimant gives written notice to the prime contractor within a 90-day period. There is no requirement for notice in the case of a party who deals directly with the prime contractor.

Under the Miller Act, first-tier subcontractors and material suppliers and second-tier subcontractors and suppliers are protected, but the payment protection of this federal statute extends no further. In addition, second-tier parties must deal with first-tier subcontractors. A subcontractor under the Miller Act has been held to mean one that performs for the prime contractor a specific part of the labor or material requirement of the project. Thus, the term subcontractor has been construed by the courts to include a party who supplied custom-made materials but did not install them. An unpaid bond claimant cannot sue on the payment bond until 90 days after the last of the labor was performed or the last of the materials was delivered. However, suit must be brought within one year after last work or delivery. Suits authorized by law are brought in the name of the United States, for the use of the party suing, in the appropriate district court. Suit is brought and prosecuted by the unpaid party's own attorney.

Several of the states have enacted "little Miller Acts." These statutes apply to projects financed by these states and establish contract bond requirements similar to those imposed by the Miller Act.

7.7 CONTRACT CHANGES

Construction contracts typically give the owner the right to make changes in the work. Because the contract comes before the bond and the bond guarantees the contract, it is commonly assumed that extension of the contract bond to include changes in the contract is provided for automatically. However, the construction contract is between the owner and the contractor, and in the event of changes, the surety is put in the position of being obligated by the terms of a contract to which it is not a party. Common law does not allow two contracting parties to bind a third without its consent. For this reason it is always advisable for the owner to obtain the written consent of the surety to any change or modification of the contract. To illustrate this point, it is standard practice that an executed consent-of-surety form be attached to each project change order. Approval of a contract change in writing is needed because of the aforementioned application of statutes of frauds to contracts of suretyship.

One aspect of the matter of contract modification is that it is possible for the surety to be exonerated from its original obligation, regardless of any provision in the construction contract that changes to the contract do not release the surety under any bond previously provided. Each such case is decided on the facts of the particular situation, and the court record is not entirely clear on this matter. However, when the changes in the contract have been such as to substantially change the manner of payment, the way the contract is to be performed, or the time for performance so as to make the contract significantly more difficult or costly to complete, the courts have given the surety relief on the basis that the contract was not the one the surety originally underwrote and agreed to be bound by. For relief in such cases it is necessary that the contract change was made without the consent of the surety and a showing is required that the change in the contract increased the risk of the surety and of the contractor substantially. As a general rule of thumb, changes in the contract that increase the amount of the contract by more than 10 percent are considered to be significant.

A second consideration in obtaining written consent of the surety for a contract change is that the surety is not obligated to provide bond for any additional or modified work unless the surety has expressly waived the right of notice. In this regard the performance bond in Figure 7.1 provides that the surety waives notice of any alteration or extension of time made by the owner. The payment bond in Figure 7.2, however, contains no waiver clause. Some bond forms used by the federal government stipulate that the surety waives notice of all extensions or modifications to the contract. However, a number

Classification			
A-1	A	B	Miscellaneous Contracts
Ash conveyors	Airfield grading	Air conditioning	Ash removal
Boiler repair	Airfield surfacing	Airport buildings	Bridges
Conveyors	Airfield runways	Aqueducts	Buildings,
Doors	Aluminum siding	Breakwaters	prefabricated
Fire alarms	Athletic fields	Buildings, erection	Culverts
Fire escapes	Beacons	and repair	Demolition
Flag poles	Ceilings, metal or	Canals	Draying
Floors, wood and	acoustical tile	Dams	Dredging
composition	Coal storage	Dikes	Garbage removal
Gas tanks	Curb and gutter	Docks	Grade eliminations
Generators	Curtain walls	Electrical work	Hauling
Guard rails	Ducts, underground	Excavation	Highways
Ironwork,	Elevators	Foundations	Maintenance
ornamental	Floodlights	Gas piping	Overpasses
Kitchen equipment	Glazing	Grain elevators	Roads
Lock gates	Greenhouses	Heating systems	Shoring
Metal windows	Machinery	Incinerators	Street paving
Parking meters	Millwork	Jetties	Structural iron and
Pipelines, oil or gas	Murals	Locks	steel
Police alarms	Parking areas	Masonry	Test borings
Radio towers	Parks	Piers	Timber cutting
Refrigerating plants	Piping, high-pressure	Pilings	Underpasses
Scaffolding	Playgrounds	Pipelines, water	Viaducts
Sidewalks	River bank protection	Plants, power	
Signal systems,	Road medians	Plants, sewage-	
railroad	Roofing	disposal	
Signs	Ski lifts	Plastering	
Stack rooms	Sprinkler systems	Plumbing	
Stand pipes	Stone, furnishing	Seawalls	
Street lighting	Storage tanks, metal	Sewers	
Tanks, gas	Tennis courts	Stone setting	
Thermostat equip-	Waterproofing	Subways	
ment	Wind tunnels	Tunnels	
Towers, water		Waterworks	
Track laying		Wells	
Traffic control		Wharves	
systems			
Weatherstripping			
Window cleaning			

Figure 7.3 Construction contract classifications for surety bond premium rates.

of these now also have a provision in them that limits the amount of changes that can be made in a bonded contract without the consent of the surety to 10 percent of the amount of the contract.

7.8 BOND PREMIUMS

For the purpose of computing contract bond premiums, construction contracts are divided into four classifications: A-1, A, B, and miscellaneous. Figure 7.3 contains a representative listing of these contract classifications.

Figure 7.4 presents the regular bureau premium rates used in conjunction with lump-sum or unit-price contracts where performance or performance and payment bonds are required. These rates include a warranty period of one year if required by the construction contract. If a longer warranty period is required, an additional premium charge is made. The rates shown in Figure 7.4 apply to the total contract amount and include all subcontracted work. Special rates apply when only payment bonds or when bonds totaling 20 percent or less of the contract price are involved.

The information given in Figures 7.3 and 7.4 is not complete and is subject to change without notice. Only the regular bureau rates are listed. In many instances downward revisions of these rates are used, these being called deviations or deviated rates. For example, contractors who have an established record of successful operation and whose financial resources meet certain standards may qualify for lower premium rates. Certain classes of work such as air conditioning systems, railroads, and privately owned public utilities can be subject to deviated premium rates. Such deviations are in a constant state of flux and vary with the surety company and with project location.

When the work can reasonably be assigned to more than one classification, the classification requiring the higher premium rate controls. Neither the classification of the contract nor the premium rate can be altered by segmenting the work or by parties other than the contractor furnishing the materials. All

Contract Price	Premium Rate per $1,000 of Contract Price for First 24 Months* (Subject to change without notice)		
	Class A-1	Class A	Class B
First $ 500,000	$6.00	$9.00	$12.00
Next 2,000,000	5.00	5.60	7.25
Next 2,500,000	4.10	4.40	5.75
Next 2,500,000	3.70	4.10	5.25
Over 7,500,000	3.30	3.70	4.80

*For construction time in excess of 24 months or 731 calendar days, increase basic premium by 1 percent per whole month.

Figure 7.4 Regular bureau rates for performance or performance and payment bonds. Lump-sum or unit-price contracts.

separate contracts are assigned the same classification as the general contract. Bond premiums for cost-plus contracts are subject to some variation from the rates of fixed-sum contracts as contained in Figure 7.4. Premium rates for cost-plus contracts with guaranteed maximum amounts are the same as for fixed-sum contracts. Cost-plus-percentage-of-cost contracts have a premium charge of 60 percent of the fixed-sum rates, excluding the contractor's fee from the final contract amount. Cost-plus-fixed-fee contracts are assessed a premium of 30 percent of the fixed-sum rates, excluding the contractor's fee.

The premiums for contract bonds are payable in advance, and the bonds are customarily delivered to the owner at the time the contract is signed. This premium payment is subject to later adjustment based on the ultimate contract amount, reflecting final work quantities on unit-price contracts and including all change orders and contract adjustments.

7.9 THE SURETY

Essentially all contractors utilize the services of national corporate surety companies whose specialties are the writing of bid bonds and contract bonds for contractors. These firms are subject to public regulation in the same manner as are insurance companies. They operate under charters and file their schedules of premium rates with designated public authorities. Because the true worth of the bond is no greater than the surety's ability to pay, the owner reserves the right to approve the surety company and the form of bond. The federal government requires that all corporate sureties proposed for use on government projects be approved by the U.S. Treasury Department.

Occasionally, on private work, the contract documents require that the contract bonds be obtained from a particular surety company. This requirement usually means the contractor must do business with an unfamiliar surety. In this situation, obtaining the bond may turn out to be a lengthy and laborious process, requiring the submission of financial reports, lists of jobs in progress, experience record, and other data of voluminous proportions that may be required to establish the contractor's record and financial standing. The professional associations of contractors and architect-engineers oppose the practice of having the surety designated by the owner and support a policy of leaving the contractor free to obtain surety bonds from a company of its choice. However, the private owner is at liberty to follow whatever practice he wishes in this regard, and the contractor must either comply with his instructions or not bid. In this regard, there are a few states with statutes that prevent an owner from requiring a contractor to obtain contract bonds from a designated surety.

On large contracts a single surety may seek protection for itself by enlisting other sureties to underwrite a portion of the contract. This is very much like reinsurance. The original surety remains completely responsible for guaranteeing the proper performance of the contract. If the bond is invoked, it is up to the original surety company to get its underwriting sureties to stand behind it in the completion of the contract.

In some instances the owner will require that the contract bonds be provided by cosureties, which means that two or more sureties split up the total contract obligation among them. On very large contracts this practice spreads the risk over the participating cosureties and correspondingly reduces the magnitude of the risk to which any one of them is exposed. This procedure also affords the owner a measurable degree of protection against possible financial default by a single surety. Occasionally it is necessary to have cosureties on large federal contracts because of limits established by the U.S. Treasury Department on the maximum amounts of single contract bonds that a given surety is authorized to execute.

7.10 INDEMNITY OF SURETY

A contract bond is not insurance for the general contractor and does not function for its protection. Under the bond the surety indemnifies the owner against default by the contractor. However, the contractor in turn must indemnify the surety against any claim that may be brought against it because of the contractor's failure to perform in the prescribed manner. Legal fees incurred by the surety because of claims under the bond are recoverable from the principal. Before the surety will provide bond service, the contractor must first sign a formal application form or contract of indemnity. This form is lengthy and contains a great deal of fine print. The net result, however, is that the contractor agrees to indemnify the surety and hold it harmless from expenses of every nature that the surety may sustain by reason of the invocation of the bond.

When the application is signed by an individual contractor or a partnership, each principal is obligated to the entire extent of his personal fortune. If a corporation makes application, only the corporation assets are pledged. However, some or all of the corporate officers, who are probably also the principal stockholders, often submit their personal contracts of indemnity to the surety in order to increase the firm's bonding capacity. Another means of providing needed financial capacity to an otherwise qualified firm is to obtain the personal guarantee or indemnity of a third-party financial backer. This is a case of where a person or entity having substantial financial resources contributes credit to the construction firm in return for a share of the profits.

7.11 INVESTIGATION BY SURETY

Before a surety will furnish an unknown contracting firm with a bid bond or contract bond, a thorough program of investigation is carried out to establish the past record and current commitments of the company. The experience, character, reputation, financial standing, equipment, integrity, personal habits, and professional ability of the firm's owners and key personnel are carefully examined. Company financial statements, both for the current period and for some years past, are subject to study and analysis. The firm's bank credit is verified, together with its relations with its sources of credit and supply. In short, a surety will issue bonds to a contractor only when the traditional three "C's" requirement is met: character, capacity, and capital. The providing of bonding services to a contractor is a highly individual matter. An exception to this general rule, however, can be found in the surety bond guarantee program that has been used by the federal government to assist small businesses and minority-owned businesses to obtain construction bonds that they could not otherwise obtain.

Once a contractor has firmly established relations with a bonding company, the contractor's bonding capacity becomes reasonably well established, and future investigations by the surety underwriter are concerned with keeping the contractor's records current and investigating the individual bond requests as they are submitted. If the contractor's work load is well below its limit and contracts of the usual variety are proposed, a bond application is generally approved without delay. However, when the maximum bonding capacity is to be approached, or when an unusually large or completely new type of construction project is proposed, approval of the bond application may require a considerably longer period of time or may not be forthcoming at all.

When the contractor makes application for a bond for a new project, it will find the surety is interested in many aspects of the work that is proposed. The following presents the usual and most important subjects of investigation:

1. The essential characteristics of the project under consideration, including its size, type, and nature. Evaluation of the hazards of construction cannot proceed until the surety is apprised of the work. Included here would be the identity of the owner and his ability to pay for the construction as it proceeds.
2. The total amount of uncompleted work the contractor presently has on hand, of both the bonded and the unbonded variety. This must, of necessity, include work that has not yet been awarded. The obvious point of concern here is to prevent the contractor from becoming overextended in regard to working capital, equipment, and organization.
3. The adequacy of working capital and the availability of credit. The contractor can assist its own cause by keeping the surety fully informed as to its activities and supplied with up-to-date financial reports. This phase of investigation can often protect the contractor against taking on a project that is too big for it to handle.
4. The amount of money the contractor "left on the table," that is, the spread between the low bid and the next highest. Competitive conditions in the construction industry are such that a spread of more than five or six percent between the two lowest bidders is generally the cause of some concern. The surety wishes to ensure that the contractor's estimating and bidding procedures are sound.
5. The largest contract amount of similar work the contractor has successfully completed in the past. Inexperience in a new field of construction has contributed to a long list of contractor failures. The surety would like the contractor to stay with the kind of work in which it is most experienced. If the contractor wishes to change to another type, the surety will urge that the first steps be small ones until the contractor acquires the necessary experience. If the contractor is not properly equipped for the new work, it must be demonstrated to the surety how the equipment problems will be solved.
6. Terms of the contract and bonds to be required, details of how payment

will be made to the contractor, retainage, time for completion, liquidated damages, and the nature of job warranties. These all influence the surety's appraisal of the contractor's ability to do the work.

7. The amount of work subcontracted and the qualifications of the subcontractors. The surety's concern here is that the prospective subcontractors possess the necessary organization, experience, and financial resources to carry out their end of the work.

After each bonded project is completed, the surety sends to the owner a request for a final report on the contractor's performance. The owner is asked to submit to the surety a statement concerning the contractor's handling of the job, changes that were made in the work, and the final total contract amount. This latter figure is used as a basis for any terminal adjustment of the bond premium.

7.12 BONDING CAPACITY

A useful concept widely used by the construction industry is that of bonding capacity. This term has no precise definition but refers to the maximum value of uncompleted work the surety will allow the contractor to have on hand at any one time. Bonding capacity is commonly obtained by multiplying a factor times either the contractor's net worth or net working capital. Such factors may vary from less than 10 to more than 30, depending on the surety's considered assessment of the individual contractor's capabilities and financial position. It is a common rule of thumb that a contractor can carry approximately $15.00 of uncompleted work for each dollar of net working capital, or $10.00 of uncompleted work for each dollar of net worth. Net working capital and net worth are discussed in Chapter 9.

The difference between a contractor's bonding capacity and its current total of uncompleted work, both bonded and unbonded, is a measure of the additional work for which the surety will bond it. There are, of course, other factors involved, such as the type of new work being considered and the size of the project. In computing the current value of a contractor's uncompleted work, the surety may decide to take into account at least part of the value of the subcontracts that are bonded.

7.13 THE SURETY AGENT

The local representative of the national surety is the surety agent. This is the person with whom the contractor must deal directly in all matters concerning bid and contract bonds. There are many advantages to be gained by the contractor in selecting a thoroughly qualified and experienced agent who has bona fide surety company affiliations that enable him to act on behalf of the surety company he represents and to provide prompt bonding service. The surety agent is a very important member of the construction circle who can contribute significantly to the success or failure of a contractor's business. The agent is a trained observer of the construction industry who has a detached point of view and whose advice is therefore particularly valuable to the contractor.

Most bonding agents are sincere and astute representatives of large corporate sureties whose businesses are solidly based on long experience and competent service. From the foregoing sections it is easy to understand how the contractor may sometimes get the impression that the surety representative is unduly meddling in its affairs or is overly limiting its volume of work. In all fairness, however, the contractor should feel fortunate that the surety is interested and alert enough to be of assistance in helping avoid the many pitfalls associated with the management of a construction firm. The contractor must realize that both are working toward the common goal of a prosperous and successful contracting business. Some surety agents are, of course, more conservative than others. It is up to the contractor to select a bonding representative who is responsive to its needs within the limits of responsible and competent service.

The bid bonds and contract bonds that are provided to the contractor seldom if ever originate directly with the corporate surety itself. These documents are prepared by the local agent. In order to verify that the agent is indeed an agent of the surety and is authorized to execute bond instruments that are binding on the surety company, it is customary that each construction surety bond include an appropriate power-of-attorney form which is attached to the bond itself.

7.14 DEFAULT BY THE CONTRACTOR

Should the contractor default, the surety is required to perform in accordance with the terms of the bond up to a maximum amount that is the face value of the bond. How the surety company elects to complete the contract is usually a matter for it to decide. There are several alternative courses of action, but the most common is for the original contractor to finish the job with the financial help of the surety and under the eye of a construction consultant hired by the surety. The surety may decide, however, to put the remaining work out for competitive bids and select a new contractor to take over, or a new contractor may be engaged on a cost-plus basis. The new contractor may contract with the surety or the owner.

Contract bonds can offer the contractor a genuine advantage when its ability to proceed has been temporarily curtailed by legal or financial difficulties. If the financial condition of the contracting firm is basically sound, the surety may choose to help it get back on its feet and into business again. It may elect to advance the contractor credit in sufficient amount for the contractor to proceed with its work. If claims have tied up the contractor's capital, the surety may furnish bonds to discharge these claims, thereby enabling the contractor to proceed on its own. Such arrangements are made privately and occur more often than might be thought.

When the contractor defaults and the surety undertakes to complete the work, the surety becomes entitled to all of the remedies the owner has against the contractor under the contract. In addition, the surety is entitled to receive from the owner the balance of the contract price, which is defined in Figure 7.1 as "the total amount payable by Owner to Contractor under the Contract and any amendments thereto, less the amount properly paid by Owner to Contractor." Provisions of the governing statute apply to this matter when a statutory bond is used. The surety may also press any claims against the owner which the defaulting prime contractor might have had. Regardless of the bond wording, the surety often finds that it must compete with other claimants for the retainage withheld by the owner.

The surety also has another right in its relationship with the owner. If the owner defaults, for example, by not making progress payments, the contractor is released from liability under the contract. Because the responsibility of the surety to the owner is the same as that of the contractor to the owner, any act of the owner that would release the contractor from its contractual obligation also releases the surety.

7.15 SUBCONTRACT BONDS

An earlier discussion pointed out that the general contractor is responsible for the job performance of the subcontractors. In addition, it can also be held liable if a subcontractor does not pay for materials, labor, or sub-subcontracts pertaining to that project. Consequently, the general contractor, to protect itself against subcontractor debt and default, frequently requires certain or all of them to provide performance and payment bonds. In this instance, the subcontractor is the principal and the prime contractor is the obligee. General contractors have found to their dismay, however, that not every subcontract bond form nor every surety company provide them with the kind of financial protection desired. It is for this reason that many prime contractors require the use of their own subcontract bond forms. Otherwise, the contractor must reserve the right to approve the surety and bond form proposed for use by the subcontractor.

A factor of major importance to be emphasized at this point is that the bond serves in no way to replace honesty, integrity, and competence on the part of the subcontractor. Bond or no bond, an inferior subcontractor means trouble. Although the bond will afford the general contractor some measure of protection against financial loss directly attributable to a particular subcontract, it does not and cannot cover expenses caused by work stoppages, delays, and disruptions of the overall construction program that inevitably result from subcontractor default.

7.16 WARRANTIES AND SURETY BONDS

A warranty is a certification, expressed or implied, that a certain aspect of a contract is, in fact, as it was declared or promised to be. Previously mentioned is the usual warranty period after the completion of construction, during which time the general contractor guarantees that the project is free from defects caused by its failure to perform. Manufacturers' or applicators' warranties are also sometimes required to guarantee the performance requirements of machinery, operating equipment, processes, or materials. A written instrument is not necessarily required in conjunction with a warranty, although specifications sometimes stipulate that a written warranty be delivered to the owner.

The true value of a warranty is, quite obviously, determined by the integrity of the firm standing behind it. In the event of a breach of warranty, the warrantor becomes liable for damages. Whether damages can actually be collected may, of course, be an entirely different matter. To ensure compliance with a warranty, a surety bond, called a maintenance bond, is sometimes required. Already mentioned is the fact that the performance bond customarily includes the usual project warranty period. Another common example of a maintenance bond is a roof bond, by the terms of which the roof is guaranteed against defects of workmanship or materials for some specified period of time. A bond of this type is executed by a corporate surety with the manufacturer of the roofing materials named as the principal. The owner and architect-engineer must be careful in this regard because some roofing "bonds" are not bonds at all, being merely the unsupported warranty of the roofing contractor or manufacturer without surety.

When a warranty bond is provided to the owner covering some aspect of the work, he must first look to the principal if defects materialize. If the principal refuses or is unable to remedy the situation, then the owner can invoke the bond, and the surety must make good the warranty as guaranteed by the bond.

7.17 MISCELLANEOUS SURETY BONDS

In addition to proposal, performance, payment, subcontract, and maintenance bonds, the contractor occasionally finds it necessary or desirable to furnish or accept several other forms of surety bonds. The most important of these are the following:

Bonds to Discharge Liens or Claims. Persons who have not received payment for labor or materials supplied to a construction project are entitled to file a mechanic's lien against the property of a private owner or against moneys due and payable to the general contractor in the case of a public contract. Other claims can also be filed. Such actions can freeze capital needed by the contractor to conduct its operations.

A surety bond in an amount fixed by an order of the court can be used to discharge a mechanic's lien. When claims for personal or property damage are involved, the bond amount is generally decided on by the owner. In any case, the bond functions as a financial guarantee to the owner and releases money that has been withheld from the contractor.

Bonds to Indemnify Owner Against Liens. The contractor may be called on to post a bond in advance that indemnifies the owner against any impairment of title or other damage that may be suffered by reason of liens or claims filed on his property. In this situation, bond is required before any such liens are filed, rather than being used to discharge liens after they are filed in the way described in the previous paragraph.

Bonds to Protect Owners of Rented Equipment and Leased Property. During construction operations, the contractor may find it desirable to rent or lease equipment, parking lots, access roads, storage installations, and similar facilities. The owner of such property often requires the contractor to post a bond that guarantees proper maintenance and payment of rental charges, and indemnifies the owner against loss, damage, or excessive wear of the property.

Fidelity Bonds. The usual fidelity bond protects the employer against dishonest acts of an employee such as theft, forgery, or embezzlement. The employee covered by the bond is the principal, and the employer is the obligee. If the employer suffers a loss proved to have been caused by the employee, the surety will reimburse the employer up to the face amount of the bond. This type of security bond is more completely discussed in Section 8.16.

Judicial or Court Bonds. When the contractor is the plaintiff in a legal action, it sometimes is required to furnish security for court costs, possible judgments, and similar financial eventualities. Such security is often provided in the form of judicial or court bonds. This legal requirement is commonly applied when the contractor institutes court proceedings in states or jurisdictions other than its own.

License Bond. Also known as a permit bond, this is a bond required by state law or municipal ordinance as a condition precedent to the granting of a contractor's license or permit. License bonds guarantee compliance with statutes or ordinances and make provision for payment to the obligee or members of the general public in the event that the licensee violates his legal or financial obligations. This topic has been discussed previously in Section 1.24.

Termite Bond. This form of bond is given by manufacturers or applicators of substances intended to prevent damage caused by termites.

Subdivision Bond. This bond, given by the developer to a public body, guarantees construction of all necessary improvements and utilities.

Self-Insurers' Workmen's Compensation Bond. This bond, given by a self-insured contractor to the state, guarantees payment of all statutory benefits to injured employees.

Union Wage Bond. This bond, given by a contractor to a union, guarantees that the contractor will pay union wages and will make proper payment of fringe benefits required by union contract.

8

CONSTRUCTION INSURANCE

8.1 RISK MANAGEMENT

Risk management may be defined as a comprehensive approach to handling exposures to loss. Any peril that can cause financial impairment to the business enterprise is the subject of risk management. The following are three steps that a contracting firm can follow in applying risk management to its business.

1. Recognize and identify the varied risks that apply to the construction process. These may arise as a consequence of contract wording, the nature of the work, site conditions, or the operation of law.
2. Decide how to protect against those risks that have been identified. If the risk cannot be eliminated by an alternative procedure or by contractual transfer to another party, the choice may be to purchase commercial insurance, to self-insure, or to assume the risk.
3. Conduct a company-wide program of loss control and prevention. Project safety programs discussed in Chapter 15 constitute an important part of this procedure.

In devising the best risk-handling program, commercial insurance is the keystone to adequate financial protection. Insurance does not eliminate the risks involved in construction contracting, but it does shift most of the financial threat to a professional risk-bearer. In addition to paying losses, insurance companies offer valuable services to the contractor in the areas of safety, loss prevention, educational and training programs, site inspections, and others. Where self-insurance is involved, it is not usually the sole risk-bearing element but works in conjunction with a commercial insurance program. Insurance is also involved with a form of risk assumption in the use of deductible amounts. This chapter is primarily directed toward a comprehensive discussion of standard commercial insurance coverages used in the construction industry.

8.2 CONSTRUCTION RISKS

Construction work by nature is hazardous, and accidents are frequent and of-ten severe. The annual toll of deaths, bodily injuries, and property damage in the construction industry is extremely high. The potential severity of accidents and the frequency with which they occur require that the contractor protect itself with a variety of complex and expensive insurance coverages. Without adequate insurance protection, the contractor would be continuously faced with the immediate possibility of serious or even ruinous financial loss.

As has been discussed previously in this book, construction projects usu-ally have in force several simultaneous contractual arrangements: between the owner and architect-engineer, between the owner and general contractor, and between the general contractor and its several subcontractors. Contracts that provide for design–construct and construction management services, and the use of separate prime contracts introduce additional features. Construed as a whole, these contracts establish a complicated structure of responsibilities for damages arising out of construction operations. Liability for accidents can de-volve on the owner or architect-engineer, as well as on the prime contractor and subcontractors whose equipment and employees perform the actual work. Construction contracts typically require the contractor to assume the owner's and architect-engineer's legal liability for construction accidents or to provide insurance for the owner's direct protection. Consequently, a contractor's in-surance program normally includes coverages to protect persons other than itself and to protect it from liabilities not legally its own.

8.3 THE INSURANCE POLICY

An insurance policy is a contract under which the insurer promises, for a con-sideration, to assume financial responsibility for a specified loss or liability. The policy contains many provisions pertaining to the loss against which it af-fords protection. Fundamentally, the law of insurance is identical with the law of contracts. However, because of its intimate association with the public wel-fare, the insurance field is closely controlled and strictly regulated by federal and state statutes. Each state has an insurance regulatory agency that admin-isters that state's insurance code, a set of statutory provisions that imposes regulations on insurance companies concerning investments, reserves, annual financial statements, and periodic examinations. Insurance companies are con-trolled as to their organizational structure, financial affairs, and business methods. In most states insurance policies must conform to statutory require-ments as to form and content.

A loss suffered by a contracting firm as a result of its own deliberate action cannot be recovered under an insurance policy. However, negligence or over-sight on the part of the contractor will not generally invalidate the insurance contract. The contractor must pay a premium as the consideration for the in-surance company's promise of protection against the designated loss. Most types of insurance require that the premium be paid in advance before the policy becomes effective. In the event of a loss covered by an insurance policy, the contractor cannot recover more than the loss; that is, a profit cannot be made at the expense of the insurance company.

Insurance companies can be organized as stock companies or as mutual companies. Stock companies are organized in a manner similar to that of a bank, and ownership is vested in stockholders. The owner of an insurance policy has no ownership in the company and assumes no risk of assessment if the insurance company encounters financial reverses.

A mutual company is one in which the policyholders constitute the members of the insuring company or association. Every policyholder of the mutual company is, at the same time, an insurer and an insured. If it happens that the premiums collected are in excess of the losses, the excess is returned to the policyholders as "dividends." By the same token, if losses outweigh income, assessments of the policyholders may be possible. State laws permit mutual companies that satisfy certain tests to limit or eliminate the assessment that can be levied against the members. Consequently, the policies of many mutual companies are nonassessable. This varies considerably with the bylaws and policies of the individual mutual company. In property and casualty insurance, a field of insurance especially important to contractors, several mutuals are among the largest companies. In life insurance probably a majority of the largest companies are mutuals.

8.4 CONTRACT REQUIREMENTS

With the many hazards that confront a construction business and the plethora of insurance types that can be purchased, one might wonder how a contractor decides just what insurance is really needed. In reality, however, the contractor quite often has no choice. For example, it is standard practice that construction contracts require the contractor to provide certain insurance coverages.

Construction contracts typically make the contractor responsible for obtaining coverages such as workmen's compensation insurance, employer's liability insurance, and comprehensive general liability insurance. Property insurance to protect the project and liability insurance to protect the owner may be made the responsibility of either the owner or the contractor, depending on the contract. Article 11 of Appendix C makes the owner responsible for obtaining both of these insurance coverages. There are, of course, many examples of special insurance being required by contract when the construction involves unusual risks or conditions. When the contract delegates specific responsibility to the contracting firm for obtaining certain insurance, it is customary that insurance certificates (see Section 8.49) must be submitted to the owner or the architect-engineer as proof that the coverage stipulated has, in fact, been provided.

As was discussed in Section 6.31, construction contracts frequently require the contractor to hold the owner and architect-engineer harmless by accepting any liability that either of them may incur because of operations performed under the contract. Most contract documents that contain such indemnity clauses are explicit in requiring the contractor to procure appropriate contractual liability insurance (see Section 8.25). Subparagraph 11.1.3 of Appendix C illustrates this point.

With regard to project insurance requirements, it is always good practice for a contractor to submit a copy of the contract documents to its insurance company before construction operations commence. The contracting firm is not an insurance expert and is not really competent to evaluate the risks and liabilities placed on it by the contract. Its insurance agents or brokers are qualified to comb the documents and advise the firm concerning the insurance needs dictated by the language and requirements of a given construction contract.

8.5 LEGAL REQUIREMENTS

Certain kinds of insurance are required by law, and the contractor must provide them whether or not they are called for by the contract. Workmen's compensation, motor vehicle, unemployment, and social security are examples of coverages required by statute. It can be argued that unemployment and social security payments made by the contractor are more in the nature of a tax than of insurance premiums in the usual sense. Nevertheless, both unemployment and social security are treated as forms of insurance for the purpose of discussion in this chapter.

The law makes the independent contractor liable for damages caused by its own acts of omission or commission. In addition, the prime contractor has a contingent liability for the actions of its subcontractors. Therefore, whether or not the law is specific concerning certain types of insurance, the contractor as a practical fact must procure several different categories of liability insurance to protect itself from liability for damages caused by its own construction operations as well as those of its subcontractors.

8.6 ANALYSIS OF INSURABLE RISKS

Aside from coverages required by law and the construction contract, it is the contractor's prerogative to decide what insurance it shall carry. Such elective coverages pertain principally to the contractor's own property or to property for which it is responsible. It is not economically possible for the contractor to carry all the insurance coverages available to it. If it purchased insurance protection against every risk that is insurable, the cost of the resulting premiums would impose an impossible financial burden on its business. The extent and magnitude of a contractor's insurance program can be decided only after careful study and consideration. If a risk is insurable, the cost of the premiums must be balanced against the possible loss and the probability of its occurrence. There are, of course, risks that are not insurable or for which insurance is not economically practicable. Associated losses must be regarded simply as ordinary business expenses.

At times, careful planning and meticulous construction procedures can minimize a risk at less cost than the premium of a covering insurance policy. Thus the contractor may choose to assume a calculated risk rather than pay a high insurance premium. A common example of assuming such a risk involves construction that is to be erected immediately adjacent to an existing structure. If the nature of the new construction is such that the existing structure may be endangered by settlement or collapse, the contractor has two courses of action open to it. As one alternative it can include in its estimate the premium for a collapse policy. Such protection is high in cost and is generally available only with substantial deductible amounts. Instead, the contractor can assume the risk without insurance protection, choosing to rely on skill and extraordinary precautions in construction procedures to get the job done without mishap. Many analogous cases can be cited with respect to pile driving, blasting, water damage, and others.

8.7 CONSTRUCTION INSURANCE CHECKLIST

Insurance coverages are complex, and each new construction contract presents its own problems. The contractor should select a competent insurance agent or broker who is experienced in construction work and familiar with contractors' insurance problems. Without competent advice, the contractor may either incur the needless expense of overlapping protection or expose itself to the danger of vital gaps in insurance coverage. The contractor can often reduce insurance costs by keeping its agent or broker advised in detail as to the nature and conduct of its construction operations.

In the long list of possible construction insurance coverages, not every policy is applicable to a given firm's operations. The following checklist is not represented as being complete, but it does contain insurance coverages typical of the construction industry.

A. Property Insurance on Project.
1. All-risk builder's risk insurance. This insurance protects against all risks of direct physical loss or damage to the project or to associated materials and job equipment caused by any external effect, with noted exclusions.
2. Named-peril builder's risk insurance. The basic policy provides protection for the project, including stored materials and job equipment, against direct loss by fire or lightning. A number of separate endorsements to this policy are available that add coverage for specific losses.
    a. Extended coverage endorsement. This covers property against all direct loss caused by windstorm, hail, explosion, riot, civil commotion, aircraft, vehicles, and smoke.
    b. Vandalism and malicious mischief endorsement.
    c. Water damage endorsement. Insurance of this type indemnifies for loss or damage caused by accidental discharge, leakage, or overflow of water or steam. Included are defective pipes, roofs, and water tanks. This does not include damage caused by sprinkler leakage, floods, or high water.
    d. Sprinkler leakage endorsement. This provides protection against

all direct loss to a building project as a result of leakage, freezing, or breaking of sprinkler installations.

3. Earthquake insurance. This coverage may be provided by an endorsement to the builder's risk policy in some states. Elsewhere a separate policy must be issued.

4. Bridge insurance. This insurance is of the inland marine type and is often termed the "bridge builder's risk policy." It affords protection during construction against damage that may be caused by fire, lightning, flood, ice, collision, explosion, riot, vandalism, wind, tornado, and earthquake.

5. Steam boiler and machinery insurance. A contractor or owner may purchase this form of insurance when the boiler equipment of a building under construction is being tested and balanced or when being used to heat the structure for plastering, floor laying, or other purposes. Unlike other property insurances listed here, this type includes some liability coverage. This policy covers any injury or damage that may occur to or be caused by the boiler during its use by the contractor.

6. Installation floater policy. Insurance of this type provides named-peril or all-risk protection for property of various kinds such as project equipment and machinery (heating and air conditioning systems, for example) from the time that it leaves the place of shipment until it is installed on the project and tested. Coverage terminates when the insured's interest in the property ceases, when the property is accepted, or when it is taken over by the owner.

B. Property Insurance on Contractor's Own Property.

1. Fire insurance on contractor's own buildings. This coverage affords protection for offices, sheds, warehouses, and stored contents. Endorsements for extended coverage and for vandalism and malicious mischief are also available.

2. Contractor's equipment insurance. This type of policy, often termed a "floater," insures a contractor's construction equipment regardless of its location.

3. Motor truck cargo policy. This insurance covers loss by named hazards to materials or equipment carried on the contractor's own trucks from supplier to warehouse or building site.

4. Transportation floater. Insurance of this type provides coverage against damage to property belonging to the contractor or others while it is being transported. It may be obtained on a per-trip, project, or annual basis.

5. Burglary, robbery, and theft insurance. This form of insurance protects the contractor against the loss of money, securities, office equipment, and similar valuables through burglary, theft, robbery, destruction, disappearance, or wrongful abstraction.

6. Fidelity bond. This surety bond affords the contractor protection against loss caused by dishonesty of its own employees.

7. Dishonesty, destruction, and disappearance policy. A comprehensive policy of this form protects against the loss of money and securities, on and off the premises, caused by dishonesty, mysterious disappearance, or destruction. It insures against dishonesty of employees, loss

of money and securities, loss of securities in safety deposit, and forgery.

8. Valuable papers destruction insurance. This policy protects the contractor against the loss, damage, or destruction of valuable papers such as books, records, maps, drawings, abstracts, deeds, mortgages, contracts, and documents. It does not cover loss by misplacement, unexplained disappearance, wear and tear, deterioration, vermin, or war.

C. Liability Insurance.

1. Employer's liability insurance. This insurance is customarily written in combination with workmen's compensation insurance. It affords the contractor broad coverage for the bodily injury or death of an employee in the course of his employment, but outside of and distinct from any claims under workmen's compensation laws.

2. Contractor's public liability and property damage insurance. This insurance protects the contractor from its legal liability for injuries to persons not in its employ and for damage to the property of others, if the property is not in the contractor's care, custody, or control, when such injuries or damage arise out of the operations of the contractor.

3. Contractor's protective public and property damage liability insurance. This protects the contractor against its liability imposed by law arising out of acts or omissions of its subcontractors.

4. Contractual liability insurance. This form of insurance is required when one party to a contract, by terms of that contract, assumes certain legal liabilities of the other party. The usual forms of liability insurance do not afford this coverage.

5. Owner's protective liability insurance. This insurance protects the owner from his contingent liability for damages arising from the operations of the prime contractor or its subcontractors.

6. Completed-operations liability insurance. This form of insurance protects the contractor from damage claims stemming from its faulty performance on projects already completed and handed over to the owner. The usual forms of liability insurance provide protection only while the contractor is performing the work and not after it has been completed and accepted by the owner.

7. Professional liability insurance. This insurance protects the contractor against damage claims arising out of design and other professional services rendered by the contractor to the owner.

D. Employee Insurance.

1. Workmen's compensation insurance. This insurance provides all benefits required by law to employees killed or injured in the course of their employment.

2. Social Security. This all-federal insurance system operated by the U.S. government provides retirement benefits to an insured worker, survivor's benefits to his family when the worker dies, disability benefits, hospitalization benefits, and medical insurance.

3. Unemployment insurance. This federal-state insurance plan provides workers with a weekly income during periods of unemployment between jobs.

4. Disability insurance. This insurance, required by some states, provides benefits to employees for disabilities caused by nonoccupational accidents and disease.

E. Automobile Insurance.

Various forms of insurance are available in connection with the ownership and use of motor vehicles. Liability coverages protect the contractor against third-party claims of bodily injury or property damage involving the contractor's vehicles or nonowned vehicles that are used in its interest. Collision insurance, together with comprehensive fire and theft coverage, indemnifies the contractor for damage to its own vehicles.

F. Business, Accident, and Life Insurance.

1. Business interruption insurance. This insurance is designed to reimburse the owner for losses suffered because of an interruption of his business.

2. Sole proprietorship insurance. A policy of this type provides cash to assist heirs in continuing or disposing of the business without sacrifice in the event of death of the owner.

3. Key man life insurance. This insurance reimburses the business for financial loss resulting from the death of a key man in the business. It also builds up a sinking fund to be available on his retirement.

4. Group life insurance. Contractors often purchase life insurance for their employees. This affords protection for each participant at a low group cost, the premium for which may be paid wholly or partly by the contractor. Additional amounts can often be purchased by the employees at their own expense.

5. Major medical insurance. Such insurance covers hospitalization and medical expenses incurred by covered employees. Policies are often written to include the families of the employees. A portion of the premium may be paid by the employer and the balance by the individuals insured.

6. Disability insurance. This insurance can provide benefits to employees for temporary or permanent disability.

7. Accidental death and dismemberment. Insurance of this type provides benefits for death or dismemberment caused by accident.

8. Corporate continuity insurance. In the event of a stockholder's death, this insurance furnishes cash for the purchase of his corporate stock. This provides liquidity for the decedent's estate and prevents corporate stock from falling into undesirable hands.

The insurance types of major importance to the construction industry will now be discussed.

8.8 PROJECT PROPERTY INSURANCE

Construction contracts make the general contractor responsible for the project until it is accepted by the owner. Consequently, it is up to the contractor to take all reasonable steps necessary to protect the project from loss or damage and to see that suitable insurance is provided for this purpose. Most construction contracts require that project insurance shall be purchased and maintained on the entire project to the full insurable value thereof, including all subcontracted work. The possibilities of property damage to the project itself and to stored materials can depend considerably on the nature of the work, its geographical location, and the season during which the work will be performed. On building construction, the loss potential is usually large and diverse and can include fire, smoke, explosion, collapse, vandalism, water, wind, freezing, and physical damage of a wide variety. Marine structures can be especially vulnerable to wind, ice, wave action, tides, water currents, collision, and collapse. In a general sense, the highway contractor faces comparatively few hazards to project property, the elements of nature probably constituting the source of greatest risk.

Depending on the terms of the contract, the project insurance may be provided by the owner or the prime contractor. Regardless of who buys this insurance, however, the intent is to protect the interests of the owner, the prime contractor, the subcontractors, and perhaps the lending institution in the event a loss occurs and to provide funds for repairs or rebuilding.

In a very real sense, project insurance must be procured that is tailored to meet the specific risks intrinsic to the work. Builder's risk policies are standard on building construction. An all-risk installation floater policy is commonly used on projects, such as water and sewer jobs, where there is little or no exposure to fire and extended coverage hazards. Policies designed for the specific construction hazards of bridges, tunnels, radio and television towers, and other special construction types are available. Highway, reclamation, and other engineering contractors often carry no project insurance, choosing to self insure the small risks and relying on "acts of God" contract clauses to protect them from major damage to the project. A typical clause of this type provides that the contractor will make good all damage to any portion of the work, except those damages due to unforeseeable causes beyond the control of and without the fault of the contractor including acts of God or extraordinary action of the elements. An act of God has been defined as a natural occurrence of extraordinary and unprecedented proportions, whose magnitude and destructiveness could not have been provided against by the exercise of ordinary foresight.

It is quite impracticable to attempt herein any comprehensive treatment of the many forms of project insurance. Builder's risk insurance, a discussion of which follows, has been selected because of its broad applicability and general similarity to other types.

On building construction, where builder's risk insurance normally constitutes the basic project policy, it is customary to deduct the costs associated with land preparation, landscaping, excavation, underground utilities, and foundations below the lowest basement floor when computing the insurable value of the structure. The rationale behind this is that these portions of the project are not directly susceptible to loss by fire or other hazards. Premium rates for builder's risk insurance can vary considerably with the type of construction and the availability of fire-fighting facilities. Premium rates are higher for unprotected areas and for the more combustible classes of construction.

There are two types of builder's risk policies, the all-risk form being the one most generally used. The named-peril form of insurance with endorsements is available but seldom utilized. The all-risk form is much broader in scope than the named-peril form and provides insurance protection for all losses not specifically excluded in the policy. Both types of builder's risk policies normally provide that the structure shall not be occupied by the owner in the course of construction without obtaining the consent of the insurance carrier. As a general rule, any loss to specifications, drawings, records, documents, accounts, deeds, currency, notes, securities, or designs is specifically excluded.

8.9 ALL-RISK BUILDER'S RISK INSURANCE

This policy, widely used for building construction projects, covers the project proper as well as temporary structures at the jobsite. Materials, equipage, apparatus, and supplies pertaining to the construction are protected while held temporarily in storage prior to delivery, while in transit to the jobsite, and after their delivery and while awaiting installation. If not otherwise covered by insurance, the contractor's tools and construction equipment are also protected while on the premises.

The policy insures against all direct physical loss or damage from any external cause to the property covered, except for stated exclusions. There are many such exclusions, however, including certain damage due to freezing, explosion of steam boilers or pipes, glass breakage, subsidence and settling, artificially generated electrical currents, rain and snow, earthquake, floods, nuclear radiation, and several others. Builder's risk policies exclude coverage for the cost of correcting faulty workmanship, materials, construction, or design, as well as loss or damage caused by error, omission, or deficiency in design, specifications, workmanship, or materials. A common omission from coverage is any loss that may occur during testing phases of the construction project. Many of these exclusions can be removed or modified for additional premium. Some policies make the insurance company liable only to the extent of the actual cash value of the property at the time of the loss. Other policies have a more liberal valuation clause that provides for the cost of replacement. Policy exclusions and format vary from one insurance company to another.

All-risk policies provide for deductible amounts that apply to various defined loss categories. The all-risk policy is very flexible and can be varied substantially. The contractor is generally able to work out specific coverage to suit its needs.

All-risk insurance can be obtained in two forms: the reporting form and the completed-value form. Both of these forms are discussed further in Section 8.11. The reporting form is a single, open-end policy that can cover all of the contractor's jobs, at least within a prescribed geographical area. New projects to be included under the policy are added automatically as they are obtained. The total insurable value of work in place for all jobs covered by the policy is reported by the contractor to the insurance company at monthly intervals. Rates for the individual projects are adjusted to suit the degree of fire protection available. Where the completed-value form is used, there is a separate policy for each job and the contractor pays a single, lump-sum premium in advance of construction.

8.10 NAMED-PERIL BUILDER'S RISK INSURANCE

Named-peril builder's risk insurance, unlike the all-risk form that protects the project against all losses except those excluded, affords coverage for only those risks specifically listed. The basic policy of this form protects building projects only against direct loss caused by fire or lightning. Such insurance ordinarily covers the cost of facilities and materials connected or adjacent to the structure insured, including temporary structures, materials, machinery, and supplies of all kinds incidental to the construction of the structure. When not otherwise covered by insurance, this policy also protects construction equipment and tools owned by the contractor or similar property of others for which it is legally responsible. All property is protected that is a part of or contained in the structure, in temporary structures, or on vehicles, or is stored on the premises adjacent to the project.

There are, of course, many possible causes of physical loss or damage to a construction project besides fire and lightning. Other significant risks can be insured against by purchasing various endorsements to the basic named-peril policy. The extended coverage endorsement provides protection against damage or loss caused by windstorm, hail, explosion, riot, riot attending a strike, civil commotion, aircraft, vehicles, and smoke. Another common endorsement covers vandalism and malicious mischief. It is to be noted that each endorsement protects against only the hazards named and that each additional endorsement is obtained by the payment of an extra premium. Many other special endorsements to builder's risk are available, including protection against water damage, sprinkler leakage, early occupancy by the owner, and earthquake.

8.11 BUILDER'S RISK PREMIUMS

There are two ways in which the premiums of builder's risk insurance can be paid, the coverage being the same in either case. One payment mode is the reporting form, which establishes the insurable value of the structure, and hence the face value of the policy, in accordance with periodic progress reports submitted to the insurance company by the contractor. This form requires the contractor to make a monthly report of the insurable value of the work in place, including all materials stored on the site. The premium for this form of policy is paid in monthly installments, the amount of each monthly premium payment being computed on the basis of the last progress report. Premium payments, therefore, are initially low but increase progressively each month as the job advances toward completion. An important consideration with the reporting form is keeping the reports accurate and up-to-date. Should a report be undervalued or overlooked, the insured stands the risk of assuming a part of the liability if a severe loss occurs.

The other payment scheme is the completed-value form. This policy is written for the full amount of the project value, which is the contract price less the cost of the foundations and other excluded work. The coverage must be written by the time the foundations are completed, and the premium is payable as a lump sum in advance. This form requires no monthly report. The premium rate for the completed-value form is usually 55 percent of the full-term rate, on the assumption that the insurable value starts from zero and increases approximately linearly during the construction period.

In general the insured is free to choose the form of premium payment that is used. At times the choice between them may be made on the basis of least probable premium cost. The premium for the completed-value form is based on a linear increase of insurable value with time. If a project's insurable value initially increases slowly but later rises rapidly, the total of the monthly premiums by the reporting form will be less than the single completed-value premium. Such a project is normally large and requires considerable time for excavating, pile driving, foundation construction, and other initial work exempted from the insurable value.

As a rule, the completed-value form of builder's risk insurance is somewhat cheaper for short-term jobs. In addition, it does not require the monthly progress reports. However, the preparation of these reports does not usually involve any real amount of additional time or expense because the essential information is already available from the monthly project pay requests.

8.12 PROVIDING OF BUILDER'S RISK INSURANCE BY THE OWNER

The standard documents of The American Institute of Architects stipulate that, unless otherwise provided, the owner shall purchase and maintain the builder's risk policy (see Subparagraph 11.3.1 of Appendix C). It is common practice, however, for the general contractor to provide this coverage, especially on public projects. By and large, most contractors would rather provide the builder's risk insurance themselves. This is true because owners, not usually being familiar with this type of insurance, sometimes procure insurance policies that do not provide all the coverage necessary or that have excessive deductible amounts.

There are, however, certain types of projects that do lend themselves well to the provision of builder's risk by the owner. One such example is a project that involves several independent contractors. Another good example is a remodeling job or an addition to an existing building. If the contractor is to provide the builder's risk insurance, there is always considerable uncertainty about protection for the existing structure. In such circumstances the contractor may make an arrangement with the owner to add the additional coverage to the owner's existing policy, although this matter can become involved from an insurance standpoint. It is much simpler and perhaps cheaper for the owner in such cases to obtain the insurance in the first place. Notwithstanding this, however, there is a contractor's-interest form of builder's risk insurance that can be purchased for alteration or remodeling projects to protect the contractor's interest only. If the owner does purchase the builder's risk insurance, the contractor must determine that the policy provides the customary construction coverage.

8.13   SUBROGATION

An important aspect of builder's risk insurance is the subrogation clause, a standard provision in many forms of insurance including builder's risk. The workings of subrogation may be illustrated by the following example. If the owner of insured property should sustain a loss to this property, the insurance company will pay the insured for the damage suffered up to the face amount of the policy. However, by the terms of the subrogation clause in the policy, the insurance company acquires the right of the insured to recover from the party whose negligence caused the loss. This process of subrogation gives the insurance company the right to sue in the insured's name for recovery of its loss. In the case where the owner provides builder's risk insurance and the general contractor's operations cause or contribute to project damage, the contractor may be exposed to action by the builder's risk carrier for recovery of its loss under the policy. Alternatively, if a subcontractor or sub-subcontractor causes or contributes to a loss on the project, this party may be subject to suit by the insurance company. If the general contractor buys the builder's risk insurance, subrogation applies to its subcontractors or sub-subcontractors if they cause a loss on the project. It is easy to see that application of the subrogation clause by the insurance company could defeat the entire purpose of the project property insurance.

One means of at least partially avoiding the undesirable effects of subrogation is to make the owner, prime contractor, and all the subcontractors and sub-subcontractors named insureds under the policy. Subrogation cannot usually be employed against parties who are insured under the policy. However, this exemption may apply only to damage that a contractor does to its own work and not to damage caused to the work of others where the contractor has no insurable interest. Another approach is for the owner and all of the contractors to waive all rights against each other for damages caused by fire or other perils. Subparagraph 11.3.6 of Appendix C illustrates this point. However, there are some builder's risk forms that provide policyholders cannot waive such rights unless written permission of the insurance company has been received. All things considered, it would appear that the naming of all parties as insureds and the waiver of rights among all parties are both desirable when either the owner or general contractor provides the builder's risk insurance.

8.14   TERMINATION OF BUILDER'S RISK INSURANCE

Builder's risk policies may be canceled on a pro rata basis at any time requested by the policyholder. If the contractor provides this insurance, the time at which it can terminate the policy is an important matter. On the one hand, the premiums are expensive, and the contractor naturally wishes the expenditure to cease at the earliest possible moment. On the other hand, the contractor cannot dispense with this protection until such time as the owner is legally responsible for the project. When the contract has been silent in this regard, the courts have repeatedly found that the contractor remains responsible for the project until the owner has made formal acceptance.

Sometimes the owner occupies the building, but his acceptance of the structure is delayed until the contractor completes some remaining work. Under these circumstances, if the contractor's operations should cause a fire, the loss to the structure would be covered by the builder's risk policy. However, the contractor may incur a liability to the owner for the latter's loss of business while the fire damage is being repaired. To protect against this possibility, the contractor may wish to take out business interruption insurance.

8.15 CONTRACTOR'S EQUIPMENT FLOATER POLICY

An equipment floater policy protects the contractor from loss or damage to its construction equipment, but does not include liability coverage of any kind. This insurance is very flexible and can be designed to fit the specific needs of the contractor. When the equipment is financed by chattel mortgage, it is a requirement of the loan that the equipment be adequately insured. The policy can be made to include equipment that is not owned but is rented, leased, borrowed, or for which the contractor is otherwise legally responsible. Many floater policies exclude coverage for licensed motor vehicles. However, this limitation can usually be removed for those vehicles that provide mobility to construction equipment.

In the usual equipment policy, each large equipment unit of high value is listed on a schedule that gives its identification, purchase price, date of acquisition, and insured value. Blanket coverage is provided for the smaller items of equipment. Equipment is normally insured for its actual cash value, which is defined as today's replacement cost less reasonable depreciation. Equipment insurance, like many other forms of property insurance, is usually sold on a 100 percent coinsurance basis. This means the contractor is required to carry insurance to the full value of the equipment. If this is not done, the contractor becomes a coinsurer with the insurance company on any loss that may occur. For example, if a contractor is carrying insurance coverage at only 80 percent of the actual value of the equipment, compensation by the insurer will be for only 80 percent of any loss, subject to overall policy limits. The contractor will have to make up the remaining 20 percent.

The protection afforded by the policy applies to the equipment while it is in the contractor's storage yard, in transit, or on a jobsite. When the total equipment value exceeds a prescribed amount, the floater policy can be made to include all equipment without the contractor providing a schedule. Each piece of equipment is automatically included as it is acquired, and no notification to the insurance company is required. An exception to this can be leased equipment where automatic coverage may require a specific policy provision to that effect. Insurance premiums are determined by annual audits. There are no standard premium rates for this type of policy. Each construction company is rated according to the type of work performed, past loss experience, company reputation, and dispersion of risk.

The policy coverage is basically determined by the hazards to which the equipment is exposed and can be made "named-peril" or "all-risk." To illustrate, the coverage may be limited to those perils most likely to happen, such as fire, transportation, upset, landslide, theft, collision, tornado, flood, explosion, windstorm, and overturn. A broad all-risk form is written that provides blanket protection against all physical loss or damage by external means that may occur to the equipment on the job, in transit, or in the contractor's yard. Deductible amounts normally apply to most or all losses. The cost of the broad form is naturally higher than that for the named risks, and the size of the deductible amounts can have a substantial effect on the premiums. The usual equipment floater policy excludes waterborne equipment, as well as damage caused by overload to certain classes of equipment.

8.16 FIDELITY BONDS

A fidelity bond indemnifies the contractor for loss of money or other property caused by dishonest acts of its bonded employees. This includes losses sustained through larceny, theft, embezzlement, forgery, and other fraudulent or dishonest acts. Under this form of surety bond, the employer is the obligee and the employee is the principal. Although a fidelity bond is not insurance, it is hardly distinguishable from insurance insofar as the employer is concerned. A fidelity bond in physical form resembles an insurance policy and is sometimes referred to as dishonesty insurance.

There are a number of ways in which fidelity bond protection can be written. The most important are described in the following.

1. Named Schedule Form. Employees named in the bond are covered.

2. Position Schedule Form. The occupants of the company positions listed are covered.

3. Blanket Position Form. A form that covers all employees of the firm without naming them or listing positions. In this form, the limit of bond liability applies for each employee.

4. Commercial Blanket Form. This form also covers all employees of the firm without naming them or listing positions. However, this is an aggregate limit form meaning that regardless of the number of employees involved in the loss, there is an aggregate and total limit of liability.

As indicated, the contractor can purchase protection on specific persons or positions such as the bookkeeper and other selected personnel. However, records almost invariably show that it is impossible to designate in advance the person who will be dishonest. On this basis a blanket form of bond may be preferable, since it covers all employees. In the event of loss under this form of policy, the contractor need only prove that it was caused by an employee or employees unknown. Under the named schedule or position schedule form, however, the dishonest act must be shown to have been that of an individual whose name is listed or who occupied a listed position. In addition, schedule forms require continuous checking as names and positions change. Other key advantages of the blanket forms are automatic additions and deletions without notice to the bonding company and no premium adjustment made during the premium year for additional employees.

8.17 BURGLARY, ROBBERY, AND THEFT INSURANCE

In addition to the possible loss of money and valuables caused by the dishonesty of its employees, the contractor can suffer similar losses at the hands of outsiders. There are a number of different policies that cover various forms of such losses. Mercantile safe burglary insurance covers loss by burglary of money, securities, and other valuables from within a contractor's safe or vault, including any damage to the premises. A paymaster robbery policy covers payroll funds in the custody of a paymaster or other payroll custodian. This insurance is available in a broad form that provides coverage for the destruction or disappearance of the funds by occurrences other than an actual holdup. The policy may be extended to cover payrolls that are unattended or otherwise not in physical custody. Safety deposit box burglary and robbery insurance covers loss of all property and securities, except money, from the contractor's safe deposit box by burglary or robbery.

A broad form of money and securities policy protects against the loss of money or securities within or away from the contractor's premises from destruction, disappearance, or wrongful abstraction. Specifically excluded from this policy is any loss caused by dishonesty of employees, war, or forgery. An office burglary and robbery policy indemnifies the insured contractor for loss by burglary and robbery of money, securities, or other property, and damage caused by forcible entry into the premises.

8.18 COMPREHENSIVE DISHONESTY, DESTRUCTION, AND DISAPPEARANCE POLICY

Contractors frequently buy comprehensive dishonesty, destruction, and disappearance insurance that combines the coverages discussed in Sections 8.16 and 8.17 into one coverage. This policy provides broad, all-risk protection against loss of money, securities, and other valuables. Commonly called a "3-D policy," this coverage is very versatile and can provide protection against loss due to the dishonesty of employees, loss of money and securities within or off the premises, damage done to the premises and equipment by thieves, loss of securities in safety deposit, and loss due to forgery. Coverage is also available for payroll robbery, burglary, and the theft of materials and office equipment.

8.19 LIABILITY INSURANCE

Liability is an obligation imposed by law. In the course of conducting its business, the contractor may incur liability for damages in any one of the following ways:

1. Direct responsibility for injury or damage to the person (not an employee) or property of third parties, caused by an act of omission or commission of the contractor.
2. Contingent liability, which involves the indirect liability of the general contractor for the acts of parties for whom it is responsible, such as subcontractors.

3. Contractual liability, whereby the contractor has assumed the legal liability of the owner, or other party, by the terms of a contract.
4. Liability that arises out of a project after the work has been completed and the structure accepted by the owner.
5. Liability that may devolve to the contractor as a result of the operation of its motor vehicles.
6. Liability to injured employees, both those covered and those not covered by workmen's compensation laws.
7. Liability that arises from design and associated professional services rendered by the contractor to the owner.

Liability insurance is also called defense coverage and serves no purpose other than to protect the contractor against claims brought against it by third parties. Insurance of this type pays the costs of the contractor's legal defense as well as paying judgments for which the contractor becomes legally liable, up to the face value of the policy. In the settlement of liability claims against the contractor, the insurance company has the right to settle as it sees fit without the approval or consent of the contractor. Liability insurance provides no protection to the contractor for loss of or damage to its own property. It is important to note that many forms of liability insurance customarily include subrogation clauses that give the insurance company the right to file suit to recover losses.

For a variety of good reasons a contractor should consider obtaining all of its liability insurance from one insurance company. This arrangement often offers the possibility of substantial premium discounts because of the larger total premium amounts involved, a better overall loss rating, and other factors. In addition, a single liability insurance carrier helps to avoid gaps and overlaps in coverage and eliminates squabbles between insurance companies about loss responsibility.

8.20 CONTRACTOR'S PUBLIC LIABILITY AND PROPERTY DAMAGE INSURANCE

The basic form of this policy, often simply called public liability insurance or premises-operations insurance, protects the contractor against its legal liability to third persons for bodily injury and property damage arising out of its own operations. Excluded from coverage are injuries to the contractor's own employees as well as all forms of automobile liability. The contractor's buildings and premises, owned or leased, are included as well as its business operations in progress anywhere in the United States. Normally included in the property damage liability coverage, either as a part of the basic policy or by broad-form property damage endorsement, is damage to the property of other contractors (including subcontractors). Consequently, if the operations of Contractor A cause damage to the work of another contractor on a construction project, the public liability insurance of Contractor A will pay the cost. Not covered, however, would be any damage that Contractor A did to its *own* work. Also normally included in this policy is an elevator liability clause that covers the insured's legal liability for bodily injury or property damage arising from ownership, maintenance, or use of elevators owned, controlled, or operated by the contractor. The term elevators, in this context, does not include material hoists. Contractors must ordinarily carry very high limits of coverage (bodily injury and property damage) for this type of insurance, either as required by contract or simply to be adequately protected. This is often done through the use of excess or umbrella liability coverage (see Section 8.43).

Standard forms of this insurance include personal coverage for individual proprietors, partners, and executive officers, directors, and stockholders of corporations while acting within the scope of their duties. The policy may be endorsed to extend coverage to company employees as additional insureds while acting within the scope of their duties. This affords protection to employees who are named as individuals in lawsuits together with their employers.

With respect to a contractor's liability for injury to third parties, the concept of "attractive nuisance" is important. The doctrine of attractive nuisance establishes a special obligation on the part of the possessor of land to children who trespass on the property. When a contractor is in possession of land, such as a jobsite, its operations frequently attract children. Because a child may not realize that he is entering on the land of another or may not appreciate the hazards involved, the contractor has an additional responsibility to protect trespassing children. The standards of due care, however, apply only to artificial conditions on the land, not to natural features such as trees, cliffs, or bodies of water. Courts usually hold that a trespassing adult is not entitled to damages for injury not willfully or negligently inflicted by the contractor. However, the situation is different with respect to children. The courts hold that a contractor who maintains a dangerous appliance or hazardous premises is expected to exercise reasonable care to eliminate the danger or otherwise to protect the children by watchmen, fences, or other protective devices.

8.21 PUBLIC LIABILITY INSURANCE PREMIUMS

Premiums for public liability insurance are based on the contractor's payrolls and, to some extent, on gross receipts. Workmen's compensation insurance is likewise based on payrolls, and it may be convenient for the same insurance company to write both policies so that one audit of the contractor's payroll records will verify premium payments for both coverages at the same time.

The premiums for both public liability and workmen's compensation insurance are adjusted up or down according to the contractor's accident experience record. When annual premium payments reach a given level, the contractor becomes eligible for experience rating. Under this process, credits or debits determined for the individual contractor are applied to the manual rates in accordance with how the premiums paid in compare with the losses paid out over a period of time. A manual rate is a standard premium charge based on a probable loss experience for a given class of risks. These rates are published in a manual and are used for general insurance cost information. A contractor whose losses are low enjoys a considerable savings on the costs of its experience-rated insurance coverages.

A possible advantage to be gained by the contractor, if the same insurance company writes its experience-rated policies, is an improved overall loss experience that may result in lower insurance premiums. An insurance company usually rates a contractor on the basis of its total insurance business with the contractor and a loss under one coverage may be counterbalanced by favorable experience on the other policies written by the same insurance carrier. These remarks also illustrate the wisdom of a contractor's continuing to do business with an insurance company with which it has established a considerable background of favorable experience.

8.22 CONTRACTOR'S PROTECTIVE PUBLIC AND PROPERTY DAMAGE LIABILITY INSURANCE

Often called contractor's contingent liability insurance, this policy protects the contractor from its contingent liability imposed by law because of injuries to persons or damage to property of others arising out of the acts or omissions of subcontractors. This includes protection for the general contractor if the subcontractor's insurance is inadequate or nonexistent. A claimant alleging damages caused by a subcontractor may sue not only the subcontractor but the prime contractor as well. This situation arises from the fact that the prime contractor exercises general supervision over the work and is responsible for the conduct of construction operations, including those of the subcontractors. Contingent liability insurance not only covers accidents arising out of operations performed for the contractor by an independent subcontractor, but also protects the contractor from liability it may incur because of any supervisory act by it in connection with a subcontractor's work. This form of liability insurance excludes workmen covered by workmen's compensation laws and property under the care, custody, or control of the contractor. If a subcontractor further subcontracts portions of its work, it will also need this form of protection.

The premiums for contractor's contingent liability insurance are derived from the subcontract amounts and do not generally vary with the work classifications. Insurance companies writing this form of insurance may require that the prime contractor obtain from its subcontractors certificates of insurance verifying that they have purchased public liability and property damage insurance to cover their own direct liability.

8.23 COMPLETED-OPERATIONS LIABILITY INSURANCE

Completed-operations liability insurance protects the contractor from liabilities arising out of projects that have been completed or abandoned. In a limited sense, the contractor is not liable for damages suffered by a third party by reason of the condition of the work after the project has been completed and accepted by the owner. According to the "completed and accepted" rule, once the owner accepts the work, he becomes the party responsible for any defects in it. When a dangerous condition exists that is obvious or readily discoverable upon inspection, and the owner allows this condition to continue, he is substituted for the contractor as the party answerable for damage to a third party. This is true even though the dangerous condition was originally created by the contractor. Nor is the contracting firm liable if it merely carries out plans, specifications, and directions given to it by another, unless the information is so obviously faulty that no reasonable person would follow it. However, there are many exceptions to these rules and the courts now increasingly hold contractors liable to anyone who might be injured because of their negligence. A contractor who creates a dangerous condition on the property of another may be held responsible to parties injured thereby, even after owner acceptance, if the parties so injured could reasonably have been expected to come into contact with the dangerous condition and provided the contractor knew or should have known of the hazard and did not exercise reasonable care to warn of the dangers.

The rule of owner responsibility does not apply to dangerous conditions of such latent nature that their existence would not be discovered by the owner during the exercise of ordinary care and usage. Indeed, courts have recently imposed liability on contractors without requiring a showing of fault in cases involving latent defects in completed projects. This is an extension to construction of the doctrine of manufacturers' or product liability whereby builders of mass-produced homes and other contractors have been held subject to strict liability for damages to third parties, even in the absence of contractor negligence. This follows from the legal theory that a producer of goods is strictly liable for injuries or damages resulting from a product defect. Consequently, a contractor's responsibility does not end with project completion. Rather, its liability for completed operations continues for the full period of the applicable state statutes of limitations.

When contractors obtain completed-operations liability insurance for the first time, they are automatically protected against loss arising from any completed project, regardless of its completion date. This is true, of course, in the absence of any specific exclusion to the contrary. The protection under this policy starts with the effective date of the insurance and continues as long as the insurance is kept in force. The date of injury or damage, however, must fall within the policy period. Protection under the policy extends only to liability to damaged parties and does not apply to damage to the work that was performed by the contractor. This exclusion can be relieved to some extent by a broad-form property damage endorsement applied to the completed-operations policy.

Another aspect of the usual completed-operations policy is that work accomplished by a joint venture is specifically not covered. If a contractor has been a member of a joint venture, it may wish to modify its completed-operations policy to cover any residual liability.

8.24 COMPREHENSIVE GENERAL LIABILITY INSURANCE

Because essentially all contractors require the same basic liability coverages, certain of these are commonly packaged together into a single comprehensive general liability policy. Practice varies somewhat as to just what coverages are included in such a policy. However, since most contractors require protection for liability arising out of their own operations, from their subcontractors, and from their completed projects, these coverages are frequently combined into a single comprehensive general liability policy. This policy also includes some limited contractual liability coverage. Despite its name, however, the comprehensive general policy is not all-inclusive in its coverage and there are many exclusions. The most important of these may be summarized as follows:

1. With minor exception, any liability the contractor assumes by contract.
2. Damage to property caused by blasting or explosion.
3. Collapse or structural injury of building or structure.
4. Damage to underground utilities.
5. Damage to property under the care, custody, or control of the contractor.
6. Personal injury.
7. Injury to the contractor's own employees.
8. Motor vehicle, watercraft, or aircraft liability of any sort.

The exclusions in items 2, 3, and 4 apply to property damage only and do not eliminate bodily injury liability coverage. All the items listed are insurable at additional cost, some as endorsements to the basic comprehensive general liability policy and some as separate policies. Contractors frequently purchase automobile liability insurance (see Section 8.31) together with their comprehensive general liability as a single policy from the same insurance company.

Most bodily injury and property damage liability insurance is written on an "occurrence" basis as defined in the policy. However, some policies are worded to protect against "accidents," and there are important differences between the two. Where occurrences are the policy basis, protection is afforded against injurious exposure as well as accidental injury or damage. The inclusion of injurious exposure is important to the contractor. If policy coverage were restricted to include only losses caused by accident, past court interpretations of this term would require that the event causing the loss be identifiable as to time and place. In other words, the event would usually have to be sudden to be covered. Policies written on an occurrence basis cover claims arising from conditions that may exist over a period of time. For example, damage caused by pile driving, leaking pipes, or dust would not ordinarily be construed as having been caused by accident. They are, however, covered under injurious exposure. Coverage is, therefore, broader when written on an occurrence basis because there is no requirement that the incident be sudden, only that it be undesigned and unexpected. Occurrence also pertains to an event resulting in bodily injury or property damage neither expected nor intended by the insured. Thus if a loss occurs, failure to correct the condition which caused the loss may permit the insurer to deny coverage for a second accident on the grounds that it logically could have been expected.

In the comprehensive general liability policy, the contractor's public liability and contractor's protective liability cover losses that occur during construction operations. However, upon completion of a project, these two coverages terminate, except for a loss that might occur because of the existence of or the hauling away of unused or scrap material remaining on the job-site after completion of construction operations. When these liability insurances cease is not always clear, although policies typically provide that field operations are completed when all work to be performed by or on behalf of the contractor has been finished or when the work has been put to its intended use. In general, therefore, termination of these two coverages depends more on physical completion of the work than on acceptance by the owner. In any event, the completed-operations insurance takes over at this general point in time. Because all three of these coverages are written by the same insurer, there is no conflict regarding when one ceases and the other begins.

Additional liability coverages that are of interest to the contractor and which are excluded from the basic comprehensive general liability policy are discussed in the following sections.

8.25 CONTRACTUAL LIABILITY INSURANCE

The liability coverages discussed to this point protect the contractor only with respect to its liability as imposed by law. In many instances in construction, however, the contractor, by terms of a construction contract, purchase-order agreement, or other form of contract, assumes the legal liability of another. This form of liability is called contractual liability and refers to the contractor's acceptance by contract of another party's legal responsibility. The other party may be the owner, the architect-engineer, third-party beneficiaries, or a material dealer.

There is an exclusion in the comprehensive general liability policy to the effect that the policy does not apply to liability assumed under a contract, except a contract defined in the policy. These defined contracts are (1) lease of premises, (2) easement agreements, (3) indemnification of a municipality required by ordinance, (4) sidetrack agreements, and (5) elevator or escalator maintenance agreements. Contractual liability coverages such as purchase-order agreement liability and hold-harmless clauses are obtainable as specific endorsements to the general policy in consideration for an additional premium. In addition, blanket contractual liability insurance that automatically covers all contractual liability is usually available from most insurance companies. In this regard, however, the contractor must beware of blanket contractual liability endorsements that exclude broad-form indemnity agreements.

Routine purchase-order agreements can impose serious contractual liability obligations on the contractor. For example, the purchase orders used by some transit-mix concrete companies contain a clause whereby the contractor agrees to hold harmless the concrete dealer from all liability that may arise out of any accident involving the transit-mix trucks or their drivers while on a jobsite. The contractor, on signing such a purchase order, assumes a legal responsibility normally belonging to the dealer. In the absence of contractual liability coverage, the contractor will be unprotected should such a loss occur.

As previously discussed in Section 6.31, the growing incidence of third-party suits against owners and architect-engineers has led to the widespread inclusion of indemnity clauses in construction contracts. By the terms of such hold-harmless agreements, the contractor assumes a liability that is not legally its own but that of another. This is contractual liability, and the contractor is not protected unless blanket contractual liability coverage is procured or each such clause is individually insured. This may be done either by a special policy or by endorsement to the comprehensive policy. For the purpose of rating hold-harmless agreements, the contractor must usually provide its insurance company with a copy of the contract section under which the hazard is assumed.

Contractual liability policies often have a number of important exclusions. A common exclusion pertains to bodily injury or property damage occurring within 50 feet of railroad property. Third-party beneficiary losses, explosion, underground damage, collapse, and other such losses may be exempted from the policy.

8.26 THIRD-PARTY BENEFICIARY CLAUSES

Contractual liability insurance policies usually state that the coverage does not apply to an action on a contract by a third-party beneficiary arising out of a project for a public authority. A third-party beneficiary contract is a contract where two parties enter into an agreement whereby one of the parties is to perform an obligation for a third party. In such a contract, if the obligated party does not perform as promised for the benefit of the third party, the third party can enforce his rights under the contract even though he is not a party to the contract. To illustrate this point, contracts with public agencies may require that the contractor be responsible for all damage to property, however caused by the construction operations. Such clauses often refer specifically to blasting. This is the natural result of the desire of governmental bodies to protect adjoining property owners from the hazards of construction operations. However, when the contract wording makes the contractor assume direct liability to third parties, a citizen property owner can assert rights as a third-party beneficiary even though he is not a party to the contract and even though the contractor was not negligent in its operations. In other words, an owner whose property has been damaged by the contractor's operations can sue for damages on the basis of breach of contract without having to prove contractor negligence. This contingency is not covered under usual comprehensive liability policies even when endorsed with the customary contractual liability coverage. Third-party beneficiary coverage can be added for an additional premium, usually on a specific exposure basis.

It is of interest to note that special provisions are now being included in some public construction contracts to limit the risk of third-party beneficiary suits. Called "no third-party liability" clauses, they provide that the contracting parties do not intend to make the public or any member thereof a third-party beneficiary under the contract, or to authorize anyone not a party to the contract to maintain a suit for bodily injury or property damage pursuant to the terms or provisions of the contract. The intent of such a clause is, of course, to remove the construction contract as a vehicle for third-party suits and to ensure that the duties, obligations, and responsibilities of the parties to the contract remain as imposed by law.

8.27 XCU EXCLUSIONS

The XCU exclusions refer to the hazards of explosion (X), collapse (C), and underground damage (U). Exclusion X excludes property damage arising out of blasting or explosions other than the explosion of air or steam vessels, piping under pressure, prime movers, machinery, or power-transmitting equipment. Exclusion C excludes property damage arising out of the collapse or structural injury to a building or structure caused by grading, excavating, filling, backfilling, tunneling, pile driving, cofferdam or caisson work, or by the moving, shoring, underpinning, or demolition of a building or structure. Exclusion U excludes property damage to wires, conduits, pipes, mains, sewers, tanks, tunnels, and any appurtenance in connection therewith beneath the surface of the ground or water, if such damage is caused by mechanical equipment used for grading, paving, excavating, drilling, filling, backfilling, or pile driving.

All three of the XCU exclusions do not necessarily pertain to every category of work. Different classifications of work are subject to one, two, or all three of the exclusions as indicated by the general comprehensive policy. The XCU exclusions can be waived for an additional premium.

The XCU exclusions delete only property damage from the basic liability policy, not bodily injury liability. In addition, the exclusions do not apply to the contractor's protective or completed-operations insurance. As long as the work associated with excavation, blasting, underpinning, pile driving, and similar operations is subcontracted, the prime contractor is protected by the subcontractor's insurance, assuming the subcontractor purchases the XCU endorsement with its general liability policy. Should the subcontractor's insurance prove to be inadequate or faulty, the prime contractor is then protected under its contractor's protective policy. If the prime contractor does this type of work with its own forces, the necessary special XCU coverage must be purchased. This form of protection can be obtained on a project or blanket basis.

At this point a few comments concerning a contractor's liability for damages caused by its blasting operations are in order. The law on contractor responsibility for blasting damage varies in different jurisdictions. Absolute or strict liability is imposed in many areas, although some require a showing of negligence. Recent years have witnessed a decided trend toward expanding the range of a contractor's liability for blasting and toward the increased application of the doctrine of strict liability. The application of absolute liability makes the contracting firm responsible for blasting damages, regardless of fault or whether it used all normal precautions and followed generally established procedures. The general rule, however, has been to prohibit recovery for damages unless those damages were caused by the concussion of the immediate blast or the injury was accompanied by physical trespass such as flying debris.

8.28 CARE, CUSTODY, OR CONTROL

A contractor's comprehensive general liability insurance contains the following exclusions:

1. Property owned, occupied by, or rented to the contractor.
2. Property used by the contractor.
3. Property in the care, custody, or control of the contractor.
4. Property over which the contractor for any reason is exercising physical control.

This exclusion means the comprehensive policy does not protect the contractor for damage done to another person's property while the contractor is working on it or with it as well as excluding coverage for damage to the insured's own property. This is consonant with the general rule that liability insurance is not intended to cover damage to one's own property. The most troublesome of these exclusions is the care, custody, or control provision and how it applies to construction projects.

The effect of these exclusions can be eased somewhat by the use of a broad form of property damage endorsement that restricts application of the exclusion to some extent. Actually, this endorsement uses more explicit language with respect to the exclusions, thereby actually covering more damage situations and providing somewhat broader protection for the contractor. Although it does not completely eliminate the care, custody, or control exclusion, it does cover more claims.

It can be readily seen that much of the property damage exempted in the comprehensive general policy is normally insured by other standard property policies. Builder's risk, equipment floater, installation floater, and other types of policies cover much of the property excluded in the liability coverage.

8.29 PERSONAL INJURY

Bodily injury, as used in insurance policies, refers to physical injury to the body. Personal injury refers to intangible harm, and an endorsement to the general comprehensive policy is available that covers personal injury liability. There are a number of reasons why this type of insurance coverage can be important to the contractor.

Personal injury liability insurance protects the contractor from liability it may incur for (1) false arrest, malicious prosecution, willful detention, or imprisonment; (2) libel, slander, or defamation of character; and (3) wrongful eviction, invasion of privacy, or wrongful entry. Such protection might be needed, for example, if the contractor caused the arrest of someone it suspected of theft or damage to its property and was sued for false arrest. It could happen that a contractor's watchman might detain someone in the course of his duties. A personal injury liability policy is normally written subject to the exclusion of the contractor's own employees, but the policy can be made to include them.

8.30 OWNER'S LIABILITY INSURANCE

The owner is responsible for procuring his own general liability insurance that applies to his normal operations and that protects him from liability that may arise because of negligent acts of himself or his employees. He may also choose to obtain owner's protective liability coverage, either as a part of his general liability policy or by having the protective coverage furnished by the general contractor.

The protective liability insurance protects the owner from injury or damage claims caused by the operations of the general contractor or any of the subcontractors. Despite the fact that the general contractor and the subcontractors are directly and legally responsible for liabilities arising out of their operations, the owner is often made a party to legal actions arising from acts or omissions connected with their construction activities. To protect the owner from this source of liability, owner's protective liability insurance is available. This insurance, with the owner as the named insured, covers his contingent liability for personal injury, including death, or property damage that may occur during the construction operations of independent contractors and subcontractors. This policy also pays legal expenses associated with the owner's defense. The owner's protective policy protects him against liability imposed by law arising out of construction operations as well as covering any liability he may incur as a result of his direction or supervisory acts in connection with the work being performed or through the omission of a duty that cannot be lawfully delegated to an independent contractor.

Where the contractor is made responsible for obtaining protective liability insurance for the owner, this insurance is normally obtained as a separate policy although, in some instances, the owner can be added as an additional insured on the contractor's comprehensive general liability policy. When a construction contract requires that the contractor provide protective liability insurance for both the owner and the architect-engineer, both parties can be named insureds on the same protective policy for an extra premium. If the construction contract contains an indemnity clause in favor of the owner and also requires that the contractor provide protective liability coverage, there is some duplication of coverage. The protective liability coverage is considered to be the primary policy, which entitles the contractor to a substantial reduction in the premium for the contractual liability coverage on that project.

8.31 AUTOMOBILE INSURANCE

The operation of automobiles exposes the contractor to two broad categories of risk, insurance for which can be written as separate policies or can be combined into a single policy. One form of risk is loss or damage to the contractor's own vehicles caused by collision, fire, theft, vandalism, and similar hazards. The other form of risk is liability for bodily injury to others or damage to the property of others caused in some way by the operation of the contractor's automobiles. Many states have statutory requirements concerning the purchase of liability insurance by owners of motor vehicles. In an insurance context, automobiles or vehicles are construed to include passenger cars, trucks, truck-type tractors, trailers, semi trailers, and similar land motor vehicles designed for travel on public roads.

Automobile insurance can be written to include bodily injury liability, property damage liability, and various forms of protection for passengers and the contractor's own vehicles. When a contractor owns more than a given number of vehicles (commonly five), he is eligible for a fleet rate, which means a reduction in overall premium cost. Premium savings can also be achieved by purchasing only fire and theft rather than comprehensive protection for vehicles that are more than some arbitrary age, such as four years old. If only a few motor vehicles are involved, collision coverage is often purchased. However, many contractors with large fleets of vehicles prefer to self-insure the collision exposure because of the high premium costs involved. Collision insurance customarily provides for a specified deductible for each collision loss occurrence. This deductible amount can be made to vary, correspondingly changing the premium rate charged. In view of recent large awards in cases of bodily injury and the claim consciousness of the public in general, the contractor must maintain high limits of motor vehicle liability coverage. As with other liability policies, the premium costs do not increase proportionately with policy limits, and the loss possibilities are very real. As has been previously mentioned, automobile liability insurance is commonly written in conjunction with a contractor's comprehensive general liability policy.

There are endorsements the contractor should consider when purchasing its motor vehicle insurance. One is a nonownership endorsement that provides protection when an employee uses his private automobile in going about the contractor's business. In this regard the contractor should not only provide the nonownership coverage but also include the employee on its policy as a named insured. Another endorsement provides hired-vehicle liability insurance that covers rented units. If the rented vehicles are already covered by automobile liability insurance, it is possible for the contractor to receive a reduction in premiums for its hired-vehicle coverage.

8.32 CONSTRUCTION EQUIPMENT INSURANCE COVERAGE

Insurance coverage for liability arising out of the operation of a contractor's construction equipment and for damage or loss to that equipment is provided in a collective way by the contractor's automobile, comprehensive general liability, and equipment floater policies. The dividing lines between these policies, insofar as equipment is concerned, are not always clear, and special arrangements can be and are often made to accommodate a contractor's specific situation.

In general, the equipment coverages of each of the three policies involved are about as follows. The automobile policy provides liability and property loss insurance for passenger cars, trucks, and similar vehicles designed for travel on public roads. Excluded from this policy, however, is mobile equipment, which is defined as self-locomoting items (a) not subject to motor vehicle registration, or (b) used entirely on the contractor's premises, or (c) designed for off-highway use, or (d) used for the sole purpose of providing mobility to construction equipment such as pumps, power cranes, air compressors, generators, and drills.

For all equipment items, mobile and nonmobile, liability insurance is provided by the contractor's comprehensive general liability policy. This coverage applies while the equipment is in use on projects and while it is being moved or transported from one job to another. Coverage for damage or loss to the equipment itself is provided by the equipment floater policy. There is a transport clause in most equipment floaters that protects equipment while it is being moved, transported, hauled, or towed.

There can be many exceptions to the general coverages just presented, however. To illustrate, the motor vehicle policy can be made to provide liability coverage for equipment when it is being carried or towed by a contractor's truck. In the case of some licensed, mobile equipment items such as truck-mounted cranes, the automobile policy may provide liability protection while the crane is in transport and the comprehensive general liability policy will apply while the crane is operating on a jobsite. The contractor with a large equipment investment will do well to seek the services of an insurance agent or broker experienced in the construction equipment field.

Many accident possibilities involving motor vehicles and construction equipment represent borderline cases between the respective coverages of the three types of insurance mentioned in the previous paragraphs. Because of possible dispute concerning the coverages between policies, it is preferable that the contractor purchase all three types from the same insurance company. Such a dispute might arise, for example, if an accident were to occur during the loading or unloading of a piece of equipment onto or off of a contractor's truck.

8.33 THE PRINCIPLES OF WORKMEN'S COMPENSATION

Before the present era of workmen's compensation laws, an employer was obligated to protect his employees only to the extent of providing a safe place to work. To obtain redress at the common law, an injured employee had to file suit against his employer and prove that the injury was due to the latter's negligence. Available to the employer were the accepted common-law defenses of contributory negligence of the injured employee, assumption of risk by the injured employee, and negligent acts of fellow employees. Under the contributory negligence doctrine, the employee could not obtain redress if he were negligent to any degree, regardless of the employer's negligence. The assumption of risk doctrine denied recovery if the worker knew or should have known about the inherent risks. The fellow servant doctrine held the employer liable for his own actions but not for those of a fellow worker of the injured party. This trinity of common-law defenses made it difficult for a disabled worker to prove employer responsibility and negligence. The process was a slow, costly, and uncertain one at best for the employee.

The social and economic consequences of this problem were instrumental in the development of workmen's compensation laws. New York enacted the first state workmen's compensation law, which became effective in 1910. Since that time, compensation legislation has been passed by the federal government, every state and territory of the United States, and each dominion of Canada. Workmen's compensation statutes apply to most private employment and much public employment (federal civil service and the military are notable exceptions).

The underlying economic principle of workmen's compensation is that the costs associated with an on-the-job injury or death of an employee, regardless of fault, is an expense of production and should be borne by the industry. The expense incurred by employers in providing suitable protection for their employees is considered as another cost of doing business and, as such, is presumably reflected in the selling price of their products or services. The fundamental objective of the compensation statutes is to ensure that an injured workman receives prompt medical attention and monetary assistance. Such support is provided by the employer and involves a minimum of legal formality.

Another basic principle is the strict liability of the employer, regardless of any fault of the injured employee. Contributory negligence of the employee, such as failure to wear a company-provided safety helmet or to conform with posted safety regulations, will not usually affect the employer's liability. As a matter of fact, about the only exceptions to the payment of compensation benefits occur when the worker deliberately inflicts the injury on himself, the injury occurs when the worker is intoxicated, the injury is sustained in the course of committing a felony or misdemeanor, or the injury is intentionally caused by a coemployee or third person for reasons not associated with the employment.

Secondary objectives of compensation laws are to free the courts from the tremendous volume of personal injury litigation, to eliminate the expense and time involved in court trials, and to serve as an instigating agent in the development of effective safety programs.

8.34 WORKMEN'S COMPENSATION LAWS

Although all of the various workmen's compensation laws embody the same general principles, they differ considerably in their working details, and no two of them are exactly alike. The contractor must be especially careful to buy proper insurance and conform with the legal requirements of the workmen's compensation laws of each state in which it works. Every law makes certain exclusions to its coverage. For example, most of the acts exclude domestic servants, farm labor, casual employees, independent contractors, and workers in religious or charitable organizations. Businesses that employ less than a specified number of employees are exempted in some of the states. Interstate railway workers and maritime employees are not covered by the compensation acts but are protected by separate federal legislation.*

Some compensation laws are designated as compulsory and others as elective. Every employer whose employees are covered by a compulsory law must accept the act and provide for the benefits specified. Failure to comply with the prescribed provisions can result in severe penalties, the payment of damages to injured workmen, and possible imprisonment. In areas in which the law is elective, employers have the option of either accepting or rejecting it. However, if an employer elects to remain outside the legislation, he risks an injured worker's civil suit for damages and simultaneously loses the three common-law defenses: assumption of risk, negligence of fellow employees, and contributory negligence. In effect, this means that all compensation laws are compulsory. In most states the presumed-acceptance principle applies, whereby in the absence of specific notice to the contrary, the employer is presumed to have accepted coverage by the compensation act. Workers in excepted or excluded employments may usually be brought within the act through voluntary action by their employers.

An injured employee who files a claim for and accepts assistance under a workmen's compensation statute ordinarily forfeits the right to sue his employer for damages. Compensation statutes usually provide that a workmen's compensation claim is the only remedy the employee has against his employer. According to the law, the employer is strictly liable and the injured employee is guaranteed benefits as prescribed by law. In return the employer avoids the danger of unpredictably large jury verdicts for bodily injury caused by employer negligence. States vary as to whether an injured employee or the workmen's compensation insurance company can sue third parties who may have caused the injury. In most states, an injured construction worker can recover from the employer's workmen's compensation insurance and then sue a third party who caused the injury. In some areas, the worker cannot collect further compensation from "fellow employees." This term is considered to include other contractors and other workers on the project. However, several states do allow coemployee suits, some of these only when the injury involved an intentional act, intoxication, a reckless act, willful and malicious conduct, unprovoked aggression, gross negligence, or when the accident involved a motor vehicle. In most states, however, an injured employee has considerable latitude in suing third persons for negligence. This contributes appreciably to the number of third-party suits filed against owners and architect-engineers. To be successful, of course, the injured employee must demonstrate in some way that the party sued was negligent and that this negligence caused or contributed to the injury.

All the states, territories, and the District of Columbia have enacted child

---

*Contractors who own watercraft, operate wharf facilities, or engage in marine construction sometimes employ workers covered by the Longshoremen's and Harbor Workers' Compensation Act or the Jones Act for Seamen. Workers covered by these federal statutes are not ordinarily protected by workmen's compensation statutes, and the usual workmen's compensation insurance does not apply. Separate insurance or an endorsement to the regular workmen's compensation policy is needed in such cases.

labor laws that regulate the conditions under which minors may be employed. Workmen's compensation laws cover legally employed minors. In some jurisdictions, double compensation or added penalties are provided in cases of injury to illegally employed minors. Minors also enjoy special benefit provisions.

8.35 ADMINISTRATION OF WORKMEN'S COMPENSATION LAWS

Workmen's compensation laws are generally administered by commissions or boards created by law. A few states provide for court administration. Statutory provisions relating to administration vary somewhat from state to state, but each of the laws contains certain regulations pertaining to its implementation. Notice to the employer of the injury by the injured workman is required within stipulated time limits, although this requirement is excused for cause. A claim must be filed by the injured workman within a statutory period. Such claims are normally settled by agreement subject to approval of the administrative body.

Review and appeal of the compensation award to the injured worker, as well as regulation of attorney fees, are provided for by the various acts. Requirements vary concerning the keeping of accident records by the employer, but all states require that the employer report injuries to a designated authority. Failure to report in accordance with the applicable statute can result in fines and, in some jurisdictions, even imprisonment. Except for preliminary reports, the contractor's insurance company usually makes the formal reports required by law except in states with monopolistic funds (see Section 8.38).

Benefits to an injured worker are only those provided by the law and approved by the administrative authority. Should the injured party not be satisfied with the award as specified by law, he can appeal to the court designated by the act. Such appeal must be initiated within a statutory period. In some jurisdictions the employee can file suit against his employer under the compensation act for alleged failure to provide safety devices.

8.36 WORKMEN'S COMPENSATION BENEFITS

All workmen's compensation laws provide various forms of benefits for the disabled employee or his dependents for job-connected injury, death, or sickness regardless of how the disability or death may have been caused. The monetary value of such benefits varies substantially from one state to another. These benefits include medical treatment, hospitalization, and income payments for the worker during his disability. Also provided are death benefits for the worker's dependents. Many jurisdictions include special benefits such as a lump-sum payment for disfigurement, rehabilitation services, and extra benefits for minors injured while illegally employed. Most statutes now stipulate that medical benefits shall be provided without limitation as to maximum total cost and that wage compensation shall be paid for as long as the worker is disabled. In those states where benefits are limited by statute, either in terms of dollars or period of time, the contractor may purchase an endorsement to its workmen's compensation insurance policy that provides "extra-legal" or additional benefits for its injured employees. Most states require a waiting period before a worker can collect certain benefits, seven days being typical. However, these states normally allow a worker who has been off the job for a specified length of time to receive benefits for the waiting period retroactively.

Four classifications of work injuries are used in conjunction with workmen's compensation benefits. These are (1) temporary-total, (2) permanent-partial, (3) permanent-total, and (4) death. Other than medical assistance, the usual benefits provided by workmen's compensation laws are quite modest, especially when compared with the worker's usual earning capacity. The idea is, of course, to give the worker sufficient monetary help so he will not become a burden to others.

The great majority of compensation cases involve injuries of the temporary-total classification, meaning the worker is unable to work temporarily but ultimately recovers fully from his injuries and returns to employment. Income benefits payable during the period of convalescence are determined as a percentage of the worker's average wages. Most states limit the minimum and maximum benefits payable weekly, as well as the number of weeks and the total dollar amount of benefit eligibility.

Permanent-partial disability connotes a continuing partial disability, although the worker is eventually able to return to work. This form of disability is classified either as a schedule injury, meaning the loss or loss of use of a thumb, eye, leg, or other member of the body; or a nonschedule injury, which is of a more general nature. Compensation acts provide additional benefits for schedule injuries.

A permanent-total disability injury prevents the worker from future employment in his former craft. Most of the state workmen's compensation laws provide that specified benefit payments shall be made for life in cases of permanent-total disability. The other statutes limit the benefits as to time, amount, or both. Rehabilitation services are utilized to train disabled people in new skills or occupations.

In the event of the accidental death of a workman, all states provide for the payment of death benefits to his family or other dependents. A few acts provide for payment of benefits to the widow for life or until remarriage, and to children until they reach a prescribed age. Most states place a limitation on the time period or total amount of such payments.

8.37 ADDITIONAL PROVISIONS OF WORKMEN'S COMPENSATION LAWS

Workmen's compensation statutes now include occupational diseases within their coverage. Provisions vary, but compensation benefits are generally the same as for other forms of disability. Many states have provided for extended periods of time during which claims may be filed as a result of certain latent, slowly developing occupational diseases. Some states have special provisions regarding silicosis, asbestosis, radiation disability, and occupational loss of hearing caused by noise.

Second-injury or special-disability funds have been developed to meet problems arising when an employee, previously injured, suffers a second injury that, together with the first, results in a combined disability much more severe than that caused by the second injury alone. Under ordinary circumstances the employer in whose employ the second injury took place must provide benefits as dictated by the total resulting disability. This condition has sometimes made employers reluctant to hire previously injured persons. Most states have now provided that the employer is responsible for payments required by the second injury only. Additional compensation to the injured employee as called for by the combined effects of his injuries comes from second-injury funds, which were created at the time the second-injury provisions were enacted into law.

8.38 WORKMEN'S COMPENSATION INSURANCE

A workmen's compensation statute requires the contractor to provide its injured workmen with all benefits as may be required by that state's law. The usual way in which a contractor does this is through workmen's compensation insurance, although almost all states allow the contractor to carry its own risk as a self-insurer if it can provide satisfactory evidence of its financial ability to do so. Monopolistic state funds have been established in a few states and the provinces of Canada. Under the laws of these jurisdictions, employers whose operations are covered by their compensation laws are required to insure in the state funds, although in some instances employers can qualify as self-insurers. Other states have competitive state funds, whereby the employer may purchase compensation insurance either from a private insurance carrier or a state fund. In the remaining states private insurance companies provide all workmen's compensation insurance.

Workmen's compensation insurance is unlimited in the policy and will pay the medical costs and provide the benefits required by law. This insurance also provides legal defense for the insured and pays any court awards in jurisdictions in which an injured worker can bring suit against his employer under the compensation act. If injury to proprietors, partners, or corporate officers is to be covered, a specific policy endorsement to that effect may be required.

The contractor must have workmen's compensation insurance in force for each state in which it works. To provide protection in those states where the contractor has no planned operations at the time its workmen's compensation policy is issued by the carrier, an "all-states" endorsement is available to extend coverage to any state except the monopolistic fund states. The broad form of this endorsement is preferable because the insurance company then stands in the insured's place rather than just reimbursing any loss. In the monopolistic fund states, separate policies must be purchased.

8.39 WORKMEN'S COMPENSATION INSURANCE RATES

The manual rates for workmen's compensation insurance are set by rating bureaus of the various states and are adjusted annually. In almost all states, premium costs are computed by multiplying each employee's total wages by the rate specified for his classification. In a few states, there is a $300 per week limitation. In these states, any excess of wages over $300 per week is not subject to assessment. Auditors of the insurance company conduct periodic audits of the contractor's payroll records to verify that sufficient premiums have been paid.

Premium rates for compensation insurance vary considerably from state to state and with the classifications of work involved. Since a higher premium rate is charged for the types of work that entail greater risks of injury, it pays the contractor to show accurately the work classifications its employees are actually performing. To cite an example, the following is an instance of base rates for compensation insurance per $100 of wages for an ironworker:

STEEL ERECTION

Doors and sash	$5.21
Interior ornamental	5.21
Structural	30.15
Dwelling, two stories	18.46
General	11.72

Should the superintendent or payroll clerk carelessly classify ironworkers as doing structural work when they actually are employed in one of the other categories, the contractor could be paying almost six times too much for this portion of its compensation insurance. The insurance company can give the contractor valuable tips on how to set up its payroll records so that no more premium payments are made than are necessary.

8.40 WORKMEN'S COMPENSATION INSURANCE RATING PLANS

There are two ways in which the annual premium the contractor pays for its workmen's compensation insurance is determined. Referred to as "rating plans," one format is the traditional guaranteed-cost plan and the second is the retrospective plan. The former is more commonly used and will be discussed first.

Workmen's compensation insurance companies have become active participants in the field of construction safety and have established merit systems whereby contractors are rated according to their accident experience. Contractors are experience-rated by various rating bureaus, according to which those contractors with good safety records are rewarded by reductions in their compensation rates, and those with poor records are penalized through rate increases. Such experience modification operates so that an employer's past loss experience is a determining factor in his present and future workmen's compensation costs. The higher the ratio of chargeable claim costs paid by the insurance company to contributions received from a given employer, the higher his premium rates are. Under this plan, workmen's compensation insurance premiums are determined by multiplying the payroll by the appropriate classification rates and then applying the contractor's individual experience modifier. This rating modifier is the means by which the contractor's insurance cost is adjusted up or down according to its loss experience. A volume discount is applied regardless of insurance losses. This discount rate increases as the amount of the annual premium increases.

Workmen's compensation insurance, as well as other liability coverages, is also sold on a retrospective basis. When the retrospective rating plan is used, the contractor pays basic premium rates during the life of the policy. Retrospective plans can be written for one or three years or for the duration of one particular project. The insurance carrier normally evaluates the contractor's losses under the policy and adjusts the premium periodically. The contractor receives a rebate if its loss experience was good and is required to pay additional premium if its loss experience was unfavorable. Compensation insurance sold on a retrospective basis stipulates both a minimum and a maximum premium rate the contractor must pay under the policy. Several different retrospective plans are available, each plan specifying a different set of minimum and maximum premium rates and providing for different possible ranges of final premium adjustment. Retrospective-rated insurance is truly a cost-plus form of insurance subject to a guaranteed maximum price but is not suitable for all contractors. Expert advice is highly desirable. As a usual rule, larger contractors with good safety programs stand to gain most from this arrangement.

8.41 WORKMEN'S COMPENSATION SELF-INSURANCE

A large majority of the compensation acts allow an employer to act as his own insurer, provided he can satisfy certain minimum financial requirements as stipulated by the various state insurance departments. Self-insurance is limited in a practical sense to big companies having such a large spread of risks that they can assume their own liability on workmen's compensation to their financial advantage. For complete self-insurance, the contracting firm must establish its own services of claim adjustment, claim investigation, safety engineering, and others similar to those furnished by insurance companies. To qualify as a self-insurer with state officials, a contractor may be required to furnish a surety bond in an amount fixed by law or the administrative agency.

Self-insurance programs have been devised between employers and insurance companies whereby the employer is self-insured up to certain maximum amounts, payments in excess of which are guaranteed by excess insurance purchased from the insurance company. Under such plans, the employer deposits a percentage (75 percent, for example) of the usual workmen's compensation insurance premiums in a bank. This establishes a fund for the payment of workmen's compensation benefits. The remainder of the premium (25 percent, in our example) is paid to the insurance company. This payment goes in part to purchase the excess insurance coverage. The remainder of the payment to the insurance company provides the contractor with the usual insurance company services pertaining to claims, inspections, engineering, auditing, accident reports, and medical and legal services.

8.42 EMPLOYER'S LIABILITY INSURANCE

Employer's liability insurance is written in conjunction with workmen's compensation insurance and affords the contractor broad coverage for bodily injury or death of an employee arising out of or occurring in conjunction with his employment but not covered by the workmen's compensation policy. Workmen's compensation laws, which provide that an injured employee shall receive specified benefits from the employer or the employer's insurer, concomitantly deny the employee the right to sue the employer at common law for an injury under the act. Because the employee is compensated for the injury, it is impossible in most jurisdictions for the injured employee to bring suit against his employer directly.

There are instances, however, when an injury to an employee may fall outside the coverage of workmen's compensation. For example, if the employee is injured through the failure of the contractor to provide safety appliances or working conditions required by state law, the employee may elect to sue the contractor for damages under common law. The employee may be entitled to recover damages from the employer if it can be proved that the latter's negligence caused the injury while the employee was eating lunch or performing other similar acts. The contractor may have an employee injured on a minor operation in another state where the contractor does not have workmen's compensation insurance in effect. An injured employee of the contractor may collect workmen's compensation benefits from the employer's insurance company and then file suit against a third party such as the owner or a subcontractor. In this event, it is usual for the owner or subcontractor immediately to make the contractor a party to the action. In each of the instances just cited, workmen's compensation insurance provides no direct protection for the contractor. However, employer's liability insurance will pay the legal costs incurred in the contractor's defense as well as any judgment, up to the face amount of the policy. For temporary exposure in a monopolistic fund state, the contractor can extend its employer's liability insurance by a "stop gap" endorsement. The purpose of this endorsement is to provide protection from suits brought under common law in these states.

The standard employer's liability insurance coverage, as provided by the present workmen's compensation rate structure, is $100,000 for each accident. The premium for this basic amount is included in the regular workmen's compensation insurance premiums. However, higher limits can be obtained by the payment of an additional premium.

8.43 UMBRELLA EXCESS LIABILITY INSURANCE

Because of the substantial hazards that its operations present and the very high awards now occasionally made in bodily injury cases, the contractor may sometimes be concerned about the adequacy of the liability insurance it carries. A common way to eliminate this question is through the means of umbrella excess liability insurance.

Although the coverage of a policy of this type is not at all standard, it serves a threefold purpose. One of these is to cover liability claims that might otherwise fall between the coverages carried under the separate basic liability policies. A second purpose is to raise the policy limits of the contractor's existing liability insurance to some value high enough to provide protection against any forseeable loss or combination of losses. The umbrella coverage does not eliminate the existing policies, it merely provides substantially greater protection for the same hazards insured against under the primary coverages. If the contractor incurs a liability beyond the limits of a primary policy, then the umbrella coverage is invoked.

The umbrella policy also provides some coverages, beyond a deductible amount, that are not contained in the usual comprehensive general liability policy. To illustrate how an umbrella policy works in this instance, suppose a contractor has no collapse insurance but does have an umbrella policy with a $25,000 self-insured deductible. If the contractor's operations cause a collapse of an adjoining building, it pays the first $25,000 of the loss, and the umbrella coverage pays the remainder up to the face amount of the policy.

Umbrella policies contain a number of exclusions. Among these are intentional property damage, damage to property owned or controlled by the insured, workmen's compensation liability, nuclear-facility hazards, and war.

8.44 WRAP-UP INSURANCE

Wrap-up insurance refers to situations where the owner of a large construction project elects to furnish certain insurance coverages for the owner, architect-engineer, prime contractor, and all subcontractors. On such projects, the contractors are required to exclude project coverages from their usual insurance programs and accept the coverages which are provided by the owner. Although practice varies somewhat, the usual procedure is for a single insurance company to provide all of the workmen's compensation and employer's liability insurance, the comprehensive general liability insurance (excluding automobile coverage), and the all-risk builder's risk insurance for the entire project.

When the single-carrier scheme is used, one of two plans is followed. Under one form, the owner buys and pays for all the insurance required. The general contractors and the subcontractors bid on the project without including the cost of the insurance in their proposals. When the second plan is used, the owner designates one insurance company from which all contractors and subcontractors must buy the designated insurance coverages at their own expense. Under this form, usually called a "designated carrier" plan, each employer is issued his own workmen's compensation and employer's liability policy and his own comprehensive general liability policy.

The prime attraction of the wrap-up form of insurance coverage is the reduced cost to the owner, both in the direct cost of the premiums and in the saving of the contractor's and subcontractor's markup on the cost of the premiums when the owner buys and pays for the insurance himself. Lower premium rates are effected through volume purchase from a single carrier. Wrap-up insurance also eliminates much of the administrative detail involved when each contractor and subcontractor on a project buys its own insurance. Under these conditions, layer upon layer of hold-harmless clauses can increase the cost of the public liability insurance. There seems to be little question that on large projects the direct cost of insurance can be reduced through use of the wrap-up insurance concept. In addition, a single carrier can provide concentrated safety inspections and quick claims service.

However, there are serious drawbacks associated with the wrap-up scheme from the viewpoint of the contractor and the subcontractors. Because any insurance dividends revert to the owner when he buys such insurance, the contractor receives no benefit from having a good safety record on the project. On the other hand, if the contractor's record on that project happens to be poor, its experience rating rises. Under state rating systems, higher premium costs will then apply to all of the contractor's subsequent work, including projects where it buys its own insurance, until such time as its rating is again adjusted.

Wrap-up plans upset the normal business arrangements that contractors have with their usual brokers and insurance companies. The contractors must deal with a different set of safety engineers, underwriters, claims people, and auditors for each different insurer. Repeated audits of the contractor's accounting records by different insurance companies can seriously disrupt a contractor's normal office routine and add to administrative costs. Wrap-up coverage may make it difficult for the contractor to obtain, at a reasonable cost, insurance for risks that are not included in the wrap-up policy. An example of this could be insurance for the contractor's plant and equipment. Contractor organizations have proposed and supported, with some success, state laws that prohibit or restrict the application of the wrap-up concept.

8.45 NONOCCUPATIONAL DISABILITY INSURANCE

Disability benefit laws in a few states require most employers to provide insurance protection for their employees for disability arising from accidents or diseases not attributable to their occupation. Disability insurance pays a weekly benefit when an eligible wage earner is disabled by an off-the-job injury or illness. Workmen's compensation insurance obviously does not apply in such instances and does not provide any benefits. Usually, a waiting period of seven days must elapse before benefits begin, thus eliminating claims for minor disabilities. All state plans provide for maximum weekly benefits and a maximum number of weeks that benefits are paid.

In some states, nonoccupational disability insurance may be obtained through either a private or a state plan. In others, all disability insurance required by law must be acquired through the state. State-administered plans are supported by payroll taxes levied against both the employer and the employee. Private plans are acceptable, particularly if underwritten by a reputable insurance carrier. Benefits provided by private plans must at least equal those in the state plan, whereas contributions from the employee must be no greater. The cost of disability insurance usually is shared by employer and employee. Some employers, however, regard these benefits as essential to their employee relations programs and underwrite the entire cost.

In some areas, the contractor's legal obligation for providing disability benefit insurance to its employees can be satisfied by contributions to union welfare funds that provide disability benefits substantially equivalent to those required by state law. Protection for nonunion employees and for union members who are not covered by such union welfare funds must be acquired through other sources.

8.46 PROFESSIONAL LIABILITY INSURANCE

Professional liability insurance provides protection from liability arising out of acts or omissions of the insured in performing design and other professional services. When the contractor's responsibilities to the owner include professional as well as construction duties, professional liability insurance must usually be obtained. While this risk applies primarily to design–construct contractors and construction managers, other contractors can have varying exposure in this area. Professional liability may devolve to the design–build contractor in two ways. Direct responsibility may arise out of design performed by in-house architect-engineers. Contingent professional liability is possible where professional design services are subcontracted to an outside architect-engineer firm or under a design–construct contract to a subcontractor. Liability exposure of a construction manager arises out of its involvement in the feasibility and design phases of the project and the supervision of work forces other than those of the construction manager itself.

When professional liability insurance is obtained, it should include contingent professional liability coverage where design work is subcontracted to outside professionals or subcontractors. The total design liability exposure of the contractor is actually covered by a combination of professional liability insurance, the comprehensive general liability policy, the umbrella excess policy, and the builder's risk policy. However, this matter can be complicated by the fact that the usual comprehensive general liability, umbrella excess, and builder's risk policies may specifically exclude professional liability protection. Insurance protection for contractors in the area of design liability is a very complex matter, and the contractor should seek expert advice in this area.

8.47 INSURANCE CLAIMS

Every loss or liability for which the contractor's insurance may be responsible must be brought to the attention of the insurance carrier and perhaps other parties through the medium of a written notice or report. State workmen's compensation statutes require that accident reports be submitted covering all compensable accidents to employees. In the event of accidental injury to a person not an employee, the contractor should, for its own protection, submit a complete report of the matter to the company carrying its comprehensive general liability insurance. Motor vehicle accidents must usually be reported both to the insurance company and to law enforcement agencies on special forms provided. Property damage is reported on a proof-of-loss affidavit, although final settlement is usually made on the basis of a detailed schedule of costs.

To ensure that claims are properly made and followed up, it is wise for the contractor to designate one person in its organization to assume complete responsibility for matters pertaining to insurance. In addition to the filing of reports and claims, this individual must be familiar with all aspects of the various insurance policies of the contractor and of the subcontractors. This person should keep a checklist of all coverages, have all new or potential construction contracts examined for insurance requirements, make necessary cancellations and renewals, keep a file of subcontractors' insurance certificates, and generally oversee the contractor's insurance program.

8.48 SUBCONTRACTORS' INSURANCE

By the terms of its subcontracts, the prime contractor normally requires each subcontractor to provide and maintain certain insurance coverages. Article 5 of Appendix L illustrates this requirement. For the most part, prime contractors look to their subcontractors for the same coverages and limits as required by the construction contract with the owner. The insurance carried by its subcontractors is a serious matter to the prime contractor. The subcontractors' coverages play an important role in the total insurance protection of the prime contractor and often, either of itself or in conjunction with the prime contractor's own insurance, provide direct protection for the prime contractor.

In cases where the insurance coverage of a subcontractor proves to be faulty or inadequate, responsibility can devolve to the prime contractor from the subcontractor. In addition, when the subcontractor does not provide insurance coverage required by law, such as workmen's compensation insurance, the general contractor may be made responsible. To illustrate, most of the states have what are called "subcontractor-under" provisions in their workmen's compensation statutes. The effect of this is that a prime contractor (or subcontractor who further subcontracts) is considered to be the statutory employer of the employees of any subcontractor that does not procure the required workmen's compensation insurance. In such an instance an injured employee of the subcontractor may be able to recover under the insurance of the prime contractor. Construction subcontracts normally provide that, if the subcontractor does not procure the insurance required, the prime contractor has the right to obtain such insurance for the subcontractor and to charge the account of the latter with the premium cost involved.

8.49 CERTIFICATES OF INSURANCE

An insurance certificate is a printed form executed by an insurance company certifying that a named insured has in force the insurance designated by the certificate. Such certificates are addressed to the party requiring the evidence of insurance and list the types and amounts of insurance the insured has purchased. These certificates note the expiration dates of the policies and contain a statement to the effect that the party in whose favor the certificate is drawn will be informed in the event of cancellation or change of the insurance described.

The general contractor is often required to submit certificates of its insurance to owners and other parties. Construction contracts usually contain the provision that the contractor shall submit suitable insurance certificates to the owner or the architect-engineer at the time the contract is signed or, at least, before field operations are started. Many building departments require proof that certain insurance is in effect before they will issue building permits.

It is standard practice for a general contractor to require suitable and up-to-date insurance certificates from all its subcontractors. A properly maintained file of such certificates will ensure that each subcontractor provides and maintains the insurance coverages and amounts required by law and subcontract. Many general contractors have a company policy that subcontractors cannot start their field operations until suitable certificates of insurance have been filed with the general contractor.

8.50 SOCIAL SECURITY

Social Security, operated by the federal government through the Social Security Administration, provides four basic types of benefits for the workers covered. Monthly cash payments are provided after a wage earner reaches a certain age and retires. When a qualified worker reaches the age of 65, hospitalization benefits are provided and supplementary medical insurance is available on application and payment of a small monthly premium. Survivor benefits are provided for the dependents of a worker should that worker die at any age. Benefits are made available to a worker who suffers a disability that renders him unable to do any substantial gainful work for which he is qualified by age, experience, training, and education for a period of 12 calendar months or longer. The Social Security Administration also administers the supplemental security income program that provides financial assistance to people who are blind, disabled, or who are 65 years of age or older and who can establish a genuine financial need. Since the passage of the original Social Security Act in 1935, benefits have been liberalized, and the number of workers covered has been increased substantially.

Employees and employers share the cost of social security by paying special taxes into a fund in the U.S. Treasury, out of which benefits are paid. The employer must contribute for each employee the same amount that is deducted from the employee's pay. The tax rates paid by employer and employee are statutory, as is the annual amount of wages subject to the tax. Both the tax rates and taxable earnings have been increased several times by congressional action.

The Social Security Administration keeps a record of each worker's wages received while in employment covered by the act. This record is maintained as a separate account for each worker, under his name and an identifying number. Each individual must have his own social security number, which he can obtain from any local social security office. Any worker can check on his social security account by writing to the Social Security Administration and asking for a statement of his wage credits.

8.51 UNEMPLOYMENT INSURANCE

Unemployment insurance is a hybrid coverage, involving both federal and state laws, that provides weekly benefit payments to a worker whose employment is terminated through no fault of his own. Each state has some form of unemployment compensation law that works in conjunction with the Federal Unemployment Tax Act. Also, each state has established its own administrative agency which works in partnership with the Bureau of Employment Security, an agency of the U.S. Department of Labor. The federal government sets minimum benefit standards for the states. Each state specifies its own qualifications, the amount of the benefit, the duration of payments, and the employment covered. Excluded from most state laws are railroad workers; domestic workers; federal, state, and municipal workers; workers in nonprofit educational, religious, or charitable organizations; agricultural workers; casual labor; and those who are self-employed.

Unemployment insurance is intended to provide workers with a weekly income to assist them during periods of unemployment. There is no intent to benefit those who cannot or will not work. Only persons who have been working for a specified period of time on jobs covered by their state unemployment compensation law, who are able and willing to work, and who are unemployed through no fault of their own are eligible to receive benefits. However, some states do provide unemployment benefits under certain circumstances to workers on strike.

All employers who come under the provisions of the state unemployment insurance laws must pay taxes based upon their payrolls up to a prescribed amount per calendar year for each employee. Employers pay all costs of the program in all but a few states. In these few, employees must also pay state unemployment taxes, although an employer in any of these states can elect to pay the employees' shares as well as the employer's own. The tax applies to a contractor's entire work force, including both field and office employees. The employer pays a part of such payroll taxes to the federal government and a larger share to the state in which the employment takes place. The federal law imposes on all covered employment a basic payroll tax whose rate is variable from year to year, depending on the financial condition of the national fund. With certain restrictions, the employer can take credit against his federal unemployment tax for his payments to the state fund and for amounts that he is excused from paying to the state because of his favorable claim experience.

Each state has an established tax rate which is adjusted up or down for a given employer, depending on his experience. A separate account is set up by the state for each employer covered. Each account is credited with tax payments made and charged with benefits paid to former employees. The extent to which the tax credits exceed the benefits charged determines by how much the employer's rate is adjusted above or below the basic tax rate. A contractor who finds it possible to offer a maximum amount of steady employment to its workers can enjoy a substantial reduction in its state unemployment tax rate. Conversely, if it has a large turnover its tax rate can be raised. The state tax rate can also vary with the solvency of the state unemployment benefit fund.

When a worker files a claim for benefits, the employee's last employer is sent a notice. If the employer replies that the worker was separated from the firm for a reason other than lack of work, the benefits may not be chargeable against that employer's account. In addition, an unemployed worker may disqualify himself from unemployment benefits by voluntarily quitting his last job without good cause, being discharged for misconduct, being directly engaged in a strike or other labor dispute, failing to apply for or to accept an offer of suitable work, and other causes. The amount of unemployment payments a worker can receive varies from state to state. The benefits are limited as to both weekly amount and the maximum number of weeks they are payable.

9

BUSINESS METHODS

9.1 GENERAL

The purpose of this chapter is to discuss selected business methods and procedures that are of particular importance to the management of a construction firm. Because of the eclectic nature of the chapter, not all the topics discussed bear a direct relationship to one another, and they are presented in no particular order. A contractor may well utilize the computer in conjunction with certain of the topics discussed. Whether manual or mechanized procedures are used, however, the basics are the same.

9.2 RECORDS

The downfall of many an otherwise well-run construction company is the lack of accurate, detailed, and current information concerning all aspects of the company's financial affairs. Financial records must be maintained to serve a variety of business and management purposes. One of the important reasons why construction contractors, like other businessmen, must keep accurate records is that such records are required by law. An assortment of governmental agencies require data pertaining to taxes, payrolls, and other company information, together with a system of accounts adequate to serve the purpose. Additionally, the contractor's records must provide the source material for obtaining indispensable support services. Financial statements and reports are required for submittal to bankers, sureties, owners, insurance companies, public agencies, lending firms, and others.

Beyond this area of legal and practical compulsion lies the realm of record keeping to serve the purposes of company management. The functions of a construction company's accounting system are not limited merely to keeping and producing the records. Although such information is basic and essential to the conduct of company operations, it is necessary to analyze and summarize such data so it can be used to best effect. If, indeed, there is anything to be gained through the conscious direction of company affairs, the success of the entire process hinges on the continuous generation of carefully selected and summarized intelligence. The accounting system must serve the purpose of providing information to assist management in controlling company operations and in utilizing its available capital to the greatest possible advantage. Without proper records it is impossible for a contractor to estimate construction costs accurately, to control costs on ongoing projects, to keep the company in a fluid cash position, to make sound judgments concerning the acquisition of equipment, and to perform the many other functions associated with the financial management of the business.

9.3 ACCOUNTING METHODS

Although the details of record keeping vary among firms, a pattern of basic accounting procedures exists that is common to the construction industry. The main theme of a contractor's accounting system centers around the determination of income and expense from each of the individual construction projects. That is to say, each contract is treated as a separate profit center. In an accounting sense, a profit center is any group of associated activities whose profit or loss performance is separately measured and analyzed. The original estimate of costs pertaining to each contract serves as a budget for that project. As costs are incurred, they are charged, directly or indirectly, against the job to which they pertain. It is customary that the keeping of a contractor's business accounts be concentrated in the home office. However, on large projects and cost-plus contracts, a subsidiary set of accounts is frequently maintained in a field office in which all record keeping pertaining to the project is done. The controlling accounts of such projects are incorporated into a master set of accounting records maintained in the contractor's central offices.

The basic accounting procedure used by a contractor can be either the cash method or the accrual method. Under the cash method, income is taken into account only when cash is actually received, and expense is taken into account only when cash is actually expended. Depreciation of equipment and other capital assets is, of course, a noncash expense. This accounting method is used principally by professional people, small businesses, nonprofit organizations, and for personal records. The pure cash method is applicable only to very small construction companies and is not often used, although a modified form that follows federal income tax regulations is sometimes utilized. If a construction company does only small jobs for cash, maintains little or no materials inventory, and owns no equipment of consequence, the modified cash method is simple and adequate. However, for most construction companies the cash method does not work to the advantage of the contractor. Income earned but not received and expenses incurred but not paid as of the end of a reporting period are not recorded or reflected in the company records. An income statement (see Section 9.9) prepared from such books will not present a realistic indication of true profit or loss. Similarly, a balance sheet (see Section 9.10) cannot reflect accurately the company's financial condition. Such financial reports are entirely inadequate for credit and other purposes.

For the majority of construction companies the accrual method is the only acceptable basic accounting procedure. Under this method income is taken into account in the fiscal period during which it is earned, regardless of whether payment is actually received. Similarly, items of expense are entered as they are incurred, whether actually paid out during the period or not. The straight accrual method is used for the accounting of short-term contracts, that is, construction contracts whose operations are completed within a single accounting period.

9.4 ACCOUNTING FOR LONG-TERM CONTRACTS

A long-term contract is defined by the Internal Revenue Service as one that covers a period in excess of one year from the date of its inception to the date it is completed and accepted. The courts have not agreed entirely with this concept, however. Numerous court decisions have ruled that a long-term contract is one that is not completed by the end of the fiscal year in which it began. This is the usual perception of a long-term contract in the construction industry. Two accounting procedures have been specifically developed for application to long-term contracts. These are called the percentage-of-completion method and the completed-contract method. The two methods differ in the manner in which costs and expenses are matched against project revenue and in the identification of the period in which the income from a job is taken into account.

Each of these methods has both tax advantages and disadvantages. It is not uncommon for a contractor to adopt the percentage-of-completion method for financial statement purposes and to report income for income tax purposes under the completed-contract method of accounting. The contractor will be well advised to seek the assistance of a competent tax accountant when selecting one or the other of these two procedures for a specific application. It should be noted that approval of the Internal Revenue Service is required before a contracting firm can make a change in its established accounting method used for tax-reporting purposes.

9.5 THE PERCENTAGE-OF-COMPLETION METHOD

The percentage-of-completion method recognizes job income as the work advances. Thus the profit is distributed over the fiscal years during which the construction is underway. This method has the advantage of recognizing income periodically on a current basis rather than irregularly as contracts are completed. It also reflects the status of contracts still in process through the current estimates of costs to complete the project or of progress toward completion.

The amount of income to be recognized in a given fiscal year from an uncompleted project can be computed in a variety of ways. A common procedure is to estimate the percentage of completion of the project as the ratio that the incurred costs to date bear to the anticipated total cost, with proper allowance for any revised estimates of costs to complete the work. Applying this percentage of completion to the total contract price and deducting the project costs incurred to date yields the net project income up to the cutoff time. The major weakness of this procedure is its dependence on estimates of costs to complete the work, which can be subject to considerable uncertainty.

It must be observed at this point that the usual monthly pay requests do not normally constitute an acceptable basis for profit allocation. In the usual case, there is little relationship between the amounts billed to owners and realized income on partially completed contracts. In some instances the contractor's early billings are in excess of the value of the work actually performed (see Section 9.23). This is done to help reimburse the contractor for move-in and start-up expenses. In other instances the architect-engineer may purposely keep the contractor's progress payments below the amounts due for the work actually performed. On a practical basis, therefore, billings to the owner do not generally provide a proper measure of partial earnings on either a lump-sum or a unit-price contract. On the other hand, costs ordinarily do.

There is considerable support for the opinion that the percentage-of-completion method most fairly presents company income when used for the preparation of financial statements. However, this is true only when the extent of job progress, contract revenues, and contract costs can be estimated with reasonable accuracy. When the lack of records, changes in the scope of the contract, or hazards inherent in the work make the estimation of completion costs subject to considerable uncertainty, the completed-contract method is usually used for financial reporting.

9.6 THE COMPLETED-CONTRACT METHOD

The completed-contract accounting method recognizes project income only when a contract is completed. During construction operations, periodic payments and costs are accumulated, but no project income is recorded until the project is completed. What actually constitutes completion of a project for income-reporting purposes has been subject to varying interpretation by the courts. Present practice tends largely to establish completion only after work finalization and acceptance by the owner, although the Internal Revenue Service takes the position that a contract is complete when it is "substantially" finished. Completion of a contract may not be delayed by the contractor if the principal purpose is to defer federal income tax.

The principal advantage of the completed-contract method is the complete deferral of income tax until the end of the contract and the collection of final payment. By the same token, however, loss recognition is also so deferred. A disadvantage of the method is that it does not reflect current performance for long-term contracts and may result in irregular recognition of income and hence, in some situations, in greater income tax liabilities. Another disadvantage of the completed-contract method is that the year of completion of a project can be subject to considerable uncertainty if the end of the project coincides with the end of the accounting period or if there are disputed claims unsettled at the end of that year. Special problems can also arise with respect to allocation of office overhead expense during years with abnormally few or abnormally small contracts in progress. The completed-contract method is used by many contractors for income tax reporting.

9.7 LONG-TERM COST-PLUS CONTRACTS

Under a cost-plus form of construction contract, it is usual that the contractor be reimbursed at regular intervals during the construction period for its expenditures and also be paid a pro rata share of its fee. Accounting procedures for recognizing income under cost-plus contracts parallel those discussed previously for fixed-amount contracts. It is generally accepted practice to recognize as income current payments on the fee in the fiscal period when they become billable if the contract is written so the risk of losing the fee is negligible. However, when there are substantial uncertainties in forecasting net income, such as with a guaranteed maximum cost contract, the operation of penalty-bonus contractual arrangements, or a provision that overhead allowed is subject to audit and adjustment, the completed-contract method may be preferable, with the entire fee being recognized as income when the contract is completed.

9.8 FINANCIAL REPORTS   CPA certified public accountant

Several forms of financial reports are derived from the company books of account. It is the purpose of such a report to group together significant facts in a way that will enable the person reading it to form an accurate judgment concerning some aspect of the company operation, such as the overall financial condition of the organization or the profit-loss results of its operations. As would be expected, much use of a company's financial reports is made by company management itself. These reports are invaluable for the advance planning of operations. They reflect the company's borrowing and bonding capacities and yield much information concerning its policies with respect to purchasing, equipment ownership, and office overhead.

In addition, these financial reports serve many important functions with respect to external agencies. Bankers, surety and insurance companies, equipment dealers, credit-reporting agencies, and clients are all concerned with the contractor's financial status and profit experience. Stockholders, partners, and others with a proprietary interest use the financial reports to obtain information concerning the company's financial condition and the status of their investment. Two financial reports of particular importance are the income statement and the balance sheet.

9.9 THE INCOME STATEMENT*

The income statement is an abstract of the nature and amounts of the company's income and expense for a given period of time, usually a quarter or full fiscal year. This statement shows the profit or loss as the difference between the income received and the expenses paid out during the period. Figure 9.1 is an income statement of the Blank Construction Company, Inc., for the fiscal year ending December 31, 19—. The following paragraphs, keyed to the lower-case letters in parentheses as they appear in Figure 9.1, explain the various items.

(a) This statement has been prepared on a completed-contract basis. Project income is the total contract value of all projects completed during the period covered by the statement. The total is obtained from a supporting schedule that shows the income figures for each completed project.

(b) Project costs include all materials, labor, equipment, subcontracts, job overhead, and other expenses that have been charged to the completed projects. A supporting schedule of job costs lists these expenses for the individual projects. Included with job costs is the office overhead expense that has been allocated to the completed projects. Office overhead is periodically distributed among the various projects in proportion to the costs incurred on them during the period.

(c) Subtracting project costs from project income yields net project income for the statement period.

(d) Other income lists the net income received from other sources.

(e) Total net income before taxes is the amount realized from all operations during the statement period.

(f) This sum represents federal and state income taxes paid or due for the statement period.

(g) Net income after taxes is the amount available for company expansion or for distribution to stockholders.

(h) This sum is the earnings accumulated as of the start of the statement period. A corporation can retain its earnings for business expansion or can distribute them as dividends to the stockholders. Tax laws provide for penalties if earnings are retained that are not actually required and are surplus to business needs. The maximum amount of earnings a corporation can retain before possibly incurring liability is set by the Internal Revenue Service and is referred to as the accumulated earnings credit.

(i) Dividends paid represent distribution of earnings made to stockholders in the form of dividends on common stock during the statement period. This construction company has 4610 shares of common stock outstanding. A dividend of $8 per share was declared and paid during the fiscal year just ended.

---

*The income statement is also known by other titles: profit and loss statement, statement of earnings, statement of loss and gain, income sheet, summary of income and expense, profit and loss summary, statement of operating results, and operating statement.

THE BLANK CONSTRUCTION COMPANY, INC.

PORTLAND, OHIO

INCOME STATEMENT

For the Year ended December 31, 19—

ITEM	TOTAL
(a) PROJECT INCOME:	$8,859,138.39
(b) Less Project Costs, including office overhead	
expense of $239,757.04	8,705,820.15
(c) Net Project Income	$ 153,318.24
(d) OTHER INCOME:	
Discounts Earned	$ 23,064.93
Equipment Rentals	23,758.93
Miscellaneous	12,882.64
Total Other Income	$ 59,706.50
(e) NET INCOME BEFORE TAXES ON INCOME	$ 213,024.74
(f) FEDERAL AND STATE TAXES ON INCOME	97,616.66
(g) NET INCOME AFTER TAXES ON INCOME	$ 115,408.08
RETAINED EARNINGS:	
(h) Balance, January 1, 19—	$ 106,127.24
(i) Dividends Paid	36,880.00
Total Retained Earnings	$ 69,247.24
(j) BALANCE, December 31, 19—	$ 184,655.32
(k) EARNINGS PER SHARE ON NET INCOME	$ 25.03

Figure 9.1   Income statement.

(j) The balance represents the total retained earnings of the corporation through the end of the report period.

(k) In a corporate form of business, it is usual for the income statement to show net earnings for the year (after taxes) per share of outstanding stock. It must be realized that an income statement has several limitations. Concentrating as it does on past events, the income statement reports only the company's profit or loss experience during the reporting period and does not show the present overall financial condition of the firm. The amounts shown are, at best, only approximations. The precision implied by the figures exists in appearance only, not in actual fact. However, assuming reasonable care is exercised in compiling the data, the results can be of great value when used with judgment. In the highly competitive and unpredictable construction industry, the profit reported is not always indicative of the management quality of the company.

9.10 THE BALANCE SHEET*

A balance sheet presents a summary of the assets, liabilities, and net worth of the company at a particular time, for example, at the close of business on the final day of a fiscal year. Balance sheets are usually prepared as of the end of each fiscal year and are of a form more or less prescribed by custom. They are universally used to describe the financial condition of business concerns. A representative example of a contractor's balance sheet is depicted in Figure 9.2.

The basic balance sheet equation may be stated as follows: assets = liabilities + net worth. The balance sheet presents in analytical form all company-owned property, or interests in property, and the balancing claims of stockholders or others against this property. The foregoing equation expresses the equality of assets to the claims against these assets. Assets are defined as anything of value, tangible or intangible. Liabilities involve obligations to pay assets or to render services to other parties. Net worth is obtained as the excess of assets over liabilities and represents the contractor's equity in the business. Here, "the contractor" is used in the broad sense meaning proprietor, partners, or stockholders.

The major headings of Figure 9.2 will now be discussed.

(a) Current assets include cash, materials, and other resources that may reasonably be expected to be sold, consumed, or realized in cash during the normal operating cycle of the business. When the business has no clearly defined cycle or when several operating cycles occur within a year, current assets (and current liabilities) are construed on a 12-month basis. This one-year rule applies to most contractors. Prepaid expenses represent goods or services for which payment has already been made and which will be consumed in the future course of operations.

(b) Noncurrent notes receivable are in the nature of deferred assets, representing the value of notes that become payable at some future date or dates.

(c) Property represents the fixed assets of the business. These assets are more or less permanent in nature and cannot readily be converted into cash, at least not in amounts commensurate with their true values to the contractor. These assets generally have a useful life of several years, although assets such as buildings, equipment, vehicles, and furnishings do wear out gradually. These assets are capitalized at their purchase prices or in accordance with appropriate cost appraisals.

(d) Accumulated depreciation represents the total decrease in value of the property as a result of age, wear, and obsolescence. Depreciation is more fully discussed in Section 9.12.

(e) Total assets is the sum of everything of value that is in the possession of or is controlled by the company.

(f) Current liabilities are debts that become payable within a normal operating cycle of the business [see Item (a)]. Presumably, payment will be made from current assets.

*The balance sheet may also be called by other names: financial statement, statement of financial condition, statement of worth, or statement of assets and liabilities.

THE BLANK CONSTRUCTION COMPANY, INC.
PORTLAND, OHIO

BALANCE SHEET
December 31, 19__

ASSETS			LIABILITIES		
(a) CURRENT ASSETS:			(f) CURRENT LIABILITIES:		
Cash on hand and on deposit	$	389,927.04	Accounts payable	$	306,820.29
Notes receivable, current		16,629.39	Due subcontractors		713,991.66
Accounts receivable, including			Accrued expenses and taxes		50,559.69
retainage of $265,686.39		1,222,346.26	Equipment contracts, current		2,838.60
Deposits and miscellaneous			Provision for income taxes		97,616.66
receivables		15,867.80	Total		$1,171,826.90
Inventory		26,530.14			
Prepaid expenses		8,490.68	(g) DEFERRED CREDITS:		
TOTAL CURRENT ASSETS		$1,679,791.31	Income billed on jobs in		
			progress at December 31, 19__		$2,728,331.36
(b) NOTES RECEIVABLE, NONCURRENT	$	12,777.97	Costs incurred to December 31,		
			19__ on uncompleted jobs		2,718,738.01
(c) PROPERTY:			Deferred Credits	$	9,593.35
Buildings	$	5,244.50	TOTAL CURRENT LIABILITIES		$1,181,420.25
Construction equipment		188,289.80	EQUIPMENT CONTRACTS, NONCURRENT		7,477.72
Motor vehicles		37,576.04	(h) TOTAL LIABILITIES		$1,188,897.97
Office furniture and equipment		13,596.18			
TOTAL PROPERTY	$	244,706.52	NET WORTH:		
(d) Less accumulated depreciation		102,722.51	(i) Common stock, 4,610 shares	$	461,000.00
NET PROPERTY	$	141,984.01	Retained earnings		184,655.32
			(j) TOTAL NET WORTH	$	645,655.32
(e) TOTAL ASSETS		$1,834,553.29	(k) TOTAL LIABILITIES AND NET WORTH		$1,834,553.29

Figure 9.2 Balance sheet.

(g) In our example, the Blank Construction Company, Inc., is using the completed-contract method of reporting income. Deferred credits represent the excess of project billings over related costs on current contracts that have not reached completion by the end of the reporting period. There is some difference of opinion concerning the proper way to treat this item. It seems reasonable to say that under most contracts the progress billings are merely advances against the amount that will be earned when the contract is completed. Consequently, the excess of progress billings over cost is a current liability representing the amount due the owner until the contract has been completed. This is the procedure followed in Figure 9.2.

(h) Total liabilities is the sum of every debt and financial obligation of the company.

(i) This is the capital stock account, showing the classes and amounts of stock that have been actually issued and paid for by the stockholders. In our example, 4610 shares of $100-par-value common stock have been purchased by the owners.

(j) This represents the net ownership interest the corporation "owes" to the stockholders. In our example, it is obtained as the sum of the common stock investments plus retained earnings (also called earned surplus). The book value of the common stock as of December 31, 19—, is obtained by dividing $645,655.32 by 4610, which gives $140.06 per common share. Book value excludes intangibles of all kinds that would have no value on liquidation.

The market value of this common stock, that is, the price for which the stock can be sold, may depart substantially from its book value. What shares are truly worth in a closely held corporation can be very difficult to determine. Yet, such a value may be needed for estate, gift, or income tax purposes. The Internal Revenue Service has issued guides to assist in valuing closely held shares.

(k) The equality of (e) to (k) illustrates that total assets = total liabilities +net worth.

The balance sheet has considerable analytical value for those who wish to determine the financial condition of a firm. It discloses the nature and composition of a company's assets and shows how these assets are financed. The sources of funds tell a great deal about the quality of management and the stability of a contractor. The balance sheet gives a good idea of the liquidity of a firm; that is, its ability to meet its short-term financial obligations. A comparison of balance sheet values over time discloses trends in the company's management policies and financial position. There are, of course, some inadequacies associated with balance sheets. The information presented applies as of a specific date and may not be representative of the normal company financial condition. Many asset values are only approximate, and the accounting method chosen for the preparation of a balance sheet can appreciably influence the data presented.

9.11 FINANCIAL RATIOS

Ratios of various kinds are frequently employed as quantitative guides for the assessment of a company's financial and earning position. For example, ratios can be used to analyze income statements and balance sheets. It is not the absolute size of the figures in these documents that is especially meaningful to the financial analyst, but rather the relationships, or ratios, among the different values. By comparing the same ratios over a series of financial reports, very significant information can be obtained regarding a company's financial performance over the years. When such ratios are compared with similar figures of other contractors, a comparative financial picture of the business is obtained.

Although many ratios are used in the interpretation of financial reports, the following ones are considered to be of major significance. The value of each ratio below is computed for the Blank Construction Company, Inc., using data from the company's income statement, Figure 9.1, and balance sheet, Figure 9.2. This company is represented to be a building contractor whose financial structure is relatively good. In brackets and just below the computation of each ratio for the Blank Construction Company, Inc., are shown ratios for general building contractors and heavy contractors (except highway and street) recently compiled by Dun & Bradstreet, Inc. The figure shown for each construction category is the median ratio for those contractors sampled.

The following data are taken from Figures 9.1 and 9.2:

Current assets	=	$1,679,791.31
Current liabilities	=	1,181,420.25
Total assets	=	1,834,553.29
Total liabilities	=	1,188,897.97
Fixed assets	=	141,984.01
Tangible net worth	=	645,655.32
Net working capital = current		
assets − current liabilities	=	498,371.06
Net profits (after taxes)	=	115,408.08
Annual volume	=	8,859,138.39

1. Current assets to current liabilities $= \dfrac{\$1,679,791.31}{\$1,181,420.25} = 1.42$ 1.8

   [Gen. Bldg. - 1.63   Heavy Const. - 1.83]

2. Net profits on annual volume $= \dfrac{\$115,408.08\ (100)}{\$8,859,138.39} = 1.30\%$

   [Gen. Bldg. - 1.87   Heavy Const. - 4.47]

3. Net profits on tangible net worth $= \dfrac{\$115,408.08\ (100)}{\$645,655.32} = 17.9\%$

   [Gen. Bldg. - 11.26   Heavy Const. - 12.52]

4. Net profits on net working capital $= \dfrac{\$115,408.08\ (100)}{\$498,371.06} = 23.2\%$

   [Gen. Bldg. - 20.08   Heavy Const. - 32.07]

5. Annual volume to tangible net worth $= \dfrac{\$8,859,138.39}{\$645,655.32} = 13.7$

   [Gen. Bldg. - 8.20   Heavy Const. - 3.31]

6. Annual volume to net working capital $= \dfrac{\$8,859,138.39}{\$498,371.06} = 17.8$

   [Gen. Bldg. - 11.26   Heavy Const. - 7.73]

7. Fixed assets to tangible net worth $= \dfrac{\$141,984.01\ (100)}{\$645,655.32} = 22.0\%$

   [Gen. Bldg. - 24.6   Heavy Const. - 54.1]

8. Current liabilities to tangible net worth $= \dfrac{\$1,181,420.25\ (100)}{\$645,655.32} = 183.0\%$

   [Gen. Bldg. - 120.7   Heavy Const. - 65.3]

9.   Total liabilities to tangible net worth $= \dfrac{\$1,188,897.97 \ (100)}{\$645,655.32} = 184.1\%$

[Gen. Bldg. - 136.8   Heavy Const. - 118.6]

The Dun & Bradstreet ratios just cited are illustrative only of contractors' financial experience in general, and the ratios for an individual contractor can and often do vary substantially from these median values. The figures also clearly reveal that most of the ratios vary considerably with the type of construction. Nonetheless, to the trained eye such financial ratios tell a great deal about the business, its financial condition, and the quality of its management. Each ratio conveys its own message, some examples of which are now given.

The ratio of current assets to current liabilities (also known as the current ratio) is the most significant short-term financial measure. It is a test of the firm's ability to meet current obligations. In the construction industry, a current ratio of 1.5 or more is generally regarded as being favorable. Net profits to tangible net worth is perhaps the most important of the long-term ratios because it reflects the efficiency with which invested capital is employed. Considering the risk, the contractor should certainly expect to earn more on its investment than it could realize from current dividend or interest rates. Annual volume to tangible net worth measures the rate of capital turnover, showing how actively the firm's capital is being put to work. If capital is turned over too rapidly, liabilities may build up at an excessive rate. If the ratio is too low, funds become stagnant and profitability suffers. In construction, this particular ratio is considerably higher than in other industries. This high figure contributes to the high failure rate among contractors and partially explains why entry into the construction industry is so easy. A construction contractor can do more work per dollar of invested capital than is possible in almost any other industry. The value of net profit on annual volume is typically less in construction than in other industries. However, this seeming low rate of return is partially explained by the respectable salaries of company principals. A significant proportion of construction profits is often taken in the form of personal salaries.

9.12 DEPRECIATION ACCOUNTING

From the day of its acquisition, most property steadily declines in value because of age, wear, obsolescence, and impaired serviceableness. This reduction in value, called depreciation, is a cost of doing business and represents the physical depletion of assets such as offices, warehouses, vehicles, accounting machines, engineering instruments, and construction equipment of all sorts. Because of the continued decline in the value of such property, its cost must be amortized over its useful life so that the contractor recovers its investment and has capital available when replacement of the property becomes necessary. The useful life of an item can be expressed in terms of time periods (months or years), operating hours, or units of production. The following paragraphs discuss methods of amortizing, or depreciating, assets, using construction equipment for the purpose of illustration.

The U.S. Treasury Department publishes a schedule of asset depreciation ranges (ADR) for construction equipment. These are suggested guidelines for the service lives to be used by contractors when depreciating various types of construction equipment for tax purposes. Lives shorter than these may be used by a contractor provided its claimed depreciation is in line with its actual replacement and modernization program.

There are different methods of computing depreciation costs for tax-reporting purposes. Each of these methods has its advantages and disadvantages, depending on many operational and financial variables of the company. Permission of the Internal Revenue Service must be obtained before an operating company can change its depreciation method used for tax-reporting purposes. For construction equipment, four different procedures are commonly utilized for the calculation of annual depreciation charges:

1. Production method.
2. Straight-line method.
3. Declining-balance method.
4. Sum-of-year-digits method.

It must be stressed that the choice of method to be used is an extremely important decision for contractors who own large fleets of equipment. The average contractor specializing in heavy and highway work may well have an equipment investment of $10 million or more, perhaps 20 times that of a building contractor with a comparable contract volume. Depreciation charges for equipment-oriented contractors account for an appreciable proportion of their annual operational expenses. The advice of an experienced tax accountant should be sought when a method is being selected for use.

With any of these methods the value of each piece of equipment is decreased annually. The sum of these reductions at any time is a depreciation reserve which, when subtracted from the initial cost of the equipment, gives its current book value. The initial cost of an asset, for depreciation purposes, includes not only the sale price but also the costs of taxes, freight, unloading, assembly, delivery, and set-up expenses. The book value represents the portion of the original cost that has not yet been amortized. As previously stated, depreciation charges are a cost of doing business and correspondingly reduce company earnings. At the same time, the depreciation reserve represents capital retained in the business, ostensibly for the ultimate replacement of the capital assets being depreciated.

Each of the aforementioned depreciation methods will now be discussed. Depreciation accounting is in a more or less continual state of flux because of changes being made in existing tax laws. For this reason the depreciation methods are discussed in general terms only. For purposes of illustration and comparison of the methods, let us consider a $16,000 ditcher having a probable service life of five years or 10,000 operating hours and an estimated salvage value of $1,000.

9.13 THE PRODUCTION METHOD

When using the production method, the annual depreciation is established in accordance with the use actually made of that piece of equipment during the accounting period. Stated in different terms, the depreciation charge is made on a cost-per-unit-of-output basis. One form of this procedure, called the production-hours method, apportions the depreciable value over the total expected hours of useful operation. The depreciable value is equal to the initial outlay less the anticipated disposal value, if any. The estimated salvage value of construction equipment should be reduced by any costs of dismantling or disposal and is frequently ignored entirely if the residual value is nominal (less than 10 percent of acquisition value) or is not subject to reliable estimate.

Under the assumption that the ditcher will have a useful service life of about 10,000 hours of field operation, a depreciation charge will be made at the rate of $1.50 for each operating hour. This value is obtained by dividing the depreciable value (initial cost minus salvage value) of $15,000 by the total estimated service life (10,000 operating hours). Figure 9.3 shows the annual depreciation charges for the production-hours method obtained by applying the depreciation rate of $1.50 per operating hour to the actual number of operating hours for each year.

In a second form of the production method, called the production-output method, the depreciable value is apportioned over the total units of production the equipment will be expected to produce during its useful life. This procedure is especially suitable to the depreciation of construction equipment whose total service-life production can be reasonably well established and whose use in the field is not uniform by periods. This procedure is applicable to production equipment such as rock crushers, aggregate plants, concrete batch plants, and asphalt paving plants. For example, the cost of an asphalt hot-mix plant can be spread over the total expected production of the plant, the depreciation rate being expressed as a fixed cost for each ton of hot-mix produced. In a similar manner, heavy-duty trucks can be depreciated on a mileage basis.

9.14 THE STRAIGHT-LINE METHOD

The straight-line method is the simplest depreciation procedure. It writes off the depreciable value of an asset at a uniform rate throughout the service life. It assumes, for record-keeping purposes, that the amount of value loss is constant year by year. Objections are sometimes raised to the use of the straight-line method on the grounds that the value decline of the capital item never actually follows such a course. It can be contended that most assets decrease in value more rapidly during the early years of their life.

In reality, few would contend that the straight-line procedure measures accurately the actual decline in asset value. In this regard, however, it must be recognized that asset appraisement is not the objective of depreciation accounting. It is concerned with the spreading of the cost of an asset over its useful life in a systematic and rational manner rather than attempting to gauge

DITCHER DEPRECIATION

Original Cost = $16,000   Service Life = 5 years
Assumed Salvage Value = $ 1,000

End of Year	Annual Operating Hours	Annual Charge	Accumulated Depreciation	Book Value
		Production — Hours Method		
0	0	$ 0	$ 0	$16,000
1	2,300	3,450	3,450	12,550
2	2,100	3,150	6,600	9,400
3	1,600	2,400	9,000	7,000
4	2,200	3,300	12,300	3,700
5	1,800	2,700	15,000	1,000
		Straight — Line Method		
0		$ 0	$ 0	$16,000
1		3,000	3,000	13,000
2		3,000	6,000	10,000
3		3,000	9,000	7,000
4		3,000	12,000	4,000
5		3,000	15,000	1,000
		Declining — Balance Method		
0		$ 0	$ 0	$16,000
1		6,400	6,400	9,600
2		3,840	10,240	5,760
3		2,304	12,544	3,456
4		1,382	13,926	2,074
5		830	14,756	1,244*
		Sum-of-Year — Digits Method		
0		$ 0	$ 0	$16,000
1		5,000	5,000	11,000
2		4,000	9,000	7,000
3		3,000	12,000	4,000
4		2,000	14,000	2,000
5		1,000	15,000	1,000

*This value cannot become less than the salvage value.

Figure 9.3 Annual depreciation of ditcher.

depreciation charges to parallel physical decline. Basically, it must be viewed as a process of allocation and not one of valuation. Figure 9.3 illustrates the workings of the straight-line method.

9.15 ACCELERATED DEPRECIATION METHODS

Both the declining-balance method and the sum-of-year-digits method give a faster write-off during the first years of equipment life than in the later years. These procedures are called accelerated depreciation methods. There is considerable opinion that a better matching of revenue and expense is achieved by applying accelerated depreciation to construction equipment because new equipment is generally more productive than old. The use of procedures that afford larger depreciation deductions during the earlier years of asset ownership may have other important advantages for a contractor. To illustrate, rapid initial depreciation causes cash to be retained in the business. Larger depreciation charges against regular income cause a reduction of income taxes for that year.

Accelerated methods provide for faster recovery of equipment value, consequently offering the contractor some measure of protection against unanticipated contingencies later in the life of the equipment, such as obsolescence, excessive maintenance, rapid wear, or need for equipment of a different size or capacity. Rapid depreciation amortizes the contractor's equipment investment more quickly and simultaneously decreases the book value at the same rate. Speedy reduction of book value leaves the contractor much freer to dispose of unsatisfactory equipment and, to some degree, can also assist in combating the problem of replacing equipment during periods of inflation.

Accelerated depreciation methods are recognized by the Internal Revenue Service (IRS) although certain restrictions on their use may apply. It is also permissible by IRS regulations for a contractor to employ different depreciation methods for financial-reporting and tax-reporting purposes. Contractors frequently use a straight-line procedure for their usual financial reporting and an accelerated method for income tax purposes.

9.16 THE DECLINING-BALANCE METHOD

With the declining-balance method, the annual depreciation charge is calculated as a constant rate times the book value. Current IRS regulations limit this rate to twice the straight-line rate that would apply for the same useful life. When this maximum rate is used, the procedure is called the double-declining-balance method, which is used in the following for purposes of illustration.

When the double-declining-balance method is used, the depreciation rate percentage is equal to 200 divided by the estimated service life in years. A piece of equipment having a five-year service life would then be depreciated at a rate of 40 percent of its undepreciated balance, or book value, each year. No salvage value is deducted from the initial asset cost when using the declining-balance method, although the book value cannot decrease below the estimated salvage value.

Figure 9.3 shows how this method works as applied to the $16,000 ditcher. It is to be noted that, when the declining-balance method is used, the book value never reduces to zero, as with the straight-line method, and a residual value always remains unrecovered at the end of the equipment's service life.

At the present writing, the IRS permits a capital asset to be depreciated by combining the double-declining-balance procedure with the straight-line method. The declining-balance method must be used initially. When the annual depreciation charges begin to taper off, the straight-line method is used for the remainder of the useful life. The IRS publishes specific regulations with regard to the year when the switchover can be made.

9.17 THE SUM-OF-YEAR-DIGITS METHOD

This method calculates the annual depreciation to be charged on the basis of the sum of the digits representing each year in the asset's life. For example, for a five-year service life, this sum would be $1 + 2 + 3 + 4 + 5 = 15$. A proportion of this total is then taken of the depreciable value for each year. To illustrate, 5/15 of the depreciable value is taken as the first year's depreciation charge; 4/15 the second year; 3/15 the third year; 2/15 the fourth year; the remaining 1/15 the last year. Figure 9.3 presents the results of applying the sum-of-year-digits method to the ditcher.

9.18 EQUIPMENT MANAGEMENT

Many contractors require extensive spreads of equipment to accomplish their construction projects. Equipment is very expensive and the contractor's investment in "iron" can easily run into the millions of dollars. Equipment makes money for the contractor but it must be carefully managed to produce the expected income. As a consequence, equipment must be managed just as any other capital investment. In the case of construction equipment, management involves making informed judgments about equipment acquisition and financing, establishing a comprehensive preventive maintenance program, and maintaining accurate and current records of equipment income and expense.

The acquisition of equipment is a major management decision and is one deserving of careful study. Considerations such as future need for the equipment, cost, size and capacity, dealer service, equipment reliability and ruggedness, warranties, risk of obsolescence, and how the new unit will fit in with presently owned equipment are all important aspects of selection. There is also the matter of how the acquisition is to be financed. New equipment can be financed in one of four different ways.

1. Renting on a short-term basis; for example, by the day, week, or month.
2. Renting on a short-term basis with an option to purchase at some future time.
3. Leasing on a long-term basis; that is, for two years or more. This can be with or without an option to purchase.
4. Direct purchase, either for cash or through some source of borrowed funds.

Each of these alternatives has its advantages and disadvantages. Detailed analysis involving discounted cash flows and rates of return on invested capital are needed to enable company management to make the wisest decision within the total context of the procurement.

A rigorous preventive maintenance program is essential to profitable equipment utilization and management. Equipment downtime can have a serious effect on project costs and schedules. A contractor's preventive maintenance program must be tailored to its own equipment specifications and the jobsite conditions where the equipment operates. Special attention must be given to those machines whose downtime would have a particularly severe effect on production and costs.

Once equipment is obtained, the contractor attempts to recover the acquisition costs by using it and realizing residual values upon its disposition. The accounting procedures used to charge construction projects with equipment costs vary substantially from one contractor to another. However, a common procedure is to establish an internal rental rate for each piece of equipment, a topic previously discussed in Sections 5.24–5.26. During the progress of work in the field, a charge equal to the internal rental rate times the number of equipment hours, weeks, or months is made against the project. At the same time, an equal, but opposite, credit is made to the ledger account of that equipment unit. This is the same equipment account previously discussed in Section 5.25 where all items of expense, exclusive of operating labor, and hours of usage are continuously maintained for that equipment item.

In essence, what the contractor is doing is to establish, at least on paper, a separate company that owns, services, and maintains all of its major equipment and rents it to the contractor at predetermined rates. This equipment accounting procedure provides a cumulative record of expense and earnings for each major equipment unit. How these figures compare provides company management with invaluable information concerning equipment utilization, maintenance, costs, and replacement. In short, the company knows which machines are paying their way and which are not. Many construction firms have a company policy that when the annual cost of an equipment item exceeds its annual earnings, it is either replaced or sold. A large construction company is apt to treat its equipment division as a separate profit center. This just says that the equipment investment is treated as a separate business, with the return on the investment required to be commensurate with the risk. Whether a contractor expects a break-even performance or a profit on its equipment investment depends on its concept of profit and how it applies to different company operations.

9.19 PROCUREMENT

Procurement is a function of first-rate importance in a contractor's organization. This activity includes a number of different actions, each playing an indispensable role in obtaining the goods and services needed for company operations. Procurement is a centralized company effort, often operating as a separate department. Collecting all such activities together is economical of both time and money. It concentrates expertise and affords protection against improper and unnecessary expenditures of many kinds. Procurement personnel occupy a position of great responsibility within a contractor's organization and perform many duties. Procurement involves the preparation and use of a number of standard document forms such as requisitions, purchase orders, and subcontracts. The principal procurement functions are discussed in the following paragraphs.

1. Purchasing. Essentially, purchasing is the obtaining of equipment, tools, materials, stores, supplies, fuel, parts, motor vehicles, and services of every sort that are needed by the office organization, the storage yard and warehouse, and the field projects. This involves the processing of requisitions, obtaining and analysis of bids, and preparation and issuance of purchase orders. The preparation of purchasing specifications is often involved with nonproject needs. When a general contract change order affects an outstanding purchase order, a suitable change must be made.

2. Expediting and Receiving. After an order has been placed, expediting contact must be maintained with the vendor to ensure timely delivery. This is especially true for project materials. The designation of a required delivery date in a purchase order is no guarantee that the vendor will deliver on schedule. Energetic follow-up action is a necessity if the most favorable delivery schedule is to be realized. When the goods are shipped, adequate arrangements must be made for their receipt, unloading, and storage.

3. Inspection. Upon receipt of the goods, they must be inspected immediately to verify quantity, quality, and other essential characteristics. In case of shortage, loss or damage in transit, or erroneous goods, immediate action is required. If quality control verification tests are called for, samples are taken for that purpose. A written receipt and inspection report is prepared and filed for each delivery.

4. Shipping. The contractor's purchase order, in most instances, designates the method of shipping the goods. This is an important aspect of procurement and requires detailed knowledge concerning the different shipping alternatives and the time and cost implications of each. In the case of strikes or other disruptive events affecting the transportation industry, last-minute changes in shipping arrangements are sometimes needed. Shipments lost or misplaced in transit must be traced and suitable claims made where goods are damaged or lost.

5. Subcontracts. The procurement function includes the preparation and processing of project subcontracts. The essential information needed to do this comes from estimating. When a general contract change order affects subcontractors, suitable changes to the subcontracts affected are prepared and processed.

9.20 CASH DISCOUNTS

Discounts are frequently offered to the contractor by material dealers and others as an inducement for early payment of bills. These are not trade discounts in the sense of adjusting a catalog price or making a concession for quantity buying. Rather, cash discounts are in the nature of a premium given in exchange for payment of an invoice before it becomes due, and the buyer is entitled to the discount only when payment is made within the time specified.

Many cash discount terms are in commercial usage; three are mentioned here. The expression "2/10 net 30" means that, if the contractor makes payment within 10 days of the invoice date, 2 percent can be deducted from the face amount of the invoice. The bill in its full amount is due and payable 30 days after the date of the invoice. Material dealers normally date their invoices the day the goods are shipped. If the customer's location is close by, he has time to check the goods before paying within the discount period. A customer who is far removed geographically, however, may have to pay before the goods are actually received if he is to take advantage of the discount. To overcome this disadvantage to distant customers, some vendors, upon request, will mark their invoices "ROG" or "AOG" to indicate that the discount period begins on "receipt of goods" or on "arrival of goods."

A cash discount provision similar to the one just discussed is termed "prox," which means the discount period is expressed in terms of a specified date in the next month after the shipments are made or billed. The expression "2/10 prox net 30" means a 2 percent cash discount is allowed if the invoice is paid not later than the tenth day of the month following the purchase. The net due date of the account is 30 days from the first of that month.

Another commonly used discount is expressed as "2/10 EOM," which indicates a 2 percent discount can be taken if the invoice is paid by the tenth day of the month after the purchases are shipped. If the buyer does not make payment within this period, the bill is considered to be due net thereafter.

Cash discounts are treated as income and appear as "discounts earned" on the income statement, Figure 9.1. On the basis of a single invoice, 2 percent may appear to be inconsequential. However, if a contractor is able to take this discount on a million dollars' worth of materials in a year's time, it has increased its earnings in the amount of $20,000. When a contractor fails to discount its bills, its sources of credit generally view this with some agitation, because it may be symptomatic of cash-flow problems (see Section 9.30) and perhaps of more serious financial ills.

The status of cash discounts is a matter of sufficient import that it should be clearly spelled out in cost-plus contracts. It is standard policy in such agreements that all cash discounts accrue to the contractor, except when the owner has advanced the latter money from which to make payments. Documents of The American Institute of Architects reflect this policy, as shown by Article 10 of Appendix J. However, some forms of cost-plus contracts stipulate that the reimbursable cost of materials shall be decreased by the amount of any cash discounts taken, regardless of who provides the funds to make payment.

9.21 TITLE OF PURCHASES

Most contractors regularly purchase appreciable quantities of construction materials. During the process of transport, delivery, unloading, and storage of these materials, a variety of losses or damage can and often does occur. When such difficulties arise, the mutual rights and responsibilities of the buyer and of the seller are largely controlled by the Uniform Commercial Code. Normally, the risk of loss or damage as related to personal property rests with the person holding title. Responsibility for loss and rights of each party as buyer or seller depend on whether title of the goods has passed from seller to buyer at the time the issue is raised. For this reason, the time at which the title of a purchase passes from the vendor to the contractor is a matter of considerable importance.

Fundamentally, title to personal property that is the basis of a sale passes from seller to buyer when the parties intend it to pass. If the sales contract either states or clearly implies at what time title is to pass, the terms of the agreement govern. Seldom, however, do the parties to a sales agreement insert specific provisions concerning the passage of title. Consequently, the Uniform Commercial Code has established a standard set of rules in this matter. In the absence of any expressed intention to the contrary, title passes in accordance with the following general rules.

Cash Sale.   Agreements that call for delivery and payment to take place concurrently are called cash sales. Title of the goods passes when the goods are paid for and delivery takes place.

On-Approval Sale.   When goods are delivered to the buyer on approval or trial, title changes when the buyer indicates acceptance, when the goods are retained beyond the time fixed for their return, or when the goods are retained beyond a reasonable time.

Sale or Return   Title passes when goods are delivered to the buyer. However, the buyer has the option to return the goods within a fixed period of time.

Delivery by Vendor.   If the purchase order requires the seller to deliver the goods to the buyer's destination and delivery is made by the seller himself, the seller retains title until the goods are delivered.

Shipment by Common Carrier.   There is a general rule that, when the goods are shipped by common carrier to the buyer, title passes to the buyer when the seller delivers the goods to the carrier for transportation. However, the following exceptions apply.

1. When the seller fails to follow shipping instructions given by the buyer, such as the buyer naming a particular carrier and the seller shipping by another.
2. When the seller is required to deliver at a particular place, such as the buyer's dock or railroad siding.
3. When the seller is required to pay freight up to a given point, as in FOB agreements.
4. When the seller is required by purchase order or custom to make arrangements with the carrier to protect the buyer, as by declaring the value of the shipment or as in CIF agreements, and fails to do so.

5. When the goods shipped do not correspond in both quality and quantity to those ordered.
6. When the seller reserves title by retaining the bill of lading.

FOB   Standing for "free on board," FOB indicates that the seller shall put the goods on board the common carrier free of expense to the buyer, with freight paid to the FOB point designated. For example, contractor's purchase orders frequently specify delivery as FOB jobsite or FOB storage yard. Under an FOB agreement, title goes to the buyer when the carrier delivers the goods at the place indicated. "FAS" signifies the same arrangement where delivery is made by ship.

CIF   Standing for "cost, insurance, freight," CIF indicates that the purchase-order price includes the cost of the goods, customary insurance, and freight to the buyer's destination. Title passes when the seller delivers the merchandise to the carrier and forwards to the buyer the bill of lading, insurance policy, and receipt showing payment of freight. C & F indicates the same shipping arrangement except no insurance need be obtained by the vendor.

COD   Meaning "collect on delivery," COD indicates that title passes to the buyer, if he is to pay the transportation, at the time the goods are received by the carrier. However, the seller reserves the right to receive payment before surrender of possession to the buyer.

9.22 PERIODIC PAYMENT REQUESTS

Construction contracts typically provide that partial payments of the contract amount shall be made to the prime contractor as the work progresses. The exact procedure followed can depend substantially on the position of the owner and the type of construction involved. For example, in housing and some other types of construction, specified payments are made at the completion of certain stages of construction. In housing, payments are due at certain points of progress, such as at completion of excavation and foundations, completion of framing, when the structure is enclosed and mechanical and electrical rough-in is installed, at completion of the interior finish, and at full completion. Each of these payments is made to the contractor by the owner or mortgagee after job inspection and receipt of proof of payment of bills from the contractor.

Payments to the contractor at monthly intervals is the more usual contract proviso. The pay requests may be prepared by the contractor, the architect-engineer, or the owner. The general conditions or supplementary conditions of the contract normally stipulate which party is to have the responsibility and authority for compiling these requests.

Each periodic pay request is based on the work accomplished since the last payment was made. Consequently, field progress must be measured or estimated at intervals, and a payment form used that clearly identifies the units of work that have been accomplished and on which the pay request is based. The owner or lending agency may require the use of a prescribed payment requisition form, or the contractor may be free to devise its own. When the pay request has been compiled, it is transmitted to the owner or lending agency, commonly through the architect-engineer or contracting officer who approves the request and certifies it for payment. Proof of payment of bills associated with the construction in the form of an affidavit or receipts, or a release of lien, must often accompany the pay request.

Approval of contractors' pay requests can be an important responsibility of the architect-engineer. In protecting the interests of his client, the owner, it is incumbent on the architect-engineer to see that the periodic payments made to contractors are reasonable measures of the work actually accomplished and that the work has been performed properly. Additionally, recent trends indicate that an architect-engineer may be held responsible if its negligence enables a contractor to divert funds. An actual example of this involved a major contract on which the contractor went bankrupt. The bonding company completed the project. The dollar value of the work in place when the bonding company took over was substantially less than the payments the contractor had received. The bonding company sued the architect-engineer for negligence in permitting overpayment to the contractor and won a sizeable indemnity.

Slow payment by the owner to the contractor continues to be a serious problem for the construction industry. Some owners will contract for a project without having completed arrangements for funding. Promoters may not have enough equity capital or permanent mortgage commitments for the full contract amount. At times owners will deliberately delay payments to the prime contractor, even though funds are available, in order to have the use of the funds themselves. Owners may at times delay payments to force the contractor into expediting the work or to extract more work than is called for by the contract documents. Public agencies are major offenders in slow-payment procedures because of the many steps involved before payment can be made. An associated problem is the slow processing of contract change orders, resulting in long delays in paying the contractor for extra work the owner himself initiated and approved.

Because of the several problems associated with the disbursement of construction funds, including late payment by the general contractor to its subcontractors and suppliers, there has been some movement toward having third parties be responsible for making payments on construction contracts. There are now private firms whose services to lending agencies include ensuring that the money moves quickly down the chain of command from lender to owner, general contractor, subcontractor, and supplier. In this way funds are not diverted and all parties to the project get paid on time.

9.23 PROJECT COST BREAKDOWNS

Construction contracts for lump-sum projects usually require that some form of cost breakdown of the project be submitted to the owner or architect-engineer by the contractor for approval before any payment request by the contractor is presented. This cost breakdown, also called a "schedule of values," is intended to serve as the basis for the subsequent monthly pay requests. Such a listing of job expenses may even be required at the time the job is bid, although it is more commonly submitted after the contract is signed. The breakdown, which is actually a schedule of costs of the various components of the structure, including all work done by subcontractors, is prepared in sufficient detail so the architect-engineer or owner can readily check the contractor's pay requests. Figure 9.4, which presents a typical monthly pay request, illustrates in the first column of figures a cost breakdown for the Municipal Airport Terminal Building. Occasionally, to satisfy an owner requirement, the specifications will stipulate the individual items for which the contractor is to present cost figures. In the absence of such instructions, it is usual to prepare the cost schedule by using the same general items as they appear in the specifications and on the final recap sheet of the estimate. This practice minimizes the time and effort necessary to prepare the breakdown figures and ensures a maximum of accuracy in the results.

Inspection of Figure 9.4 shows that the cost figures given for the individual work classifications do not check with those indicated on the recap sheet, Figure 5.7. The explanation for this discrepancy is that construction equipment, job overhead, markup, bond, and tax (often referred to collectively as job burden), are not shown on the cost breakdown as line items but have been incorporated back into the listed job costs. The total of these five amounts has not been distributed on a completely pro rata basis, however. Rather, items of work completed early in the construction period, such as excavation, concrete, reinforcing steel, and forms, have been increased proportionately more. This procedure, called "front-end loading," serves the purpose of helping to reimburse the contractor for the initial costs of moving in, setting up, and commencing operations, for which a specific pay item is seldom provided. Such moderate unbalancing of cost figures assists the contractor financially and minimizes its initial investment in the owner's project.

It should be pointed out that a cost breakdown compiled for payment purposes cannot be used for the pricing of either extra work to or deductions from the contract. Items of general expense have been prorated into the various pay items whether or not they apply directly to a given item of work. Additionally, as has been discussed, these cost breakdowns are usually unbalanced to some degree.

9.24 PAYMENT REQUESTS FOR LUMP-SUM CONTRACTS

Figure 9.4 illustrates the form of a typical pay request for a lump-sum building contract. Usually prepared by the contractor, it includes all subcontracted work as well as that done by the contractor's own forces. For each work classification that it does itself, the contractor estimates the percentage completed and in place. From invoices submitted by the subcontractors, suitable percentage figures are entered for all subcontracted work. These percentages are shown in column 4 of Figure 9.4. The total value of each work classification is multiplied by its percent completion and these figures are shown in column

THE BLANK CONSTRUCTION COMPANY, INC.
1938 Cranbrook Lane
Portland, Ohio

PERIODIC ESTIMATE FOR PARTIAL PAYMENT

Project ___Municipal Airport Terminal Building___ Location ___Portland, Ohio___

Periodic Estimate No. ___4___ for Period ___September 1, 19—___ to ___September 30, 19—___

Item No.	Item Description	Total Cost (1)	Completed to Date (2)	Cost to Complete (3)	Percent Complete (4)
1	Clearing and Grubbing	$ 14,909	$ 14,909	$ —	100
2	Excavation and Fill	44,749	38,037	6,712	85
3	Concrete and Forms				
	Footings	71,915	51,060	20,855	71
	Grade Beams	72,131	41,836	30,295	58
	Beams	43,690	4,369	39,321	10
	Columns	14,113	4,234	9,879	30
	Slab	253,749	30,450	223,299	12
	Walls	39,455	14,204	25,251	36
	Stairs	28,734	—	28,734	0
	Sidewalks	23,622	—	23,622	0
4	Masonry	486,566	14,597	471,969	3
5	Carpentry	37,772	—	37,772	0
6	Millwork	85,634	—	85,634	0
7	Steel and Misc. Iron				
	Reinforcing Steel	85,590	44,507	41,083	52
	Mesh	18,800	2,820	15,980	15
	Joist	92,953	—	92,953	0
	Structural	178,779	30,392	148,387	17
8	Insulation	25,688	—	25,688	0
9	Caulk and Weatherstrip	5,741	—	5,741	0
10	Lath, Plaster, and Stucco	196,240	—	196,240	0
11	Ceramic Tile	22,290	—	22,290	0
12	Roofing and Sheet Metal	206,712	8,268	198,444	4
13	Resilient Flooring	26,085	—	26,085	0
14	Acoustical Tile	32,305	—	32,305	0
15	Painting	96,091	—	96,091	0
16	Glass and Glazing	71,853	—	71,853	0
17	Terrazzo	91,976	—	91,976	0
18	Miscellaneous Metals	79,642	—	79,642	0
19	Finish Hardware	55,178	—	55,178	0
20	Plumbing, Heating, Air-Cond.	622,064	55,985	566,079	9
21	Electrical	392,160	19,608	372,552	5
22	Clean Glass	3,228	—	3,228	0
23	Paving, Curb, and Gutter	81,724	—	81,724	0
		$3,602,138	$375,276	$3,226,862	
24	Change Order No. 1	5,240	—	5,240	0
	Total Contract Amount	$3,607,378	$375,276	$3,232,102	10.4
A	Cost of Work Performed to Date	$375,276			
B	Materials Stored on Site (Schedule Attached)	67,699			
C	Total Work Performed and Materials Stored	$442,975			
D	Less 10% Retainage	44,297			
E	Net Work Performed and Materials Stored	$398,678			
F	Less Amount of Previous Payments	180,369			
G	Balance Due This Payment	$218,309			

Figure 9.4 Periodic estimate for partial payment.

2. To the total completed work is added the value of all materials stored on the site, but not yet incorporated into the work. The cost of stored materials includes that of the subcontractors and is customarily set forth in a supporting schedule. From the total of work in place and materials stored on site is subtracted the retainage. This gives the total amount of money due the contractor up to the date of the pay request. From this is subtracted the amount of progress payments already made. The resulting figure gives the net amount now payable to the prime contractor.

Although the pay request procedure for lump-sum contracts has been in general use for many years, it has one serious defect. As shown in Figure 9.4, the project is divided for payment purposes into relatively few work classifications, most of which are extensive and often extend over appreciable portions of the construction period. This situation can make it difficult to estimate accurately the percentages completed of the various work categories. Actual measurement of the work quantities accomplished to date is the key to accurate percentage figures, but this can become very laborious and therefore most of the percentages are established by visual appraisal or other approximate procedure. Contractors want these estimates to be fair representations of the actual work achieved, but understandably, they do not want them to be too low. Hence most of their percentage estimates are apt to be on the generous side.

This circumstance continues to produce vexing problems for both the contractor and the owner. If it is difficult for the contractor to estimate the completion percentages accurately, it is at least equally difficult for the architect-engineer or owner to check these reported values. This presents the architect-engineer with a difficult problem because, in the interest of its client, it must make an honest effort to see that the monthly payments made to the contractor are reasonably representative of the actual progress of the job. Architect-engineers are at times casual with the processing of a pay request, feeling that a delay in payment will offset the generous nature of the completion percentages provided by the contractor. Although effective, this is often at odds with contract provisions regarding payment. It also ignores the fact that the progress payments are being reduced by the prescribed retainage.

9.25　PAYMENT REQUESTS FOR UNIT-PRICE CONTRACTS

Payment requests under unit-price contracts are based on actual quantities of each bid item completed to date. The determination of quantities accomplished in the field is done in several different ways depending on the nature of the particular bid item. When cubic yards of aggregate, tons of asphaltic concrete, or bags of portland cement are set up as bid items, these quantities are usually measured as they are delivered to the work site. Delivery tickets or fabricator's certificates are used to establish tons of reinforcing or structural steel. Other work classifications, such as cubic yards of excavation, lineal feet of pipe, or cubic yards of concrete, are measured in place or computed from field dimensions. Survey crews of the owner and of the contractor often make their measurements independently and adjust any differences.

Many owners use their own standard forms for monthly pay estimates. On unit-price contracts, the owner usually prepares the pay request and sends it to the contractor for checking and approval before payment is made. On unit-price contracts that involve a substantial number of bid items, each monthly pay request is a sizable document consisting of many pages. In essence, the total amount of work accomplished to date on each bid item is multiplied by its corresponding contract unit price. All of the bid items are totaled and the value of materials stored on the site is then added. From this total is subtracted the prescribed retainage. The resulting figure represents the entire amount due the contractor for its work to date. The sum of all prior progress payments that have already been paid is then subtracted, this yielding the net amount of money payable to the contractor for its work that month.

9.26 PAYMENT REQUESTS FOR COST-PLUS CONTRACTS

Negotiated contracts of the cost-plus variety provide numerous methods for making payments to contractors. In many cases, contractors furnish their own capital, receiving periodic reimbursements from owners for costs incurred. Other contracts provide that owners will advance money to the contractors for the purpose of meeting their payrolls and paying other expenses associated with the work. One scheme is where the contractor prepares estimates of expenses for the coming month and receives the money in advance. Then, at month's end, the contractor prepares an accounting of the actual expenses. Any difference between the estimated expenses and actual expenses is adjusted with the next monthly estimate. Other contracts provide for zero-balance or constant-balance bank accounts where checks are written by the contractor and funds are furnished by the owner. In any event, the contractor must make periodic accountings to the owner of the cost of the work, either to receive direct payment from the owner or to obtain further advances of funds.

A periodic payment to the contractor under a cost-plus type of contract is not usually based on quantities of work performed but on expenses incurred by the contractor during the preceding pay period. Consequently, such pay requests consist primarily of the submission of cost records. Copies of invoices, payrolls, vouchers, and receipts are submitted in substantiation of the contractor's claims. In addition to cost records of payments made by the contractor to third parties, the periodic pay requests customarily include equipment expense and a pro rata share of the negotiated fee. If the contractor is furnished funds by the owner to pay for construction costs, the owner is credited with all cash discounts.

Because of the sensitive nature of cost reimbursement, it is common practice to maintain a separate set of accounting records for each cost-plus project. When the size of the project is substantial enough to justify it, a field office is established where all matters pertaining to payroll, purchasing, disbursements, and record keeping for the project are performed. Project financial records are either routed through the owner's representative or are available for inspection at any time. This procedure does much to eliminate misunderstandings and facilitates the final audit. Cost-plus contracts with public agencies customarily impose special conditions on the contractor pertaining to form of payment, payment application, affidavits, and preservation of project records.

9.27 FINAL PAYMENT

The procedural steps leading up to acceptance of the project and final payment by the owner vary with the nature of the work and the specific provisions of the contract. In building construction, the process typically commences when the contractor, having achieved substantial completion, requests a preliminary inspection. The owner or his authorized representative, in company with general contractor and subcontractor personnel, inspect the work. A list of deficiencies to be completed or corrected is prepared and the architect-engineer issues a certificate of substantial completion. After the deficiencies have been remedied, a final inspection is held and the contractor presents its application for final payment. If the work is determined to be complete and acceptable, the architect-engineer issues a certificate for final payment which includes retainage still withheld by the owner. Under a lump-sum form of contract, the final payment is the total contract price less the sum of all progress payments previously made. With a unit-price contract, the final total quantities of all payment items are obtained, and the final contract amount is determined. Final payment is again equal to the total contract price less the sum of all previous payment installments made by the owner.

Construction contracts typically require that the general contractor's request for final payment be accompanied by a number of different documents. For example, releases or waivers of lien executed by the general contractor, all subcontractors, and material suppliers are common requirements on privately financed jobs. An affidavit that the releases and waivers furnished include all parties who might be entitled to lien may also be required. Other contracts call for an affidavit certifying that all payrolls, bills for materials, payments to subcontractors, and other indebtedness connected with the work have been paid or otherwise satisfied. Construction contracts frequently require the contractor to provide the owner with as-built drawings, various forms of written warranties, maintenance bonds, and literature pertaining to the operation and maintenance of job machinery. Consent of surety to final payment is an almost universal prerequisite. Subparagraph 9.9.2 of Appendix C is an example of a contractual provision concerning documentation required for final payment.

If there are unpaid construction costs still outstanding at the time of request for final payment, the contractor must report the parties to whom money is owed and the amounts due. When the owner has been informed of sums payable on the project, either by affidavit from the contractor or by direct notice from the unpaid party, the owner may become liable for these debts if sufficient funds are not withheld from the general contractor to cover them. If the general contractor certifies payment of all bills and the owner has not received any notification of debt, the owner is generally entitled to rely on the general contractor's statement and make final payment. There can be exceptions to this rule, however, in those states whose lien statutes impose unlimited liability on the owner's property. In such cases, the owner must require releases of lien from all possible claimants.

9.28 PAYMENTS TO SUBCONTRACTORS

When the owner makes payment to the general contractor, it is expected that the general contractor will then make prompt payment to its subcontractors. Payment to subcontractors, although a routine process for the general contractor, is often a troublesome one and requires attention and care. Indeed, the wise general contractor will apply many of the same precautions used by the owner in paying the general contractor to its paying of the subcontractors. It does seem clear, however, that the safeguards applied will depend considerably on the subcontractor and the contractor's past experience with that subcontractor.

The general contractor must check each monthly pay request from a subcontractor to ensure that it is a fair measure of work actually performed. The prime contractor does not wish to allow its subcontractors to be overpaid any more than the owner wants to overpay the general contractor. To assist in this regard, the prime contractor frequently requires the subcontractors to submit cost breakdowns for use in checking their requisitions.

When the general contractor makes payment to a subcontractor, it is understandable that the general contractor wants assurance that the subcontractor is meeting its financial obligations. In this regard, the contractor should use certain precautions to protect itself from the hazard of a subcontractor's failure to pay its bills and meet its payrolls. An affidavit that the subcontractor is keeping its accounts current can be helpful, although it must be recognized that not all such certifications are bona fide or accurate. Waivers of lien or further corroboration in the form of receipts and other satisfactory evidence of payment can be required.

There continue to be instances where the general contractor, after making payment to a subcontractor, is required to pay that subcontractor's workmen and suppliers. Subcontracts frequently provide that the general contractor can withhold payments from a subcontractor if it appears the subcontractor is not paying its bills, although such a clause may make the contractor liable to third persons as third-party beneficiaries unless care is used in the subcontract language.

Payment is frequently made to a subcontractor without specification by the general contractor as to how the payment shall be applied by the recipient. Some caution may be in order here if the two parties have dealings on more than one job. In many jurisdictions, if the general contractor does not designate, the subcontractor may apply the payment to a job other than the one for which payment is being made. This follows from a generally recognized rule of law that when one who owes two or more separate debts makes a payment, he has the right to designate which shall be credited. If he fails to do so, the creditor can choose. If proper application of payment is important, the general contractor can indicate on the check the application of proceeds desired and include a letter of transmittal that specifically states the purpose for which the payment is being made. If a general contractor makes payment to a subcontractor's supplier by a check payable to both, specific instructions on how the payment should be credited by the payees should be provided.

There is another side of the coin, however. A not uncommon situation is when payment to the subcontractor is delayed because the prime contractor diverts the funds to other areas of its own business. This may follow from the general contractor's attempt to perform a volume of work in excess of its financial capability, resulting in payments from the newer jobs being used to pay the bills of older work. The subcontractor is often reticent to press for payment, not wanting to antagonize the general contractor. In the final analysis, however, the subcontractor can seek relief in different ways including the general contractor's payment bond, a suit for breach of contract, and filing either a mechanic's lien or stop order, depending on whether the owner is private or public. There is a fourth avenue for unpaid subcontractors and material dealers in those states with construction trust-fund statutes. A construction trust-fund statute declares that the funds paid to a prime contractor by the owner are trust funds held by the contractor as trustee for the benefit of its subcontractors and material suppliers. Such statutes afford these parties an additional avenue of recovery from the general contractor.

It occasionally happens that the general contractor will incur or assume an expense that is chargeable to a subcontractor. The general contractor may pay some of the subcontractor's bills; temporarily provide the subcontractor with labor, equipment, materials, or facilities; or perform cleanup or hauling services on behalf of the subcontractor. The usual way for the general contractor to recoup these expenses is to subtract them from payments made to the subcontractor. Such a deduction is called a "backcharge" and may lead to a dispute between the general contractor and the subcontractor who is backcharged, unless the amount involved is discussed and preferably agreed to in advance.

9.29   PAYMENTS TO MATERIAL SUPPLIERS

Payments by the contractor to its material suppliers are made in accordance with the terms of the applicable purchase order or usual commercial terms. Payment for materials is not normally dependent on any disbursements made by the owner to the general contractor but is due and payable in full 30 days after invoice date, receipt of materials, end of the month in which delivery was made, or other stipulation. Such payment terms frequently include provisions for cash discounts to encourage early payment by the contractor.

With reference to the payment for materials, most contractors have established some system of checking and control. To illustrate a typical procedure, a purchase order is issued for all project materials of any consequence. Such a purchase order not only specifies the cost of the materials but also serves as an internal control for the contractor. As deliveries are made, the materials are inspected, quantities are verified, and any quality control or acceptance tests are performed. After inspection and testing, a receiving report is sent to the disbursement section of the contractor's office listing the materials received, identifying the vendor, and describing any shortages, damage, or variations from specified quality. No payment of the vendor's invoice is made until a receiving report has been received from the project or the contractor's warehouse attesting to proper delivery and until the invoice amount has been proved against the purchase order. Backcharges, as discussed in the preceding section, can also apply to material vendors.

9.30 CASH FLOW

Many contractors have gotten into deep financial trouble when they suddenly and unexpectedly ran out of cash to pay their bills. Although it can happen to any business, cash flow is one of the major causes of failure for small construction firms. Cash flow refers to a contractor's income and outgo of cash. The net cash flow is the difference between disbursements and income over a period of time. A positive cash flow indicates that cash income is exceeding disbursements and a negative cash flow signifies just the opposite. Cash is the fuel that runs the business, and a contractor must maintain a cash balance sufficient to meet payrolls, pay for materials, make equipment payments, and satisfy other financial obligations as they become due.

In itself, however, cash is not a productive asset; it must be invested in some way to make it productive. Therefore, management wants to have enough funds readily available for needs as they occur but to keep excess cash suitably invested. By the same token, if a contractor must borrow funds on which to operate, it wishes to recognize this need as early as possible and does not want to borrow any more than is necessary. The timing of cash flow is especially important in determining a firm's working capital requirements, and the future cash position of a construction firm is of great importance to its management.

9.31 CASH FORECASTS

A cash forecast is a schedule that summarizes the estimated cash receipts, estimated disbursements, and available cash balances for some period into the future. The forecast, starting with a known beginning cash balance, estimates cash income and disbursements, either on a weekly or monthly basis, yielding an estimated cash balance at the beginning of each period. For the most part, only short-term future cash predictions are optimally useful for contractors because of the unpredictability of new contract acquisition as well as other financial matters generally. These figures will indicate when the available cash balance will be below minimum needs and when it will be above. This, in turn, gives advance warning that additional funds must be obtained by borrowing or that excess funds will be available for investment or company growth.

The preparation of a cash forecast begins with the collection of detailed information regarding future cash income and expenditures. For a construction contractor this must usually first be done for each individual project based on its proposed progress schedule. The resulting cash-flow figures (the net result of cumulative cash income and outgo on a time basis) are then combined with the company's general and administrative disbursements to develop a total cash-flow forecast for the given planning period. The computational process is complicated somewhat by the fact that both project earnings and expenses occur on a discrete basis, and these are normally not on the same time frequency. To illustrate, project income is received in some regular and periodic way such as monthly payments. Project expenses are extremely variable in their payment patterns. Payrolls are on a weekly basis, whereas disbursements to material dealers and subcontractors are on a monthly basis. Payment of taxes, insurance premiums, and certain equipment expenses are largely uncoordinated with the physical progress of the work.

When done properly, the preparation of a cash forecast is a complex and time-consuming operation, but the resulting management information can make the expense and effort well worthwhile. After the forecast estimates are made, they must be reviewed periodically to determine their continued validity. When material differences develop, new estimates are prepared and the forecast revised.

9.32 THE MECHANIC'S LIEN

A mechanic's lien is a right created by law to secure payment for work performed and materials furnished in the improvement of land. This statutory right attaches to the land itself in much the same way as a mortgage. The purpose of a lien statute is to permit a claim on the premises where the value or condition of real property has been increased or improved and where suitable payment has not been made by the owner. Following the lead of Maryland in 1791, every state has enacted some form of mechanic's lien law. These laws are similar in general import but differ considerably in detail. Lien laws are strictly construed by the courts, and full compliance with all provisions of the local statute is mandatory.

Lien laws are based on the theory of unjust enrichment and are designed to protect workmen, material suppliers, and, under certain conditions, general contractors and subcontractors. Rights accruing to general contractors and subcontractors are the same, but there are differences in the notices required of the two parties. For a general contractor to obtain a lien, a usual requirement is that the owner agreed to have the work done and to pay for it. In some states, a written contract must exist. In case of default by a private owner on a construction contract, the general contractor actually has two remedies available. It can file suit against the owner for breach of contract, or it can exercise its right of lien. In some cases the contractor takes both courses of action.

Architect-engineers are given lien rights by some state statutes. However, for such a lien right to be available, work usually has to commence on the property. Design service, in and of itself, may not be sufficient unless construction has actually proceeded. In some states only services that directly benefit the land, such as supervision, apply; design itself is not included. Because of wide variations in state laws, generalizations about the lien rights of design professionals can be badly misleading, and the intricacies of individual state statutes must be investigated.

Public property is not subject to a statutory lien, although some lien laws provide that a worker, material supplier, or subcontractor who has not been paid by the prime contractor can file a notice of lien claim or "stop notice" with the owner. A stop notice advises the owner that the party has not been paid, and the owner is required to hold up payments to the prime contractor until the debt is satisfied. Where permitted by statute, this same provision applies to private owners or lenders as well.

Although lien laws are beneficial, there are many drawbacks associated with them. For the claimant to thread his way through the intricacies of the law requires a number of time- and money-consuming measures. Recovery is technical, cumbersome, and often unsatisfactory for all concerned.

The lien rights of parties dealing with the general contractor depend on whether the state in which the work was performed has a lien statute based on the New York system or the Pennsylvania system. Under the New York system, the amount such a party can collect is limited to the amount due the general contractor from the owner. If nothing is due when the lien is filed, the lienor may not look to the owner of the premises for payment. However, in states that abide by the Pennsylvania system, the owner is not so protected. The unpaid party has a right to file a lien even if the entire contract amount has been paid to the prime contractor. Correspondingly, the owner may be forced to pay for some of the contract work twice should the general contractor fail to pay his subcontractors or material suppliers. Withholding payment from the general contractor in amount sufficient to cover open accounts and waivers of lien afford the owner protection against such possible double payment. In some states with the Pennsylvania system, an owner can limit liens to the contract price by filing a copy of the contract and posting a surety bond.

9.33 FILING A LIEN CLAIM

For any person who contracts directly with the owner to be able to obtain a lien, the statutes typically require him to record a notarized claim for public record with a county authority within the statutorily prescribed time. In most jurisdictions, this lien claim is considered sufficient if it names the owner, describes the project, contains appropriate allegations as to the work performed or materials furnished, and states the amount and from whom it is due. The time for filing differs from state to state, but general contractors typically have 60 days after the filing by the owner of a notice of project completion or, if no notice is filed, 90 days after the physical completion of the work that is the subject of the lien. Many states require, in addition to the filing of a lien claim, that written notice also be given to the owner or his agent within a specified time. In some cases, notice to the owner must precede the filing of the claim. The advice of a local attorney can be invaluable when filing a claim of lien.

Laborers, subcontractors, or material dealers who contract with the general contractor rather than directly with the owner are also entitled to liens, but the statutory requirements for these people are often different from those for the general contractor. They must not only file a notice of lien for the public record, but also are usually required to give notice in writing to the owner or his agent. The time limit for such filing is prescribed by statute and is frequently different from that required of people contracting directly with the owner. In addition, the statutory time for filing a claim may begin with the date when the last labor or materials were furnished rather than with project completion.

9.34 FORECLOSURE OF LIEN

Once the lien claim is filed and no payment has been made, proceedings must be brought to enforce the lien, usually within 90 days after filing. The court procedure by which the claim is judicially determined is known as a foreclosure action; it varies from state to state but is highly technical and requires the services of a lawyer. If the evidence substantiates the claim under the mechanic's lien and the court finds that there is a sum of money due and owing, the court can order the property to be sold and the proceeds used to satisfy the indebtedness. The sale of the property is made at auction by the sheriff, and a certificate of sale is executed to the successful bidder. The holder of the lien is paid from the proceeds of the sale. Most lien statutes give the original owner a prescribed period of time to redeem his property on payment of the judgment, interest, and costs.

The priority of claims is a matter treated by the pertinent state statute. A mortgage subsequent to the construction contract is usually inferior to a mechanic's lien that arises out of work done under the contract. Mortgages outstanding at the time of the contract may or may not be inferior, depending on the law. As a general rule, different liens stand on an equal footing with one another regardless of the order in which they are filed. An exception is the preferred status usually given to workmen and material dealers.

9.35 WAIVER OF LIEN

The right to lien may be waived in a number of ways, depending on the particular statute. Construction contracts sometimes include a clause whereby the general contractor agrees not to file or place any liens against the owner's premises and waives its right in this regard. After the contract has been signed, this clause is binding on the contractor. The courts have long held that a contractor, by the terms of a contract, may waive its right of lien. However, some states have enacted legislation that makes the waiver of a mechanic's lien void as against public policy, hence unenforceable.

A broader form of waiver of lien sometimes used in construction contracts provides that no liens shall be filed by the general contractor or any subcontractor or material dealer. This proviso can be made binding on the subcontractors and material dealers provided it is permitted by the statute involved and provided certain actions are taken by the general contractor, these actions depending on the requirements of the law. For example, the contractor may be required to give timely notice of the waiver before the purchase orders and subcontracts are signed, either by direct notification in writing to the subcontractors and material dealers or by making the waiver agreement a matter of public record. Additionally, the subcontracts and purchase orders may have to include a clause that expressly provides for the waiver of lien by the subcontractors and material dealers. This is a consequence of the fact that lien statutes often provide that the right of lien may be waived only by an express agreement in writing specifically to that effect.

Rather than the contractor waiving its right of lien through the medium of a contract clause to that effect, it may be required to submit a signed waiver form to the owner as a condition of receiving final payment or perhaps each time that it requests a progress payment. Similar waivers are often also required from the subcontractors and material vendors. Standard waiver of lien forms are available from suppliers of commercial and business forms and are widely used for the purpose.

9.36 ASSIGNMENTS

Practically all rights arising out of contracts are assignable. A common contractual right is that of receiving payment in exchange for the performance of a stipulated contractual duty. Assignment means the transfer of such a right from the party to whom the right belongs by contract to a third party. There are certain limitations to the general policy that permit people entitled to receive money to assign their rights to others, but they seldom apply to construction contracts. To provide security for a loan or to pay off an impatient creditor, a contractor may assign funds to become due under a construction contract to a lending institution or the creditor. After notice of the assignment has been given to the owner, he must make payments to the assignee. Basically, the assignee acquires the same but no greater rights than his assignor had, and the owner is required only to make payments as required by contract. The owner cannot be placed in a worse position than he would have been if the assignment had not been made. For example, payment by the owner to the assignee can be conditioned by the contractor's failure to perform properly.

A party to a contract can assign his rights under the contract without the consent of or notice to the other contracting party (the debtor). Freedom of assignment, however, can be regulated by the terms of the contract itself. Construction contracts usually contain provisions that expressly forbid the owner or the general contractor from assigning either the contract as a whole or funds due or to become due under the contract to third parties without the written consent of the other (see Subparagraph 7.2.1 of Appendix C.)

Subcontract forms used by general contractors customarily include a restriction of assignment (see Article 3.9 of Appendix L). The general contractor is often requested by subcontractors to approve an assignment of moneys due or to become due under a subcontract. By assignment of such funds to a bank, for example, the subcontractor can receive a loan from the bank to finance its operations. It is important to note that the failure of the general contractor to honor a subcontractor's assignment of funds has resulted in liability of the general contractor to the assignee. On receipt of notice of an assignment, the general contractor generally has the duty to pay the assignee whether it has accepted the assignment or not. This follows from the Uniform Commercial Code, which provides if an assignment is made as security for a loan, consent of the other party is not required despite any contract clause to the contrary.

PROJECT MANAGEMENT

10.1 THE NEED FOR PROJECT MANAGEMENT

The essential focus of a construction company is its field projects. These are, after all, the heart and soul of any such business enterprise. If a project is to be constructed within its established budget and time schedule, close management control of field operations is a necessity. Project conditions such as technical complexity, importance of timely completion, resource limitations, and substantial costs put great emphasis on the planning, scheduling, and control of construction operations. Unfortunately, the construction process, once it is set into motion, is not a self-regulating mechanism and requires expert guidance if events are to conform to plans.

It must be remembered that projects are one-time and largely unique efforts of limited time duration which involve work of a nonstandardized and variable nature. Field construction work can be profoundly affected by events that are difficult, if not impossible, to anticipate. Under such uncertain and shifting conditions, field construction costs and time requirements are constantly changing and can seriously deteriorate with little or no advance warning. Skilled and unremitting management effort is not only desirable, it is absolutely imperative for a satisfactory result.

10.2 PROJECT ORGANIZATION

All construction projects require some field organization, although large jobs will obviously require considerably larger organizations than smaller jobs. Terminology differs somewhat from one construction firm to another and organizational patterns vary, but the following description is more or less representative of current trade practice.

The management of field construction is customarily run on a project basis, with a project manager being made responsible for all aspects of the work. Project management cuts across functional lines of the parent organization, and the central office acts in a service role to the field projects. Working relations with a variety of outside organizations including architect-engineers, owners or owner representatives, subcontractors, material and equipment dealers, labor unions, and regulatory agencies are an important part of guiding a job through to its conclusion. Project management is directed toward pulling together all the diverse elements involved into a going venture with the common objective of project completion.

The form and extent of a project's organization depend on the nature of the work, size of the project, and type of construction contract. A firm whose jobs are not particularly extensive will have essentially all office functions, such as accounting, payroll, and purchasing, concentrated in its main or area office. Only larger projects can justify the additional overhead necessary to carry out the required office tasks in a field office on the jobsite. Extensive projects frequently support a substantial field management team, the extent of this depending on the nature of the work, its geographical location, and the type of contract. For example, a large cost-plus contract might well have all associated office functions performed at the project site. A project management staff is customarily developed along much the same lines as the contractor's main operating organization.

10.3   THE PROJECT MANAGER

The project manager organizes, plans, schedules, and controls the field work and is responsible for getting the project completed within the time and cost limitations. He acts as the focal point for all facets of the project and brings together the efforts of those organizations having inputs into the construction process. He coordinates matters relevant to the project and expedites project operations by dealing directly with the individuals and organizations involved. In any such situation where events progress rapidly and decisions must be consistent and informed, the specific leadership of one person is needed. Because he has the overall responsibility, the project manager must have broad authority over all elements of the project. The nature of construction is such that he must often take action quickly on his own initiative, and it is necessary that he be empowered to do so. To be effective the project manager must have full control of the job and be the one voice that speaks for the project. Project management is a function of executive leadership and provides the cohesive force that binds together the several diverse elements into a team effort for project completion.

Large projects normally will have a full-time project manager who is a member of the firm's top management or who reports to a senior executive of the company. The manager may have a project team to assist him or he may be supported by a central office functional group. When smaller contracts are involved, a single individual may act as project manager for several jobs simultaneously.

The project manager must have expertise and experience in the application of specialized management techniques for the planning, scheduling, and control of construction operations. These are procedures that have been developed specifically for application to construction projects and are those discussed in Chapters 11 and 12. Because much of the project management system is often computer-based, the project manager must have access to adequate computer support services.

10.4 THE PROJECT SUPERINTENDENT

Project management and field supervision are quite different responsibilities. The day-to-day direction of project operations is handled by a site supervisor or project superintendent. His duties involve supervising and directing the trades, coordinating the subcontractors, working closely with the owner's or architect-engineer's field representative, checking daily production, and keeping the work progressing smoothly and on schedule. He is responsible for material receiving and storage, equipment scheduling and maintenance, job safety, and job records and reports.

Centralized authority is necessary for the proper conduct of a construction project, and the project manager is the central figure in that respect. Nevertheless, some freedom of action by the field superintendent is required in field construction work. In practice, construction project authority is wielded much as a partnership effort with the project manager and project superintendent functioning much as allied equals. Notwithstanding this, however, top company management must make a clear delegation of authority to the project manager and must also make sure the duties and responsibilities of each are explicitly understood.

On large projects, the field staff normally includes a field engineer who reports either to the project manager or the project superintendent. The engineer is assigned such responsibilities as project scheduling, progress measurement and reporting, progress billing, shop drawings, keeping of job records and reports, cost studies, testing, job engineering and surveys, safety and first aid, and payrolls.

Project Engineer

10.5 ASPECTS OF PROJECT MANAGEMENT

In general terms, project management might be described as the judicious allocation and efficient usage of resources to achieve timely completion within the established construction budget. The resources required are money, manpower, equipment, materials, and time. Considerable management effort is required if the contractor is to meet its construction objectives. The achievement of a favorable time-cost balance by the careful scheduling and coordination of labor, equipment, and subcontractors, and the maintaining of a material supply to sustain this schedule requires endeavor and skill. The project organization must blend the quantitative techniques of scientific management together with the subjective ingredients of experienced judgment and intuition into an effective and efficient operating procedure. Astute project management requires at least as much art as science—as much human relations as management techniques.

The details of a job management system depend greatly upon the contractual arrangements with the owner. Basic to any contractor's project management system, however, is the control of project time and cost during the construction period. Before field operations begin, a detailed time schedule of operations and a comprehensive construction budget are prepared. These constitute the accepted time and cost goals that will be used as a flight plan during the actual construction process. After the project has been started, monitoring systems are established that measure actual costs and progress of the work at periodic intervals. The reporting system provides progress information that is measured against the programmed targets. Comparison of field costs and progress with the established plan quickly detects exceptions that must receive prompt management attention. Data from the system can be used to make corrected forecasts of costs and time necessary to complete the work. In addition to the close management of project time and costs, any system of project control must include various actions known collectively as "project administration." Chapters 11 and 12 of this text discuss the essentials of time and cost control of field construction. This chapter describes a number of essential aspects of project administration.

10.6 PROJECT ADMINISTRATION

Project administration refers to those actions that are required to achieve the established project goals. These involve duties that may be imposed by the construction contract or that are required by good construction and business practice. The efficient handling, control, and disposition of contractual and administrative matters in a timely fashion are of paramount importance for a smooth-running job.

Specifically, project administration refers to those practices and procedures, usually routine, that keep the project progressing in the desired fashion. Whatever is required to provide the project, efficiently and in a timely manner, with the materials, labor, equipment, and services required lies within the general jurisdiction of project administration.

10.7 PROJECT MEETINGS

On larger projects it is common practice for the owner and the architect-engineer to meet with contractor project personnel before field operations actually begin. The purpose of this meeting is to establish important ground rules and to call the contractor's attention to certain critical areas. It permits key personnel of both sides to be introduced to each other and allows the parties to arrive at clear understandings concerning a variety of mutual concerns such as construction site surveys, submittal of shop drawings, sampling and testing, project inspection, the authority of the owner field representative, payment requests, handling of claims, change orders, and similar matters.

The agenda of such meetings usually includes reminders to the contractor concerning insurance certificates, required permits, cost breakdowns, lists of subcontractors, the construction schedule, the schedule of owner payments, and other actions required before the start of construction. Such contractual provisions as the completion date, liquidated damages, bonus clauses, safety program, and extensions of time may also be discussed. This pre-construction meeting gives the contractor the opportunity to raise questions and clear up misunderstandings.

Once construction has started, regular jobsite meetings are standard practice on larger projects. Usually held once a week, such meetings are run by the project manager with representatives of the owner, material vendors, and subcontractors attending. Minutes of these meetings are kept and copies distributed to interested parties. At such meetings, job progress is discussed with trouble spots being identified and corrective action planned. These regular meetings are very valuable in the sense that all parties concerned are kept fully informed concerning the current job status and impressed with the importance of meeting their own obligations and commitments. Such face-to-face exchanges are invaluable in resolving misunderstandings and quickly getting to the real source of project problems and difficulties.

10.8 SCHEDULE OF OWNER PAYMENTS

A common contract provision requires the contractor to provide the owner with an estimated schedule of monthly payments that will become due during the construction period. This information is needed by the owner so that cash will be avilable as needed to make the necessary periodic payments to the contractor. Because the owner must often sell bonds or other forms of securities to obtain funds with which to pay the contractor, it is important that the anticipated payment schedule be as accurate a forecast as the contractor can make it.

The most accurate basis for determining such payment information is the project time schedule (see Section 11.23) that is established by the contractor either before field operations start or very early in the construction period. By establishing the total costs associated with each scheduled segment of the project and by making some reasonable assumption concerning how construction cost varies with time (a linear relationship is often assumed), a reasonably accurate prediction of the value of construction in place at the end of each month can be made. By taking retainage into account. these data can be reduced to an estimated schedule of owner monthly payments. These values may have to be subsequently revised as the project progresses.

10.9 SHOP DRAWINGS

The working drawings and specifications prepared by the architect-engineer, although adequate for job pricing and general construction purposes, are not suitable for the fabrication and production of many required construction products. Manufacture of the necessary job materials and machinery often requires that the contract drawings be amplified by detailed shop drawings* that supplement, enlarge, and clarify the contract design. Such descriptive technical submissions are prepared by the producers or fabricators of the materials and are submitted to the general contractor and thence to the architect-engineer for approval before the item is supplied. This procedure also applies to materials provided by subcontractors, in which case shop drawings are submitted to the general contractor through the subcontractor. Shop drawings are required for almost every product that is fabricated away from the building site. To illustrate, in building construction, shop drawings must be prepared for everything from reinforcing steel and glazed tile to millwork and finish hardware. The general contractor must sometimes prepare shop drawings covering work items or appurtenances it has designed and will fabricate.

10.10 APPROVAL OF THE SHOP DRAWINGS

When shop drawings are first received from a supplier, the contractor is responsible for checking them carefully against the contract drawings and specifications. The shop drawings are then forwarded to the architect-engineer for examination and approval. The checking and certifying of these drawings is properly the responsibility of the architect-engineer, since they are basically a further development and interpretation of the design, and the final verification is the designer's responsibility. Paragraph 4.12 of Appendix C describes the workings of a typical approval procedure.

A sufficient number of copies of each drawing must be provided so there are enough for distribution to all interested parties. After the approved drawings have been returned by the architect-engineer, the contractor notes the nature of any comments or corrections and routes copies as necessary, returning at least one copy to the supplier. Occasionally the shop drawings must be redone and resubmitted to the architect-engineer. Because the material supplier must receive a copy of the approved shop drawings before placing the order into production, it is important that the submittal-approval-return process be expedited as much as possible. Otherwise, materials may not be delivered to the project by the time they are needed. It is usual for the contractor to establish some form of check-off and reminder system with regard to shop drawings to guard against oversight or delay in the approval process.

It is important to note that only a qualified approval is given to shop drawings by architect-engineers. Such approval relates only to conformance with the design concept and overall compliance to the contract drawings and specifications. Checking does not include quantities, dimensions, fabrication methods, or construction techniques. Approval of shop drawings by the architect-

---

*The term "shop drawings" includes such fabrication, erection, and setting drawings; manufacturer's standard drawings or catalog cuts; performance and test data; wiring and control diagrams; schedules; samples; and decriptive data pertaining to material, machinery, and methods of construction as may be necessary to carry out the intent of the contract drawings and specifications.

engineer does not relieve the contractor of its responsibility for errors or inadequacies in the shop drawings or for any failure to perform the requirements and intent of the contract documents. Approval of shop drawings does not authorize any deviation from the contract unless the contractor gives specific notice of the variance and receives express permission to proceed accordingly. However, as long as there is no explicit disagreement between the shop drawings and the construction contract, approval is usually binding in the event of a subsequent dispute over design requirements. For obvious reasons it is important that contractors carefully check their project shop drawings. They cannot act merely as go-betweens with suppliers and architect-engineers. An interesting aspect of approved shop drawings is that they are not usually considered to be contract documents, and they do not modify or extend the obligation of either party to the construction contract.

10.11 QUALITY CONTROL

Quality control during field construction has to do with ensuring the work is accomplished in accordance with the requirements specified in the contract. The architect-engineer establishes the criteria for construction, and the quality control program checks contractor compliance with those standards. A field quality control program involves inspection, testing, and documentation for the control of the quality of materials, workmanship, and methods. On a given project, the quality control program may be administered by the architect-engineer, the owner, consultants, or the prime contractor.

By the terms of most of the usual design contracts between owner and architect-engineer, the designer accepts only very limited responsibility for field operations. For example, Subparagraph 1.5.4 of Appendix A provides that the designer will make periodic site visits but will not assume any responsibility for continuous on-site inspections unless provided for under a supplementary agreement. If the architect-engineer does agree to administer the project quality control program, it will assign a full-time employee to the project to provide administrative and surveillance services. Laboratory and field testing may be done by this person, or a commercial testing laboratory may be engaged to perform the specialized work.

Many government agencies and corporate owners establish their own internal quality control programs to monitor their ongoing construction projects. Where there is a continuous construction program or where very large and complex projects are involved, the owner may establish a functional department within his overall organization that acquires trained personnel and develops standards for quality control application in the field.

Owners sometimes hire specialized consultants to provide field quality control services. In this context, a consultant is a technical firm such as a testing laboratory, specialized consulting engineer (not the project designer), or construction mangement firm that provides quality control services. Such consultants are frequently employed on extremely complex construction or on projects that have specialized and highly technical quality control requirements. These firms are entirely independent of the architect-engineer and contractor and are usually hired before the design is completed to provide advance quality assurance input and advice.

During recent years, some public and private organizations have required prime contractors to take a more active role in the control of project quality by making them develop and manage their own quality control programs. Construction contracts with these owners require the contractor to maintain a job surveillance system of its own and to perform such inspections that will assure the work performed conforms to contract requirements. The contractor is required to maintain and make available adequate records of such inspections. Under such contracts, the contractor is required to provide significant and specific inspection and documentation to satisfy both itself and the owner that the work being performed meets the contract requirements. In the usual case, the contractor is required to report on a daily basis the construction progress, problems encountered, and corrective action taken, and to certify that the completed work conforms to the drawings and specifications. The owner is responsible for final inspection and may inspect at any other time deemed necessary to ensure strict compliance with the contract provisions.

The last aspect of quality control is the final inspection, field acceptance testing, and start-up of the facility. The prime contractor, appropriate subcontractors, and manufacturer's representatives start up the project equipment and systems with the owner checking all control and instrument operation. This includes simulation of operating and emergency conditions. The facility is turned over to the owner with a complete set of job files, shop drawings, maintenance and operating manuals, and as-built drawings.

10.12 EXPEDITING

Construction is of such a nature that the timely delivery of project materials is of extreme importance. If required items are not available when needed, the contractor can experience major difficulties because of the disruption of the construction schedule. Such delays are expensive, awkward, and inconvenient, and every effort must be made to avoid them. When purchase orders are written, delivery dates are designated which, if met, will ensure that the materials will be available when needed. These dates are established on the basis of the project progress schedule and must necessarily make allowance for the approval of shop drawings.

Unfortunately, the contractor cannot assume that the designation of delivery dates in its purchase orders or the securing of delivery promises from the sellers will automatically ensure that the materials will appear on schedule. To obtain the best service possible, a series of follow-up actions, referred to as expediting, are taken after each material order is placed to keep the supplier constantly reminded of the importance of timely delivery. Expediting may be a jobsite function or the construction firm may provide all of its construction projects with a centralized expediting service. A full-time expediter is sometimes required on a large project. On work where the owner is especially concerned with job completion, or where certain material deliveries are crucial, the owner will often participate with the contractor in cooperative expediting efforts.

A necessary adjunct to the expediting function is the maintaining of a check-off system or log where the many steps in the material delivery process are recorded. Starting with the issuance of the purchase order, a record is kept of the dates of receipt of shop drawings, their submittal to the architect-engineer, receipt of approved copies, return of the approved drawings to the vendor, and delivery of the materials. Because shop drawings from subcontractors are submitted for approval through the general contractor, the check-off system will include materials being provided by the subcontractors. This is desirable because project delay can be caused by any late material delivery, regardless of who provides the material. This same documentation procedure is followed for samples, mill certificates, concrete-mix designs, and other submittal information required. General contractors sometimes find it necesssary on critical material items to determine the manufacturer's production calendar, testing schedule if required, method of transportation to the site, and data concerning the carrier and shipment routing. This kind of information is especially helpful in working the production and transportation around strikes and other delays.

Each step in the approval, manufacture, and delivery process is recorded and the status of all materials is checked frequently. At intervals, a material status report is forwarded to the project manager for his information. This system enables job management to stay current on material supply information and serves as an early-warning device when slippages in delivery dates seem likely to occur.

The intensity with which the delivery status of materials is monitored depends on the nature of the materials concerned. Routine materials such as sand, gravel, brick, and lumber usually require little follow-up. Critical made-to-order items, whose late delivery would badly cripple construction operations, must be closely monitored. In such cases, the first follow-up action would be taken weeks or months in advance of the scheduled delivery date. This action, perhaps a letter showing order number, date of order, and delivery promise, would request specific information on the anticipated date of shipment. Return answers to such inquiries can be very helpful. If a delay appears forthcoming, strong and immediate action is necessary. Letters, telegrams, telephone calls, and personal visits, in that order, may be required to keep the order progressing on schedule.

10.13 DELIVERIES

In addition to working for the timely delivery of materials, the expediter is also usually responsible for their receipt, unloading, and storage. In general, deliveries are made directly to the projects to minimize handling, storage, insurance, and transportation costs. However, there are often instances where it is preferable or necessary to store materials temporarily at off-site locations until they are needed on the job.

When notice of a material delivery is received, suitable receiving arrangements must be made. Advance notice of shipments is provided directly by the vendors or through bills of lading or other shipping papers. If a shipment is due on a jobsite, notice is given to the project superintendent. If suitable unloading equipment is not available on the site, such equipment must be scheduled or the project superintendent be authorized to obtain whatever may be required. If a shipment is to be made to the contractor's storage yard or warehouse, the person in charge must be advised and unloading equipment scheduled if required. In the case of rail-car shipments, unloading must be accomplished before expiration of the free unloading period or the car goes under demurrage. Demurrage is a daily charge made by the railroad until the car is released by the contractor.

The scheduling of material deliveries to the jobsite can be especially important on some projects. For example, consider the delivery of structural steel to a building project in a downtown city area. On projects of this type, storage space is extremely limited and deliveries must be carefully scheduled to arrive in the order needed and at a rate commensurate with the advancement of the structure. Such projects obviously require very careful scheduling of deliveries and the close cooperation of the contractor and the steel supplier. An additional factor is the routing of the trucks through the city streets, often at off-hours, and arranging for the direction of traffic around the vehicles during the delivery and unloading operations. Arrangements for necessary permits, police escorts, labor, and unloading equipment must be made in advance. On such projects, many material deliveries are not made directly to the site but to temporary storage facilities owned or rented by the contractor. When such off-site storage is used, deliveries to the jobsite are made in accordance with short-term job needs.

10.14 RECEIVING

Delivery of job materials is usually made directly to the jobsite. There are times, however, when it is either undesirable or impossible to accept shipments at the project. Construction in congested urban areas is an instance already mentioned. Another example is early delivery of items that would be susceptible to damage, loss, or theft if stored on the job for extensive periods of time. Whenever possible, such materials are stored in the contractor's yard or warehouse until they are needed.

Truck shipments may be made by common carriers or the vendor's own vehicles. In either case, the material must be checked for damage as it is being unloaded and quantities checked against the freight bill or vendor's delivery slip. Observed damage must always be noted on all copies of the freight bill and be witnessed by the truck driver's signature. The receiver should not sign the delivery slip or freight bill until he has checked the quantity delivered against that indicated.

When shipment is made by rail car, the contractor advises the carrier where it desires the car to be spotted as soon as the contractor is advised of the car number. The shipment should be checked after the car is placed for unloading and any visible damage reported to the railroad claim agent. In case of damage, unloading must be deferred until the shipment has been inspected and proper notations made on the bill of lading. A claim for damage or loss is submitted to the freight claim agent on the standard form of the carrier.

This claim must be accompanied by the original bill of lading, the receipted original freight bill, the original or a certified copy of the vendor's invoice, and other information in substantiation of the claim. Should damage be such that it is not visible and cannot be detected until the goods are unpacked, the contractor must make its claim at that time on the carrier's special form which is used for concealed damage. Rail shipments of less-than-carload (LCL) may be such that the contractor must pick up the material at the freight depot, or it may be delivered to his project or yard by truck. Delivery depends on the FOB point designated by the purchase order.

The party who receives a shipment on behalf of the contractor should immediately transmit the covering delivery ticket, freight bill, or bill of lading to the contractor's office. Information pertaining to damage or shortage and location of material storage should be included.

Although most purchase orders include freight charges in their face amounts, material vendors do not always prepay freight charges. As a result, materials often arrive at the contractor's location with freight charges collect. The contractor usually has an account with the carrier that allows it to receive the goods without having to pay the transportation charges at the time of delivery. However, in the case of common carriers, Interstate Commerce Commission regulations require that freight charges be paid within a short time of delivery. Therefore, it is important that freight bills for collect shipments be transmitted immediately to the contractor's office for payment. Where the purchase order amount included freight, it is usual for the contractor to pay the freight charges and back-charge the account of the vendor.

10.15 INSPECTION OF MATERIALS

Inspection of delivered goods for quantity and quality should preferably be done concurrently with their unloading and storage. This is not always possible, however, and often would involve an objectionable delay in the unloading and release of transporting equipment. The checking of the package count as shown by the freight bill or delivery ticket should always be done, with any variations being indicated on the bill or ticket. However, the quantities of different items and quality often must be verified at the first opportunity after receipt. To do this, the party making the inspection obtains copies of the covering purchase order and approved shop drawings. A thorough check of the delivered items is made to verify both item quantities and quality. This verification can become quite laborious in some cases, but can pay big dividends in minimizing later job delays because of missing, faulty, or erroneous materials. The project inspector will frequently participate and assist in this inspecting process. Inspection of material deliveries must be done with reasonable promptness so that there is time to take any corrective measures that may prove necessary.

After the delivered materials have been reconciled with the shop drawings and purchase orders, the inspector will normally file a brief inspection report with the company's procurement section. This report shows the purchase order number, the date of inspection, the material, the location of storage, the quantity, remarks, and the signature of the inspector. The inspection report verifying receipt of the proper count and quality clears the order and authorizes payment to the vendor. With partial shipments, several inspection reports may be required to clear the entire order.

Inspection duties at times involve the sampling of various kinds of construction materials. Construction contracts may require the laboratory testing of certain materials as proof of quality. Thus inspectors must be acquainted with the standard methods of sampling sand, gravel, bulk cement, asphalt, reinforcing steel, and other construction commodities. Another aspect of materials inspection is the obtaining of certification of quality or the results of laboratory control tests from the manufacturer or producer. Submittal of the manufacturer's certification of quality may enable the contractor to avoid duplicate acceptance testing.

10.16 SUBCONTRACTOR SCHEDULING

As important members of the field construction team, subcontractors have an obligation to pursue their work in accordance with the project schedule established by the prime contractor. Failure of a subcontractor to commence its operations when required and to pursue its share of the work diligently can be a serious matter for the general contractor. Consequently, the scheduling of subcontractors deserves and receives appropriate action by the general contractor. A practice followed by many contractors is to notify each subcontractor by letter two weeks or more before the subcontractor is expected to move onto the project and commence operations. Subcontractors must be given adequate time to plan their work and make the necessary arrangements to start their operations. Follow-up telephone calls are made if needed. In the interest of good subcontractor relations, the project manager should not schedule a subcontractor to appear on the site until the job is ready and the subcontracted work can proceed unimpeded. After the subcontractor is on the job, its progress must be monitored to ensure its operations are keeping pace with the overall project time schedule. If its work falls behind, the project manager may reasonably instruct the subcontractor to take appropriate measures to accelerate its progress.

With regard to the general matter of subcontractor scheduling, the form and content of the subcontract can be very important. A carefully written document with specific provisions regarding conformance with time schedule, material orders, and shop drawing submittals can strengthen the project manager's hand in keeping all aspects of the project on schedule. In this context, a common problem is the failure of a subcontractor to order major materials in ample time to meet the construction schedule. General contractors occasionally find it advisable to monitor their subcontractors' material purchases. This can be accomplished by including a subcontract requirement that the subcontractor submit unpriced copies of its purchase orders to the general contractor within ten days after execution of the subcontract. In this way, the general contractor can oversee the expediting of key materials provided by subcontractors along with its own.

10.17 RECORD DRAWINGS

A common general contract requirement is that the contractor must maintain and prepare one set of full size contract drawings marked to show various kinds of "as-built" information. These drawings show the actual manner, location, and dimensions of all work as actually performed. This involves marking a set of drawings to show details of work items that were not performed exactly as they were originally shown, such as changed work, changed site conditions, and variations in alignment or location. In addition, details and exact dimensions are given for those work items that were not precisely located on the original contract drawings. Depths, locations, and routings of electrical service and underground piping and utilities are examples of this point. The set of record drawings is prepared by the contractor as the work progresses and is turned over to the architect-engineer or owner at the end of the project.

10.18 DISBURSEMENT CONTROLS

To coordinate the actions of the company accounting office with the project, it is necessary to implement a system of disbursement controls. These controls are directed toward controlling payments made to vendors and subcontractors and require that no such payments be made without proper approval from the field. The basic purpose of disbursement control is twofold: first, to ensure that payment is made only up to the value of the goods and services received to date and second, to see that total payment does not exceed the amount established by the purchase order or subcontract.

Payments made for materials are based on the terms and conditions of covering purchase orders. Copies of all job purchase orders are provided to the project manager for his use and information. Purchase order disbursement by the accounting office is conditioned on the receipt of a signed delivery ticket or receiving report from the jobsite. Suitable internal controls are established to ensure that total payments do not exceed the purchase order amount. Any change in purchase order amount, terms, or conditions is in the form of a formal written modification with copies sent to the jobsite.

Disbursements to subcontractors follow a similar pattern. Because there are no delivery tickets or receiving reports for subcontractors, all subcontractor invoices are routed for approval through the project manager, who has copies of all the subcontracts. The project manager determines if the invoice reflects actual job progress and approves the invoice or makes appropriate changes. General contractors normally withhold the same percentage from their subcontractors that owners retain from them. If the subcontractor bills for materials stored on site, a common requirement is that copies of invoices be submitted to substantiate the amounts billed. Any change to a subcontract is accomplished by a formal change order.

10.19 JOB RECORDS

To serve a variety of purposes, a documentation system is needed on each project that will produce a comprehensive record of events that transpired during the construction period. The extent to which this is done and the job records that are maintained is very much a function of the provisions of the construction contract and the size, complexity, and risks inherent in the work. It is up to the contractor to establish what records are appropriate for a given project and to see that they are properly kept and filed.

The original estimating file and the contract documents are obvious basic job records. During the construction phase, periodic progress reports, cost reports, the job log, correspondence, minutes of job meetings, time schedules, subcontracts, purchase orders, field surveys, test reports, progress photographs, shop drawings, and change orders are routinely maintained as a permanent project record. On larger projects additional records on manpower, equipment, operating tests, back charges, piledriving and welding records, progress evaluation studies, and others are kept.

10.20 THE DAILY JOB LOG

A job log is a historical record of the daily events that take place on the jobsite. The information to be included is a matter of personal judgment, but the log should include everything relevant to the work and its performance. The date, weather conditions, job accidents, numbers of workers, and amounts of equipment should always be noted. It is advisable to indicate the numbers of workers by craft and to list the equipment items by type. A general discussion of the daily progress, including a description of the activities completed and started and an assessment of the work accomplished is important. Where possible and appropriate, the quantities of work put into place can be included. The job log should be maintained in a hardcover, bound booklet or journal. Pages should be numbered consecutively in ink with no numbers being skipped. Every day should be reported and all entries made on the same day as they occur. No erasures should be made, and each diary entry should be signed immediately under the last line of the day's entry.

The diary should list the subcontractors who worked on the site together with the workers and equipment provided. Note should be made of the performance of subcontractors and how well they are conforming to the project time schedule. Material deliveries received must be noted together with any shortages or damage incurred. It is especially important to note when material delivery dates are not met and to record the effect of such delays on job progress and costs.

The diary should include the names of visitors to the site and facts pertinent thereto. Visits by owner representatives, the architect-engineer, safety inspectors, union representatives, and people from utilities and government agencies should be documented and described. Meetings of various groups at the jobsite should be recorded, including the names of people in attendance, problems discussed, and conclusions reached.

Complete diary information is occasionally necessary to substantiate payment for extra work and is always needed for any work that might involve a claim. The daily diary should always include a description of job problems and what steps are being taken to correct them. The job log is an especially important document where disputes result in arbitration or litigation. To be accepted by the courts as evidence, the job diary must meet several criteria. The entries in the log must be original entries made on the dates shown. The entries must have been made in the regular course of business and must constitute a regular business record. The entries must be original entries, made contemporaneously with the events being recorded and based on the personal knowledge of the person making them. Where the above criteria have been met, the courts have generally ruled that the diary itself can be entered as evidence, even if the author is not available to testify.

10.21 CLAIMS

Construction contracts typically require the contractor to advise the owner in writing, within a prescribed period of time, concerning any event that will result in delay and/or additional cost. The actual claim against the owner follows and would include a detailed description of the job condition underlying the claim, identification of the contract provisions under which the claim is made, details of how the condition has caused the extra cost and/or delay, and a summary of the increased costs and extension of time requested.

A claim dispute arises when the contractor gives the owner written notice that a condition exists that requires additional contract cost or time and the owner refuses to recognize it by initiating a suitable change order. Most claims result from (1) changes, (2) changed conditions, (3) delay or interference, (4) acceleration, (5) errors or omissions in design, (6) suspension of the work, (7) variations in bid-item quantities, or (8) rejection of or-equal substitutions. Almost any extra cost or time caused to the contractor by the action or inaction of the owner or the owner's agent can be a valid basis for a claim against the owner. Refusal by the owner to recognize the claim does not authorize the contractor to refuse to continue. However, the contractor should proceed with the work in dispute only after filing a written protest with the owner.

Settlement of claims under the contract will ultimately be made in accordance with the provisions of the contract or as provided by law. This may be either in the courts, by arbitration, or through contract appeal boards. Regardless of the nature of the tribunal, however, the contractor often does not prevail, not because the claim has no merit but because of the contractor's inability to prove actual damages. This inadequacy is usually a direct result of the fact the contractor did not maintain adequate records to support its position. Records that are sufficient for project management are not necessarily adequate for proof of a claim.

In the final analysis, the successful settlement of a disputed claim depends largely on painstaking documentation. Preparation of a claim may be based on the routine project records compiled during the construction process. However, these records may not contain the detailed information needed to substantiate such a demand. For this reason, the contractor may be well advised to maintain a special set of records that pertain specifically to the matter in dispute. The standard dictum of "put everything in writing" applies here. It is to be noted in this regard that, to be effective, documentation must be prepared during the construction process. Records created after the fact will not generally receive consideration.

10.22 ACCELERATION

Acceleration refers to the owner's directing the prime contractor to accelerate its performance so as to complete the project at an earlier date than the current rate of work advancement will permit. If the job has been delayed because of the fault of the contractor, the owner can usually order the contractor to make up the lost time without incurring liability for extra costs of construction. However, if the owner directs acceleration under the changes clause of the contract so that the work will be completed before the required contract date, the contractor can recover under the contract change clause for the increased cost of performance plus a profit.

There is another instance of acceleration, however, where additional costs may or may not be recoverable by the contractor, this often being referred to as constructive acceleration. This is a case where the owner either fails or refuses to grant the contractor a requested extension of time and still insists the work be completed within the original contract period. In such an instance, the contractor must be able to demonstrate certain key elements before being granted relief for the constructive change of acceleration. A constructive change is one where the owner or his representative has acted in a manner that has the same effect as a formal change to the contract under the changes clause. Along with the directive to accelerate, there must have been either actual or implied threats to take action if the completion date is missed. There must have been an excusable delay for which the contractor requested an extension of time and this request was denied. In addition, the contractor must have completed the contract on time and actually have incurred extra costs.

11

PROJECT PLANNING AND SCHEDULING

11.1 INTRODUCTION

Time is an important aspect of job control. If a construction project is to proceed efficiently and be completed within the contract time, the work must be carefully planned and scheduled in advance. Construction projects are complex, and a large job will involve literally thousands of separate operations. If these tasks were to follow one another in single file order, job planning and scheduling would be relatively simple, but this is not the case. Each operation has its own time requirement, and its start depends on the completion of certain preceding operations. At the same time, many tasks are independent of one another and can be carried out simultaneously. Thus a typical construction project involves many mutually dependent and interrelated operations that, in total combination, comprise a tangled web of individual time and sequential relationships. When individual task requirements of materials, equipment, and labor are superimposed, it becomes obvious that project planning and scheduling is a very complicated and difficult management function.

The traditional basis for the planning and scheduling of construction projects has been the bar chart. Admittedly, this graphical representation of work versus time is a useful and convenient device for depicting an established schedule of construction operations and recording its progress. However, the bar chart (see Section 11.24) falls far short of being an adequate tool for project planning, and the resulting construction schedule is based more on the contractor's experience and intuition than on any rational analysis of the work to be performed. The bar chart has a major weakness in that it does not show the interrelationships and interdependencies among the various phases of the work, and there is no way of determining which operations actually control the overall progress of the project. Modern conditions of intense competition, soaring costs, tight time schedules, and increasing complexity of construction procedures have lent great impetus to the development of more efficacious ways of planning and scheduling construction projects.

11.2 THE CRITICAL PATH METHOD

The critical path method (CPM) is a relatively recent management innovation developed especially for the time management of construction projects. CPM involves the analysis of the sequential and time characteristics of projects by the use of networks. It is a widely used procedure for construction time control, and contractors are now frequently required by contract to apply network methods to the planning and scheduling of their field work. Complete and comprehensive time management systems have been developed based on the CPM procedure. The reader is referred to books that describe these systems in detail.* This chapter confines itself to a discussion of basics.

CPM is a project management system that provides a basis for informed decision making on projects of any size. It provides information necessary for the time scheduling of a construction project, guides the contractor in selecting the best way to shorten the project duration, and predicts future manpower and equipment requirements.

The procedure starts with project planning. This phase consists of identifying the elementary items of work necessary to achieve job completion, establishing the order in which these work items will be done, and preparing a graphical display of this planning information in the form of a network.

The scheduling phase that follows requires an estimate of the time required to accomplish each of the work items identified. With the use of the network, computations are then made that provide information concerning the time-schedule characteristics of each work item and the total time necessary to achieve project completion.

The procedure just described may suggest that project planning must follow a definite step-by-step order of development. In actual practice this is not the case, the three planning steps proceeding more or less simultaneously with one another. However, for purposes of discussion, the three steps described will be treated separately in the order mentioned.

The computations associated with project scheduling are simple additions and subtractions. Manual computation is easy and logical but can become tedious and very time-consuming on large projects. For this reason many contractors utilize computers to produce their project schedules and to update them occasionally during construction in the field. The computer's role in the process is merely to perform the many arithmetic operations in a faster and more economical manner than would be possible with human effort alone. For a full understanding of the method and a thorough appreciation of the data generated, however, it is necessary that the practitioner be familiar with how the calculations are made. In addition, there are many important applications

*CONSTRUCTION PROJECT MANAGEMENT, second edition, Clough and Sears, Wiley-Interscience, New York, 1979.

of CPM in which manually developed data are adequate and usual. Consequently, the following discussion of planning and scheduling is based on manual procedures.

11.3 GENERAL CONSIDERATIONS

Although some preliminary study of a project may take place during the estimating or negotiation process, it is usual that the detailed planning be started immediately after the contractor has been awarded the construction contract. A characteristic of CPM is that its effectiveness depends on the accuracy of the input information and the skill and judgment with which the generated data are used. Consequently, the development and application of a project plan and schedule must be made the responsibility of people who are experienced in and thoroughly familiar with the type of field construction involved.

A project plan is usually the result of a study made by a group of people. Key persons involved with the job such as estimators, field supervisors, and the project manager are usual participants. A consultant might also be involved, particularly when the contractor's own staff planners have no background of practical CPM application. Because the prime purpose of CPM is to produce a coordinated project plan, key subcontractors should also be brought into the planning. Their input information is vital to the development of a workable construction schedule. Normally, the prime contractor sets the general timing reference for the project. The individual subcontractor then reviews the portions of the plan relevant to its own work and needed alterations are made.

The basic procedure followed by such a planning group is to "talk the project through" from start to finish. In so doing, the job is subjected to careful, detailed advance planning. It is usual that the network diagram be developed in rough form as the job is dissected into its basic elements and the sequential order of construction operations is discussed. It is often helpful to list the major operations of the project and to use them to develop a preliminary diagram. This diagram can serve as a basis for discussion and as a basic framework for the subsequent development of a fully detailed network. It is important that the plan prepared be the one the contractor actually expects to follow. This means, therefore, that the people preparing the plan must have authority to make decisions concerning methods, procedures, equipment, and manpower.

11.4 PROJECT PLANNING

Planning is the devising of a workable scheme of operations to accomplish an established objective when put into action. Besides being the most time-consuming and difficult aspect of the job management system, planning is also the most important. It requires an intimate knowledge of construction methods combined with the ability to visualize discrete work elements and to establish their mutual interdependencies. If planning were to be the only job analysis made, the time would be well spent. It involves a depth and thoroughness of study that gives the construction team an invaluable understanding and appreciation of job requirements.

For purposes of planning, the project must first be broken down into elemental, time-consuming activities. An activity is a single discrete work step in the total project. The extent to which the project is subdivided into activities depends on a number of practical considerations, but the following factors should be taken into account:

1. Different areas of responsibility, such as subcontracted work, which are distinct and separate from that being done by the prime contractor directly.
2. Different categories of work as distinguished by craft or crew requirements.
3. Different categories of work as distinguished by equipment requirements.
4. Different categories of work as distinguished by materials such as concrete, timber, or steel.
5. Distinct and identifiable subdivisions of structural work such as walls, slabs, beams, and columns.
6. Location of the work within the project which necessitates different times or different crews to perform.
7. Owner's breakdown for bidding or payment purposes.
8. Contractor's breakdown for estimating purposes.

The activities chosen may represent relatively large segments of the project or may be limited to small steps. For example, a concrete slab may be a single activity, or it may be broken down into the erection of forms, placing of reinforcing steel, pouring of concrete, finishing, curing, and stripping of forms. If the activity breakdown is too gross, the job plan developed will not yield information in sufficient detail to be optimally useful. However, if the subdivision of the work is carried to the other extreme, the excessive detail tends to obscure the truly significant planning factors. Experience and observation of practices used by other contractors are probably the best tutors in this regard.

As the separate activities are identified and defined, the sequential relationships among them are determined. This is referred to as "job logic" and consists of the established time order of construction operations. When the time sequence of activities is being determined, restraints must be recognized and taken into consideration. Restraints are practical limitations of one sort or another that can influence or control the start of certain activities. For example, an activity that involves the placing of reinforcing steel obviously cannot start until the steel is on the site. Hence the start of this activity is restrained by the time required to prepare and approve the necessary shop drawings, fabricate the steel, and deliver it to the job. In like manner, the start of an activity may depend on the availability of labor, subcontractors, equipment, completed construction drawings, owner-provided materials, and other required inputs.

Failure to consider such constraints can be disastrous to an otherwise adequate job plan.

Some restraints are shown as time-consuming activities. For example, the preparation of shop drawings and the fabrication and delivery of job materials are material restraints that require time to accomplish and are depicted as activities on project networks. Restraints are also shown in the form of dependencies between activities. If the same crane is required by two activities, the equipment restraint is imposed by having the start of one activity depend on the finish of the other.

11.5 PRECEDENCE NOTATION

As the identification of activities proceeds and their logic is established, the resulting job plan is depicted graphically in the form of a network. There are two commonly used symbolic conventions used to portray network activities. One, called "precedence notation," depicts each activity as a rectangular box. The other shows each activity as an arrow and is called "arrow notation." Both kinds of diagrams are widely used, but precedence diagrams are considered to have some important advantages. For this reason, precedence notation will be used herein.

In drawing a precedence network, each time-consuming activity is portrayed by a rectangular figure. The dependencies between activities are indicated by dependency or sequence lines going from one activity to another. The identity of the activity and a considerable amount of other information pertaining to it are entered into its rectangular box. This matter will be developed further as the discussion progresses.

11.6 THE PRECEDENCE DIAGRAM

The preparation of a realistic precedence diagram requires time, work, and experience with the type of construction involved. The management data extracted from the network can be no better than the diagram itself. The diagram is the key to the entire time control process. When the network is first being developed, the planner must concentrate on job logic. The only consideration, at this stage, is to establish a complete and accurate picture of activity dependencies and interrelationships. Restraints that can be recognized at this point should be included. The time durations of the individual activities are not of concern during the planning stage.

Each activity in the network must be preceded either by the start of the project or by the completion of a previous activity. Each path through the network must be continuous with no gaps, discontinuities, or dangling activities. Consequently, all activities must have at least one activity following, except the activity that terminates the project. With the usual project network, it is not possible to have the finish of one activity overlap beyond the start of a succeeding activity. When such a condition exists, the work must be further subdivided. It follows, therefore, that a given activity cannot start until *all* those activities immediately preceding it have been completed.

Each activity is given a unique numerical designation with the numbering proceeding generally from project start to finish. Usual practice is that numbering is not done until after the network has been completed. Leaving gaps in the activity numbers is desirable so spare numbers are available for subsequent refinements and revisions. In this text, activities are numbered by multiples of ten. It is standard practice, as well as being a requirement of many computer programs, that precedence diagrams start with a single opening activity and conclude with a single closing activity.

11.7 EXAMPLE PROBLEM 1

To illustrate how the three planning steps of activities, job logic, and network go together, consider a simple project consisting of only a few steps, such as the construction of a heavy reinforced concrete slab on grade. This will be called Example Problem 1. Let us assume the job to be subdivided into eight activities as listed in Figure 11.1. The activity "Procure reinforcing steel" will be recognized as a job restraint.

The job logic or the time-sequence relationships among the activities of Example Problem 1 must now be determined. The sequence of operations will be the following.

The procurement of reinforcing steel, the excavation, and the building of forms are all opening activities that can proceed independently of one another. Fine grading will follow excavation, but forms cannot be set until both the excavation and form building have been completed. The placing of reinforcing steel cannot start until fine grading, form setting, and steel procurement have all been carried to completion. Concrete will be poured after the steel has been placed, and finishing will be the terminal operation, proceeding after the concrete pour.

EXAMPLE PROBLEM 1: REINFORCED CONCRETE SLAB ON GRADE		
Activity	Symbol	Activity Immediately Following
Excavate	EX	FG, SF
Build Forms	BF	SF
Procure Reinforcing Steel	PS	PR
Fine Grade	FG	PR
Set Forms	SF	PR
Place Reinforcing Steel	PR	PC
Pour Concrete	PC	FC
Finish Concrete	FC	- - -

Figure 11.1 Example Problem 1, job logic.

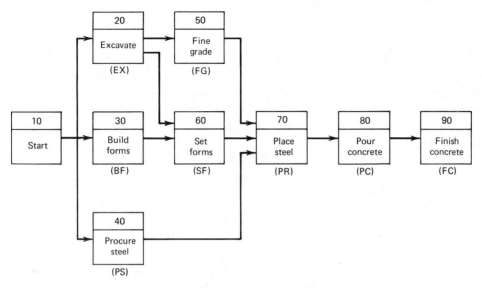

Figure 11.2   Example Problem 1, precedence diagram.

In elementary form, this is the kind of information generated as a project is talked through.

As has already been mentioned, the diagram is normally drawn concurrently with the development of the job plan, and the operational sequence is not otherwise recorded or listed. There are occasions, however, when it is useful to list the job logic.

Job logic can be listed in several different ways, no one procedure necessarily being any better than another. A common way of expressing the necessary activity sequence is to list for each activity those following activities that can start immediately after, and only after, the given activity is finished. This is a sufficient system for enumerating job logic and will be used herein. Figure 11.1 presents the previously discussed job logic of Example Problem 1. The precedence diagram describing the prescribed sequence of activities is shown in Figure 11.2.

Figure 11.2 is a complete project plan in graphical form. It shows what is to be done and in what order it will be accomplished. The diagram serves as the construction flight plan and points the way on a daily basis. A precedence network of the project is an extraordinarily efficient medium for communication among owner, architect-engineer, contractor, subcontractors, and other members of the construction team.

11.8 THE NETWORK FORMAT

A horizontal diagram format has become standard in the construction industry, although vertically oriented diagrams are sometimes used in conjunction with the construction of high-rise structures such as multistory buildings and chimneys. The general synthesis of a network is from start to finish; from project beginning on the left to project completion on the right. The sequential relationship of one activity to another is shown by the dependency lines between them and the essence of the network is the manner in which the constituent activities are joined together into a total operational pattern. In the usual precedence diagram, the lengths of the lines between activities have no significance because they indicate only the dependency of one activity on another. Arrowheads are not always shown on the dependency lines because of the obvious left-to-right flow of time. However, arrowheads are shown here for additional clarity.

During initial diagram development, the network is sketched emphasizing activity relationships rather than the appearance or style of the diagram. The first version is apt to be a freewheeling conglomeration of sweeping curves, random direction lines, and many sequence-line crossovers. Corrections, revisions, and erasures are plentiful. The rough appearance of the first version of the diagram is of no concern; its completeness and accuracy are. The finished diagram can always be put into a more tidy form at a later date.

The activity symbols appearing in Figures 11.1 and 11.2 are used only as a convenience in depicting the job logic. The use of symbols on the precedence network in lieu of writing out an abbreviated description of each activity is not recommended. For instance, "PR" might have been used for "Place reinforcing steel" or "PC" for "Pour concrete" in Figure 11.2. Even mnemonic codes such as this make a diagram difficult to read and the user must spend a considerable amount of time consulting the symbol listing to check the identity of activities. Networks are much more intelligible and useful when each activity is clearly identified on the diagram.

When a working precedence diagram is being prepared, the scale and spacing of the activities deserve attention. If the scale is too big and activities are widely spaced, the resulting network is likely to become so large that it is unmanageable. On the other hand, a small scale and overly compact makeup render the diagram difficult to read and inhibit corrections and modifications. With experience and observation of the work of others, practitioners will soon learn to adjust the scale and structure of their diagrams to the scope and complexity of the project involved. It must not be forgotten that the network is intended to be an everyday tool, used and consulted by a variety of people. Emphasis here is not on drafting elegance but on producing the most realistic and intelligible form of network possible.

Dependency lines that go backward from one activity to another should not be used. Backward, in this context, means going from right to left on the diagram which is against the established direction of time flow. Backward sequence lines are confusing and increase the chances of unintentional logical loops being included in the network. A logical loop involves the impossible requirement that an activity be followed by an activity that has already been accomplished. Logical loops are, of course, completely illogical, but they can be inadvertently included in large and complex networks if backwardly directed sequence lines are permitted. Crossovers occur when one dependency line must cross over another to satisfy job logic. Careful layout will minimize the number of crossovers, but there are usually some that cannot be avoided. Any convenient symbol can be used to indicate a crossover.

11.9 PROJECT SCHEDULING

In the discussion thus far, the vital concern has been with the identification of activities and their sequential constraints. Before the procedure can be carried further, attention must be given to time requirements. This is called the "scheduling phase" of CPM. The essential input information for project scheduling is the network diagram and a time duration estimate for each activity.

Project scheduling is accomplished by carrying out a series of simple computations that yield valuable project control information. An overall project completion time is first established. Times or dates within which each activity must start and finish if the established project completion date is to be met are then determined.

11.10 ACTIVITY DURATIONS

The first step in the scheduling process is to estimate the time necessary to carry out each activity. These durations are generally expressed in working days and do not include nonproductive periods such as holidays and weekends. Other units of time may be used (e.g., hours, weeks, or shifts). The only requirement is that the same unit of time be used throughout. The process of estimating activity durations will usually result in a certain amount of network refinement and activity redefinition.

Much of the value of the planning and scheduling process depends on the accuracy of the estimated activity durations. The following rules constitute the basis for this important step.

1. It is important that each activity be evaluated independently of all the others. For a given activity, assume that materials, labor, equipment, or other needs will be available when required. If there is reason to believe that this will not be so, then the use of a preceding restraint may be in order.
2. For each activity, assume a normal level of manpower and/or equipment. Exactly what "normal" is in this context is difficult to define. Based on experience, customary and relatively standard crew sizes and equipment spreads have emerged as being efficient and economical. In short, a normal level is intrinsic to the situation and is optimum insofar as expedient completion and minimum costs are concerned. At times normal may be dictated by the availability of labor or equipment. If shortages are anticipated, this factor must be taken into account.
3. A normal workday or week is assumed. Overtime or multiple shifts are not considered unless this is standard procedure or part of a normal work period. Around-the-clock operations are usual in many tunnel jobs, for example, and overtime is routinely utilized on highway projects during the summer months to beat the approaching cold weather. Some labor contracts guarantee overtime work as a part of the usual workday or workweek.

4. Activity durations must be estimated without regard to any predetermined contract completion date. Otherwise there is apt to be, consciously or unconsciously, an effort made to fit the activities within the total time available. The only consideration pertinent to estimating an activity time is how much time will be required to accomplish *that* activity, and that activity alone.

5. Use consistent time units throughout. It must be remembered, for example, that when working days are used, weekends and holidays are not included. Certain job activities such as concrete curing or plaster drying carry through nonworking days, and their times must be adjusted accordingly. Material delivery times are invariably given in terms of calendar days. A delivery time of 30 calendar days translates into approximately 21 working days.

Quantitative information is available to assist in the estimation of activity durations; for example, the estimating sheets compiled when the job was priced will yield the man-hours or equipment-hours for each activity. By making assumptions concerning the size of work crew or number of equipment units, activity durations can be computed. Time estimates of good accuracy can often be made informally. Off-the-cuff time estimates made by experienced construction supervisors usually prove to be surprisingly accurate. In any event, it is important that someone experienced in and familiar with the type of work involved estimate the activity times. Up to a point, the accuracy of duration estimates can be improved by subdividing the work into smaller units.

11.11 TIME CONTINGENCY

When the activity times are being estimated, it is assumed that the work will progress reasonably well. However, Murphy's Law is especially applicable to construction, and some safety factor must be included in project time estimates to make allowance for possible project disruptions of many kinds. Practice varies with regard to adding contingency allowances to the estimated durations of activities. There is general agreement, however, that a contingency allowance should not be applied to individual activities to allow for unforeseen project delays caused by fires, accidents, equipment breakdowns, strikes, late material deliveries, difficult site conditions, floods, legal delays, and the like. It is generally impossible, in such cases, to predict in advance which activities will be affected and by how much. A suitable contingency is better included in the time required for overall project completion (see Section 11.15).

How allowances for time lost because of inclement weather are handled depends on the type of work involved. When projects, such as most highway, heavy, and utility work, are affected by the weather, the entire project is normally shut down. As a result, it is usual that an estimated number of days to be lost because of weather be added to the overall project duration.

On buildings and other work that can be protected from the weather, allowances for time lost are commonly added to the durations of those activities susceptible to weather delay. It is relatively easy to establish average time losses for particular activity types during the season that they will be underway. After a job of this type passes a certain stage, it is not often completely shut down by bad weather. Although some parts of the project may be at a standstill, others can still proceed. Consequently, a better overall job of scheduling will possibly result if allowances for weather are included in the durations for those activities likely to be involved rather than adding a weather contingency to the entire project.

11.12 EXAMPLE PROBLEM 2

To illustrate subsequent discussion, let us now consider a somewhat more involved construction project which will be Example Problem 2. Although much simplified from an actual construction project, it will serve to illustrate the essential workings of the scheduling procedure.

The project to be discussed is a natural gas compressor station involving a prefabricated metal building, compressor foundation, standby butane fuel system, and compressor appurtenances such as piping, electrical services, and controls. Figure 11.3 presents the job logic and the estimated activity dura-

EXAMPLE PROBLEM 2: COMPRESSOR STATION			
Activity	Symbol	Duration (working days)	Activity Immediately Following
Grading	GR	7	CF, EXB
Granular Fill	GF	21	CC, BP, SF
Run-in Electrical	RE	42	EC
Procure Building	PB	28	ERB
Compressor Foundation	CF	8	CC, BP, SF
Excavate Butane	EXB	17	BUS
Butane Storage	BUS	34	PS
Connect Cooling	CC	11	PS
Bolts and Plates	BP	6	SC
Slab and Footings	SF	14	SC, ERB
Piping Systems	PS	15	CP
Set Compressor	SC	13	SCP
Electrical to Compressor	EC	20	SCP
Erect Building	ERB	29	WB
Wire Building	WB	17	TC
Set Control Panel	SCP	14	CP
Connect Piping	CP	10	TC
Test and Clean	TC	4	—

Figure 11.3   Example Problem 2, job logic and activity times.

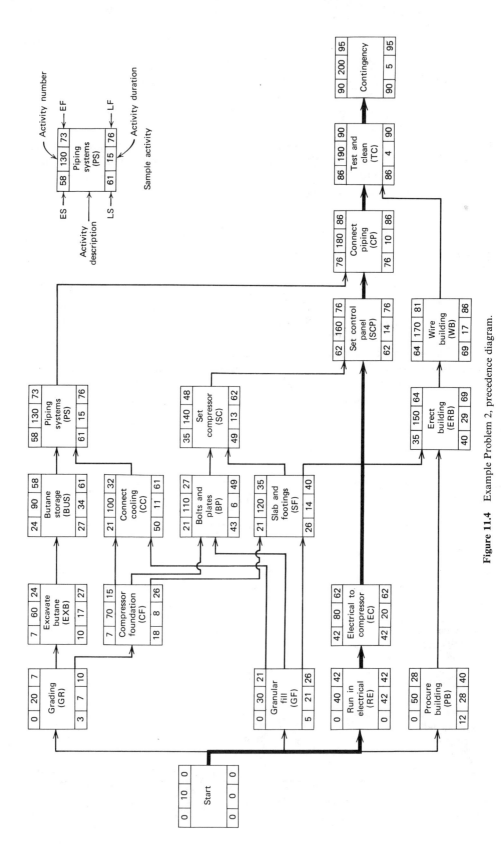

Figure 11.4 Example Problem 2, precedence diagram.

tions, and Figure 11.4 shows the completed precedence diagram. The sample activity in Figure 11.4 shows that each activity duration in terms of working days is shown in the lower, central part of the activity box. The identifying number of each activity is located in the upper, central part of the activity box. The other numerical values shown with the activities and the contingency activity at the right terminus of Figure 11.4 will be discussed subsequently.

11.13 NETWORK COMPUTATIONS

After a time duration has been estimated for each activity, some simple and step-by-step computations are performed. The purpose of these calculations is to determine (1) the overall project completion time and (2) the time brackets within which each activity must be accomplished if this completion time is to be met. The network calculations involve only additions and subtractions and can be made in different ways, although the data produced are comparable in all cases. The usual procedure is to calculate what are referred to as "activity times."

The calculation of activity times involves the determination of four limiting times for each network activity. The "early start" or "earliest start" (ES) of an activity is the earliest time the activity can possibly start allowing for the times required to complete the preceding activities. The "early finish" or "earliest finish" (EF) of an activity is the earliest possible time by which it can be completed and is determined by adding that activity's duration to its early start time. The "late finish" or "latest finish" (LF) of an activity is the very latest it can finish and allow the entire project to be completed by a designated time or date. The "late start" or "latest start" (LS) of an activity is the latest possible time it can be started if the project target completion date is to be met and is obtained by subtracting the activity's duration from its latest finish time.

The computation of activity times can be performed manually or by computer. When calculations are made by hand, they are normally performed directly on the network itself. When the computer is used, activity times are generally printed out in tabular form. When making the initial computations, it is usual that activity times be expressed in terms of *expired* working days. Commensurately, the start of the project is customarily taken to be at time zero.

11.14 EARLY ACTIVITY TIMES

The compressor station network shown in Figure 11.4 is used here to describe how the manual calculation of activity times is performed directly on a precedence diagram. Such networks are exceptionally convenient for the manual calculation of activity times and afford an excellent basis for describing how such calculations are made. The computation of the early start and early finish times is treated in this section. The determination of the late activity times will be described subsequently. The early time computations proceed from project start to project finish, from left to right in Figure 11.4, this process being referred to as the "forward pass." The basic assumption for the computation of early activity times is that every activity will start as early as possible. That is to say, each activity will start just as soon as the last of its predecessors is finished.

The ES value of each activity is determined first, with the EF time then being obtained. Reference to Figure 11.4 shows that activity 10 is the initial activity. Its earliest possible start is, therefore, zero elapsed time. As explained by the sample activity, shown in Figure 11.4, the ES of each activity is entered in the upper left of its activity box. The value of zero is commensurately entered at the upper left of activity 10 in Figure 11.4. The EF of an activity is obtained by adding the activity duration to its ES value. Activity 10 has a duration of zero. Hence the EF of activity 10 is its ES of zero added to its duration of zero, or a value of zero. EF values are entered into the upper right of activity boxes, and Figure 11.4 shows the EF of activity 10 to be zero. Activity 10 calculations are trivial, but the use of a single opening activity is customary with precedence diagrams and is required by some computer programs.

Figure 11.4 shows that activities 20, 30, 40, and 50 can all start after activity 10 has been completed. In going forward through the network, the earliest that these four activities can start is obviously controlled by the EF of the preceding activity. Since activity 10 has an EF equal to zero, then each of the following four activities can start as early as zero. Consequently, zero values appear in the upper left of activities 20, 30, 40, and 50. The EF value of each of these four is obtained by adding its ES of zero to its respective duration. For example, the EF value of activity 20 is zero plus 7 or a value of 7. Thus a 7 appears in the upper right of activity 20.

Continuing into the network, neither activity 60 nor 70 can start until 20 has been completed. The earliest that 20 can be finished is at the end of the seventh day, so the earliest that activity 60 or 70 can be started is after the expiration of seven working days (or at a time of 7). The ES of activities 60 and 70 are indicated to be 7 in Figure 11.4. The EF of activity 60 will be its ES value of 7 plus its duration of 17 or a value of 24. In like fashion, activity 70 will have an EF value of 15.

The start of activities 100, 110, and 120 all must await the completion of *both* activities 30 and 70. The earliest that activity 70 can be finished is 15, but activity 30 cannot be completed until 21. Because activities 100, 110, and 120 cannot start until *both* activities 30 and 70 are finished, the earliest possible start times for each of the three activities is time 21. Activities 100, 110, and 120 are examples of what is called a merge activity. This is an activity whose start depends on the completion of two or more antecedent activities. The computation rule for a merge activity on the forward pass is that its earliest possible start time is equal to the latest (or largest) of the EF values of the immediately preceding activities.

The forward-pass calculations consist of only repeated applications of the few simple rules just discussed. Working methodically in step-by-step fashion, the computations in Figure 11.4 proceed from activity to activity until the end of the project, activity 190, is reached.

11.15 PROJECT DURATION

Reference to Figure 11.4 discloses that the early finish time for the last work activity (190) is 90 elapsed working days. For the job logic established and the activity durations estimated, it will require 90 working days to complete the work. What this says is that if a competent job of planning has been done, if activity durations have been estimated with reasonable accuracy, and if everything goes reasonably well in the field, project completion can be anticipated in 90 working days or about $(7/5)$ x 90 = 126 calendar days.

The matter of contingency must again be considered at this point. Although allowances for lost time caused by inclement weather can be ascribed to the affected activities in some cases, an overall weather contingency is more appropriate in other cases (see Section 11.11). In addition, some provision must be made for general project delays caused by a variety of troubles, mischances, oversights, difficulties, and job casualties. Many contractors will, at this stage of the project scheduling, plan on a time overrun of five to ten percent and add this to the overall projected time requirement for the entire work. The percentage actually added must be based on a contractor's judgment and experience. In Figure 11.4 a contingency of five working days has been added to the diagram in the form of the final contingency activity 200. Adding an overall contingency of five working days gives a probable job duration of 95 working days. In the contractor's way of thinking, 95 working days represents a more realistic estimate of actual project duration than does the value of 90. If everything goes as planned, he will probably finish the job in about 90 working days. However, if the usual difficulties arise, he has allowed for a 95-working-day construction period.

Whether the contractor chooses to add in a contingency allowance or not, now is the time to compare the computed project duration with any established project time requirement. Following along with Example Problem 2 and assuming the contingency of five working days will be used, the figure of $(7/5)$ x 95 = 133 calendar days is compared with the required completion date. If the compressor station must be completed in 120 calendar days, then the contractor will have to consider ways in which to shorten the time duration. If a construction period of 140 calendar days is admissible, the contractor will feel reasonably confident that this requirement can be met and no action to shorten the work is required. How to go about decreasing a project's duration is the subject of Section 11.20.

The probable project duration of 133 calendar days is a valuable piece of information. For the first time, the contractor has an estimate of overall project duration that can be relied on with considerable trust.

11.16 LATE ACTIVITY TIMES

Project calculations now "turn around" on the value of 95 working days and a second series of calculations is performed to find the late start (LS) and the late finish (LF) times for each activity. These calculations, called the "backward pass," start at the project end and proceed backward through the network, going from right to left in Figure 11.4. The late activity times now to be computed are the latest times at which the several activities can be started and finished with project completion still achievable in 95 working days. The supposition during the backward pass is that each activity finishes as late as possible without delaying project completion. The LF value of each activity is obtained first and is entered into the lower right portion of the activity box. The LS, in each case, is obtained by subtracting the activity duration from the LF value. The late start time is then shown at the lower left.

The backward pass through Figure 11.4 is begun by giving activity 200 a LF time of 95 as shown in the lower right of the activity box. The LS of an activity is obtained by subtracting the activity duration from its LF value. Activity 200 has a duration of 5. Hence, the LS of activity 200 is its LF of 95 minus its duration of 5, or a value of 90. This value of 90 is entered at the lower left of activity 200 in Figure 11.4.

Figure 11.4 shows that activity 190 immediately precedes activity 200. In working backward through the network, the latest that activity 190 can finish is obviously controlled by the LS of its succeeding activity 200. If activity 200 must start no later than day 90, then activity 190 must finish no later than that same time. Consequently, activity 190 has an LF time equal to the LS of the activity following (200) or a value of 90. Activity 190, with a duration of 4, has a LS value of 90 minus 4, or 86. These values are shown on activity 190 in Figure 11.4.

Figure 11.4 shows that activity 190 is preceded by two activities: 170 and 180. The LF of each of the two antecedent activities is set equal to the LS of activity 190 or day 86. Subtracting the activity durations from the LF values yields the LS times. The LS times for activities 170 and 180 are, correspondingly, equal to 69 and 76.

Some explanation is needed when the backward pass reaches a burst activity, which is one that has more than one activity immediately following it. In Figure 11.4, activity 120 is the first such activity reached during the backward pass, this activity being followed immediately by activities 140 and 150. To obtain the late finish of activity 120, the late start of the immediately succeeding activities are noted. These are obtained from Figure 11.4 as 49 for activity 140 and 40 for activity 150. Keeping in mind that activity 120 must be finished before either activity 140 or 150 can begin, it is obvious that activity 120 must be finished by no later than day 40. If it is finished any later than this, the entire project will be delayed by the same amount. The rule for this and other burst activities is that the LF value for such an activity is equal to the earliest (or smallest) of the LS times of the activities following.

The backward-pass computations proceed from activity to activity until the start of the project is reached. All that is involved are repetitions of the rules just discussed.

11.17 TOTAL FLOAT

Examination of the activity times appearing in Figure 11.4 discloses that the early and late start times (also early and late finish times) are the same for certain activities and not for others. The significance of this is that there is time leeway in the scheduling of some activities and none at all in the scheduling of others. This leeway is a measure of the time available for a given activity above and beyond its estimated duration. This extra time is called float, two classifications of which are in general usage: total float and free float.

The total float of an activity is obtained by subtracting its ES time from its LS time. Subtracting the EF from the LF gives the same result. Once the activity times have been computed on the precedence diagram, values of total float are easily computed and may be noted on the network if desired, although this has not been done on Figure 11.4. Referring to Figure 11.4, the total float for a given activity is found as the difference between the two times at the left of the activity box or between the two at the right. The same value is obtained in either case. An activity with zero total float has no spare time and is, therefore, one of the operations that controls project completion time. For this reason, activities with zero total float are called "critical" activities. The second of the common float types, free float, will be discussed in Section 11.19.

11.18 THE CRITICAL PATH

In a precedence diagram, a critical activity is quickly identified as one whose two start times at the left of the activity box are equal. Also equal are the two finish times at the right of the activity box. Inspection of the activities in Figure 11.4 discloses that there are seven activities (10, 40, 80, 160, 180, 190, and 200) that have total float values of zero. Plotting these on the figure discloses that these seven activities form a continuous path from project beginning to project end, this chain of critical activities being called the "critical path." The critical path is normally indicated on the diagram in some distinctive way such as with colors, heavy lines, or double lines. Heavy lines are used in Figure 11.4.

Inspection of the network diagram in Figure 11.4 shows that numerous paths exist between the start and end of the diagram. These paths do not represent alternate choices through the network. Rather, each of these paths must be traversed during the actual construction process. If the time durations of the activities forming a continuous path were to be added for each of the many possible routes through the network, a number of different totals would be obtained. The largest of these totals is the critical or minimum time for overall project completion. Each path must be traveled, so the longest of these paths determines the length of time necessary to complete *all* of the activities in accordance with the established project logic.

If the total times for all of the network paths in Figure 11.4 were to be obtained, it would be found that the longest path is the critical path already identified using zero total floats and that its total time duration is 95 days. Consequently, it is possible to locate the critical path of any network by merely determining the longest path. However, this is not usually a practical procedure. The critical path is normally found by means of zero total float values. Although there is only one critical path in Figure 11.4, more than one such path is always a possibility in network diagrams. One path can branch out into a number of paths, or several paths can combine into one. In any event, the critical path or paths must consist of an unbroken chain of activities from start to finish of the diagram. There must be at least one such critical path, and it cannot be intermittent. A break in the path indicates an error in the computations. On the compressor station, 5 of the 18 activities (exclusive of start and contingency), or about 28 percent, are critical. This is considerably higher than for most construction networks because of the small size of the project. In larger diagrams, critical activities generally constitute 20 percent or less of the total.

Any delay in a critical activity automatically lengthens the critical path. Because the length of the critical path determines project duration, any delay in the finish date of a critical activity, for whatever reason, automatically prolongs overall project completion by the same amount. As a consequence of this, identification of the critical activities is an important aspect of job scheduling because it pinpoints those job areas that must be closely monitored at all times if the project is to be kept on schedule.

11.19 FREE FLOAT

Free float is another category of spare time. The free float of an activity is found by subtracting its early finish time from the earliest start time of the activities directly following. To illustrate how free floats are computed, consider activity 20. Figure 11.4 shows the earliest finish time of activity 20 to be 7. The activities immediately following are 60 and 70, and the earliest start time for both is 7. The difference between the earliest finish time of activity 20 (day 7) and the earliest start time of the activities immediately following (day 7) is zero. Hence, activity 20 has a free float of zero. Another example could be activity 70. Figure 11.4 shows the earliest finish time of activity 70 to be at time 15. The activities immediately following are 100, 110, and 120, and their earliest start time is 21. Thus, activity 70 has a free float of $21 - 15 = 6$ days. As an alternate statement of procedure, the free float of an activity can be obtained by subtracting its EF (upper right) from the smallest of the ES values (upper left) of those activities immediately following.

11.20 LEAST-COST PROJECT SHORTENING

If the 95-working-day construction time is not satisfactory, the contractor must reexamine the operational plan. It may be, for example, that the construction contract for Example Problem 2 stipulates a completion time of less than 95 working days. In the absence of some analytical procedure such as CPM, there is no way in which those activities that truly control total project time can be identified. Lacking this, the usual impulse is to accelerate most, if not all, of the job operations when the construction time for a project has to be reduced or when a project is falling behind schedule. This procedure is neither necessary nor economical, because there is seldom any need to speed up any but the critical job activities. On large construction networks, only 10 to 20 percent of the activities normally prove to be critical. Consequently, most job activities are floaters; that is, more time is available for them than is required by their estimated durations. There is nothing to be gained by diminishing even further the durations of such noncritical activities.

Because the critical path determines the overall job duration, the only way in which job duration can be shortened is to reduce the length of the critical path, either by revising the job logic or by shortening some of the individual critical activities, or perhaps both. The first action is to determine the possibility of performing some of the critical activities in parallel with one another rather than in series. Sometimes a localized reworking of job logic will make possible a shortening of the critical path. One caution is needed at this point: any revision of the network diagram may result in the appearance of new critical paths. Thus when the critical path is being shortened, the network must be continuously reexamined to ensure that all critical operations have been identified.

If the time reduction achieved by reworking the job logic is not sufficient, the durations of the critical activities themselves must be examined with the intent of decreasing certain of them. This shortening of critical activities cannot be done arbitrarily. It must be both practicable and feasible in the field.

Assume that it has become necessary to shorten the job duration of Example Problem 2 by means of reducing the times of the critical activities. For this purpose, there are only five critical activities that are susceptible to possible shortening. Activity 10, Start, is already at zero duration and activity 200, Contingency, must remain at five days. Consequently, each of the five critical activities (40, 80, 160, 180, and 190) must first be studied to ascertain whether it can be shortened and, if so, what the additional direct cost will be. When necessary, most activities can be completed in less than the normal time. Speed-up actions do increase the direct cost of the activity, however, because they involve the additional expense of overtime work, multiple shifts, more equipment, payment of premiums for quick delivery of materials, larger but less efficient crews, and similar speed-up actions. There is a point, of course, beyond which the duration of an activity cannot be compressed further.

Suppose the contractor on the compressor station of Example Problem 2 ascertains that it can shorten activity 40 (Run-in Electrical) by as much as three days by sending in another pole-setting truck crew. The additional expense of moving the extra truck and crew to and from the jobsite will be a single lump sum of $2000 whether one, two, or three days are saved. Activity 80 (Electrical to Compressor) can be diminished by one day if the electricians work overtime. The additional cost of premium time will amount to $400. Activity 160 (Set Control Panel) can be shortened by one day if the manufacturer's installation engineer does part of his work at night. This will require special lighting that will cost about $600 to install and remove. There does not seem to be any feasible way in which activity 180 (Connect Piping) can be shortened significantly. Activity 190 (Test and Clean) is already scheduled as a 24-hour-a-day operation and cannot be shortened.

The contractor's analysis discloses, first of all, that the original critical path can be shortened by a maximum of five days. This means that the probable job duration can be decreased from 95 to 90 days, assuming any new critical paths created during the process can also be shortened. The project duration cannot be reduced by more than five days, however, because the original critical path remains a critical path during the entire shortening process.

The critical activities to be shortened will be selected on the basis of least cost. The exact activities selected will depend on the number of days the project is to be reduced. In our Example Problem 2, the time shortening pattern and cost for each of the first three days of reduction will be as follows:

Step	Project Duration (working days)	Critical Activities Shortened	Total Direct Cost
0	95	—	—
1	94	Activity 80—1 day ($400)	$400
2	93	Activity 80—1 day ($400)	
		Activity 160—1 day ($600)	$1000
3	92	Activity 40—3 days ($2000)	$2000

This now reduces the original critical path to 92 days. In doing so, a path consisting of activities 10, 20, 60, 90, 130, 180, 190, and 200, whose total length is also 92 days, now becomes a second critical path. To shorten the project further, it is now necessary to compress both critical paths simultaneously. This could be done by diminishing either activity 180 or 190 because they are common to both critical paths. However, it has already been established that neither of these can be reduced. Further project shortening, therefore, depends on reducing concurrently the two parallel branches of the critical path. Suppose activity 90 is the cheapest way to shorten the new critical loop, and it can be decreased by as much as three days at an extra cost of $300 per day. The pattern and cost for the next two days' reduction will be:

Step	Project Duration (working days)	Critical Activities Shortened	Total Direct Cost
4	91	Activity 40—3 days ($2000)	
		Activity 80—1 day ($400)	
		Activity 90—1 day ($300)	$2700
5	90	Activity 40—3 days ($2000)	
		Activity 80—1 day ($400)	
		Activity 160—1 day ($600)	
		Activity 90—2 days ($600)	$3600

This now reduces the probable project duration to 90 working days, and this is the most that can be achieved. When there is more than one critical path (our example now has three), all critical paths must be shortened simultaneously to achieve any overall project reduction. Since the original critical path has now been decreased to an irreducible minimum, no further project shortening is possible.

At this point, the contractor must make a management decision concerning how much project shortening it is willing to buy. To answer this question, the consequences of running over on contract time must be weighed against the costs of shortening the project.

11.21 TIME-SCALED NETWORKS

The project network shown in Figure 11.4 shows activities located in the general order of their accomplishment but is not plotted to a time scale. Project diagrams can be and frequently are drawn to a horizontal time scale to serve a variety of useful purposes. When drawing a time-scaled diagram, two time scales can be used: one in terms of expired working days and the other as calendar dates. One scale is, of course, convertible to the other. Figure 11.5 is the time-scaled plot of Example Problem 2 plotted to a working-day scale with the calendar months also indicated. This is the same network diagram as Figure 11.4, and the two portray the same logic and convey the same scheduling information.

Figure 11.5 is obtained by plotting, for each activity, its ES and EF values. The horizontal distance between is equal to the estimated time duration of that activity. Although arrowheads are shown in this figure, they actually are not needed and do not appear in most time-scaled diagrams. Vertical solid lines indicate sequential dependence of one activity on another. When an activity has an early finish time that precedes the earliest start of activities following, the time interval between the two is, by definition, the free float of the activity. Free floats are shown as horizontal dashed lines in Figure 11.5, and a time-scaled plot of this type automatically yields to scale the free float of each activity. When an activity has no free float, no dashed extension to the right of that activity appears. The horizontal dashed lines also represent total float for groups or strings of activities, a topic that is discussed in Section 11.22. The circles shown in Figure 11.5 represent the ES times of the activities that immediately follow. The numerical values in these circles are the ES times in terms of expired working days.

The time-scaled diagram has some distinct advantages over a regular network for certain applications because it provides a graphical portrayal of the time interrelationships among activities as well as their sequential order. Such a plot enables one to determine immediately which activities are scheduled to be in process at any point in time and to detect quickly where time and resource problems exist. These networks are very convenient devices for checking daily project needs of labor and equipment and for the advance detection of conflicting demands among activities for the same resource. Such a diagram is also an effective way to record and analyze the progress of the work in the field. As of any given date, progress can be plotted on each activity that is in process in accordance with its percentage of completion. The resulting plot shows at a glance which activities are on, ahead of, or behind schedule.

11.22 SIGNIFICANCE OF FLOATS

The float of an activity represents potential scheduling leeway. When an activity has float time, this extra time may be utilized to serve a variety of scheduling purposes. When float is available, the early start of an activity can be delayed, its duration can be extended, or a combination of both can occur. To do a proper job of scheduling noncritical activities, it is important that the practitioner understand the workings of float times.

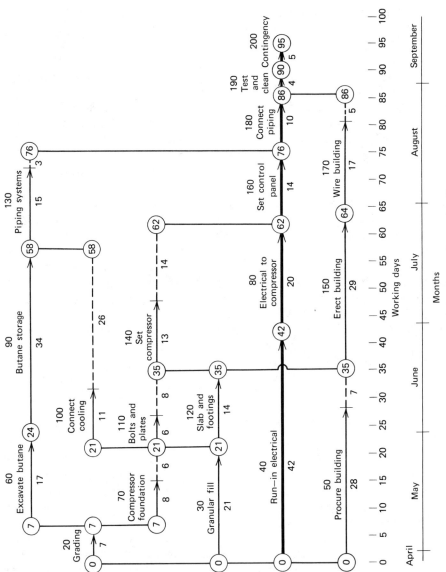

Figure 11.5 Example Problem 2, time-scaled diagram.

296

The total float of an activity is the maximum time that its actual completion date can go beyond its earliest finish time and not delay the entire project. It is the time leeway available for that activity if the activities preceding it are started as early as possible and the ones following it are started as late as possible. If all the total float of any one activity is utilized, a new critical path is created. The free float of an activity is the maximum time by which its actual completion date can exceed its earliest finish time and not affect either overall project completion or the times of any subsequent activities. If an operation is delayed to the extent of its free float, the activities following are not affected, and they can still start at their earliest start times.

A time-scaled network illustrates and clarifies the nature of total and free float. In Figure 11.5 the critical path can be visualized as a rigid time spine extending through the diagram. The horizontal dashed lines (float) can be regarded as compressible connections between activities. Activities, singly or in groups, can move to the right (be delayed), provided the correct dependency relationships are maintained and there is flexibility (float) present.

To illustrate the nature of total float, consider the rigid subassembly in Figure 11.5 that consists of activities 20, 60, 70, 90, and 130. Reference to Figure 11.5 shows that not only is the free float of activity 130 equal to 3 days but also the total float of each of the activities 20, 60, 90, and 130 is the same 3 days. The diagram also shows that if any one of these four activities is allowed to consume an additional 3 days, the succeeding activities are pushed along to the right, and all the float in that path is gone. Consequently, it becomes a second critical path. The result would be the same if 3 additional working days were to be consumed by any combination of activities 20, 60, 90, and 130. Figure 11.5 clearly illustrates that total float is usually shared by a string or aggregation of activities, and the additional time it represents is associated with the group.

The significance of free float is equally obvious from a study of Figure 11.5. For example, consider activity 100 that has a duration of 11 days and a free float of 26 days. Figure 11.5 shows there is a total of 37 days in which to accomplish activity 100. Within its two boundaries, activity 100 can be delayed in starting, have its duration increased, or a combination of the two without disturbing any other activity. For all practical purposes, activity 100 can be treated as an abacus bead 11 units in length that can be moved back and forth on a wire 37 units long. The free float of an activity, therefore, is extra time associated with that activity alone and which can be used or consumed without affecting the early start time of any succeeding activity.

11.23 THE EARLY-START SCHEDULE

If the network calculations are made by computer, the printout will give activity start and finish times as calendar dates. However, if manual computations are made directly on the network, as in Figure 11.4, the numerical data obtained must be translated into calendar dates at which the activities are scheduled to start and finish. This calendar date schedule is customarily based on the early start and finish times of each activity. If modifications to the computed early times have resulted from project shortening, this could be incorporated into the schedule.

Thus far project times have been expressed only in terms of elapsed working days. Conversion of elapsed working days into calendar dates is easily done with the aid of a calendar on which the working days are numbered consecutively, starting with number 1 on the anticipated start date and skipping all weekends, holidays, and vacation periods. It is assumed that the start date of Example Problem 2 will be Monday, April 29.

When making up a job calendar, the true meaning of elapsed days must be kept in mind. To illustrate, the early start of activity 90 in Figure 11.4 is shown to be 24. This means that this activity can start *after the expiration* of 24 working days, so the *starting date* of activity 90 will be the calendar date numbered 25 (June 3). There is no such adjustment for early finish dates. In the case of activity 90, its early finish time from Figure 11.4 is 58, which indicates that it is finished by the end of the 58th working day. This is indicated by the calendar date numbered 58 (July 19).

Figure 11.6 is the early-start schedule for Example Problem 2, with the activities being listed in the order of their starting dates. Also indicated are critical activities and free float values, information that is very useful to field supervisors. Not only is this form of operational schedule very useful to the contractor but it can also satisfy the usual contract requirement of providing the owner and/or architect-engineer with a projected timetable of construction operations.

As a matter of practicality on larger projects, it is not usually desirable to work up or print out a job schedule for more than a month or two in advance.

EXAMPLE PROBLEM 2: COMPRESSOR STATION				
Activity	Scheduled Duration (working days)	Free Float	Scheduled Starting Date (early start)	Scheduled Completion Date (early finish)
20	7	0	April 29, 19—	May 7, 19—
30	21	0	April 29, 19—	May 28, 19—
*40	42	0	April 29, 19—	June 26, 19—
50	28	7	April 29, 19—	June 6, 19—
70	8	6	May 8, 19—	May 17, 19—
60	17	0	May 8, 19—	May 31, 19—
100	11	26	May 29, 19—	June 12, 19—
120	14	0	May 29, 19—	June 17, 19—
110	6	8	May 29, 19—	June 5, 19—
90	34	0	June 3, 19—	July 19, 19—
140	13	14	June 18, 19—	July 5, 19—
150	29	0	June 18, 19—	July 29, 19—
*80	20	0	June 27, 19—	July 25, 19—
130	15	3	July 22, 19—	August 9, 19—
*160	14	0	July 26, 19—	August 14, 19—
170	17	5	July 30, 19—	August 21, 19—
*160	10	0	August 15, 19—	August 28, 19—
*190	4	0	August 29, 19—	September 4, 19—

*Critical activities

Figure 11.6   Example Problem 2, early-start schedule.

There is no need for a voluminous schedule extending to project completion when much of this schedule will probably be rendered obsolete by subsequent changes and updatings.

11.24 BAR CHARTS

Previously discussed was the fact that the bar chart is not a good planning device for construction. Nevertheless, the bar chart is very useful in other respects, and its unsurpassed visual clarity makes it a very valuable medium for displaying job schedule information. It is immediately intelligible to people who have no knowledge of CPM or network diagrams. It affords an easy and convenient way in which to monitor job progress, check delivery of materials, schedule equipment and crews, and record project advancement. For these reasons, bar charts will undoubtedly continue to be widely used in the construction industry.

Fortunately, it is now possible to prepare bar charts on a more rational basis, avoiding their intrinsic weaknesses and incorporating the strengths and advantages of CPM. This is possible by recognizing that a time-scaled diagram *is* a form of bar chart. Conventional bar charts can be quickly derived from the arrow diagram or project schedule. Activities may not always be the most desirable basis for bar chart preparation or usage. Simpler diagrams with fewer bars and showing larger and more comprehensive segments of the work may be more suitable for ordinary job applications. In such a case it is an easy matter to combine strings or groups of activities into a single bar-chart item.

11.25 RESOURCE SCHEDULING

If a project schedule is to be workable, it must be sustained by adequate resources as they are needed on the project. In a construction context "resources" include materials, subcontractors, labor crews, and equipment. The calendar date job schedule provides reasonably accurate information concerning when such resources will be required. It is now the responsibility of job management to follow through and see that they are available accordingly.

From the job schedule now established, the contractor can determine the calendar dates by which the various materials must be delivered to the jobsite. Allowing some time as a factor of safety, the purchase orders can specify reasonable delivery times. In a similar manner, a calendar schedule of subcontractor operations can be abstracted. Subcontractors can be notified well in advance of when their crews must be on the job. Advance information regarding materials and subcontracted work can be of inestimable value to the contractor's procurement people and expediters in their efforts to provide timely and adequate support to the project.

The job schedule can serve a somewhat different purpose with respect to labor crews and equipment. Not only does the contractor now know when the resources will be required on the job, but also conflicting demands for the same resource can be detected. For example, from the calendar date schedule for the job activities it is possible to determine total job labor and equipment requirements on a day-to-day basis. This information can prove to be valuable in that it can disclose hitherto unsuspected peaks and valleys in labor demand and conflicts between activities for the same equipment item. It is often possible to smooth manpower demand or remove equipment conflicts by utilizing the floats of noncritical activities.

As a simple illustration, suppose the start of activity 140 in Example Problem 2 required the use of the same heavy-duty crane already engaged on activity 90. Two such cranes on the project simultaneously would be difficult and expensive to arrange. Figure 11.5 shows that the start of activity 140 can be deferred by as much as 14 days. Rescheduling the start of activity 140 undoubtedly will remove the equipment conflict. As can be seen, a time-scaled plot of the precedence diagram can be valuable and convenient for the scheduling of construction equipment.

The procedure just discussed can be applied to labor demands also. The main objective here is to detect in advance where several activities may concurrently require workmen of the same craft, thereby causing pronounced peaks in labor demand. Sharp fluctuations in labor needs on a project are undesirable and often impossible to arrange. The judicious rescheduling of activities by the use of float times can do much to alleviate impractical fluctuations in labor demand. In addition, it assists management in recognizing manpower needs well ahead of time where shortages of certain crafts or highly specialized labor may be a factor.

11.26 PROGRESS MONITORING

After the operational plan and calendar schedule have been prepared, the work can now proceed with the assurance that the entire project has been thoroughly studied and analyzed. A plan and schedule have been devised that will provide specific guidance for the conduct of the work in the field. However, it is axiomatic that no plan is ever perfect, nor can the planner possibly anticipate every future job circumstance and contingency. Problems arise every day that could not have been foreseen. Adverse weather, material delivery delays, labor disputes, equipment breakdowns, job accidents, and other conditions can and do disrupt the original plan and schedule. Thus, after construction operations commence, there must be continual evaluation of field performance as compared with the established schedule. Considerable time and effort are expended to check and analyze the progress of the job and to take whatever action that may be required either to bring the work back on schedule or to modify the schedule to reflect changed job conditions.

How often field progress should be measured and evaluated depends on the degree of time control that is considered to be feasible for the particular work involved. Fast-paced projects using multiple shifts may demand daily progress reports. On the other hand, large-scale jobs such as earth dams that involve only a limited range of work classifications may use a reporting frequency of a month or even longer. However, for most jobs, progress reporting is normally done on a weekly basis.

11.27 PROGRESS ANALYSIS

The analysis of job progress is concerned primarily with determining the effect that reported job progress has on the established project completion date and any intermediate time goals that have been identified. When a progress report is received from the field, the time status of the activities, completed and in progress, is compared with the project schedule. Of particular importance in this regard are the critical activities. If a critical activity has been started late, a setback in project completion will occur unless the delay can be somehow made up by the time that the activity is completed. If a critical activity has been finished late, the overall completion date is delayed commensurately. The analysis of job progress data establishes the current time status of the project and where the work is ahead or behind schedule.

11.28 SCHEDULE UPDATING

Updating is primarily concerned with the effect that schedule deviations and changes in the job plan have on the portions of the project yet to be constructed. One of the nice features of CPM is its ability to incorporate changes as the project goes along. No project plan is ever perfect, and deviations from the original program are inevitable. Consequently, occasional updating of the plan is necessary during the construction period. Updating involves revising the schedule of the work yet to be accomplished to reflect the effect of changes that have occurred as work on the project has progressed. The objective is to determine corrected information regarding activity start and finish dates, floats, and critical activities and to establish a revised project completion date.

How often the project schedule should be updated depends on how often it is needed. If the progress of the work has followed the plan closely and if the schedule has held up reasonably well, there is really no need for a network updating. If there have been a number of significant deviations from the plan, then an updating is probably in order. Attempting to make the project fit an obsolete schedule can undo much of the good realized from formalized planning and scheduling. When the accumulated changes start to render the network and its schedule ineffective, it is time to revise it.

11.29 FAST-TRACKING

CPM is used as a planning and scheduling device for a broad range of industrial applications in addition to field construction operations. Included among these are the planning and design of construction projects. Many state highway departments, for instance, use CPM methods for this purpose. Under ordinary circumstances the design is accomplished and finalized before field construction starts. As such, the planning and scheduling of the design and of the construction phases are done essentially independently of one another. However, the two are closely coordinated under a design–construct or construction-management type of contract in which phased construction or fast-tracking is utilized. This topic has been previously discussed in Section 1.12.

Fast-tracking refers to the planned coordination and overlapping achievement of the design and construction of a project, as opposed to the traditional single file completion of each step before the next is initiated. For the process to be truly effective, however, the best possible cooperative efforts of the architect-engineer and the contractor or construction manager are required. It is also necessary that the owner have a well-thought-out program of requirements and be able to make immediate design decisions. CPM is an especially useful device to achieve a truly integrated and practicable fast-track schedule. No new principles are involved, but the planning, design, and construction of the project must be merged into a single concerted course of action.

12

PROJECT COST
ACCOUNTING

12.1 INTRODUCTION

Project cost accounting refers to the retrieval of field production rates and costs from ongoing projects. This information is used in two important ways. One is to develop labor and equipment production and cost data in a form suitable for estimating the cost of future work. The ability to estimate construction costs accurately is a key element in the success of any contracting firm. The other application of cost accounting data is to assist in keeping the construction costs of a project within its established budget. Regardless of the type of contract with the owner, it is important that the contractor exercise the maximum control possible over field costs during the construction period. A functioning and reliable cost system plays a vital role in the proper management of a construction project.

How the costs of an ongoing construction project are obtained and used is a variable with its size and character. A large, complex job requires a detailed reporting and information system to serve project management needs. Simpler and less elaborate procedures are sufficient for smaller and simpler projects. In any event, the only justification for the expense of a project cost system is the value of the management data it provides. If the information produced is not used or if it is not supplied in a usable form or timely fashion, then the system has no real value and its cost cannot be justified. Properly designed and implemented, project cost accounting is an investment rather than an expense.

This chapter discusses the methods involved in project cost control and the estimating feedback process. Although the details of how these actions are actually accomplished vary substantially from one construction firm to another, the ensuing treatment can be regarded as being reasonably typical of present practice within the construction industry.

12.2 PROJECT COST CONTROL

Project cost control actually begins with the preparation of the original cost estimate and the subsequent construction budget. Keeping within the cost budget and knowing when and where job costs are deviating are two factors that constitute the key to profitable operation. As the work proceeds in the field, cost accounting methods are applied to determine the actual costs of production. The costs as they actually occur are continuously compared with the budget. Field costs are obtained in substantial detail because this is the way jobs are originally estimated and also because excessive costs in the field can be corrected only if the exact causes can be isolated. To learn that construction costs are going over the budget is not helpful if it is impossible to identify where the trouble is occurring.

Cost reports are prepared at regular time intervals. These reports are designed to serve as management-by-exception devices, making it possible for the contractor to determine the cost status of the project and to pinpoint those work classifications where expenses are excessive. In this way, management attention is quickly focused on those job areas that need it. Timely information is required if effective action against cost overruns is to be taken. The detection of excessive costs only after the work has been finished leaves the contractor with no possibility of taking corrective action.

12.3 DATA FOR ESTIMATING

As discussed in Chapter 5, when the cost of a project is being estimated, many elements of cost must be evaluated. Labor and equipment expenses, in particular, are priced in the light of past experience. In essence, historical production records are the only reliable source of information available for estimating these two job expenses. The company cost system provides a reliable and systematic way of accumulating labor and equipment productivity and, costs for use in estimating future jobs.

With regard to feedback information for cost estimating purposes, there is some variation in practice about the form of production data that is recovered. It can be argued that production rates are fundamental to the estimating of labor and equipment costs. However, as seen in Chapter 5, costs per unit of production, or "unit costs," are widely used for estimating labor and equipment expense because of the convenience of their application. Such unit costs are, of course, determined from production rates and hourly costs of labor and equipment. Unit costs can be kept up-to-date by being adjusted for changes in hourly rates and production efficiencies. Information generated for company estimating purposes can, therefore, be in terms of labor and equipment production rates, unit costs, or both. For estimating purposes, the feedback system must be designed to produce information in whatever form or forms are compatible with company needs and procedures.

12.4 ACCOUNTING CODES

It is customary that an identifying code designation be assigned to each individual account of a contractor's accounting system. The coding systems used by contractors are not standardized, although suggested schemes have been proposed by various technical and professional groups. Contractors often use their own individual coding systems, which they have tailored to suit their particular operations. Alphabetic, decimal, and mixed cost codes are in use. The code used in this text is a decimal code and may be considered as typical of construction accounting practice. An advantage of a decimal system is that it is expandable to admit any level of detail desired.

Appendix M contains an abbreviated list of typical ledger accounts in common use by contractors. The asset accounts as included in the general ledger are identified by whole numbers from 10 through 39. General ledger liability accounts are designated by 40 through 49, net worth by 50 through 69, income accounts by 70 through 79, and expense accounts by 80 through 99. Subaccounts are assigned a distinctive decimal number, the first part of which identifies the general ledger or control account under which the subaccount exists. For example, consider the account 14.105. The whole number 14 indicates a Property, Plant, and Equipment account. The first decimal number, 0.1, indicates that it is associated with Real Estate and Improvements. The last two numbers identify the specific piece of property to which this account applies.

In Appendix M, Project Expense is designated by account number 80.000. This is where every item of expense chargeable to a particular job is recorded. This major category of Project Expense is often subdivided into two major subdivisions: project work accounts and project overhead accounts. A subsidiary ledger is maintained for each of these two project cost classifications. Each of these major subdivisions has an extensive internal breakdown into detailed items of cost. An abridged list of cost accounts for each of these is shown in Figure 12.1. The project cost breakdown shown is not intended to be complete nor to apply to all categories of construction. The cost accounts shown in Figure 12.1 would best apply to building construction. Once established, the code number for a given cost account remains the same. It is used consistently throughout the company and does not vary from one project to another. The use of "general" or "miscellaneous" cost accounts is avoided.

12.5 THE RECORDING OF PROJECT COSTS

A basic accounting principle for construction contractors is that project costs are recorded by job. Cost keeping must of necessity be a function of each project, and profit or loss is evaluated at the individual job level. When a new construction contract is obtained, a set of separate cost accounts is established for that project. A ledger sheet is set up for each cost item that pertains to the project. Each job expense is then posted to the appropriate cost account. Labor, materials, supplies, equipment charges, subcontract payments, overhead costs, and all such expenditures are charged to the appropriate project

	MASTER LIST OF PROJECT COST ACCOUNTS Subaccounts of General Ledger Account 80.000 PROJECT EXPENSE		
Project Work Accounts .100-.699		Project Overhead Accounts .700-.999	
100	Clearing and grubbing	700	Project administration
101	Demolition	.01	project manager
102	Underpinning	.02	office engineer
103	Earth excavation	701	Construction supervision
104	Rock excavation	.01	superintendent
105	Backfill	.02	carpenter foreman
115	Wood structural piles	.03	concrete foreman
116	Steel structural piles	702	Project office
117	Concrete structural piles	.01	move-in and move-out
121	Steel sheet piling	.02	furniture
240	Concrete, poured	.03	supplies
.01	footings	703	Timekeeping and security
.05	grade beams	.01	timekeeper
.07	slab on grade	.02	watchmen
.08	beams	.03	guards
.10	slab on forms	705	Utilities and services
.11	columns	.01	water
.12	walls	.02	gas
.16	stairs	.03	electricity
.20	expansion joint	.04	telephone
.40	screeds	710	Storage facilities
.50	float finish	711	Temporary fences
.51	trowel finish	712	Temporary bulkheads
.60	rubbing	715	Storage area rental
.90	curing	717	Job sign
245	Precast concrete	720	Drinking water
260	Concrete forms	721	Sanitary facilities
.01	footings	722	First-aid facilities
.05	grade beams	725	Temporary lighting
.07	slab on grade	726	Temporary stairs
.08	beams	730	Load tests
.10	slab	740	Small tools
.11	columns	750	Permits and fees
.12	walls	755	Concrete tests
270	Reinforcing steel	756	Compaction tests
.01	footings	760	Photographs
.12	walls	761	Surveys
280	Structural steel	765	Cutting and patching
350	Masonry	770	Winter operation
.01	8-in. block	780	Drayage
.02	12-in. block	785	Parking
.06	common brick	790	Protection of adjoining
.20	face brick		property
.60	glazed tile	795	Drawings
400	Carpentry	796	Engineering
440	Millwork	800	Worker transportation
500	Miscellaneous metals	805	Worker housing
.01	metal door frames	810	Worker feeding
.20	window sash	880	General clean-up
.50	toilet partitions	950	Equipment
560	Finish hardware	.01	move-in
620	Paving	.02	set-up
680	Allowances	.03	dismantling
685	Fencing	.04	move-out

Figure 12.1 Master list of project cost accounts.

from the original documents such as payrolls, invoices, freight bills, and equipment cost reports.

12.6 LABOR AND EQUIPMENT COSTS

Job costs associated with materials, subcontracts, and most project overhead items are of a reasonably fixed nature, and cost control of these kinds of expenses is effected in the main by disbursement controls applied to purchase orders and subcontracts. Barring oversight or mistake, these costs are determined with reasonable exactness when the job is estimated, and such costs are seldom subject to much fluctuation during the construction period. For this reason costs of this type are normally allocated only to general cost accounts in the project expense ledgers and are not subjected to detailed cost analysis. The cost information available from the general accounting system pertaining to materials, subcontracts, and job overhead is normally in sufficient detail and timely enough for job cost monitoring.

Labor and equipment costs, however, are subject to considerable uncertainty. These two categories of job expense can vary substantially during the construction process. The inherent variability of labor and equipment costs explains the fact that contractors' cost analysis systems usually concentrate on these two items of expense. The general system of accounts will not provide the needed cost information concerning these two expense types either in suitable detail or in sufficient time. Accordingly, the discussion of project cost accounting in this chapter is limited to these two categories of job expense which most concern the contractor. The process of determining at regular intervals how much work is being put into place in relation to the costs of labor and equipment being supplied is described in the following sections.

12.7 COST ACCOUNTING

To accumulate and analyze labor and equipment cost and production information from an ongoing project, special records and detailed accounts are required. The maintenance of such detailed job records is referred to, in the aggregate, as "project cost accounting." Cost accounting is not independent of the contractor's general system of accounts. Rather, it is in the nature of an elaboration of the basic project expense accounts. Knowledge of total costs is not sufficient. Cost performance has to be expressed in terms of the detailed makeup of the total cost. Cost accounting involves the continuous determination of labor and equipment costs, together with the work quantities produced, the analysis of these data, and the presentation of the results in summary form.

It is thus seen that project cost accounting differs from the usual accounting routines in that the information gathered, recorded, and analyzed is not entirely in terms of dollars and cents. Construction cost accounting is necessarily concerned not only with costs, but also with man-hours, equipment-hours, and the amounts of work accomplished. The systematic and regular checking of costs is a necessary part of obtaining reliable, time-average production information. A system that evaluates field performance only spasmodically in the form of occasional "spot checks" does not provide trustworthy feedback information, either for cost control or for estimating purposes.

Project cost accounting must strike a workable balance between too little and too much detail. A too general system will not produce the detailed costs necessary for meaningful management control. Excessive detail will result in the objectives of the cost system being obliterated by masses of data and paperwork as well as needlessly increasing the time lag in making the information available. The company cost system must be tailored to suit the contractor's own particular mode of operation. The detail used in this book is reasonably typical of actual practice in the industry.

A project cost accounting system supplements field supervision; it does not replace it. In the final analysis, the best cost control system a contractor can have is skilled, experienced, and energetic field supervision. It is important that field supervisors realize that project cost accounting is meant to assist them by the early detection of troublesome areas. Trade and site supervisors are key members of the team and without their support and cooperation, the job cost system cannot and will not perform satisfactorily.

For purposes of both estimating and cost accounting it is necessary that the project be broken down into the same elementary work classifications that are identified by the same cost code numbers. Reference to Figures 5.3 and 5.4 shows that the estimator identifies each work item by its code designation when he sets down the results of his quantity takeoff. Throughout the construction process, a continuous record is kept of the actual costs of production of these same work items. From estimate to project completion, the same work breakdown and cost code numbers apply.

12.8 THE PROJECT BUDGET

For purposes of cost control, a project or control budget is prepared at the beginning of field operations. To do this, the contractor must extract pertinent information concerning labor and equipment from the original project estimate. This budget information may be in the form of labor and equipment hours, unit costs, or both. For the discussions in this chapter, it is assumed the budget is prepared in terms of the total estimated work quantity, unit cost of labor and/or equipment, and total labor and/or equipment cost for each work cost code involved on the job. This budget then serves as the yardstick against which the actual field costs are measured during the progress of the work.

Contractors vary as to whether the budget costs include or do not include indirect labor costs and overhead costs. Practice in this regard depends on the detailed estimating and cost accounting procedures being used. What is important is that the budget and field costs be expressed on exactly the same basis and include the same items of cost. In this way, valid comparisons can be made between estimated and actual costs of production. In this text, the project cost budget and labor cost reports are all in terms of direct labor costs only and do not include indirect labor expense.

12.9 COST ACCOUNTING REPORTS

Summary labor and equipment cost reports must be compiled often enough so that excessive project costs can be detected while there is still time to remedy the situation. Cost report intervals are very much a function of project size, the nature of the work, and the type of construction contract involved. Obviously, there must be a balance struck between the cost of generating such reports and the value of the management information received. Daily cost reports are sometimes prepared on complex projects involving multiple shifts. It is not the usual job, however, that can profit from such frequent cost reporting. Some very large projects that involve relatively uncomplicated work classifications find intervals of a month or even longer to be satisfactory. For most construction projects, however, cost reports are needed more often than this.

It is generally agreed that weekly labor and equipment cost reports are about optimum for most construction operations, and this is the basis for discussion here. The cutoff time can be any desired day of the week, although contractors will often match their cost system to their usual payroll periods and also to their time monitoring system. The procedures involved with labor cost accounting will be discussed first. Following this will be a treatment of equipment cost accounting methods.

12.10 LABOR TIME CARDS

The source document for both payroll purposes and labor cost accounting is the labor time card. It is used to report the hours of labor time for each worker and the work categories to which the labor applied. Figure 12.2 shows a typical daily labor time card, and Figure 12.3, a weekly labor time card. Which of the two is used depends on company preference and policy.

When daily time cards are used, a time card is filled out each day. The head of each card provides for entry of the project name and number, date, weather conditions, and the name of the person preparing the card. The body of the time card provides for the name, badge number, and craft of each worker covered. Hours are reported as regular time (RT) or overtime (OT) as the case may be. For each worker listed, several slots are provided for the distribution of his hours to specific cost codes. Absolute perfection in time distribution is not possible; nevertheless, the need for care and reasonable accuracy cannot be overemphasized.

In Figure 12.2 it can be noted that the foreman's time is charged to account 701.03, which is an overhead account, and to accounts 240.05 and 240.08, which are work accounts. This is illustrative of the usual practice that

THE BLANK CONSTRUCTION COMPANY, INC.
DAILY LABOR TIME CARD

Project: Municipal Airport Terminal Bldg.
Date: August 18, 19—
Project No. 881
Weather: Cloudy-windy
Prepared by: R. D. Jones

Badge Number	Name	Craft	Time Classif.	Hourly Rate	701.03	240.01	240.05	240.07	240.08	240.51	240.91	Total Hours	Gross Amount
316	Jones, Richard D.	CF	RT	$15.00	4		2		2			8	$120.00
			OT										
109	Adams, Claud	CM	RT	14.25						6	2	8	114.00
			OT										
422	Chavez, S. C.	CM	RT	14.25				4		4		8	114.00
			OT										
461	Womac, C. T.	CM	RT	14.25		4	2			2		8	114.00
			OT										
247	Craig, O. N.	L	RT	10.95		4	4					8	87.60
			OT										
356	Johnson, Clyde	L	RT	10.35				4	4			8	82.80
			OT										
393	Newmann, Stan	L	RT	10.35		4	4					8	82.80
			OT										
211	Prong, Glace	L	RT	10.35				4	4			8	82.80
			OT										
Total Labor Cost					$60.00	$142.20	$143.70	$139.80	$112.80	$171.00	$28.50		$798.00

Figure 12.2 Daily labor time card.

310

THE BLANK CONSTRUCTION COMPANY, INC.
WEEKLY LABOR TIME CARD

Name ___ Womac, C. T. ___
Badge No. ___ 461 ___
Week Ending ___ August 22, 19— ___

Craft ___ CM ___

Project ___ Municipal Airport Terminal Bldg. ___
Project No. ___ 881 ___
Prepared by ___ R. D. Jones ___

Cost Code	Time Classif.	Hourly Rate	Mon.	Tues.	Wed.	Thurs.	Fri.	Total Hours	Total Cost
240.01	RT	$14.25	4	8	8			20	$285.00
	OT								
240.05	RT	14.25	2				4	6	85.50
	OT								
240.07	RT	14.25				4		4	57.00
	OT								
240.08	RT	14.25				4		4	57.00
	OT								
240.51	RT	14.25	2				4	6	85.50
	OT								
	RT								
	OT								
Total Hours	RT		8	8	8	8	8	40	
	OT								
Gross Amount			$114.00	$114.00	$114.00	$114.00	$114.00		$570.00
Weather			Cloudy–Windy	Clear–Windy	Clear	Clear–Hot	Cloudy		

Figure 12.3 Weekly labor time card.

311

a foreman's time spent in supervisory duties is charged to an overhead account and the time that a foreman spends working with his tools is charged to the appropriate work cost codes. As shown by Figure 12.2, the foreman spent four hours supervising his crew, two hours in pouring grade beams, and two hours pouring beams.

If weekly time cards are used, a separate card for each individual worker is prepared to record the hours he worked that week. Although the arrangement of the weekly time card is different from that of the daily time card, it presents the same information concerning the individual tradesman. The hourly rates shown on the time cards in Figures 12.2 and 12.3 are base wage rates only and do not include any indirect labor costs, such as payroll taxes, insurance, or fringe benefits. Labor unit costs derived from such time card information obviously do not include any indirect labor costs either, a subject previously discussed in Section 5.22. In this text, all labor unit costs are derived from base wage rates only.

Where the contractor is using computer-based payroll and cost accounting procedures, computer punch cards are often used for labor time reporting. The contractor's office or data processing service prepares time cards that are prepunched and preprinted with standard information such as employee name and number. The person filling out the cards on the project need add only the hours and cost codes.

12.11 TIME CARD PREPARATION

The distribution of each worker's time to the proper cost accounts normally is done by the foreman because he is in the best position to know how each person's time is actually spent. It is usual that this information is first recorded in the foreman's pocket time book with the nature of the work performed by the individuals of his crew often being described by words rather than by cost code numbers. The foreman identifies the craft and position of each worker by using any simple letter or numerical code the contractor may decide on. For example, in Figure 12.2, CF indicates concrete foreman, CM means cement mason, and L indicates laborer. Ordinarily, the foreman enters the hourly rate for each worker because it is not unusual for an individual during a week, or even in a single day, to be employed on work requiring different rates of pay. However, the extension of hours and wage rates into totals is usually done by others.

The importance of accurate and honest time reporting cannot be overemphasized. On the basis of the allocation of labor (and equipment) time to the various account numbers, cost and production information is generated. If this information is inaccurate or distorted, it not only is unusable, it can be seriously misleading when used for estimating or cost control purposes. Loss items must be identified as such without any attempt at cover-up by charging time to other cost accounts.

It may well be that the formal time card is filled in by someone other than the foreman, such as the field engineer, field superintendent, or project manager. One reason for this is the fact that paperwork is usually the least desirable part of a foreman's job. The field engineer, for example, can collect the foremen's time books at some convenient time each day and fill out the individual time cards, adding the necessary information for payroll and cost accounting, such as employee number and work cost codes, and perhaps make the extensions.

Even if a weekly time card is used, it is preferable that the information be entered on a daily basis. Setting down the time and its cost distribution as the work progresses will eliminate the practice of the foreman letting the matter go until the end of the week and then trying to enter the information from memory. It is probable that completing the labor record on a daily basis will improve the accuracy of the distribution. It is for this and other reasons that many contractors favor the use of daily time cards.

12.12 MEASUREMENT OF WORK QUANTITIES

To determine labor production rates or unit costs, it is necessary to obtain not only the hours and costs involved but the amount of each elementary work classification that has been accomplished. On some types of work it may be feasible and convenient for the foreman to include with his daily time card the work quantities accomplished for that day. It is more common practice, however, that work measurements be made at the end of each weekly payroll period, which is the basis for discussion here. The items of work measured must be identical with the standard cost code work classifications. Although labor costs are now being discussed, the weekly measurement of work quantities includes all work items performed, whether accomplished by labor, equipment, or a combination of both. Consequently, the same weekly work quantity determination serves for both labor and equipment cost accounting.

Work quantities can be obtained in a variety of ways depending on the nature of the work involved and company management methods. Direct field measurement on the job, estimation of percentages completed, computation from the drawings, and obtaining of quantities from the estimating sheets or CPM activities are all used. Direct measurement in the field is common, either in terms of total units achieved to date or units accomplished that week. This procedure is easily applied to projects that involved few cost code classifications. Many heavy, highway, and utility contracts are of this type. There are often instances in which quantities can be obtained with reasonable accuracy by applying estimated percentages of completion to the work amount totals. Although not as accurate as direct measurement of actual quantities, this procedure can yield useful information as long as accurate determinations are made occasionally, such as for monthly pay requests. Total work quantities of all types should be checked back occasionally against the estimating sheets as a check on the overall accuracy of quantity reporting.

The field measurement of work quantities on projects that involve many cost code classifications can become a substantial chore. Most building and industrial projects entail substantial numbers of different cost classifications. One convenient procedure in such cases is to mark off and dimension the work advancement in colored pencil on a set of project drawings reserved for that purpose. The extent of work put into place can be indicated as of the end of each day or each week as desired. By using different colors and dating succes-

THE BLANK CONSTRUCTION COMPANY, INC.
WEEKLY QUANTITY REPORT

Project _____Municipal Airport Terminal Building_____   Project No. _____881_____

Week Ending _____August 22, 19—_____   Prepared by ___R. D. Jones___

Cost Code	Work Description	Unit	Total Last Report	Total This Week	Total To Date
240.01	Concrete, footings	c.y.	479	196	675
240.05	Concrete, grade beams	c.y.	208	208	416
240.07	Concrete, slab	c.y.	595	65	660
240.08	Concrete, beams	c.y.	0	60	60
240.51	Concrete, trowel finish	s.f.	50,595	2,865	53,460
240.91	Concrete, curing, slab	s.f.	50,595	2,865	53,460

Figure 12.4   Weekly quantity report.

sive stages of progress, work quantities can be determined from the drawings or estimating sheets as of any date desired.

The field measurement of work quantities can be and often is done by the field supervisors. However, on large projects and especially those with many work codes, it may be desirable that the field engineer or project manager carry out this function because of the time and effort required. Weekly reports of work done are submitted on standard forms such as that in Figure 12.4. Prepunched and preprinted cards with cost codes, work description, and unit of measure already entered can be used for reporting work quantities when cost reports are prepared by computer.

12.13 FORMS OF LABOR REPORTS

Weekly labor reports can be prepared on either a man-hour or cost basis. That is, labor productivity can be monitored in terms of either man-hours per unit of work (production rates) or cost per unit of work (unit prices). Which of these is used is a function of project size, type of work involved, and project management procedures. Where man-hour control is used, a budget of man-hours per unit of work is prepared. In this regard, total man-hours are usually used with no attempt to subdivide labor time by trade specialty, such as so many hours of carpenter time and so many of ironworker time. Total man-hour estimates are based on an "average" crew mix for each work type.

During field operations, actual man-hours and work quantities are obtained. This makes possible a direct comparison of actual to budgeted productivity. Such an approach, of course, reflects productivity but not cost. It is simple to implement and avoids many problems associated with labor cost analysis. One such problem occurs on projects that require long periods of time to complete. Wage rates may be increased several times during the life of such a job. When the project cost is first being estimated, educated guesses are made about what the wage raises will likely be. What this means is that labor costs on long-term jobs are often estimated without exact knowledge of the wage rates that will actually apply during the construction process. As a result of this, actual labor wage rates during the work period may turn out to be different from those rates used in preparing the original control budget. Thus labor costs produced by the project cost system are not directly comparable to the budget. To make valid cost comparisons, it is necessary to either adjust these costs to a common wage-rate basis or work in terms of man-hours rather than labor cost.

In the foregoing case, man-hours can serve as a very effective basis for labor cost control. However, for most construction applications, labor cost analysis is more widely applied than is man-hour analysis. For this reason, the labor cost reports discussed here are all based on costs.

12.14 WEEKLY LABOR COST REPORTS

Once a week, labor costs obtained from the time cards are matched to the work quantities produced. The results of this analysis are summarized in a weekly labor cost report, two different forms of which are illustrated in Figures 12.5 and 12.6. These labor reports classify and summarize all labor costs incurred on the project up through the effective date of the report (August 22). The labor costs in these two reports are direct labor costs only and do not include indirect labor costs. These reports have the objective of providing job management with detailed information concerning the current status of labor costs and of indicating how these costs compare with those estimated. Both of the labor report forms are designed to identify immediately those work classifications having excessive labor costs and to give an indication of how serious those overruns are. Labor cost reports vary considerably in format and content from one construction company to another, although all such report forms are designed to convey much the same kind of management information.

Figure 12.5 is a weekly cost report for concrete placement that summarizes labor costs as budgeted, for the week being reported, and to date. Not all cost report forms include costs for the week being reported. These values can be of significance, however, in indicating downward or upward trends in labor costs. The labor report form in Figure 12.5 involves work quantities as well as labor expense and yields unit costs for each work type. Unit prices, obtained by dividing the total labor cost in each work category by the respective total quantity, enable direct comparisons to be made between the actual costs and the costs as budgeted. In Figure 12.5, the budgeted total quantity, budgeted total labor cost, and budgeted unit cost for each work type are taken from the project budget. The other quantities and labor costs are actual values, either for the week reported or to date.

When the total quantity of a given work item has been completed, its to-

THE BLANK CONSTRUCTION COMPANY, INC.
WEEKLY LABOR COST REPORT

Project ___Municipal Airport Terminal Building___

Week Ending ___August 22, 19——___

Project No. ___881___

Prepared by ___W. W. Smith___

Cost Code	Work Description	Unit	Quantity			Direct Labor Cost			Labor Unit Cost			To Date		Projected	
			Budget	This Week	To Date	Budget	This Week	To Date	Budget	This Week	To Date	Saving	Loss	Saving	Loss
240.01	Concrete, footings	c.y.	1,040	196	675	$ 4,680	$ 927	$ 3,243	$4.50	$4.73	$4.80		$203		$ 312
240.05	Concrete, grade beams	c.y.	920	208	416	5,796	1,247	2,519	6.30	6.00	6.06	$ 100		$ 221	
240.07	Concrete, slab	c.y.	2,772	65	660	16,632	450	4,473	6.00	6.92	6.78		515		2,162
240.08	Concrete, beams	c.y.	508	60	60	3,200	381	381	6.30	6.35	6.35		3		25
240.51	Concrete, trowel finish	s.f.	128,000	2,865	53,460	32,000	717	12,477	0.25	0.25	0.23	1,069		2,560	
240.91	Concrete, curing, slab	s.f.	180,000	2,865	53,460	5,400	86	1,286	0.03	0.03	0.024	321		1,080	

Figure 12.5 Weekly labor cost report (#1).

THE BLANK CONSTRUCTION COMPANY, INC.
WEEKLY LABOR COST REPORT

Project ___ Municipal Airport Terminal Building ___

Week Ending ___ August 22, 19— ___

Project No. ___ 881 ___

Prepared by ___ W. W. Smith ___

Cost Code (1)	Work Description (2)	Unit (3)	Total Quantity Budgeted (4)	Total Quantity To Date (5)	Percent Completed (6)	Budgeted Direct Labor (7)	Budgeted Labor To Date (8)	Actual Labor To Date (9)	Cost Difference (10)	Deviation (11)
240.01	Concrete, footings	c.y.	1,040	675	64.9	$ 4,680	$ 3,037	$ 3,243	−$206	1.07
240.05	Concrete, grade beams	c.y.	920	416	45.2	5,796	2,620	2,519	101	0.96
240.07	Concrete, slab	c.y.	2,772	660	23.8	16,632	3,958	4,473	− 515	1.13
240.08	Concrete, beams	c.y.	508	60	11.8	3,200	378	381	− 3	1.01
240.51	Concrete, trowel finish	s.f.	128,000	53,460	41.8	32,000	13,376	12,477	899	0.93
240.91	Concrete, curing, slab	s.f.	180,000	53,460	29.7	5,400	1,604	1,286	318	0.80
	Totals to Date						$24,973	$24,379	$ 594	0.98

Figure 12.6 Weekly labor cost report (#2).

date and projected saving or loss figures are obtained merely by subtracting its actual total labor cost from its estimated total cost. When a work item has been only partially accomplished, the to-date saving or loss of that work item is obtained by multiplying the quantity in place to date by the underrun or overrun of the unit price. The projected saving or loss for each work type can be obtained in different ways. In Figure 12.5, it is determined by assuming the unit cost to date will continue to completion of that work type. Multiplying the total estimated quantity by the underrun or overrun of the unit price to date yields the projected saving or loss figure.

The projected saving and loss figures shown in Figure 12.5 afford a quick, informative summary of how the project is doing insofar as labor cost is concerned. Those work types with labor overruns are identified together with the financial consequences if nothing changes. Some labor cost reports indicate the trend for each cost code, that is, whether the unit cost involved has been increasing or decreasing. This information can be helpful in assessing whether a given cost overrun is improving or worsening and in evaluating the efficacy of cost reduction efforts.

Figure 12.6 is an alternative form of weekly labor cost report that presents in somewhat different form the same weekly cost information for concrete placement. This figure shows actual and budgeted total labor costs to date for each cost classification. The budgeted total quantities and total labor cost for each cost code are obtained from the project budget. The actual work quantities and labor costs to date are cumulative totals for each work classification obtained from the time cards and weekly quantity reports. Column 10 of Figure 12.6 shows the cost difference as column 8 minus column 9, with a positive difference indicating that the cost as estimated exceeds the actual cost to date. Hence in column 10, a positive number is desirable; a negative number, undesirable. The deviation is the actual cost to date (column 9) divided by the budgeted cost to date (column 8). A deviation of less than one indicates that labor costs are within the budget, whereas a deviation of more than one indicates a cost overrun.

Although column 10 does indicate the magnitude of the labor cost variation for each cost code, it does not indicate the relative seriousness of the cost overruns. The deviation is of value in this regard because it shows the relative magnitude of labor cost variance. For those work types not yet completed, the cost differences listed in column 10 of Figure 12.6 do not always check exactly with the to-date saving and loss values of Figure 12.5. These small variations are caused by the rounding off of numbers and are not important.

12.15 EQUIPMENT EXPENSE

Cost keeping for equipment, especially on engineering construction, is an important cost accounting application. Equipment constitutes a substantial proportion of the cost of such projects, and the need to analyze equipment costs parallels the need for the determination of labor costs. The costs associated with large pieces of construction equipment are substantial, inherently variable, and deserving of a comprehensive record-keeping system. The objectives of equipment cost accounting are the same as those discussed for labor costs.

Management requires timely information for effective project cost control, and estimators need data to use for future bids. Only major equipment items merit detailed cost study, however. Lesser equipment such as power saws, concrete vibrators, and hand-operated soil compactors are normally charged to a project on a flat-rate or lump-sum basis and do not require detailed cost analysis.

Labor wage rates are, in almost all instances, fixed by or related to local labor agreements. No such determination exists for equipment costs, however, and it is up to the contractor to establish its own equipment expense rates. In the case of rental or leased equipment, the rental or lease rates are known, but the contractor must still establish the field operating costs. In the case of contractor-owned equipment, both ownership and operating costs must be determined. As discussed in Section 5.23, ownership expense includes depreciation and investment costs. Investment expense includes the costs of interest, insurance, taxes, and storage. Operating costs are on-the-job expenses such as tire replacement, tire repairs, mechanical repairs and parts, fuel, oil, grease, and, possibly, operating labor. Some contractors prefer to regard the labor associated with equipment operation as a labor expense rather than as an equipment cost. Others include the labor cost as a part of equipment operating expense. There are some cost accounting advantages in treating equipment-operating labor as any other labor cost and not lumping it together with equipment costs proper. This is the basis for discussions in this text.

When contractors estimate new work, they must obtain the most accurate values possible of the ownership and operating expenses of the various equipment types that will be required. Figure 5.5 illustrates a widely used procedure for estimating such costs. For most items of operating equipment, ownership, lease, or rental expense is combined with operating costs into a total cost per operating hour. When the project is under construction, it is the purpose of the cost accounting system to determine the number of hours each equipment type is in operation and the project work accounts to which these hours apply. With the use of the hourly equipment rates previously established for estimating purposes, equipment costs are periodically determined for each work type. When these costs are matched with work quantities accomplished, equipment unit costs of production are obtained. Thus once the hourly equipment rates are determined, equipment cost accounting proceeds very much like labor cost accounting and similar kinds of cost reports are produced.

As a general rule, equipment expense is directly chargeable to a single project work account. However, there are occasions when this is not true, and equipment costs must be accumulated in a suspense account until such time as they can be distributed to the proper cost accounts. An example of this might be a central concrete-mixing plant, consisting of many separate equipment items, that is producing concrete for several different cost accounts. Equipment expenses of this type must be collected into a suspense account and be periodically distributed equitably to the appropriate cost accounts on the basis of the quantities involved.

The internal rental rates used to charge equipment time to projects are based on time-average ownership and operating expenses that actually vary over the service life of the equipment. To illustrate, investment costs decrease and repair costs increase with equipment age. However, the use of lifetime average costs is the only way to have each project bear its proper share of the ultimate total expense associated with any particular equipment item. When equipment rental rates are assessed to a job, this is an all-inclusive charge. Correspondingly, the costs of fuel, lubrication, maintenance, repairs, and other such equipment expenses are not charged to the job on which they are actually incurred, but to the applicable equipment accounts.

12.16 EQUIPMENT TIME CARDS

Because equipment costs are expressed as a time rate of expense, time reporting is the starting point for equipment cost accounting. Equipment time is kept in much the same way as labor time. Where major equipment items are involved, a common procedure is to have the equipment supervisor make out daily or weekly equipment time cards that include each equipment item on the job. Equipment time cards are separate from and in addition to the operators' time cards. This procedure has merit because, by using different time cards, separate reportings are available for payroll and labor cost accounting purposes and for equipment cost accounting. In addition, equipment items such as pumps and air compressors may not have full-time operators and could be overlooked otherwise. Figure 12.7 is a typical daily equipment time card.

The equipment time card performs the same cost accounting function as the labor time card. By allocating equipment times to the proper cost codes, it is possible to determine the equipment costs chargeable to the various work types. Accuracy of time allocation to cost codes is just as important for equipment as it is for labor if reliable information is to be obtained for purposes of cost control and estimating. Even if a weekly time card is used, it is preferable that the time information be entered on a daily basis. Just as with labor time, distributing the equipment time daily is conducive to better accuracy. Someone other than the field supervisor, such as the field engineer, may fill out the time card and enter the budget rates of the individual equipment items reported and make the cost extensions.

Figure 12.7 records equipment time as working time (W), repair time (R), and idle time (I). Excessive equipment idle time (it may be difficult to get this item reported honestly) may indicate field management problems such as too much equipment on the job, lack of operator skill, improper balance of the equipment spread, or poor field supervision. Appreciable repair time can indicate inadequate equipment maintenance, worn-out equipment, severe working conditions, or operator abuse. On the other hand, substantial unproductive time can also be caused by job accidents, inclement weather, unanticipated job problems, or unfavorable site conditions.

12.17 EQUIPMENT COST REPORTS

Once each week, equipment costs are matched with the corresponding quantities of work produced. Work quantities are derived from the weekly quantity report previously dealt with in Section 12.12. By following the same process as has already been described for labor, a weekly equipment cost report is pre-

THE EXCELLO COMPANY, INC.
DAILY EQUIPMENT TIME CARD

Project _____ Holloman Taxiways _____
Date _____ October 16, 19— _____
Weather _____ Warm–clear _____
Project No. _____ 8112 _____
Prepared by _____ J. Brown _____

Machine No.	Machine	Rate per Hour	Cost Code			Total Hours			Total Cost
			101.05	103.07		W	R	I	
16	Bottom dump hauler	$23.71	8			7	1	0	$ 189.68
12	Bottom dump hauler	23.71	8			8	0	0	189.68
17	Bottom dump hauler	23.71	8			8	0	0	189.68
48	2 c.y. shovel	41.20		8		8	0	0	329.60
7	HG-11 tractor	19.84	4	4		7	0	1	158.72
21	Air compressor	7.25		8		8	0	0	58.00
	Total Cost		$648.40	$466.96					$1,115.36

Figure 12.7 Daily equipment time card.

THE EXCELLO COMPANY, INC.
WEEKLY EQUIPMENT COST REPORT

Project _____ Holloman Taxiways

Week Ending _____ October 18, 19—

Project No. _____ 8112

Prepared by _____ J. Brown

Cost Code	Work Description	Unit	Quantity			Equipment Cost			Equipment Unit Cost			To Date		Projected	
			Budget	This Week	To Date	Budget	This Week	To Date	Budget	This Week	To Date	Saving	Loss	Saving	Loss
101.05	Excavation, hauling	c.y.	127,000	12,200	113,680	$68,580	$6,343	$60,255	$0.54	$0.52	$0.53	$1,137		$1,270	
103.07	Excavation, common	c.y.	127,000	12,200	113,680	48,260	5,488	54,564	0.38	0.45	0.48		$11,368		$12,700
145.11	Base course, spreading	ton	79,500	7,360	77,420	43,725	4,343	43,359	0.55	0.59	0.56		774		795
250.03	Concrete, production	c.y.	23,625	2,090	4,139	81,270	6,917	14,070	3.44	3.31	3.40	166		945	
254.01	Concrete, hauling	c.y.	23,625	2,090	4,139	8,269	795	1,656	0.35	0.38	0.40		207		1,181
258.02	Concrete, lay-down	s.y.	90,000	8,360	16,556	44,100	3,846	7,282	0.49	0.46	0.44	828		4,500	

Figure 12.8   Weekly equipment cost report.

pared. An equipment cost report summarizes all equipment costs incurred on the project up through the effective date of the report. Either of the two cost report forms previously used for labor (Figures 12.5 and 12.6) can be used for equipment. Figure 12.8 is a frequently used format for weekly equipment cost reports. As can be seen, this figure is very similar to the weekly labor cost report form shown in Figure 12.5. Figure 12.8 serves to inform project management in a quick and concise manner how equipment costs are faring and about those work items where costs are going over the budget. The equipment report form in Figure 12.8 presents both work quantities and equipment expense and yields actual unit costs for each work type. Comparing the estimated with the actual equipment unit costs discloses where equipment costs are overrunning the project budget. The budgeted total quantity, budgeted total equipment cost, and budgeted unit cost for each work type are taken from the project budget. The other quantities and equipment costs are actual values, either for the week reported or to date. The to-date and projected saving and loss values capsulize the equipment cost experience on the project up through the report date of October 18.

12.18 COST INFORMATION AND FIELD SUPERVISORS

There is considerable difference of opinion over whether detailed cost information should be divulged to lower-level field supervisors. The statement is made, for example, that craft foremen will be tempted to charge labor or equipment time incurred on operations showing losses to other cost codes on which performance has been good. It can also be said that confidential cost information may be compromised or that a foreman may tend to relax when he knows that his costs are within the estimate. There is some truth to all of these.

On the other hand, it is well recognized that the only way a cost system can succeed is with the support and cooperation of the field supervisors. They try to achieve the best possible performance and expect to receive credit, and perhaps a bonus, if they beat the estimated costs. Field costs are very much involved when companies enter into profit-sharing or incentive plans with their supervisors. Probably the best answer is to provide craft foremen with the total amounts of cost overruns without specifying the estimated or actual unit costs.

12.19 COST CONTROL

The weekly cost reports make it possible for company management to assess quickly the cost status of the project and to pinpoint those work areas where expenses are proving to be excessive. In this way management attention is quickly focused on those work classifications that need it. If the project expense information is developed promptly, it may be possible to bring the offending costs back into line. In actual fact, of course, project cost control starts when the job is first priced because this is when the control budget is actually established. No amount of management expertise or corrective action can salvage a project that was priced too low in the beginning. In this regard, there likely will always be some work classifications whose actual costs will exceed those estimated. The project manager is primarily responsible for getting the total project built for the estimated cost. If some costs go over, hopefully they can be counterbalanced by savings in other areas.

Having identified where production costs are excessive, project management must then decide just what to do about them, if indeed anything can be done. For certain, the hourly rates for labor and equipment are not controllable by management. The only real opportunity for cost control resides in the area of improving production rates. This element of work performance can, to a degree, be favorably influenced by skilled field supervision, astute job management, energetic resource expediting, and the improved makeup of labor crews and selection of equipment.

Any efforts to improve field production must be based on detailed knowledge of the pertinent facts. If the cause of excessive costs cannot be specifically identified, then a satisfactory solution is not likely to be found. It is impossible to generalize on this particular matter, but certainly the treatment must be gauged to the disease. In the usual case, full cooperation between the field supervisors and project management is needed before any real cost improvement can be realized. Field supervisors play a key role in implementing corrective procedures. There are no precise guidelines for reducing excessive project costs. The effectiveness of corrective procedures depends largely on the ingenuity, resourcefulness, and energy of the people involved.

Production costs are frequently high early in the construction process, but tend to become lower as the work progresses. This is a "learning curve" phenomenon where costs decrease as job experience and familiarity are gained. In a general way, production costs usually tend to decrease as the job goes along because crew members learn how to work as a team, become familiar with the job, and find out what their foremen expect of them.

12.20 INFORMATION FOR ESTIMATING

Estimating requires production rates and unit costs that are a balanced time average of good days and bad days, high production and low production. For this reason, information for estimating is normally not recovered from the cost accounting system until after project completion, or at least not until all the work type being reported has been finished. In so doing, the best possible time-average rates will be obtained. Permanent files of cost and productivity information are maintained on paper, on microfilm, or by computer, providing the estimator with immediate access to data accumulated from prior projects.

Both production rates and unit costs are available from the project cost accounting system. To be of maximum value in the future, however, it is important that such productivity data be accompanied by a description of the project work conditions that applied while the work was being done. Knowledge of the work methods, equipment types, weather, problems, and other job circumstances will make the basic cost and productivity information much more useful to an estimator. Such written narrative becomes a part of the total historical record of each cost account.

12.21 COMPUTER APPLICATION

Computers are widely used by construction contractors in conjunction with their project cost systems. Because cost accounting can become laborious and time-consuming, even for relatively small operations, the computer has advantages of economy, speed, and accuracy over manual methods. In addition, the computer provides a cost system with flexibility and depth that manual systems often cannot match. This does not mean that job costs cannot be developed satisfactorily by hand. Many small contractors have completely adequate manual cost systems. Experience indicates, however, that few contractors of substantial size are able to generate manually field cost reports in time to serve a genuine cost-control purpose. Contractors often find that manual methods serve well as generators of estimating information, but not for cost control. A common experience in this regard is that the project is finished before the contractor knows the profit status of the work. It is only being realistic to recognize that most contractors find computer support to be a necessary part of their project cost accounting system.

Supporting computer software is now very flexible in the sense that just about any cost report or information can be generated that project management may desire. Cost reports in the same general formats as those presented in this chapter are produced by several current computer programs. The programs commonly used by contractors actually perform a whole series of cost accounting and financial accounting functions. After the input of cost and production information, the computer generates payroll checks, keeps payroll records, maintains the equipment accounts, and performs other functions as well as producing a variety of productivity and cost reports and project cost forecasts.

12.22 EQUIPMENT CHARGES TO PROJECTS

The usual procedure for charging equipment costs to construction projects has been previously discussed in Section 9.18. There are, however, some aspects of equipment charges that require special arrangements. For example, some equipment expenses are not included in the usual hourly rental rates. The costs of move-in, erection, dismantling, and move-out are fixed costs that cannot be incorporated into time rates of expense. Such costs are normally charged to appropriate job overhead accounts and are not included in the hourly or monthly equipment rates.

With regard to equipment charged to the project at an hourly rate, how idle and repair time is handled is a matter of company policy. Several different procedures are followed. Probably the most common approach is to charge the project at the established internal rates for the full working day for each piece of equipment on the job. However, credit is given for repair time and for idle time caused by weather and other uncontrollable causes. The significance of this procedure is that the job is charged for all equipment on the site whether it is used or not. This policy can materially assist in controlling underusage of equipment. Where standby equipment units are purposely kept on a job to handle emergencies, the project is usually charged only with the ownership expense involved.

Where the project is charged with extensive periods of equipment idle time, it is probably best to charge the individual cost accounts with net operating hours, plus ordinary or usual idle time. The excessive idle time can be charged to a special overhead account. The reason for this accounting maneuver is that the equipment cost charged to a given work item is used to compute summary information for purposes of cost control and estimating. The use of the special account avoids having excessive idle time distort the reported equipment unit costs.

13

LABOR LAW

13.1 INTRODUCTION

The employment of labor by the contractor is subject to the provisions of an imposing array of both federal and state statutes. These laws have such an important bearing on the conduct of a contracting business that the construction manager must have at least a general grasp of their workings and implications. The purpose of this chapter is to discuss the important features of the principal federal statutes that apply to the employment of construction workers. Federal laws are discussed because of their wide applicability and the fact that most state labor statutes are patterned after federal law. Emphasis is placed more on the broad implications of these laws than on the intricacies of case studies.

For purposes of discussion in this chapter, those statutes pertaining to labor-management relations are considered first (Sections 13.2 through 13.24). Following are the federal laws pertinent to equal employment opportunity (Sections 13.25 through 13.27). The last sections of this chapter discuss labor-standards legislation and other topics.

13.2 HISTORY OF LAW OF LABOR RELATIONS

In the early days of this nation the right of working people to associate together for their mutual aid and protection was severely restricted. Unions were strictly curtailed and were sometimes referred to as "unlawful conspiracies." Up until the 1930s the law of labor relations was created principally by the courts. In the almost complete absence of applicable statute law, employer and union complaints were adjudicated primarily in accordance with common law. In general the courts tended to grant employers relief from unionizing activities but refused to assist unions against employers on the grounds that there was no precedent for this in the common law. Court injunctions were widely used to negate the usual union weapons of strikes, picketing, and boycotts. At the same time, there was no comparable judicial instrument available to the unions to assist their organizing efforts.

This state of affairs continued until 1932, although the Sherman Antitrust Act of 1890 did provide the beginnings of a statutory basis for labor-management policy. This act made statutory provisions against the restraint of trade. It was primarily intended to limit the growth of business cartels, and it is debatable whether it was ever intended to apply to labor unions. However, the U.S. Supreme Court ruled in 1908 that labor organizations were covered by the provisions of the act. The Sherman Act provided a broad new basis for the use of court injunctions against unions and also placed another effective weapon of unions in jeopardy, that is, the boycott. The Supreme Court ruled that a union could be sued for damages suffered by reason of a boycott and that the union and its members were individually liable. In summary, it can be reported that the law was strongly discriminatory against labor unions until the advent of the New Deal in the early 1930s.

During the intervening years Congress has passed a series of major federal labor statutes that contain positive and detailed statements of national labor policy. Under the law today the right of workers to form or join unions and to take concerted action to improve their economic condition is guaranteed, and the exercise of that right is protected. National labor policy today is the sum of the policies and provisions contained in the major federal labor relations statutes: the Norris-LaGuardia Act (1932), the National Labor Relations Act (1935), the Labor Management Relations Act (1947), and the Labor-Management Reporting and Disclosure Act (1959).

The following sections discuss these major pieces of federal labor-management legislation. The discussion of these laws concentrates on those provisions of the law that pertain especially to the construction industry.

13.3 THE NORRIS-LAGUARDIA ACT

In 1932 Congress enacted the Norris-LaGuardia Act, which strictly limits the power of the federal courts to issue injunctions against union activities in labor disputes and protects the right of workers to strike and picket peaceably. Also called the Anti-Injunction Act, this statute makes it very difficult for an employer to secure injunctions in a federal court against union activities in labor disputes. Although the act itself pertains only to federal courts, many states have enacted similar injunction-control legislation.

Although it is difficult for private parties to obtain federal court injunctions against peacefully conducted labor action, injunctions are available to certain government agents under modern labor relations statutes. In this regard, however, injunctions are issued against only those union activities that are in violation of the law or that imperil national health or safety. In addition, the U.S. Supreme Court has decreed that an employer can obtain a federal court injunction against a striking union that is violating a no-strike arbitration pledge in its labor contract.

The Norris-LaGuardia Act also expressly prohibits "yellow-dog contracts" and makes them unenforceable in federal courts. Designed to discourage union membership, such employment contracts provide that a job applicant will not be hired until he promises not to join a union during his tenure of employment and to renounce any existing membership. Such contracts were widely used by many industrial employers prior to the passage of the Norris-LaGuardia Act.

13.4 THE NATIONAL LABOR RELATIONS ACT

The National Labor Relations Act (NLRA), also known as the Wagner Act, was passed by Congress in 1935. Enacted in an atmosphere of depressed business conditions and extensive unemployment, the central purpose of the Wagner Act was to protect union-organizing activity and to foster collective bargaining. Employers are required to bargain in good faith with the properly chosen representatives of their workers. Employers are forbidden to practice discrimination against their employees for labor activities or to influence their membership in any labor organization. These and other unfair labor practices were defined as they pertained to employers. However, no such prohibitions were applied to employees or unions in their relations with employers. Enforcement of the law was vested in a National Labor Relations Board, which the Wagner Act created. Under the shelter of this piece of legislation, union strength and membership increased enormously during the 1937–1945 period. A number of state acts followed, which were more or less patterned after the federal law.

It was almost inevitable that the sudden removal of the traditional restraints would result in union excesses. Starting in about 1938, public opinion concerning organized labor became increasingly antagonistic, with the mounting incidence of union restrictive practices, wartime strikes, and criminal activities by some labor leaders. Congressional resentment against organized labor's high-handed actions resulted in the War Labor Disputes Act (Smith-Connally Act) of 1943. However, the provisions of this act proved to be largely ineffective. If nothing else, however, the act reflected the mounting popular sentiment for the enactment of positive union-control legislation. During this period several state legislatures passed statutes that regulated and curbed union activities. By 1947, 37 states had passed some form of labor-control legislation.

13.5 THE LABOR MANAGEMENT RELATIONS ACT

In 1947 Congress passed the Labor Management Relations Act, commonly known as the Taft-Hartley Act. This was the first federal statute that imposed comprehensive controls over the activities of organized labor. It amended the earlier National Labor Relations Act in several important respects and added new provisions of its own. The National Labor Relations Board was reconstituted, and its authority was redefined. Section 7 of the Taft-Hartley Act established the basic right of every worker to participate in union activities or to refrain from them, subject to authorized agreements requiring membership in a union as a condition of employment. To protect such rights, Section 8 of the Taft-Hartley Act defined unfair labor practices for both employers and labor organizations. The act established the Federal Mediation and Conciliation Service, gave the President of the United States certain powers regarding labor disputes imperiling national health or safety, and restricted political contributions by labor organizations and business corporations (see Section 13.24).

In contradistinction to the Wagner Act, the Taft-Hartley Act was designed to curtail the freedom of action of unions in several different and important ways. Although the act reiterates a national labor policy of encouraging and assisting collective bargaining, the provision is made that the public interest must prevail in the conduct of labor affairs. The provisions of the Taft-Hartley Act have had far-reaching effects on the labor-management scene.

Experience with the provisions of the act quickly revealed several imperfections and shortcomings. Some features of labor employment peculiar to the construction industry proved to be inadequately covered. However, the act's extremely controversial nature and the appreciable strengths of both its backers and opponents made revision of the law a very touchy and difficult matter.

13.6 THE LABOR-MANAGEMENT REPORTING AND DISCLOSURE ACT

In 1959 Congress passed the Labor-Management Reporting and Disclosure Act, also known as the Landrum-Griffin Act. This act established a code of conduct for unions, union officers, employers, and labor relations consultants. In addition, it guaranteed certain rights to rank-and-file union members, and imposed stringent controls on union internal affairs. The principal thrust of this law was to safeguard the rights of the individual union member, to ensure democratic elections in unions, to combat corruption and racketeering in unions, and to protect the public and innocent parties against unscrupulous union tactics. Under the act, reports pertaining to union organization, finances, activities, and policies are required from unions, union officials and employees, employers, labor relations consultants, and union trusteeships. It was made illegal for an employer to pay or lend money to any labor representative or union of his employees (see Section 13.23).

Additionally, the Landrum-Griffin Act amended the National Labor Relations Act and the Taft-Hartley Act. The 1959 law enumerated additional union unfair labor practices and remedied several inadequacies of the Taft-Hartley Act with respect to pressures that unions and their agents can legally apply to employers and their employees. So-called "hot-cargo" labor agreements (see Section 13.10, item 6) were forbidden, with an exception made for the construction industry. Limitations were applied to organizational picketing of employers by labor organizations. Most of the restrictions on union-security agreements in the construction industry were removed, and union hiring halls (see Section 13.16) were made lawful.

13.7 COVERAGE OF THE NATIONAL LABOR RELATIONS ACT

The National Labor Relations Act, as amended by the Taft-Hartley Act and the Landrum-Griffin Act, plays a dominant role in national labor relations policy. Its declared purpose is to state the recognized rights of employees, employers, and labor unions in their relations with one another and with the public, and to provide machinery to prevent or remedy any interference by one with the legitimate rights of another. In particular, it protects employees in the free exercise of their right to join or not to join a union, to bargain collectively through representatives of their own choosing, and to act together with other employees for mutual aid and protection. Any violation of these rights, whether by management or by labor representatives, is declared to be an unfair labor practice.

The National Labor Relations Act applies to employers and employees engaged in interstate commerce or the production of goods for such commerce. Interpretation by the courts as to what constitutes interstate commerce in the construction industry has been so broad that the act's authority extends to almost all construction work of any consequence. The act specifically excludes the following employers and employees from its coverage:

1. Exempted Employers.
   a. The United States.
   b. States and their political subdivisions.
   c. Wholly owned government corporations.
   d. Federal reserve banks.
   e. Employers subject to the Railway Labor Act.
   f. Labor organizations (when not acting as employers).
   g. Officers or agents of labor organizations.
2. Exempted Employees.
   a. Employees of exempted employers as listed in Item 1 above.
   b. Agricultural laborers.
   c. Domestic servants.
   d. Individuals employed by their parents or spouses.
   e. Independent contractors.
   f. Supervisors.

Certain selected portions of the National Labor Relations Act, as amended, have been selected for discussion. These are presented in Sections 13.8 through 13.24.

13.8 THE NATIONAL LABOR RELATIONS BOARD

Administration of the National Labor Relations Act is the responsibility of the National Labor Relations Board (NLRB), which is composed of five members, and the general counsel of the board. Members of the NLRB are appointed by the President of the United States, with consent of the Senate, for terms of five years. The general counsel is appointed by the President, with consent of the Senate, for a term of four years. The NLRB has two primary functions: (1) to establish, usually by secret-ballot elections, whether groups of employees wish to be represented by designated labor organizations for collective-bargaining purposes, and (2) to prevent and remedy unfair labor practices.

Much of the day-to-day work of investigating and processing charges of unfair labor practices and of handling representation proceedings has been delegated by the board to the various NLRB regional offices located in major cities throughout the nation. The board has given its regional directors final authority in election cases, subject to limited review. In unfair labor practice cases the board acts much like an appellate court to determine if an unfair labor practice actually exists and how such practices should be remedied. The board does not ordinarily become involved until an investigation has been conducted and recommendations made by a regional office. The general counsel has largely independent authority in the prosecution of unfair labor practices. Responsible for general supervision over the regional NLRB offices and for most of the administrative routine of the agency, the general counsel has broad and direct authority to seek injunctions against unfair labor practices.

By statute the NLRB exercises its powers over all enterprises whose operations affect interstate commerce. It does not act, however, on every case over which it could exercise jurisdiction. Rather, the board restricts its attention to a case load it can handle expeditiously and within its budgetary limitations. The Landrum-Griffin Act authorized the NLRB to limit its cases to those whose effect on commerce is, in the board's opinion, substantial. As a guide to when it will exercise its power, the board has established minimum measures of the annual volume of business that must be involved before the NLRB will accept the case. These standards are expressed in terms of the gross dollar volume of the employer's sales and purchases that cross state lines and vary for different areas of industry. The Landrum-Griffin Act further provided that state and territorial courts and agencies can assume jurisdiction over labor disputes the NLRB declines to hear.

13.9 REPRESENTATION ELECTIONS

The National Labor Relations Act requires that an employer bargain with the representative selected by a majority of his employees but does not stipulate a selection procedure. The only requirement is that the representative clearly be the choice of the majority. The representative may be an individual or a labor union but cannot be a supervisor or other representative of the employer.

For the employees to select a majority representative, it is usual for the regional office of the NLRB to conduct representation elections. However, such an election can be held only when a petition has been filed by the employees, by an individual or a labor organization acting in their behalf, or by an employer who has been confronted with a claim of representation from an individual or labor organization. The present rule is that a union may secure an election to determine the wishes of the employees if it can show at least 30 percent of the eligible employees have indicated they desire such representation. The board, through its regional office, supervises every step in the election procedure.

In a representation election, the employees are given a choice of one or more bargaining representatives or no representative at all. To be chosen, a labor organization must receive a majority of the valid votes cast. The NLRB will certify the choice of the majority of employees for a bargaining representative only after a secret-ballot election.

Section 7 of the Taft-Hartley Act has a free speech provision that establishes the employee's right to hear the arguments of both labor and management. The expressing or disseminating of any views, arguments, or opinions by either side does not constitute an unfair labor practice as long as it contains no threat of reprisal or force or promise of benefit. Within these limitations, an employer who wants to stay nonunion can state his opinions to his employees.

13.10 EMPLOYER UNFAIR LABOR PRACTICES

Under the National Labor Relations Act, as amended, an employer commits an unfair labor practice if he:

1. Interferes with, restrains, or coerces employees in the exercise of rights protected by the act, such as their right of self-organization for the purposes of collective bargaining or other mutual assistance [Section 8(a)(1)].
2. Dominates or interferes with any labor organization in either its formation or its administration or contributes financial or other support to it [Section 8(a)(2)]. Thus "company" unions that are dominated by the employer are prohibited, and employers may not unlawfully assist any union financially or otherwise.
3. Discriminates against an employee in order to encourage or discourage union membership [Section 8(a)(3)]. It is illegal for an employer to discharge or demote an employee or to single him out in any other discriminatory manner simply because he is or is not a member of a union. In this regard, however, it is not unlawful for employers and unions to enter into compulsory union-membership agreements permitted by the National Labor Relations Act. This is subject to applicable state laws prohibiting compulsory unionism.
4. Discharges or otherwise discriminates against an employee because he has filed charges or given testimony under the act [Section 8(a)(4)]. This provision protects the employee from retaliation if he seeks help in enforcing his rights under the act.
5. Refuses to bargain in good faith about wages, hours, and other conditions of employment with the properly chosen representatives of his employees [Section 8(a)(5)]. Matters concerning rates of pay, wages, hours, and other conditions of employment are called mandatory subjects, about which the employer and the union must bargain in good faith, although the law does not require either party to agree to a proposal or to make concessions.
6. Enters into a hot-cargo agreement with a union [Section 8(e)]. Under a hot-cargo agreement, the employer promises not to do business with or

not to handle, use, transport, sell, or otherwise deal in the products of another person or employer. This unfair labor practice can be committed only by an employer and a labor organization acting together. A limited exception to this ban on hot-cargo clauses is made for the garment industry and the construction industry.

13.11 UNION UNFAIR LABOR PRACTICES

Under the National Labor Relations Act, as amended, it is an unfair labor practice for a labor organization or its agents:

1. a. To restrain or coerce employees in the exercise of their rights guaranteed in Section 7 of the Taft-Hartley Act [Section 8(b)(1)(A)]. In essence Section 7 gives an employee the right to join a union, to assist in the promotion of a labor organization, or to refrain from such activities. This section further provides that it is not intended to impair the right of a union to prescribe its own rules concerning membership.
   b. To restrain or coerce an employer in his selection of a representative for collective-bargaining purposes [Section 8(b)(1)(B)].
2. To cause an employer to discriminate against an employee in regard to wages, hours, or other conditions of employment for the purpose of encouraging or discouraging membership in a labor organization [Section 8(b)(2)]. This section includes employer discrimination against an employee whose membership in the union has been denied or terminated for cause other than failure to pay customary dues or initiation fees. Contracts or informal arrangements with a union under which an employer gives preferential treatment to union members are violations of this section. It is not unlawful, however, for an employer and a union to enter an agreement whereby the employer agrees to hire new employees exclusively through a union hiring hall as long as there is no discrimination against nonunion members. Union-security agreements that require employees to become members of the union after they are hired are also permitted by this section.
3. To refuse to bargain in good faith with an employer about wages, hours, and other conditions of employment if the union is the representative of his employees [Section 8(b)(3)]. This section imposes on labor organizations the same duty to bargain in good faith that is imposed on employers.
4. To engage in, or to induce or encourage others to engage in, strike or boycott activities, or to threaten or coerce any person, if in either case an object thereof is:*
   a. To force or require any employer or self-employed person to join any

*Section 8(b)(4) permits "publicity," other than picketing, provided such action does not have the effect of inducing a work stoppage by neutral employees. Such publicity can notify the public, including the customers of a neutral employer, that a labor dispute exists concerning a primary employer's products being distributed by the neutral.

labor or employer organization, or to enter a hot-cargo agreement that is prohibited by Section 8(e) [Section 8(b)(4)(A)].

b. To force or require any person to cease using or dealing in the products of any other producer or to cease doing business with any other person [Section 8(b)(4)(B)]. This is a prohibition against secondary boycotts, a subject discussed further in Section 13.17. This section of the National Labor Relations Act further provides that, when not otherwise unlawful, a primary strike or primary picketing is a permissible union activity.

c. To force or require any employer to recognize or bargain with a particular labor organization as the representative of his employees that has not been certified as the representative of such employees [Section 8(b)(4)(C)].

d. To force or require any employer to assign certain work to the employees of a particular labor organization or craft rather than to employees in another labor organization or craft, unless the employer is failing to conform with an order or certification of the NLRB [Section 8(b)(4)(D)]. This provision is directed against jurisdictional disputes, a topic discussed in Section 13.21.

5. To require of employees covered by a valid union shop membership fees that the NLRB finds to be excessive or discriminatory [Section 8(b)(5)].

6. To cause or attempt to cause an employer to pay or agree to pay for services that are not performed or not to be performed [Section 8(b)(6)]. This section forbids practices commonly known as featherbedding.

7. To picket or threaten to picket any employer to force him to recognize or bargain with a union:
   a. When the employees of the employer are already lawfully represented by another union [Section 8(b)(7)(A)].
   b. When a valid election has been held within the past 12 months [Section 8(b)(7)(B)] .
   c. When no petition for a NLRB election has been filed within a reasonable period of time, not to exceed 30 days from the commencement of such picketing [Section 8(b)(7)(C)]*.

The National Labor Relations Board has ruled that discrimination by a labor union because of race is an unfair practice under the Taft-Hartley Act. Discrimination on the basis of race in determining eligibility for full and equal membership and segregation on the basis of race has been found unlawful by the NLRB. This has made possible the filing of unfair labor practice charges against a union because of alleged racial discrimination. A union found guilty of such practices faces cease-and-desist orders, as well as possible rescission of its right to continue as the authorized employee representative.

---

*Subparagraph (c) does not apply to picketing or other publicity for the purpose of truthfully advising the public that an employer does not have a union contract or employ union labor, unless it has the effect of inducing employees of persons doing business with the picketed employer not to pick up, deliver, or transport goods or not to perform services. If the purpose is purely informational, it can be continued indefinitely. If the purpose is organizational, it cannot be continued more than 30 days without a petition for an election.

13.12 CHARGES OF UNFAIR LABOR PRACTICES

Charges of unfair labor practices can be filed by an employer, a union, or an individual worker. These charges must usually be filed in the NLRB regional office that serves the area in which the case arose within six months from the date of the alleged unfair activity. After charges are filed, field examiners investigate the circumstances, and a formal complaint is issued if the charges are found to be well-grounded and the case cannot be settled by informal adjustment.

When a complaint is issued, a public hearing is held before a trial examiner whose findings and recommendations are served on the parties and are sent to the NLRB in Washington, D.C. If no exceptions are filed by either party within a statutory period, the examiner's judgment takes the full effect of an order by the NLRB. If exceptions are taken, the NLRB reviews the case and makes a decision.

If an employer or a union fails to comply with an order of the NLRB, the board has no statutory power of enforcement of its own but can petition the appropriate United States Court of Appeals for a decree enforcing the order. If the court issues such a decree, failure to comply may be punishable by fine or imprisonment for contempt of court. Parties aggrieved by the order may seek judicial review.

13.13 REMEDIES

When the NLRB finds that an employer or a union has engaged in an unfair, labor practice, it is empowered to issue a cease and desist order and to take such affirmative action as deemed necessary to erase the effects of the unfair practice found to have been committed. The purpose of the board's orders is remedial, and it has broad discretion in fashioning remedies for unfair labor practices. Typical affirmative actions ordered by the NLRB include reinstatement of persons discharged, reimbursement of wages lost, or refund of dues or fees illegally collected.

The law provides that whenever a charge is filed alleging certain unfair labor practices relating to secondary boycotts, hot-cargo clauses, or organization or recognition picketing, the preliminary investigation of the charge must be given first priority. The board or the general counsel is authorized to petition the appropriate federal district court for an injunction to stop any conduct alleged to constitute an unfair labor practice. If the preliminary investigation of a first-priority case reveals reasonable cause to believe the charge is true, the law requires that the general counsel seek such injunctive relief or temporary restraining order as seems proper under the circumstances.

13.14 UNION-SHOP AGREEMENTS

The National Labor Relations Act outlaws the closed shop but permits the establishment of a union shop. A closed shop requires that a worker be a member of the appropriate union at the time he is hired. Under a union shop a new employee need not be a union member at the time of employment, but he must join within a stipulated period to retain his job. Therefore, a union-security agreement (an agreement providing for compulsory union membership) cannot require that applicants for employment be members of the union to be hired but can stipulate that all employees covered by the agreement must become members of the union within a certain period of time. This grace period cannot be less than 30 days after hiring, except in the building and construction industry, in which a shorter grace period of seven days is permissible.

Union-shop agreements often provide for the checkoff of union dues, an arrangement where the employers deduct dues from their employees' wages and pay the money withheld to the union. Under the NLRA, such checkoff is permitted only by written assignment of each employee, and such assignment is not to remain in effect for a period of more than one year or beyond the end of the current collective agreement, whichever occurs sooner. A mandatory checkoff is illegal and is an unfair labor practice on the part of both the employer and union.

Section 14(b) of the Taft-Hartley Act provides that the individual states have the right to forbid negotiated labor agreements that require union membership as a condition of employment. In other words, any state or territory of the United States may, if it chooses, pass a law making a union-shop labor agreement illegal. This is called the "right-to-work" section of the act, and such state laws are termed right-to-work statutes. At the present writing 20 states have such laws in force.* It is interesting to note that most of these state right-to-work laws go beyond the mere issue of compulsory unionism inherent in the union shop. Most of them outlaw the agency shop, under which workers, in lieu of joining a union, must pay as a condition of continued employment the same initiation fees, dues, and assessments as union members. Some of the laws explicitly forbid unions to strike over the issue of employment of nonunion workers.

13.15 PREHIRE AGREEMENTS

Section 8(f) of the National Labor Relations Act allows an employer engaged primarily in the building and construction industry to sign a union-shop agreement with a union that has not been elected as the representative of his employees. The agreement can be made before the contractor has hired any employees and will apply to them when they are hired. This proviso is called the "prehire" provision of the act. Contractors frequently make prehire agreements with construction unions at the outset of large projects as a means of trying to stabilize labor costs and securing a ready-made source of skilled manpower.

The union's right to maintain and enforce such a labor contract can be challenged at any time, however. The law provides that the employer can test

*Alabama, Arizona, Arkansas, Florida, Georgia, Iowa, Kansas, Louisiana, Mississippi, Nebraska, Nevada, North Carolina, North Dakota, South Carolina, South Dakota, Tennessee, Texas, Utah, Virginia, and Wyoming now have right-to-work legislation in effect.

by a representation vote the majority status of the union. Until the union can show it has the support of a majority of the employees, the contractor is free to ignore the prehire bargaining agreement or unilaterally change the terms and conditions of employment. The courts have ruled that prehire agreements are not mandatory subjects of bargaining and unions cannot strike or picket to obtain them. In addition, the law does not give the union any right to enforce such an agreement, either by picketing or in the courts, unless the union can demonstrate that it represents the majority of all the company's employees at the company's various jobsites. Hence, no enforceable agreement exists prior to a representation election.

13.16 UNION HIRING HALLS

The NLRA provides that in the building and construction industry a labor agreement can require the contractor to notify the union of opportunities for employment or to give the union an opportunity to refer qualified applicants to the contractor. The agreement may also specify minimum training or experience qualifications for employment or provide for priority in job referrals based on length of service with the employer, in the industry, or in the particular geographical area. However, hiring-hall agreements that give priority to employees who previously worked for employers subject to collective bargaining agreements with the union have been found illegal under the National Labor Relations Act.

Contracts or informal arrangements with a union under which an employer gives preferential treatment to union members are illegal. It is not unlawful, however, for an employer and a union to enter into an agreement whereby the employer agrees to hire new employees exclusively through a union hiring hall, as long as there is no discrimination against nonunion members in favor of union members. Both the agreement and the actual operation of the hiring hall must be nondiscriminatory; job referrals must be made without reference to union membership. The employer must not discriminate against a nonunion employee if union membership is not available to that employee under the usual terms or if membership is denied to that employee for reason other than nonpayment of union dues and fees. Hiring-hall provisions in labor contracts usually give the contractor the right to reject any applicant and the right to obtain employees from other sources when the union is unable to supply a sufficient number of qualified people.

The courts have held that a lawful hiring hall is a mandatory subject of bargaining between employers and unions and that a union can strike and picket to press its demands for an exclusive nondiscriminatory hiring-hall referral system. It is important to note that contractors who were parties to labor agreements providing for exclusive union-hall referrals have in certain circumstances shared the union's guilt when the system has been found guilty of using discriminatory practices. The fact that a contractor has delegated the selection of its employees to the union does not relieve it from liability if the union (as the contractor's agent) acts in an unlawful manner. The courts have also ruled that union hiring halls are permissible in states with right-to-work laws as long as the hiring-hall agreement expressly states that union membership is not to be considered in job referrals.

Construction hiring halls or referral systems that discriminate against minorities are illegal under civil rights statutes, even though they may appear to be legal under the National Labor Relations Act. The matter of contractor responsibility has become particularly troublesome with regard to the hiring of workers from minority groups. In this regard, the NLRB has held that hiring is a management responsibility and cannot be delegated to a union. The contractor, not the union, must be the judge of a worker's competence, and a worker's access to a construction job cannot be conditioned on his ability to pass a union examination. If a contractor hires or retains a worker the union will not accept, the union is liable for the consequences if it strikes to force that worker off the job.

13.17 SECONDARY BOYCOTTS

A primary boycott arises when a union, engaged in a dispute with an employer, exhorts that firm's customers and the public to refrain from all dealings with that employer. A secondary boycott occurs if a union has a dispute with Company A and attempts to exert pressure on it by causing the employees of Company B to stop handling or using the products of Company A or otherwise forces Company B to stop doing business with Company A. The primary employer, in this case Company A, is the employer with whom the union has the dispute. Company B is the neutral secondary employer and hence the name secondary boycott.

Secondary boycotts have a long and turbulent legal history. Illegal at common law, secondary boycotts were ruled to have been forbidden by the Sherman Act in a decision by the U.S. Supreme Court. This prohibition was reversed by the Norris-LaGuardia Act and the Wagner Act, which gave unions almost complete immunity from liability for damages arising out of secondary boycotts. The pendulum has since returned almost to its original position, with the National Labor Relations Act forbidding secondary boycotts.

Secondary boycotts in construction can assume many forms, and the dividing line between a legal primary boycott and an illegal secondary boycott is sometimes hazy and difficult to establish. The wording of the law is strictly construed in determining the legality of a boycott action. A form of secondary boycott that is of extraordinary importance to the construction industry will now be discussed.

13.18 COMMON-SITUS PICKETING

A common situs is a given location such as an industrial plant or a construction project at which several different employers are simultaneously engaged in their individual business activities. A dispute between one of these employers and a union is likely to involve the other neutral employers, especially if picketing is involved. Decisions of the NLRB and of the courts have evolved some rules for establishing whether such common-situs picketing constitutes an illegal secondary boycott action.

In an attempt to give effect to both the union's right to picket the primary employer and the right of the secondary employer to be free from disputes that are not his own, the NLRB established in 1950 the Moore Dry Dock tests, which determine when a union may picket a common site without committing an illegal secondary boycott. These rules are:

1. That the picketing be limited to times when the employees of the primary employer are working on the premises.
2. That the picketing be limited to times when the primary employer is carrying on his normal business there.
3. That the picket signs clearly indicate the identity of the primary employer with whom the union is having the dispute.
4. That the picketing be carried on reasonably close to where the employees of the primary employer are working.

Common-situs picketing occurs frequently in the construction industry, often as the result of the general contractor awarding work to a nonunion subcontractor. The resulting picketing causes the employees of the union contractors on the project to refuse to cross the picket line. In 1951 the U.S. Supreme Court decided the Denver Building and Construction Trades Council case. This dispute involved a general contractor whose employees were union members and a nonunion subcontractor. The project was picketed and shut down by the construction unions which demanded that the subcontractor be discharged. The U.S. Supreme Court found that the unions were guilty of an illegal secondary boycott because the object of the picketing was to force the general contractor (secondary employer) to quit doing business with the nonunion subcontractor (primary employer).

Subsequent to the Denver Building Trades case, the "separate gate" doctrine was developed. On multiemployer construction sites one gate is reserved and marked for the primary contractor involved in the labor dispute and another gate for the neutral contractors not involved. On the basis of the Moore Dry Dock standards traditionally applied to common-situs picketing and the Denver Building Trades case, the NLRB and the courts now hold that picketing of the gate reserved for the contractor directly involved in the dispute is permissible, but that the unions cannot picket separate gates used by employees of neutral secondary contractors if the effect is to keep them off the project. In essence this decision says that construction contractors at the site are separate and distinct employers and union picketing must be limited to the primary contractor involved in the dispute. Recent years have seen a sustained attempt by organized labor to prevail upon Congress to amend the National Labor Relations Act to permit unrestricted picketing of construction sites. To date, however, these efforts have been to no avail.

The U.S. Supreme Court, in the General Electric case (1961), decided the matter is different, however, when construction work is being done at an industrial plant by a construction contractor. The court ruled that picketing by plant strikers of gates reserved exclusively for contractor personnel can be banned only if there is a separate, marked gate set apart for the contractor, if the work being done by the contractor is unrelated to the normal operations of the industrial company, and if the work is of a kind that will not curtail normal plant operations. The courts have ruled that the General Electric decision does not apply to a prime contractor and subcontractors at the usual construction site.

13.19 SUBCONTRACTOR AGREEMENTS

The Landrum-Griffin Act made it an unfair labor practice for an employer and a union to enter into an agreement whereby the employer agrees to refrain from handling the products of another employer or to cease doing business with any other person. As has already been pointed out, such a contract provision is called a hot-cargo clause. However, the construction industry was exempted from the ban under certain circumstances. Under the construction industry proviso to Section 8(e) of the National Labor Relations Act, contractors can agree to restrictions on subcontracting or can agree not to handle certain products, as long as the restrictions relate to the contracting or subcontracting of work to be done at the site. This construction industry exemption has led to the widespread use of two forms of hot-cargo clauses in construction labor contracts. One of these is the subcontractor agreement, a subject discussed in this section. The other is the work preservation clause that is discussed in the following section.

Subcontractor agreements typically require the general contractor to award work only to those subcontractors who are signatory to a specific union labor contract or who are under agreement with the appropriate union. Such agreements do not extend to supplies or other products produced or manufactured elsewhere and delivered to the construction site. The NLRB and the courts have ruled that construction unions may strike to obtain subcontractor clauses in their labor contracts if no secondary boycott is involved. To illustrate, picketing to induce a general contractor to accept a subcontractor clause is legal, but picketing is illegal as a secondary boycott if it is designed to force a neutral general contractor to stop doing business with an existing and identified nonunion subcontractor. The courts have held that unions cannot enforce subcontractor clauses by threats, coercion, strikes, or picketing but that such labor contract provisions can be judicially enforced by the courts through breach-of-contract suits. Self-enforcing clauses have been ruled to be illegal and unenforceable. These are where the contracting firm agrees that if it violates the subcontract agreement, the union can take action against it such as picketing, refusing to provide workmen, or canceling the labor contract between them.

By its decision in the Connell case (1975), the U.S. Supreme Court ruled that a subcontractor agreement can be subject to federal antitrust laws where it is not within the context of a collective bargaining relationship and is not restricted to a particular jobsite. The court further ruled that where these conditions are not met, such an agreement is a hot-cargo clause prohibited by Section 8(e) of the National Labor Relations Act and does not fall within the construction industry exemption. However, the Connell case has proven to be less than definitive in providing specific guidance as to just what constitutes a legal subcontracting clause. Recent decisions by different U.S. Courts of Appeals have disagreed on the issues of whether a subcontractor clause can apply to more than one jobsite and to sites where no union workers are employed.

13.20   WORK PRESERVATION CLAUSES

The work preservation clause is the second form of hot-cargo provision ruled to be permissible under the construction industry exemption. In accordance with a U.S. Supreme Court decision made in 1967 concerning the installation of precut doors in Philadelphia, construction unions may legally obtain labor agreements to bar the use of prefabricated products in construction in order to preserve their customary on-site work, and can enforce them by strikes and picketing. Commonly called "work preservation clauses" or "prefabrication clauses," these provisions ban the use of prefabricated construction products manufactured off the site—products that eliminate work normally done on the project itself. Examples of these construction products are precut and prefitted wooden doors, precut pipe insulation, prefabricated trusses, and prepackaged boilers. Such product boycotts have been construed to fall within the construction industry exemption from the ban on hot-cargo clauses and can be legal work preservation clauses in a labor agreement if the prefabricated products replace work customarily and traditionally performed on the site by the union members. Prefabrication clauses are now commonplace in construction labor contracts.

The NLRB applies the "right-of-control test" as one factor in determining the legality of prefabrication clauses. Under this test, if a prefabricated product is specified by the architect-engineer and/or owner, then the contractor is required by the terms of the construction contract to provide the materials as specified and has no control over product selection. In such a case the union cannot refuse to handle and install the product, whether or not a work preservation clause exists. On the other hand, if the construction contract does not specify the use of a prefabricated product but the contractor, on its own volition, decides to use such materials, then the union is entitled to enforce the work preservation clause and to strike and picket the offending contractor to block the use of the prefabricated materials. Application of the right-to-control test to determine when a work preservation dispute becomes an illegal secondary boycott was upheld by the U.S. Supreme Court in the Enterprise case (1977).

13.21   JURISDICTIONAL DISPUTES

A jurisdictional dispute can arise when more than one union claims jurisdiction over a given item of work. The dispute is between unions and the employer is caught in the middle. The construction industry, in particular, is plagued with this type of dispute, primarily because each of its many craft unions regards its type of work as a proprietary right and jealously guards against any encroachment of its traditional sphere by other unions. Lines of demarcation between the various jurisdictions are sometimes indistinct, and the development of new products and methods often brings with it jurisdictional clashes between unions whose members claim exclusive right to the work assignment. When disputed work is at issue, the contractor must assign the work to one of the unions involved. This often prompts the other union or unions to strike or resort to some other form of work stoppage or slowdown, and the contractor finds itself with a jurisdictional dispute on its hands.

There are different procedures that can be followed by the construction contractor when jurisdictional disputes occur. One of these is to utilize voluntary machinery established by and within the construction industry itself, a subject discussed in the next section. Another is to take the case to the NLRB as an unfair labor practice. However, the NLRB will not hear a jurisdictional dispute where voluntary methods for settlement have been agreed upon until after such procedures have been exhausted.

Section 8(b)(4)(D) of the NLRA makes a jurisdictional strike a union unfair labor practice unless the employer is failing to conform with an order of the NLRB. An accusation under this section, however, is handled differently from charges involving other types of unfair labor practices. After an unfair labor practice charge involving a jurisdictional dispute is filed with the NLRB, a preliminary investigation is held to ascertain if there is reasonable cause to believe that the charge is true. If so, the NLRB schedules a hearing on the case within 10 days. It may also seek an injunction to halt the strike or work stoppage in the interim. If, within the 10 days, the rival unions settle their dispute or agree on a method for adjusting it, the hearing is canceled. If they do not, the board is empowered and directed to hold a hearing on the dispute in accordance with Section 10(k) of the Taft-Hartley Act and to make an assignment of the work under contention. If the determination is not obeyed, the charge is processed in the same manner as any other unfair labor practice. Jurisdictional awards made by the NLRB are limited in scope to the disputes from which they arise. The Taft-Hartley Act gives private parties damaged by a jurisdictional strike, picketing, or other coercion the right to sue the union or unions involved.

13.22 VOLUNTARY JURISDICTIONAL SETTLEMENT PLANS

Because the Taft-Hartley Act does not require the NLRB to rule on jurisdictional matters when the disputants have agreed to voluntary methods of adjustment, the construction industry has set up a number of voluntary plans, one national in scope and many strictly local in coverage. The first national plan to settle jurisdictional disputes in the construction industry was established shortly after the Taft-Hartley Act was enacted. In 1948 the National Joint Board for the Settlement of Jurisdictional Disputes was formed by specialty contractor associations, the Associated General Contractors of America, and the AFL building trades unions.

After many ups and downs over the years, the National Joint Board was disbanded in 1973 and was replaced by the Plan for Settlement of Jurisdictional Disputes in the Construction Industry. Parties to this plan are the AFL-

CIO Building and Construction Trades Department, the National Constructors Association, the Associated General Contractors of America, and a number of specialty contractor associations. Under this plan an Impartial Jurisdictional Disputes Board decides jurisdictional disputes referred to it. Procedural rules of the plan require the contractor to assign disputed work in accordance with jurisdictional agreements and decisions of record such as those contained in the "Green Book" and the "Gray Book."* If no such agreement or decision exists, the work must be assigned according to prevailing trade or area practice. If there has been no local precedent, the contractor must use its best judgment in assigning the work.

After the assignment is made by the contractor, no change can be effected unless directed by the Impartial Jurisdictional Disputes Board or by agreement between the unions involved. The plan provides that there shall be no strikes or work stoppages arising out of any jurisdictional dispute. Jurisdictional awards made by the Impartial Jurisdictional Disputes Board apply only to that project on which the dispute occurs. Decisions of the Impartial Jurisdictional Disputes Board can be appealed to the Joint Administrative Committee. Participation in the national jurisdictional plan is voluntary with each contractor or employer bargaining unit, and unions may not force contractors to use it. The plan provides financial penalties for building trades unions whose locals use strikes, slowdowns, or picket lines to protest work assignments. Recently, both management and labor have expressed dissatisfaction with the operation of the Impartial Jurisdictional Disputes Board and have suggested that a new approach may be desirable.

In addition to the national plan just discussed, there are many local areas that have set up machinery for the voluntary settlement of jurisdictional disputes. Several large metropolitan areas maintain their own boards for settling jurisdictional strikes. Local plans and procedures of wide variety are used over the country. Without doubt, a large proportion of the jurisdictional disputes in the construction industry are settled at the local level.

The U.S. Supreme Court in the Texas Tile case (1971) held that the contractor is a party to a jurisdictional dispute as well as the rival unions. The effect of this is that any private arrangement between unions providing for the settlement of their jurisdictional disputes must include the employer if the employer is to be bound by it. Otherwise, such an agreement between the unions does not prevent a contractor from referring the matter to the NLRB and will not bar the NLRB from hearing the dispute and assigning the contested work.

13.23 PAYMENTS TO EMPLOYEE REPRESENTATIVES

Section 302 of the Taft-Hartley Act prohibits any employer or association of employers from paying, lending, or delivering money or other thing of value

---

*The "Green Book," approved by the Building and Construction Trades Department, AFL-CIO, and the "Gray Book," published by the Associated General Contractors of America, contain permanent jurisdictional agreements reached between unions which are pertinent to the construction industry.

to its employees or their representatives. This includes labor unions or officers thereof and any employee or group of employees if the purpose of the payment is to influence the right of employees to organize and bargain collectively. Specifically permitted by this section, however, are various fringe benefits such as health, welfare, pension, vacation, holiday, and annuity payments, as well as apprenticeship plans and prepaid legal services for which employer contributions are permissible and over which unions are given some control. Such contributions must be paid to a trust fund, and the trustees must consist of an equal number of labor and management representatives.

13.24 POLITICAL CONTRIBUTIONS

The Taft-Hartley Act makes it unlawful for certain organizations, including corporations and labor organizations, to make a contribution or expenditure in connection with the election of federal officials. A labor organization is defined to be any organization in which employees participate and which exists for the purpose of dealing with employers concerning grievances, labor disputes, wages, rates of pay, hours of employment, or conditions of work.

Despite these restrictions, however, corporations and labor unions are very active on the American political scene through the medium of Political Action Committees (PACs). Under federal campaign laws and Federal Election Commission regulations, both corporations and labor unions are permitted to establish PACs at the national, state, and local levels. Within limitations established by law, these organizations are authorized to solicit and receive personal contributions from individuals and disburse these to selected candidates for political office. PACs established by corporate businesses and trade associations use contributed funds to help elect candidates philosophically in tune with management viewpoints. On the other hand, PACs associated with labor unions contribute large sums toward the election of candidates who support the causes of organized labor.

In addition, labor unions maintain local "education" funds through the AFL-CIO Committee on Political Education (COPE). This falls under federal law that allows labor unions to spend unlimited amounts of their own monies on communications to members and their families on any subject. These funds support get-out-the-vote efforts through the publication and distribution of union newsletters that make specific candidate endorsements. They also finance meet-the-candidates sessions where union-endorsed candidates may speak and may solicit funds and volunteer labor. While the campaign laws allow business corporations to conduct similar education activities with their management employees and stockholders, this is not commonly done.

13.25 THE CIVIL RIGHTS ACT OF 1964

In passing the Civil Rights Act of 1964, Congress confirmed and established certain basic individual rights pertaining to voting; access to public accommodations, public facilities, and public education; participation in federally assisted programs; and opportunities for employment. Title VII of this act, Equal Employment Opportunity, prohibits discrimination in employment or union membership. It is an unlawful practice for an employer (1) to refuse to hire or to discharge any individual or otherwise discriminate against him regarding conditions of employment because of his race, color, religion, sex, or national origin, or (2) to limit, segregate, or classify employees in any way that would deprive the individual of employment opportunity or adversely affect his status as an employee because of race, color, religion, sex, or national origin.

Administration and enforcement of the Civil Rights Act is made the responsibility of the Equal Employment Opportunity Commission (EEOC), which the act created. The responsibility of the commission is to assure that consideration for hiring and promotion is based on ability and qualifications, without discrimination. Title VII prohibits discriminatory practices on the part of employers, employment agencies, labor organizations, and apprenticeship or training programs. The Civil Rights Act applies to interstate commerce and covers employers and labor organizations. There are several exemptions from the act, some of which are local, state, and federal agencies; government-owned corporations; Indian tribes; and religious organizations. The law requires that employers, labor unions, employment agencies, and joint labor-management apprenticeship committees keep such records and submit such reports as the EEOC may require. Special rules apply in states that have their own enforceable fair employment practice laws.

The Equal Employment Opportunity Act of 1972 amended the Civil Rights Act of 1964 in several important respects and expanded its coverage substantially. It authorized the EEOC for the first time to go directly to court for temporary restraining orders and for permanent injunctions against unlawful discrimination. This is in addition to other remedies such as reinstatement or hiring with back pay and appropriate affirmative action directives. The Civil Rights Act of 1964 now covers joint labor-management committees for apprenticeship and other training programs. The coverage of the act was expanded to include employers of 15 or more employees and unions that operate hiring halls or referral systems or that have 15 or more members (25 in the original Civil Rights Act of 1964). Also created was the position of general counsel to act along much the same lines as the general counsel under the NLRB. The general counsel of the EEOC has authority to bring civil court actions against patterns and practices of employment discrimination in interstate commerce.

13.26 EXECUTIVE ORDER 11246

Issued in 1965, Executive Order 11246 applies to contracts and subcontracts exceeding $10,000 on federal and federally assisted construction projects. Federal and federally assisted construction is not normally construed to include projects involving federal assistance that is in the nature of a loan guarantee or insurance. Contractors are prohibited from discriminating against any employee or applicant for employment because of race, color, religion, or national origin. The contractor must take positive action to ensure that applicants are employed, and that employees are treated during employment, without discrimination. Affirmative action must be taken by contractors on projects covered by the executive order to increase the level of minority representation in their workforces. Actions pertaining to employment, promotion, transfer, recruitment, layoff, rates of pay, training, and apprenticeship must not be discriminatory. Executive Order 11375 (1968), as an extension of Executive Order 11246, applies to federal and federal-aid contracts and prohibits discrimination against any employee because of sex.

Executive Order 11246, administered by the Office of Federal Contract Compliance Programs (OFCCP), U.S. Department of Labor, states that each federal contracting agency shall be primarily responsible for obtaining compliance with the provisions of the order. In addition, each administering agency is made responsible for compliance by the recipients of federal financial assistance. Federal agencies have compliance officers whose duties are to ensure adherence to the objectives of the order, including compliance reviews. A compliance review is a procedure used to check an ongoing contract. The contractor is required to give information to show that it is complying with the nondiscriminatory requirements of its contract, including affirmative action.

The OFCCP has issued nationwide affirmative action goals for hiring minorities and women in the construction trades. The minority goals, set for both major metropolitan areas and rural areas, are equal to the percentages of minorities in the civilian workforce in those areas. The goals apply when a contractor has a federal or federally assisted construction contract or subcontract in excess of $10,000 and are expressed as a percentage of the total hours worked by the contractor's entire onsite construction workforce employed in the relevant area. The covered contractor is required to make every effort to meet the goals for minority and female participation on *all* of the contractor's work, federally financed and otherwise, regardless of the location of the federal contract.

In the event of noncompliance with OFCCP rules, the contract may be canceled or suspended, and the contractor can be declared ineligible for further government or federally assisted construction contracts. Additionally, the OFCCP has the authority to withhold progress payments from contractors that are in violation of Executive Order 11246. This authority stems from the government's right to suspend payment when a contractor fails to comply with any requirement of the contract. Compliance reports from contractors are required, and the general contractor must include suitable provisions concerning compliance with the order in its subcontracts and purchase orders.

13.27 THE AGE DISCRIMINATION IN EMPLOYMENT ACT

The Age Discrimination in Employment Act of 1967 prohibits arbitrary age discrimination in employment. This act protects individuals 40 to 65 years old from age discrimination by employers of 25 or more persons in an industry affecting interstate commerce. Employment agencies and labor organizations are also covered.

By the terms of the act it is against the law for an employer to:

1. Fail or refuse to hire, to discharge, or to otherwise discriminate against any individual as to conditions of employment because of age.
2. Limit, segregate, or classify his employees so as to deprive any individual of employment opportunities or to adversely affect his status as an employee because of age.
3. Reduce the wage rate of any employee in order to comply with the act.

The prohibitions against discrimination because of age do not apply when age is a bona fide occupational qualification, when differentiation is based on reasonable factors other than age, when the differentiation is caused by the terms of a bona fide seniority system or employee benefit plan, or when the discharge or discipline of the individual is for good cause.

Employers must post an approved notice of the Age Discrimination in Employment Act in a prominent place where employees can see it and must maintain records as required by the Secretary of Labor. The act is enforced by the Secretary of Labor, who can make investigations, issue administrative rules and regulations, and enforce its provisions by legal proceedings.

13.28 THE DAVIS-BACON ACT

The Davis-Bacon Act (1931), as subsequently amended, is a federal law that determines the wage rates, including fringe benefits, that must be paid workers on all federal construction projects and on a host of federally assisted jobs. The law applies to contracts in excess of $2000 and states that the wages of workers shall not be less than the wage rates specified in the schedule of prevailing wages as determined by the Secretary of Labor for similar work on similar projects in the vicinity in which the work is to be performed. The general contractor and subcontractors are required to pay at least once a week all workmen employed directly on the site of the work at wage rates no lower than those prescribed. Full payment must be made, with the exception of such payroll deductions as are permitted by the Copeland Act (see Section 13.29). The law's purpose is to protect the local wage rates and local economies of each community and presumably to put union and nonunion contractors on a more nearly equal footing competitively in bidding on federal and federally assisted projects.

For purposes of defining the coverage of prevailing wages prescribed for a given project, the work site is defined as being limited to the physical place or places where the construction called for will remain and to other adjacent or nearby property used by the contractor or subcontractors which can reasonably be included in the "site" because of proximity. Fabrication plants, batch plants, borrow pits, tool yards, and the like are considered to be a part of the work site, provided they are dedicated exclusively or nearly so to performance of the contract and are so located in relation to the actual construction location that it would be reasonable to include them. Exempt from the work site definition, and therefore from Davis-Bacon wage rates, are permanent offices, branch plants, or fabrication plants of a contractor or subcontractor whose locations and continuance are governed by its general business operations.

The Davis-Bacon Act is administered by the U.S. Department of Labor. Contractors whose projects are covered must keep certain records, file periodic reports, and comply with various regulations with respect to the use of apprentices. Violation of Davis-Bacon requirements constitutes a breach of contract and exposes the contractor and subcontractors to government compliance action and suits by employees. Restitution is secured for workers found to have been underpaid, and penalties are assessed for violations of the overtime requirements. Violators can be denied the right to bid on other federal or federal-aid projects. The act does not provide for judicial review of Labor Department wage determinations, but a Wage Appeals Board operating under delegated authority from the Secretary of Labor has been established to hear appeals from findings or decisions of the Davis-Bacon Division. Contracting agencies have primary responsibility for Davis-Bacon enforcement, because obligations under the act become a part of the construction contract. Many of the states have enacted "little Davis-Bacon" statutes that pertain to state and municipal construction projects.

Recent years have seen the Davis-Bacon Act come under attack from a number of different quarters. The General Accounting Office of the federal government has recommended that the act be repealed and demands for its reform have been voiced by many responsible parties including Congressional committees. Much of the criticism is directed at the administration of the act and the procedures that are followed in determining prevailing wages. There are many allegations that prevailing wages as determined are usually too high, this resulting in excessive construction costs and more inflation. In many instances, prevailing wage rates are merely equated to local union pay scales. In addition, it has been stated that the Davis-Bacon Act eliminates competition between union and nonunion contractors and imposes costly and burdensome reporting requirements on contractors.

13.29 THE COPELAND ACT

As passed in 1934, and since amended, the Copeland Act makes it a punishable offense for an employer to deprive anyone employed on federal construction work or work financed in whole or in part by federal funds of any portion of the compensation to which the employee is entitled. Other than deductions provided by law, the employer may not induce "kickbacks" from his employees by force, intimidation, threat of dismissal, or any other means whatever. This portion of the Copeland Act is commonly known as the Anti-Kickback Law, violation of which may be punished by fine, imprisonment, or both. Regulations issued by the Secretary of Labor allow the contractor or subcontractor to make additional deductions from wages, provided that the prior approval of the Department of Labor is obtained by showing that the proposed deductions are proper. Union dues may be deducted by the employer if such holdback is consented to by the employee and is provided for in a collective bargaining agreement.

The law stipulates that payroll records shall be maintained and reports submitted by contractors as the Department of Labor may require. The Copeland Act covers all construction projects on which Davis-Bacon prevailing wages apply. The contracting agency is responsible for enforcing compliance with the act.

13.30   THE FAIR LABOR STANDARDS ACT

First enacted by Congress in 1938 and since amended several times, the Fair Labor Standards Act, also known as the Wage and Hour Law, contains minimum wage, maximum hours, overtime pay, equal pay, and child-labor standards. Workers whose employment is related to interstate commerce or consists of producing goods for interstate commerce are covered without regard to the dollar volume of business.

The Fair Labor Standards Act provides for a minimum wage for all employees covered, this minimum wage having been steadily increased during recent years. Also required is payment of an overtime rate of one and one half times the regular hourly rate of pay for all hours worked in excess of 40 hours in any workweek. However, payment of overtime is not required for more than eight hours per day, nor is there a limit set on the number of hours that may be worked in any one day or during any one week. The law does not require premium pay for Saturday, Sunday, or holiday work, or vacation or severance pay.

An employer who violates the wage and hour requirements is liable to his employees for double the unpaid minimum wages or overtime compensation plus associated court costs and attorney's fees. Willful violation of the law is made a criminal act, and the errant employer can be prosecuted. Several classes of employees are exempted from coverage under the act, such as bona fide executive, administrative, and professional employees who meet certain tests established for exemption.

The Fair Labor Standards Act, as amended by the Equal Pay Act of 1963, provides that an employer must not discriminate on the basis of sex by paying employees of one sex wages at rates lower than he pays employees of the other sex for doing equal work on jobs requiring comparable skill, effort, and responsibility and performed under similar working conditions. Pay differentials can be justified by a seniority system, merit system, piecework pay system, or other system based on factors other than sex.

The basic minimum age for employment covered by the act is 16 years except for occupations declared to be hazardous by the Secretary of Labor, to which an 18-year minimum age applies. Construction per se is not designated as hazardous, but specified work assignments such as truck driving, wrecking and demolition, and power tool operation are so designated.

13.31   THE CONTRACT WORK HOURS AND SAFETY STANDARDS ACT

In 1962 Congress passed the Contract Work Hours and Safety Standards Act, also known as the Work Hours Act of 1962. This act, as subsequently amended, applies to federal construction projects and to projects financed in whole or in part by the federal government. The main requirement of this law is that every mechanic and laborer shall be paid at a rate not less than one and one half times the basic rate of pay for all hours worked in excess of eight hours per day or 40 hours per week. In the event of violation, the contractor or subcontractor responsible is liable for unpaid wages to the employees affected and for liquidated damages to the federal government. Willful violation of the Work Hours Act is punishable by fine or imprisonment. The enforcement of the law and the withholding of funds from the contractor to secure compliance with the act are made the responsibility of the government agency for which the work is being done.

13.32 THE HOBBS ACT

Also known as the Anti-Racketeering Act, the Hobbs Act, which was enacted in 1946, makes it a felony to obstruct, delay, or affect interstate commerce by robbery or extortion. To attempt or conspire to do so is also made a felony. Robbery is defined as the unlawful taking or obtaining of personal property from a person against that person's will by means of actual or threatened force or violence. Extortion is defined as the obtaining of property from another, with that person's consent, induced by the wrongful use of actual or threatened force, violence, or fear. The underlying motive behind the act was to put an end to the use of threats, force, or violence by union officials to obtain payments from employers under the guise of recompense for services rendered. Prosecution of violators is placed in the hands of the U.S. Department of Justice.

Extortion by unions and union officials has been practiced in many guises against contractors, payments being required as a condition of avoiding "labor trouble." These payments have been concealed behind many subterfuges such as "gifts," "commissions," "equipment rentals," and "services." The courts have held that under the Hobbs Act extortion extends to attempts by a union or union officials to obtain money from an employer in the form of wages for imposed, unwanted, and superfluous services. Violence or threats of violence need not be involved if such attempts involve fear of economic loss, injury to employees, or damage to equipment. However, by its 1973 decision in the Enmons case, the U.S. Supreme Court held that the Hobbs Act does not extend to extortion connected with legitimate labor disputes.

13.33 THE NATIONAL APPRENTICESHIP ACT

In 1937 Congress passed the National Apprenticeship Act, authorizing the establishment of the Bureau of Apprenticeship and Training of the U.S. Department of Labor. The bureau has the responsibility to encourage the establishment of apprenticeship programs and to help improve existing ones, but it does not conduct them itself. One of its prime objectives is to promote cooperation between management and organized labor in the development of such programs. Since its establishment the bureau has provided technical assistance in developing and improving apprenticeship and other industrial training programs and has set up minimum standards for the registration of local apprenticeship programs. The bureau works closely with state apprenticeship agencies, trade and industrial education institutions, and management and labor. Through its field staff it cooperates with local employers and unions in developing apprenticeship programs to meet specific needs.

To implement apprentice training on a local level, many states have passed laws that provide for the establishment of state apprenticeship councils. These state agencies function cooperatively with the Bureau of Apprenticeship and Training and are made up of labor, management, and public representatives, often with the addition of members from the state labor departments and others. Using the standards recommended by the bureau as a guide, the councils have established detailed standards and procedures to which apprenticeship and training programs in the state are expected to conform. The state council becomes a part of the national apprenticeship program by securing recognition of its standards and procedures by the Bureau of Apprenticeship and Training.

13.34 APPRENTICESHIP PROGRAMS

Traditionally, apprenticeship programs in the construction industry have been a joint effort between union-shop contractors and the AFL-CIO building trades unions. A National Joint Apprenticeship Committee for each craft is responsible for developing suggested national training standards for that craft. The actual employment and training of apprentices is essentially a local matter supervised by local joint apprenticeship committees that are made up of both contractor and labor union representatives. However, in 1971, the Bureau of Apprenticeship and Training approved national apprenticeship standards for the employees of open-shop contractors. This was the first time a unilateral apprenticeship program was approved on a national basis and placed under the direct supervision of employers only. Consequently, independent programs can be registered when the sponsors decline to participate in existing local AFL-CIO training programs. For local programs to be approved by and registered with a state apprenticeship council or with the Bureau of Apprenticeship and Training, the operating standards adopted by the local body must meet certain criteria. In general, a construction trade requires two to four years of on-the-job training and a minimum of 144 hours a year of related classroom instruction. It is important to note than an apprentice cannot work on a project covered by the Davis-Bacon Act unless he is part of an approved training program.

Federal regulations designed to increase minority-group participation in apprenticeship by requiring unions and contractors to go beyond passive nondiscrimination have been in effect for some years. These regulations require that apprenticeship programs take in enough minority-group apprentices to make the racial composition of the local labor force reflect the racial mix of the community. In addition, special apprenticeship programs are in operation to accelerate the entry of women into the building trades. Unions and/or contractors operating apprenticeship programs must have a positive affirmative action plan covering all phases of the program including recruitment, admission, and selection. Compliance reviews are made, and noncompliance can lead to deregistration of the apprenticeship program or prosecution under Title VII of the Civil Rights Act of 1964.

The U.S. Department of Labor administers the Apprenticeship Outreach Program through the Bureau of Apprenticeship and Training. This program, operating almost exclusively within the periphery of the building and construction trades, reaches out to minority youth, introducing them to registered craft apprenticeship programs. Minority youths are recruited from a variety of sources, interviewed, given basic tests to determine aptitude and general ability, and tutored to prepare them for apprenticeship examinations. Various groups, financed by the Labor Department, sponsor Outreach programs in cooperation with the local building and construction trades councils of the AFL-CIO. Outreach not only assists getting minority youth into apprenticeship programs but also helps to upgrade the skills of minority construction workers.

14

LABOR RELATIONS

14.1 THE CONSTRUCTION WORKER

The average construction worker has no fixed relationship with any one contractor, and his tenure of employment with a given employer is normally indefinite and temporal. He is tightly bound to his occupation and is only loosely associated with any given construction company. He may work for several different employers during the course of a year and is known as a carpenter or cement mason rather than as the employee of any particular firm. Although these generalities are much less true for workers in the specialty trades, such as electricians and plumbers, than they are for carpenters, ironworkers, and others in the basic trades, they still typify employment in the construction industry. The construction worker is in the atypical position of being a skilled artisan with no permanent place of employment. This circumstance is a manifestation of the fact that the projects of a typical construction company are short-term, variable as to location, and demand different combinations of trade skills. Any contractor experiences fluctuating requirements for manpower as new jobs are started and existing ones are completed. Economical operation dictates that most of a contractor's workforce be drawn from a local pool of manpower as needs dictate.

Although the popular conception of a construction worker is that of a "migratory bird," the geographical mobility of such people is variable with the craft and the individual circumstances of employment. Crews on large industrial construction projects and on highways, pipelines, bridges, transmission lines, tunnels, and other engineering construction are necessarily mobile. Millwrights, boiler makers, pipe fitters, pile drivers, and structural ironworkers follow their specialties over wide geographical areas, moving from project to project. Building construction tradesmen are more apt to find continuous or relatively continuous employment within a given locality. A number of considerations such as home ownership, family ties, schools, pensions, and other factors make many construction workers reluctant to move their places of residence. This, among other reasons, leads to a certain amount of movement into and out of the construction industry itself. There are many tradesmen who find jobs elsewhere in the economy when local job opportunities in construction for their particular craft skills are limited.

Although reliable data are difficult to obtain, available evidence suggests that something more than 50 percent of all construction workers are union members. Union membership varies by type of construction, geographical area, and craft. Residential construction, one of the largest single segments of the industry, is probably the least organized, with an estimated 80 percent of its workers being nonunion. Large industrial and nonresidential building construction are probably the most fully organized branches of the industry, with heavy and highway work following. Geographically, union strength in the construction industry is greatest in the major cities and least in suburban and rural areas and in the southeastern states in general. Proportionately, union membership in the construction trades has not changed appreciably in more than 30 years. Many tradesmen who carry union cards conveniently alternate between union and nonunion jobs, depending on current job opportunities.

Some contractors work open-shop, others operate under union-shop contracts. An open-shop contractor is one whose employment of workers is not governed by the terms of a union contract but who hires tradesmen, union or nonunion, from the open labor market. A union-shop contractor is one whose labor relations are controlled and regulated by labor contract. No attempt is made here to argue the merits of either system. Considering the country as a whole, it is probable that the employment of construction workers for approximately half of the nation's contractors is controlled by union labor contracts. Relations with labor unions are, therefore, simply an everyday fact of life for many construction concerns. Even for open-shop contractors, union wage scales more or less set the prevailing wage patterns in most localities, and the craft unions provide an element of stability and cohesiveness to an otherwise uncertain area of manpower supply. All contractors, union and nonunion, are influenced by local union labor policies and work rules. Any discussion of labor relations in the construction industry must, of necessity, devote considerable attention to the construction unions.

14.2 ENTRY INTO THE BUILDING TRADES

Completion of a formal apprenticeship program is by no means the only method of entry into the skilled building trades. There are many alternative informal training opportunities, and available data indicate that only about 20 percent of the union journeymen in the construction industry today ever went through formal apprenticeship training. Experience in the armed forces, summer work, manpower training programs of diverse sorts, learning on nonunion projects, and "permit" work are all common examples of possible alternatives to the serving of a formal apprenticeship. Such informal routes may feed into journeyman upgrading programs or directly to journeyman status.

The importance of apprenticeship varies considerably with the trade and locality, being much more prominent in the mechanical trades and the large metropolitan areas. Formal apprenticeship has been criticized as overtraining the worker and being unnecessarily comprehensive in scope. There is no question that such programs provide highly skilled workers as well as a core of lead journeymen, foremen, supervisors, and even contractors.

14.3 RELATIONS WITH UNIONIZED TRADESMEN

As a general rule relations between the contractor and its unionized employees are casual and impersonal. The tradesman is referred to a project by his union, and his job performance is largely judged by his foreman, a fellow union member. There is little direct contact between the construction worker and the contractor employer. Personal relations are almost nonexistent because rates of pay, holidays, overtime, and other conditions of employment are not negotiated with the employee directly but with his union and are dictated by labor contract.

When he reports to a project, the tradesman is assumed to know his trade and is expected to perform any task that is assignable to his craft. He receives the same rate of pay as his fellow craftsman, regardless of their relative skills and abilities to produce. By and large his loyalties are to his union, whose well-being he closely associates with his own. He is not especially "company minded" because he has no fixed relationship with any given contracting firm.

As a result of these circumstances pertaining to the employment of unionized tradesmen, a contractor's labor relations are conducted almost entirely with the craft union locals whose members the contractor employs and not with the workers themselves. Labor relations for a union-shop contractor consist primarily, therefore, not of interpersonal relationships but of labor contract negotiation and administration.

14.4 THE ROLE OF THE UNIONS

Despite personal feelings in the matter, there is no denying the fact that unions make an important contribution to the operation of the construction industry. The unions have a stabilizing influence on a basically unstable area of business, an influence which, from the point of view of the contractor, has both its good and bad points. Through the medium of negotiated labor contracts, fixed wage rates are established, and much uncertainty associated with the employment of labor is removed. The unions provide a pool of skilled and experienced labor from which the contractor can draw as its needs dictate. In addition, they help to police the industry on both sides of the fence, serving to maintain discipline among their own members and helping to control the entry and actions of irresponsible, fly-by-night contractors.

Belonging to a union can offer many compensations to the working person. A prime consideration is the leading role organized labor takes in raising wages and establishing some form of job security for its members. Although the bargaining power of the individual worker is weak, that of an organization of workers can be very strong. The workers are secure in their belief that their union will protect them from unfair treatment and will exert every effort to improve their situation. They enjoy a sense of belonging to a group with the common purpose of mutual help, and through their elected union representatives they have a voice in the determination of their wages and conditions of employment.

It is true that some union members belong, not of their own volition, but as a matter of necessity to keep their jobs. However, available evidence incontrovertibly indicates that most union members are not unwilling captives of organized labor but belong because it is their desire. Of course, not all members are enthusiastic unionists. Many of them do not participate actively in union affairs or even attend meetings regularly. However, this passive attitude cannot be interpreted as a lack of union patriotism.

14.5 UNION HISTORY

Unions are worldwide, and had their origins at least 150 years ago. However, organized labor has become a stable, responsible element in America only within the past half century. In 1886 the first enduring union association was founded after a long era of repeated failures. The American Federation of Labor (AFL) was organized in that year with Samuel Gompers as its first president. Since the time of its inception, the AFL has traditionally been identified with the skilled craft worker. The AFL is a loose confederation of many sovereign national unions, each of which remains free to manage its own internal affairs. The construction trades were charter members of the AFL.

At the time the AFL was founded, semiskilled and unskilled factory workers were largely unorganized. The Knights of Labor, an organization of diverse membership, had made some progress in that direction but was waning rapidly. Nevertheless, the general sentiment of the AFL was not concordant with organization of the industrial worker. In 1905 the Industrial Workers of the World (IWW) was organized to fill the void left by the AFL's failure to act. The IWW advocated the elimination of capitalism, engaged frequently in violent strikes, and declined rapidly in the post-World War I era. Once again no central force existed for the organization of the mass-production worker, although several independent industrial unions were leading somewhat precarious existences.

Enactment of the Norris-LaGuardia and Wagner Acts in the 1930s encouraged and assisted union activity. The AFL began to extend a lukewarm welcome to some industrial unions but relegated them to a second-class position known as "federal locals." This scheme enjoyed no great success because the AFL craft unions, apprehensive that they would be obliterated in a huge mass of industrial workers, pressed demands on the federal locals for jurisdiction over members who were engaged in craft occupations. In 1935, the disgruntled leaders of eight industrial unions associated with the AFL formed the Committee for Industrial Organization for the avowed purpose of organizing the mass-production industries. This committee was organized within the AFL and sought to induce it to assume a dual personality, representing both craft and industry. This was nothing short of treason in the eyes of the AFL hierarchy, and the committee was summarily ordered to disband or be expelled from the federation. In 1936, the AFL executive council suspended the Committee for Industrial Organization. In 1938, after two fruitless years of attempts at reunification, the committee became the Congress of Industrial Organizations (CIO), an association of autonomous industrial unions, with John L. Lewis of the United Mine Workers as its first president.

In the following years, the AFL and the CIO engaged in a bitter struggle for the leadership of American labor, and both sides came to recognize the need for reconciliation. However, personal animosities and conflicts of interest proved difficult to resolve, and the merger was delayed for many years. Not until 1955 were the two groups able to patch up their differences and reassociate.

14.6   AFL-CIO CRAFT UNIONS

A large proportion of the organized construction workers belong to one of the international unions that form the Building and Construction Trades Department of the AFL-CIO. These unions are listed in Figure 14.1. Not included is the International Brotherhood of Teamsters. This is an independent union

---

AFL–CIO CONSTRUCTION UNIONS

1. International Association of Bridge, Structural and Ornamental Iron Workers
2. International Association of Heat and Frost Insulators and Asbestos Workers
3. International Brotherhood of Boilermakers, Iron Shipbuilders, Blacksmiths, Forgers and Helpers
4. International Brotherhood of Electrical Workers
5. International Brotherhood of Painters and Allied Trades
6. International Union of Bricklayers and Allied Craftsmen
7. International Union of Elevator Constructors
8. International Union of Operating Engineers
9. Laborers International Union of North America
10. Operative Plasterers' and Cement Masons' International Association of the United States and Canada
11. Sheet Metal Workers International Association
12. Tile, Marble, Terrazzo, Finishers and Shopmen International Union
13. United Association of Journeymen and Apprentices of the Plumbing and Pipe Fitting Industry of the United States and Canada
14. United Brotherhood of Carpenters and Joiners of America
15. United Union of Roofers, Waterproofers and Allied Workers

---

Figure 14.1   International unions affiliated with the Building and Construction Trades Department, AFL-CIO.

that includes construction truck drivers as a part of its membership. As used here, the term "international" refers to the fact that such unions have jurisdiction over some members in Canada and Mexico. At the present writing, the total membership of the 15 building trades unions listed in Figure 14.1 is approximately 3,300,000.

At this point it may be appropriate to define the two major types of labor organizations, the craft union and the industrial union. The craft union follows jurisdictional lines and is made up exclusively of workers in a single craft, trade, or occupation. Its members may be and often are employed in different industries. All the building and construction trades unions are of this type. An industrial union includes everyone who works in a particular industry, regardless of what their individual job responsibilities may be. For example, the United Automobile Workers of America (UAW) encompasses everyone involved with the production of motor vehicles including machinists, pattern makers, upholsterers, welders, painters, and janitors.

14.7 OTHER CONSTRUCTION UNIONS

The vast majority of unionized construction tradesmen belong to the 15 AFL-CIO construction trade unions and the International Brotherhood of Teamsters. However, there are significant numbers of construction workers who are represented by other unions. In some coal-mining areas of the country, workers engaged in highway and certain heavy construction belong to the United Mine Workers union. There are a number of small independent unions that represent limited numbers of construction workers in a few localities. Independent, in this context, means the unions involved are not affiliated with the AFL-CIO. Most of these unions are organized along industrial lines without trade jurisdictions. One of the largest union competitors to the AFL-CIO craft unions is the Construction Industry Conference of the United Steelworkers of America.

Formerly referred to as District 50, this conference represents construction workers in many localities across the United States and Canada. Most of the contractors who have labor contracts with locals of this conference are engaged in heavy and highway construction or are specialty contractors. This Conference does not divide its members into separate crafts or jurisdictions, and contractors can shift their workers about from one occupational classification to another as long as the workmen are able to perform the work and are paid the wages established for it. Charters of the locals forbid strikes without approval of the international union, members can transfer freely from one local to another, and union membership is open to anyone the contractor hires. Labor contracts with these locals specifically give contractors control over the management of the work, assignment of the workers, source of the construction materials, and use of tools. Grievance machinery that can terminate in binding arbitration is written into all labor contracts.

In terms of total numbers, the AFL-CIO craft unions represent a very large majority of the organized construction workers. For this reason the discussion in the following sections pertains principally to those unions.

14.8 THE LOCAL UNION

The basic unit of a construction union is the "local," each local exercising jurisdiction over an assigned geographical area such as a borough, city, county, or state. The local serves as a headquarters and is responsible for all union activities of that craft within its boundaries. The geographical area assigned to a given local can vary widely from craft to craft. For example, there may be an appreciable number of carpenter's locals with a given state but only a few locals of the plumber's union. Over the continental United States there are approximately 12,000 locals of the AFL-CIO building and construction trades unions.

Each local union elects its own officers, these usually consisting of a president, vice-president, secretary-treasurer, and sergeant at arms. These officials may or may not be salaried; a typical arrangement is that they are paid only for time actually spent on union business. An executive board or some equivalent body may also be established, which is concerned primarily with the admission of new members, discipline, financial matters, and contract negotiation. The day-to-day affairs of the local are managed by a paid business agent who is elected from the membership. The local is represented on each project by a job steward, who is expected to see that union rules are observed and to report any violations or grievances to the business agent. The steward for each project is appointed by the local.

The locals of the same international union are commonly grouped together on a district, state, county, or city basis. For example, carpenter's locals within designated regions band together to form District Councils of Carpenters, which serve to coordinate the activities of all member locals into a unified approach to common problems.

Locals of the different construction unions unite on a regional basis to form building and construction trades councils. These councils are formed of delegates from each member local and serve as agencies for regional cooperation. An important function of these councils is to present a united front to employers during periods of collective bargaining.

14.9 LOCAL UNION AUTONOMY

The local is chartered by its international union and is subject to the constitution and bylaws of the parent organization. Beyond this, however, the decentralized nature of construction requires that locals possess a high degree of independence and freedom of action. Construction locals typically have the authority to negotiate their own labor agreements and to call strikes without the formal approval of their international unions. There are exceptions to this generality, however, and the parent unions can restrain their recalcitrant locals from reckless resort to work stoppages and strikes. Locals cannot, with impunity, defy their international union and strike at will. Locals do have considerable autonomy, but parental authority can be lawfully applied when the occasion demands. Each local has its own set of bylaws that governs the election of local officers, ratification of labor contracts, conduct of meetings, payment of dues, expulsion of members, and other union business.

The legislative body of local unions is the membership. Elected officers have only the responsibility for carrying out union rules as ratified by majority vote of the members. Correspondingly, all matters of union policy, labor contract terms, decisions to strike, and other basic issues must be approved by the voting membership.

14.10   UNION WORK RULES

Most local unions promulgate work rules that pertain to the employment of the locals' members. These rules vary considerably from craft to craft and with geographical area. Some of the work rules represent concessions gained from employers in past collective bargaining and cover a wide range of issues such as work hours per day, jurisdiction of work, multiple shifts, overtime and holidays, apprentices, prohibition of piecework, paydays, reporting time, foremen, crew size, safety provisions and devices, tools, job stewards, drinking water, and a variety of others. Contractors and others often criticize certain aspects of union work rules as being restrictive labor practices that unduly increase costs of production, unnecessarily prolong construction time, and interfere with the contractor's management prerogatives. Unions defend their working rules on the grounds that they make the employment of their members more secure, defend against the loss or subversion of hard-won gains, protect the members from unfair or arbitrary treatment by the employer, and ensure the safe employment conditions to which their members are entitled.

Without doubt, many work rules do function to preserve legitimate union rights and prerogatives. Unfortunately, many locals have shown little restraint in their imposition and enforcement of work rules. Union insistence on the use of unnecessary men on the job, prohibitions on the use of labor-saving methods and tools, inflexible application of overtime requirements, and other restrictive practices have substantially lowered productivity and commensurately raised construction costs in many sections of the country, especially the large metropolitan areas. Overlong coffee breaks, requirements that skilled workers do unskilled jobs, strict trade jurisdictions, flagrant featherbedding, excessive nonproductive time, limitations on the daily production of a tradesman, and the requiring of unnecessary work or the duplication of work already done are examples of work rules that seriously concern owners and contractors.

There is evidence, however, that the work-rule picture is changing. High construction labor costs are resulting in changes to faster and cheaper ways to build. The shift to dry-wall, precast concrete, and prefabricated steel buildings illustrates this point. In addition, the tremendous increase in open-shop operations in many parts of the country is resulting in a noticeable easing of work rules by many construction union locals.

14.11 THE BUSINESS AGENT

The local union elects a business agent, sometimes called a business manager, to conduct its business affairs and to serve as its representative to outside agencies. He is paid at the top rate of his craft for his services. The business agent is the key man in a building trades local. As the full-time spokesman of the local, he has the primary responsibility for "policing the trade" within the local's assigned geographical area. He refers members to jobs, has considerable authority in determining who joins the union, and is clearly a powerful man insofar as local contractors and union members are concerned. Business agents are typical of unions, such as the building trades, that deal with many different employers and in which union members are employed in scattered groups.

Large locals often have more than one full-time business agent; small locals may have one of their number act as part-time business agent. The business agent is the contractor's only direct contact with a local. The agent helps negotiate agreements, enforces them, ameliorates grievances, protects the union's work jurisdiction, and serves as a general go-between for the local members and the employers. The agent directs strikes and other concerted activities undertaken by the local. To the contractor, the business agent *is* the local union.

The business agent has substantial authority to make decisions for the local and bears almost the entire responsibility for management of its affairs. Obviously such freedom of action is a necessary condition in the construction industry where jobs are scattered and can be of short duration. To be effective, the business agent must be empowered to act quickly without having to await approval of a higher body. In a complex and shifting environment, he seeks to protect the interests of the local and its members. Even though he can act freely, however, he must remain responsive to members of his local because he is elected by them, and his job depends on how well he pleases his constituency.

The business agent, in many respects, is the middleman of the industry who strives to reconcile the conflicting demands of the employers on the one side and the union rank and file on the other. Contractors look to the business agent to find qualified people for their jobs and to curb recalcitrant workers. Members of the local depend on him for such services as finding jobs for them and settling disputes with employers. He must manage his local's office, keep necessary records, and supervise the finances. He must also visit the jobs and check employer compliance with union work rules and contract provisions.

11ᵗʰ hour bargining

14.12 COLLECTIVE BARGAINING

Labor contracts between construction contractors and labor unions are obtained by a process of collective bargaining and constitute the essential basis for labor-management relations in the construction industry. These contracts, as negotiated between the two parties, establish wages, hours, and other essential conditions of employment. As existing labor agreements expire, negotiations are conducted to arrive at new, mutually acceptable contracts.

The National Labor Relations Act requires management and labor to bargain in good faith with one another. Failure to do so by either party is an unfair labor practice. The law does not require that concessions be made, or even that the two sides come to an agreement. The law does not actually define what is meant by good-faith bargaining, although decisions of the NLRB and the courts identify it as a duty to approach negotiations with an open mind and a real intention of reaching an agreement. Lack of good faith on the part of the employer may be indicated by ignoring a bargaining request, anti-union activities, failing to appoint a bargaining representative with power to reach an agreement, delaying and evasive tactics, attempting to deal directly with employees during negotiations, refusing to consider each and every proposal, failing to respond with counter-proposals, and refusing to sign an agreement.

14.13 PATTERNS OF BARGAINING

Some contractors choose to conduct their own labor relations independently and on their own by dealing directly with the unions. In so doing, such contractors can negotiate a contract solely as it pertains to them. Freelance operation offers the real advantage of independence and freedom of action. On the other hand, there is the decided disadvantage that the independent contractor does not enjoy the benefits of collective strength and experience gained through affiliation with an association of contractors.

The most prevalent pattern of labor negotiations in the construction industry is where an association of contractors bargains with a union or unions on behalf of all of its members. Collective bargaining between contractors and the construction unions has a long history and is a well-established practice in this country. Although the makeup of the bargaining units of both sides is variable, a reasonably stable pattern of negotiations has emerged. Local associations of general contractors typically negotiate with the locals of the basic trades: carpenters, cement masons, laborers, operating engineers, and construction teamsters. In some areas ironworkers are also included. The basic trades may choose to bargain as individual locals, as a group of locals affiliated with the same international union, or through organizations of locals of different unions such as building trades councils. The resulting labor contracts are generally referred to as "master labor agreements," because they cover the employment of several different crafts. The resulting agreements apply only within the geographical jurisdiction of the participating local unions and pertain only to the craft or crafts involved.

In general the specialty trades unions do not subscribe to the principle of master labor agreements. Although locals of these crafts associate to form district councils and other cooperative units, these units have not been widely utilized for bargaining purposes. Local specialty contractor groups generally negotiate with the individual union locals whose members they employ. For example, a city or regional association of painting contractors will negotiate with the appropriate local or locals of the International Brotherhood of Painters and Allied Trades. In a similar manner locals of plumbers, plasterers, roofers, electricians, and the other specialty trades bargain with organizations of their contractor counterparts.

14.14 THE BARGAINING PROCESS   give & take

As indicated in the previous section, usual practice is that contractors participate in collective bargaining through a trade association such as a local branch of the Associated General Contractors of America. A common procedure is for the actual negotiations to be conducted by a committee of contractors, legal counsel, and association staff. The unions appoint the members of their negotiating team, normally officers of the locals, district councils, or building trades councils. Representatives of the international unions frequently assist the local unions and participate in the negotiations.

Invariably, the labor side demands more than it actually expects to obtain. The resulting sessions assume somewhat the characteristics of a poker game, involving a complex strategy of offer and counter offer. By bluff, bluster, and compromise, the contest slowly and carefully picks its way toward the ultimate settlement. The experienced negotiator will hold back some concessions that are almost sure to be made eventually but that are useful to "horse trade" in obtaining the final agreement. It takes experience and skill to recognize the psychological moment at which the most propitious deal can be made. As a general rule, union negotiators do not have final authority to consummate binding agreements but must obtain ratification by the union membership. It is not unusual for the members to refuse the proposed settlement, sending the negotiators back to the bargaining table.

A fatal mistake that management must constantly guard against is entering into negotiations without thorough preparation and expert advice. It must be remembered that the primary purpose of unions is to establish favorable working conditions for their members. The periodic negotiations with employers are the focal point for union activity and are not taken lightly by union representatives. By the same token, these negotiations are a serious matter for management also, and unsophisticated bargaining by poorly prepared management representatives can only lead to the contractors being outmaneuvered and outpointed by their union adversaries, who spend a great deal of their time in constant association with labor-management matters. The highly specialized art of labor negotiation has led to the increasing practice of hiring professional assistance for this purpose. As a matter of fact, most large contractors and many of the contractor associations have labor relations specialists on their staffs.

14.15 LABOR AGREEMENTS

Once a settlement is reached between contractors and labor unions, a written instrument is prepared that contains the essentials of the agreement and is signed by the parties. This instrument, referred to as a "labor agreement" or "labor contract," is binding on both parties. The life of a labor agreement may run from one year to as many as five years, a multiyear contract normally providing for periodic pay increases or containing reopener clauses. As a general rule employers favor the long-term contract, because it has a stabilizing influence and removes much of the uncertainty associated with the bidding of future work.

A contractor who is a member of a bargaining unit is normally bound by any labor agreement that is negotiated unless the contractor withdraws its bargaining consent with adequate, unequivocal, written notice prior to the commencement of negotiations. However, once bargaining has started, a contractor cannot withdraw without consent of the union side or a showing of unusual circumstances. Most contractor associations require their members to execute a "designation as bargaining agent" agreement that appoints the association as the sole bargaining representative. The point here is that a contractor cannot make its participation in multiemployer negotiations contingent on a favorable outcome.

Bargaining agreements in the construction industry, although not nearly as long and involved as the average industrial union contract, nevertheless have many provisions covering a wide range of subjects. Such contracts stipulate wages, hours, fringe benefits, overtime, and a wide variety of working conditions. Most agreements are negotiated to cover a particular category of construction. For example, separate agreements are negotiated for the same geographical area between the same unions and different contractor groups to cover building, heavy, highway, and residential construction. Most agreements provide for the settlement of jurisdictional disputes, a job referral system, apprenticeship, and grievance procedures. A large proportion of construction labor agreements contain no-strike and no-lockout pledges. Many of the no-strike clauses are conditional, however, permitting strikes under certain conditions such as after the grievance procedure has been exhausted or when the employer is in noncompliance with the provisions of the agreement.

Section 8(d) of the National Labor Relations Act provides that the party desiring to terminate or modify an existing labor contract covering employees in an industry affecting interstate commerce must do the following:

1. Serve written notice on the other party of the termination or modification 60 days before the termination date or date of modification.
2. Offer to negotiate a new or modified contract.
3. Notify the Federal Mediation and Conciliation Service and any similar state agency that a dispute exists, if no agreement has been reached 30 days after the notice was served.
4. Continue to live by the existing contract terms without resort to slowdowns, strike, or lockout until expiration of the 60-day notice period or until expiration of the contract, whichever is later.

Any employee who engages in a strike within the 60-day period loses his status as an employee of the struck employer and is no longer protected by the National Labor Relations Act. This means, for example, that the striking employee is not entitled to reinstatement.

14.16 GEOGRAPHICAL COVERAGE OF AGREEMENTS

In terms of geographical coverage, labor contracts in the construction industry run the gamut from the entire country down to an individual project. However, such agreements are predominantly local in coverage, their application ranging from cities to entire states. There are, however, a few instances where labor agreements are negotiated to cover large areas or even the entire United States. A few construction unions, or specialty crafts, negotiate agreements that apply nationally. As one example, a national agreement (exclusive of New York City, which has a separate contract) is negotiated between the National Elevator Industry, Inc. (NEII), an employer group, and the International Union of Elevator Constructors. The rate of pay of an elevator constructor journeyman is based primarily on the average of the four highest paid trades in the area where he is working. The general aspects of wages, hours, and work rules are prescribed by the national agreement. There is another form of national agreement that pertains to several crafts that is discussed in Section 14.17. Regional agreements covering several states are sometimes negotiated by individual contractors or contractor associations. Boilermakers' agreements are normally on a multistate, regional basis.

Project agreements have become very common in recent years. A project agreement is a labor contract that applies only to a specific large construction project. These agreements may be negotiated by the contractor, the owner, or both. These contracts have the goal of constructing projects efficiently, on time, and with reasonable and predictable labor costs. Such labor contracts have proliferated because of the inadequacies of local bargaining and local agreements. Project agreements attempt to avoid work stoppages, ensure adequate supplies of labor, obtain reasonable work rules, and stabilize wage rates. Agreements of this type have contributed to animosities among contractor groups because of their detrimental impact on local bargaining. Owner-negotiated project labor contracts have provoked much controversy between contractor associations and owners. Contractors maintain that such agreements undermine the traditional role of contractors in industry labor negotiations and further fragment collective bargaining in an already badly segmented industry. This has the effect of negating industry efforts to restructure and improve collective bargaining in construction (see Section 14.18).

14.17 NATIONAL AGREEMENTS

The term "national agreement" has come to refer to a nationally applicable labor contract such as those negotiated between certain of the AFL-CIO international building trades unions and the National Constructors Association (NCA), National Tank Fabricating and Erector Contractors, Pipe Fabricators Institute, and certain individual contractors. A national agreement provides for the employment of the members of a given union anywhere within the United States and often in Canada. Employers who sign national agreements are mainly large industrial contractors who perform work in widely scattered localities. Rather than attempt to become parties to the myriad of local labor contracts in force about the country, these contractors use the national agreement as a convenient method of handling their labor relations. The international construction trades unions do not negotiate these national agreements indiscriminately but limit them to a restricted number of selected general and specialty contractors.

Under a national agreement, a joint labor-management committee is created which has broad powers to administer the agreement. Wages and fringe benefits are those rates locally negotiated. However, many national agreements provide that where local wages and conditions are considered to be contrary to the best interests of the industry, the union and contractor can renegotiate them. Typically, these agreements contain explicit work rules and specifically ban slowdowns, standby crews, and featherbedding practices. Hours of work, overtime, shifts, and holidays are spelled out and made uniform.

Most national agreements contain a no-strike, no-lockout clause backed by a formal grievance procedure that can terminate in final and binding arbitration. A clause to protect the integrity of local bargaining by permitting the owner or contractor to shut a project down during a local collective bargaining strike is often included. However, some national agreements contain a retroactive pay provision. By the terms of such a clause, the contractor agrees not to shut the job down in the event of a local strike and the union agrees to keep working. The national contractor pays retroactively any increase in wages and fringe benefits subsequently agreed on during the local bargaining. National agreements, especially those with retroactive pay provisions, can seriously disrupt and undercut local labor negotiations.

14.18 CONTRACTOR BARGAINING STRENGTH

The discussion in the preceding sections disclosed that, in the main, labor negotiations in the construction industry can be best described as being local in character and enormously fragmented. This, as well as other factors, results in a considerable disparity of bargaining strength between the labor-management adversaries with the contractors occupying the disadvantaged position. Contractors who elect to negotiate individually are often very vulnerable and give the unions the opportunity to divide the contractor ranks by wresting special concessions from the independent operator. Area negotiation by contractor associations does offer some advantage of management solidity, but even such group action under the present prevailing pattern of bargaining still places contractors in a basically weak position.

Badly disadvantaged by uncoordinated bargaining, fragmentation, disunity, and economic vulnerability, union contractors have gradually been forced into labor contracts that badly erode their freedom of job management and that establish high wage rates that have lost all association with productivity. A contractor with continuing overhead expense, equipment costs, and contract completion deadlines is in no position to withstand a prolonged strike stemming from an impasse in labor negotiations. In contrast, labor union members usually have no difficulty in finding work in adjoining local areas not on strike, on nonunion jobs, or even temporarily outside the construction industry. Strike funds provided by unions and unemployment benefits paid by some states to strikers tilt the power imbalance even further.

Within recent years many attempts have been made to strengthen the employers' bargaining position in the construction industry. Thus far these efforts have met with varying degrees of success. Associations of contractors have been formed for the specific purpose of developing programs to improve the conduct of collective bargaining and to develop a better balance of power in labor negotiations. Some of these organizations serve to improve and co-ordinate management's collective bargaining efforts on an industry-wide basis by providing existing bargaining units with professional guidance, legal assistance, wage and contract information, and other technical expertise.

Other efforts are being made in several parts of the country to band contractors together from all segments of the industry for mutual aid and protection in collective bargaining. The ultimate objective is to achieve multiemployer, multiassociation, multitrade, coordinated bargaining on a wide-area basis resulting in multiyear contracts with uniform contract expiration dates. This is being approached through the voluntary consolidation of local contractor bargaining groups into larger units on a citywide, multicity, statewide, or regional basis. The individual contractor, or contractor group, assigns its bargaining rights to the association that engages in multiemployer, broad-area bargaining on behalf of all of its members. There is an inherent weakness to this latter approach, however. According to present law, any employer or union can withdraw unilaterally and unconditionally from such multiemployer-multiunion bargaining as long as notice of withdrawal is timely and unequivocal. The NLRB has ruled, however, that a union cannot refuse to bargain with a wide-area coordinated employer bargaining unit. To do so is to restrain and coerce employers in the selection of their bargaining representative, which is an unfair labor practice.

14.19   THE FEDERAL MEDIATION AND CONCILIATION SERVICE

Established by the Taft-Hartley Act in 1947, the Federal Mediation and Conciliation Service is an independent agency of the federal government that is charged with the responsibility of assisting employers in interstate commerce and labor organizations in promoting labor peace. As discussed in Section 14.15, employers or unions who wish to modify or terminate existing collective bargaining agreements must serve notice on the other party 60 days before the effective date of these changes. Should the matter not be resolved with 30 days, notice is required to the Federal Mediation and Conciliation Service and any similar state agency having jurisdiction. The service is automatically notified when negotiations threaten to lead to a dispute.

Mediation by the Federal Mediation and Conciliation Service is a voluntary process, and absolute neutrality in labor disputes is a guiding principle of the service. The parties to a dispute are encouraged to settle their differences by themselves, but either side may request a mediator's assistance at no charge. The aim of the service is reconciliation of conflicting views without intervention or dictation into the affairs of either party. Mediators cannot compel either side in a labor dispute to do anything, but they bring experience and dispassionate advice to the bargaining table to help the disputing parties reach a mutually acceptable area of agreement. Thus the Federal Mediation and Conciliation Service functions to keep the parties bargaining, offers helpful suggestions, and otherwise assists in the ultimate achievement of collective bargaining agreements. In no way altered, however, is the fact that the resulting agreement is both the product and the responsibility of the signatories.

The service employs a staff of mediators who are located in regional offices over the country and who are recruited from both management and labor. Through the years the Federal Mediation and Conciliation Service has established a reputation for impartiality and devotion to duty while occupying a most difficult role. As a general rule, it concentrates its energies on the resolution of disputes that have an appreciable impact on interstate commerce. In a dispute that threatens to cause a substantial interruption to commerce, the Federal Mediation and Conciliation Service may enter the dispute either on its own motion or on the request of one or more of the parties to the dispute. In disputes that imperil the national health or safety, the service must intervene regardless of the effect on commerce. Should the parties agree to arbitrate their differences, the service will furnish them with a list of qualified arbitrators.

14.20 EMPLOYER LOCKOUTS

Employer lockouts are sometimes called strikes in reverse. A lockout occurs when an employer or, more commonly, an association of employers close their establishments against employees during negotiations and cease operations until a settlement has been reached. A strike is the withholding of workmen from the contractor until a settlement is reached. A lockout is an employer device to withhold employment from his workers until an agreement is reached.

Labor's right to strike or to engage in concerted activity for bargaining purposes is protected. The law does not, however, give employers the parallel right to engage in a lockout. Federal labor law does not expressly permit or forbid lockouts, and they are legal under certain circumstances that are not spelled out in the law. The National Labor Relations Act gives workers the right to bargain collectively and makes it an unfair labor practice for employers to interfere with that right. Hence it is up to the NLRB and the courts to decide whether a particular lockout amounts to "interference." The guideposts to legality are vague, and contractors who engage in protective lockouts assume the risk of having to prove that the circumstances justified their actions. Competent legal advice is highly desirable for contractors who contemplate such action.

Over the years some guidelines have emerged that are useful in judging the legality of a lockout action. Generally speaking, the following circumstances could justify a lockout.

1. Unusual economic circumstances exist such that a strike would cause substantial loss to the contractor.
2. A strike would pose a serious threat to the public health and welfare.
3. A lockout by members of an employer association is necessary to prevent "whipsawing" by the union.

4. A union is engaging in selective strikes against individual members of an employers' association.

In addition, the U.S. Supreme Court has held that the lockout is a valid counter-weapon after an impasse has been reached in a bargaining dispute, even if the employer is not anticipating an immediate strike. The lockout cannot be used in situations in which there is anti-union intent but only in support of a legitimate bargaining position as an economic tool to counter a union's strike weapon. The NLRB has held that the fact that multitrade contractor groups are acting in concert does not make a lockout illegal as long as the interests of the employers involved are sufficiently interwoven to justify their taking concerted action in their common interest. The courts have ruled that, in the usual case, an employer may not continue to operate by using labor replacements while involved in a lockout during contract negotiations. However, the use of temporary replacements during a lockout is lawful under certain circumstances, such as where the union uses whipsawing tactics.

14.21 WAGES AND HOURS

Details associated with wages and hours are generally the crucial bargaining issues; negotiations on other matters are more in the nature of skirmishes around the edge of the major area of battle. Actually, wages and hours are broad subjects, covering such considerations as fringe benefits, overtime rates, show-up time, premiums for high work, travel time, subsistence, apprentice wage rates, and cost-of-living wage escalators.

It is universal practice for overtime rates to be required for all hours worked in excess of eight per day, 40 per week, or for work done on Saturdays, Sundays, or holidays. However, standard workweeks of 35 or 30 hours, and even of 25 hours, have been established in some metropolitan areas. Many of the agreements that provide for such shortened workweeks also include additional guaranteed hours of employment per week at overtime rates. Overtime rates are usually one and one half or two times the regular rate of pay, depending on the craft and the labor agreement involved.

Travel time and subsistence pertain to projects located in remote areas. Most bargaining agreements establish machinery to determine the status of a given project. If a job is found to have a remote classification, special rules apply regarding contractor-furnished transportation, field camp facilities, payment of subsistence, and travel reimbursement.

The wage rates of construction workers have risen steadily since the late 1930s and have far outstripped those in other industries. This upward trend gives every appearance of being a permanent condition when one considers the underlying philosophy of the trade unions. Labor believes that constantly rising wages are necessary to provide increased consumer buying power in an expanding economy. In the eyes of the average union member, the success of his union is measured in terms of a constant increase in wages and other benefits. Fringe benefits add further to already high construction costs. Long-term labor contracts tend to perpetuate the upward spiral. Automatic increases preclude any voluntary curbs on wage escalation, and negotiations by the various labor unions have become a game of "catching up" with the other fellow.

Although resisted by many contractors, cost-of-living adjustment (COLA) clauses in construction labor contracts have been used increasingly in recent years. Unions are attempting to protect the purchasing power of their members by negotiating contracts that automatically adjust wage rates for inflation. The most commonly used measure of the rate of inflation is the Consumer Price Index (CPI) published every month by the Bureau of Labor Statistics for selected cities and the country as a whole. Changes in the CPI, expressed in either index points or percentages, are the basis for calculating periodic wage changes.

14.22 FRINGE BENEFITS

Labor contracts customarily require the contractor to provide its employees with a number of pay augmentations known collectively as fringe benefits. These benefits vary somewhat from one contract to another and include such things as pension plans, profit sharing, health and welfare funds, insurance, paid vacations, paid holidays, employee education, legal aid funds, annuities, sick leave, bonuses, supplemental unemployment payment plans, apprenticeship programs, and other types of benefits. Recently reported national figures indicate that employer contributions to fringe benefits for construction workers now average almost 30 percent of direct payroll. This figure can vary considerably with geographical area and the type of work involved. Fringe benefits required by labor contracts are paid by the contractor into special trust funds jointly administered by both unions and contractors.

A common procedure in collective bargaining in the construction industry is for the unions to first negotiate a package settlement and then to distribute this settlement between direct wages and fringe benefits. In most cases the union reserves the right to allocate a part of the settlement either to wages or to a cents-per-hour contribution to designated fringe benefits.

Pension and welfare plans provided by private employers in interstate commerce to their employees fall under the Employee Retirement Income Security Act (ERISA) passed by Congress in 1974. A pension plan is any company program that provides retirement income to employees or results in a deferral of income by employees until the termination of employment or beyond. Included are profit-sharing, stock bonus, disability retirement, and thrift plans. A welfare plan provides hospital, medical, surgical, sickness, accident, disability, death, vacation, or other benefits. ERISA does not require any employer to establish a plan, but those that do must meet certain standards. The chief purpose of this act is to protect the interests of workers and their beneficiaries. Workers are not required to satisfy unreasonable age and service requirements before becoming eligible for a company pension plan. ERISA ensures that money will be available to pay benefits when they are due and that plan funds are handled prudently. The reporting and disclosure pro-

visions of the act require the employer to advise employees and their beneficiaries of their rights and obligations under the company benefit plan. Employers bear the responsibility of ensuring that benefit promises to their employees are fulfilled.

14.23 INDUSTRY ADVANCEMENT FUNDS

For many years, labor and management in the construction industry have
joined together in cooperative programs designed to advance some aspect of
construction. The funds established for this purpose are called "industry advancement funds" or "industry promotion funds." Generally, such a fund is
created by labor contract and is financed by negotiated hourly contributions
paid by the participating contractors into a fund established for that purpose.
The employer's association acts as the administrator of the fund, which is used
to finance programs in education, public relations, safety, equal employment
opportunity, market development, labor relations, building codes, specification writing, and other relevant areas. Although there are a few industry funds
on a national level, such arrangements are predominantly local in coverage.

The courts have ruled that Section 302 of the Taft-Hartley Act prohibits
joint administration of industry advancement funds by labor and management
but that such funds are legal as long as they are managed by employers alone.
However, a 1978 amendment to the Comprehensive Employment and Training Act permits union participation in the trusteeship of certain organizations
in the name of labor-management cooperation. Uncertainty presently exists as
to whether this amendment permits building trades unions to share in the control of construction advancement funds. The National Labor Relations Board
has ruled that such funds are not mandatory subjects of bargaining. Consequently, neither the contractor association nor the union can force the other
to bargain on the industry fund issue or to accept such a provision in a labor
agreement.

14.24 ADMINISTRATION OF THE LABOR CONTRACT

The signing of the negotiated contract does not close the matter of contractor-
union relationships. Rather, the status of the agreement merely changes from
that of bargaining to one of interpretation and administration. A carefully
considered and clearly worded contract will do much to minimize contract
misunderstandings.

Labor agreements in the construction industry typically contain procedures for the settlement of disputes that may arise during the life of the contract. When a dispute occurs that cannot be resolved by a conference of the
steward, business agent, superintendent, and any other party directly involved,
the grievance procedure set forth in the agreement is followed. This procedure

often provides for meetings between successively higher echelons of contractor and union officials, during which time no work stoppage is to occur. Should the grievance procedure not resolve the dispute, arbitration of the matter may or may not be provided for in the labor agreement. Historically, unions have resisted the concept of binding arbitration, preferring to remain free to select their own course of action to suit the situation. Nevertheless, many construction labor contracts now provide for the arbitration of contract disputes with no resort to strikes or lockouts.

It is worthwhile to note that an agreement to arbitrate can be enforced against either contractors or unions by a federal court injunction. The law on this matter with respect to management has been long established. In 1970, the U.S. Supreme Court reversed an earlier ruling that the Norris-LaGuardia Act barred federal court injunctions against labor unions and ruled that injunctions could be issued on complaint of employers to enforce no-strike agreements to arbitrate.

Because administration of the contract is an everyday requirement for the contractor, a carefully selected individual is usually appointed to handle the labor relations for the firm. When a specific person is assigned to this duty, even if only as one of several responsibilities, he can become versed in the complexities of labor law and the provisions of the local labor agreements. He becomes, in fact, the labor relations specialist for the company. In this way company labor policy is at least consistent and informed. The contractor is obligated to live up to the terms of the agreement, and must be insistent that the unions do likewise.

14.25 DAMAGE SUITS

Section 301 of the Taft-Hartley Act provides that suits for violation of labor contracts between an employer and a labor union representing employees in an industry affecting interstate commerce can be brought in the federal district court having jurisdiction. The law provides that any labor organization or employer subject to the Taft-Hartley Act is bound by the actions of its agents, and that a labor organization may sue or be sued as an entity and in behalf of the employees whom it represents. Any monetary judgment so obtained against a union is enforceable only against the organization as an entity and not against any individual member. The U.S. Supreme Court has ruled that an employer can sue a union for damages resulting from violation of a no-strike arbitration agreement but may not sue union officials or members individually.

Damage suits for losses suffered as a result of unlawful strikes, picketing, or other such union actions may also be brought in federal courts under the provisions of the Taft-Hartley Act. Section 303 establishes the right of an employer who is injured in his business or property by a secondary boycott, hot-cargo agreement, jurisdictional dispute, or other unfair labor practice to sue the responsible union or unions for damages. Any injured party, not just the employer against whom a strike or boycott is called, is entitled to bring suit for damages sustained through unlawful union action.

14.26 PREJOB CONFERENCES

When employment conditions pertaining to a given project are out of the ordinary (that is, the conditions are not clearly provided for in the area agreement), it is usual practice for a prejob conference to be held between the contractor and the union or unions involved. This meeting may be held before bidding, so as to establish standard conditions for the bidding contractors, or just before the start of field operations. In either case the underlying motive is to achieve a meeting of minds between the employer and the unions regarding job conditions of employment. For example, consider a job that is to be located in a remote area. The contractor wishes to man the job in the most economical manner possible, and the unions require that fair standards be maintained. If the project is near a town of any size, the contractor may decide to quarter its workers there and furnish transportation back and forth to the site. If this solution is not feasible, temporary barracks or trailers may be provided at the project. At any rate, the contractor and the unions must arrive at a mutually acceptable understanding relative to quarters, subsistence, and transportation. In addition, the locals having jurisdiction must be checked to determine whether they have enough workers available to do the job. If not, arrangements must be made to bring workers in from the outside, either by the local or by the contractor.

There is some contractor resistance to the prejob-conference concept, based on the belief that such meetings encourage the unions to make exorbitant demands. Objection or acceptance seems to be a matter of opinion and company policy. However, available evidence seems to indicate that unions are much more reasonable at this stage than they are later on when they detect conditions that lead them to believe the contractor is trying to get away with something. Obtaining advance union acceptance of its procedure makes it much easier for the contractor to devote its energies to getting the job done without having to combat active union resistance and interference.

14.27 OPEN-SHOP CONTRACTING

The open-shop "movement" has made tremendous strides during recent years. Despite arson, job violence, sabotage, threats, harassment, picketing, and vandalism, an increasingly large proportion of construction of all types is being performed by open-shop contractors. It is estimated that approximately 60 percent of the annual construction volume is now accounted for by nonunion contractors. The Associated Builders and Contractors, Inc. (ABC), an association of contractors who operate open-shop (or "merit-shop" as favored by the ABC), is now a national organization with many state chapters. Approximately 50 percent of the members of the traditionally union-oriented Associated General Contractors of America (AGC) now operate open-shop or double-breasted (see Section 14.28).

By and large, open-shop contractors do not assume an anti-union posture. Although they generally pay somewhat less than union scale, continuity of employment provides the nonunion worker with an annual income much the same as or better than that of his union counterpart. Open-shop contractors provide fringe benefits that compare favorably with those established by union contract. These contractors employ union and nonunion workers alike, believing that every tradesman should have the right to belong or not to belong to a labor organization. They oppose union make-work rules and strict trade jurisdictions, and maintain that every contractor should have the right to deal with any other contractor or business firm, union or nonunion, as it sees fit. The open-shop concept defends the right of the contractor to manage its projects. For example, open-shop contractors decide for themselves the size of work crews and to what jobs a worker can be assigned. They are free to use prefabricated materials and are not subject to jurisdictional disputes, featherbedding, forced overtime, and work slowdowns. They pay workers according to their ability and performance.

14.28 DOUBLE-BREASTED OPERATION

A number of construction companies operating under labor agreements have found that their labor costs have made them noncompetitive for work of certain categories or in some geographical areas. These contractors have found it advantageous to form and operate a second company that is open-shop. The original union-shop firm continues to function in areas in which unions are strongly entrenched, and the nonunion firm does business where open-shop work has become established. The NLRB has approved such "double-breasted" operations when certain conditions are met.

The main criterion for double-breasted operation is that there be two separate and independent business entities, each of whose labor policies are conducted without interference from the other. This requires that the two companies be independent in their day-to-day operations and in their labor relations policies. The usual procedure has been to establish separate corporations (not wholly owned subsidiaries or divisions) with separate management, field supervision, work forces, equipment, and financing.

The NLRB, with court approval, has held that common ownership does not, in itself, dictate an illegal status of dual operations when there is no common control of labor policies. The general rule is that two commonly owned corporations in a double-breasted operation are considered to be two separate entities under the National Labor Relations Act if there is no direct control from one company over the other with regard to day-to-day operations or labor relations. The principal consideration is that independence of labor policy makes the two act as different employers and, as such, employees of each constitute a separate bargaining unit. When each company is a separate and independent entity, each is entitled to the same protection against secondary boycotts and to the same protection from each other's labor controversies. Competent legal advice is a necessity for any contractor contemplating double-breasted operation or changing from union to nonunion status.

14.29 OPEN-SHOP REFERRAL CENTERS

A problem faced by open-shop contractors is obtaining adequate manpower. The construction unions and union hiring halls are obviously not available to these companies. To assist in this regard, employee referral systems have been established in several cities about the country that provide a central source of open-shop personnel. Operated for the most part by either the Associated Builders and Contractors (ABC) or the Associated General Contractors of America (AGC), these open-shop "hiring halls" provide any contractor with names and other information about registered individuals with specific construction skills. These centers maintain a file of names, skills, experience, and addresses of people who register for job placement. Advertisements in local newspapers and other media advise the community that the referral service is available. Contractors obtain names from the center and make their own contacts, conduct their own interviews, and do their own hiring. In addition to referral centers, various open-shop training programs are being conducted by ABC and individual contractors.

15

PROJECT SAFETY

15.1 THE COST OF CONSTRUCTION ACCIDENTS

Construction is, by its very nature, an extremely hazardous undertaking. The annual toll of accidents in the construction industry is high, in terms of both cost and human suffering. The consequences of construction accidents are not expressible in terms of dollars alone. Money loses much of its significance when bodily injury and death are involved. Nevertheless, the financial consequences of accidents are an important matter to the construction industry and for the individual contractor.

Job accidents impose on the construction industry a tremendous burden of needless and avoidable expense. It has been estimated that the total cost of construction accidents may amount to as much as $8 billion per year. It is true that insurance can be purchased to protect the contractor from certain direct expenses resulting from construction accidents. For example, workmen's compensation insurance provides hospital and medical care, subsistence payments, rehabilitation, and other benefits provided by law for injured workers and their families. However, construction accidents can and do involve substantial costs that are not insurable or that are not included in the usual construction insurance coverage. These are sometimes spoken of as "hidden" or "indirect costs" whose total is normally several times greater than the direct expenses paid by insurance. Examples of uninsured costs are first aid expenses, damage or destruction of materials, cleanup costs, idle machine time, unproductive labor time, spoiled work, schedule disruptions, loss of trained manpower, work slowdowns, poor public relations, administrative and legal expense, lowered employee morale, and other expensive side effects.

15.2 SAFETY LEGISLATION

The advent of the industrial revolution in the United States was marked by a simultaneous proliferation of unsafe working conditions. Although employers had certain common-law duties toward their employees, such as to provide a safe place to work, the safeguards to health and safety that are taken for granted today were not customary at that time. As a matter of fact, there was no general acceptance of the notion that employers should be concerned with the welfare and safety of employees while they were on the job. Certainly no concerted effort was made to render working conditions less hazardous. It was believed by management that most work accidents were caused by the carelessness of the employees themselves and that it was the worker's responsibility to avoid accidents. However, the grim loss of life, limb, and livelihood aroused the public conscience, and the latter half of the nineteenth century witnessed a gradual change in the general attitude toward work safety.

The primary responsibility for statutory job-safety requirements in the United States has traditionally rested with the state governments. In 1867 Massachusetts took the first legal step toward remedying the dangerous working conditions so characteristic of the time. During the years following many of the states enacted laws pertaining to working conditions in factories, mine safety, machinery operation, inspection of steam boilers and elevators, and fire protection. In 1911 Wisconsin established a state industrial commission which was authorized to develop and issue rules and regulations in the field of industrial health and safety that would have the force of law. All states have now enacted some form of legislation that gives designated state agencies general rule-making authority in the areas of occupational health and safety.

15.3 STATE SAFETY CODES

Details of safety codes vary somewhat from state to state, but there is a trend toward greater uniformity in the provisions of the regulations through adoption of the standards of nationally recognized codes developed by such bodies as the American National Standards Institute. Although some states have established comprehensive safety standards applicable to all employments, the tendency has been to develop special codes for particular industries, operations, or hazards. Some specific hazards are regulated in a large majority of the states. Codes now relate to boilers, construction, elevators, mechanical power transmission, cranes and derricks, fire protection, floors and stairways, illumination, sanitation, ventilation, electrical hazards, explosives, ladders, spray painting, welding, and other areas of potential danger. All states require the provision of first aid and protective equipment. State safety codes make the employer and his supervisory personnel responsible for compliance with the code and for suitable safety instruction to the worker. In turn the employee is required to make use of safeguards provided for his protection and to conduct his work in conformance with the established safety rules.

Each state industrial commission or labor department has jurisdiction over every employment and place of employment within its state and is authorized to enforce and administer established codes and rules pertaining to the safety and protection of workers. The commission is vested with full power and authority to establish and enforce necessary and reasonable rules and regulations for the purpose of implementing the state law. In general whenever the commission finds that any employment or place of employment is not safe or that employees are not being adequately protected, it is empowered to order the employer to rectify the situation and to furnish safety devices and other safeguards reasonably required. Violation of a state safety code is punishable by fine and/or imprisonment as provided by the applicable statute. As discussed in the next section, a number of federal statutes are now preeminent over state law in matters of health and safety in certain employments unless the state has adopted health and safety requirements that meet federal standards.

15.4 FEDERAL HEALTH AND SAFETY ACTS

Several federal statutes now establish health and safety standards in a variety of occupational areas such as nuclear energy, mining, and transportation. Two federal laws have been enacted that impose safety standards on the construction industry. These are the Construction Safety Act of 1969 and the Williams-Steiger Occupational Safety and Health Act (OSHA) of 1970.

The Construction Safety Act of 1969 applies to construction projects financed in whole or in part by federal funds. It prohibits contractors from requiring tradesmen to work under conditions that are unsanitary, hazardous, or dangerous, as determined under standards issued by the Secretary of Labor. Before the provisions of this act could be effectively implemented, however, the 1970 legislation was passed that applies to employers in interstate commerce, including those in construction. The safety regulations promulgated for the Construction Safety Act have been included with the construction regulations for the Occupational Safety and Health Act.

On all private and essentially all federal construction, enforcement action is taken and penalties are assessed under OSHA. If the job is federally funded, however, it is possible for a contractor to be penalized under the Construction Safety Act. Penalties under the act consist primarily of contract cancellation. Blacklisting from additional federal contracts for a period of up to three years is also possible if the safety and health violations are flagrant. However, the contractor does not face double jeopardy in the sense that action cannot be brought against it under both the Construction Safety Act and OSHA.

15.5 THE OCCUPATIONAL SAFETY AND HEALTH ACT (OSHA)

OSHA established the first nationwide program for job safety and health by directing the Secretary of Labor to set safety and health standards for all industry. Every employer in interstate commerce has a twofold obligation to provide employment and a place of employment that are free of recognized health and safety hazards, and to comply with OSHA standards. Employers must also keep and preserve stipulated records of recordable occupational injuries and illnesses. Exempted from the act, however, is industry already under the jurisdiction of those federal agencies that have statutory authority to establish their own safety regulations. Examples of this are mining and rail transportation. Consequently, all construction performed at both surface and underground mines falls under the Mine Safety and Health Act (1977). This act is administered and enforced by the U.S. Bureau of Mines. Commercial aggregate producers are similarly covered.

OSHA established the position of Assistant Secretary for Occupational Safety and Health to take charge of standards and enforcement. Also established was a National Advisory Committee on Occupational Safety and Health to assist with the devising and establishing of standards. An independent Occupational Safety and Health Review Commission appointed by the President, with the advice and consent of the Senate, was created to enforce the standards and to hear appeals. The commission is assisted by an organization of trial examiners and supporting staff similar to the system used by the National Labor Relations Board.

Under OSHA as originally enacted, safety inspectors were authorized to conduct unannounced site inspections to see if employers were complying with safety standards. In the event of such a surprise visit, employers were required to admit the inspectors to their places of business. This was changed in 1978 by a U.S. Supreme Court decision to the effect that such visits are unconstitutional because they violate the prohibition against unreasonable search. Under this ruling, inspections are still made but admission of the OSHA inspector must be voluntary with the employer or the inspector must first obtain a search warrant. However, warrants are available as long as the inspection is conducted as part of OSHA's general administrative plan for enforcement of the act.

An employer representative and an employee representative have the right to accompany the inspector during his rounds of the premises. When a violation exists, a citation is issued describing the nature of the violation, the amount of any civil penalty imposed, and a reasonable time in which to correct the situation. A copy of the citation must be posted in a location near the site of the violation. The employer has 15 days in which to contest the citation. If a citation is contested, the review commission holds a formal hearing. Enforcement of the commission's orders or review of its decision is handled by the appropriate U.S. Court of Appeals. If a citation is not contested, it becomes final.

Employees or their representatives can demand inspections of their employers' premises by making direct complaints to the Labor Department in writing. If it is determined that there are reasonable grounds to believe a violation or danger exists, a special inspection is made as soon as possible. Also, during the course of any inspection any employee or employee representative can notify the inspector of any violations that may exist. The act provides that employees may not be discharged or discriminated against in any way for filing safety and health complaints or otherwise exercising their rights under the act. In 1980 the U.S. Supreme Court ruled that employees have the right to refuse to perform tasks they reasonably believe could result in serious injury or death.

If imminent danger to safety or health is noted, the inspector is required to promptly notify the employer, the employees, and the Secretary of Labor. If the imminent danger is not eliminated, the Secretary of Labor is required to seek a temporary injunction in a federal district court to shut down that part of the operation where the danger exists.

The act provides for mandatory civil penalties against employers of up to

$1000 for each serious violation and for optional penalties of up to $1000 for each nonserious violation. Penalties of up to $1000 per day may be imposed for failure to correct violations within the proposed time period. Any employer who willfully or repeatedly violates the act may be assessed penalties of up to $10,000 for each such violation. Criminal penalties are also provided for in the act. Any willful violation resulting in the death of an employee, upon conviction, is punishable by a fine of not more than $10,000 or by imprisonment for not more than six months, or by both. Conviction of an employer after a first conviction doubles these maximum penalties. The law requires workers to observe applicable health and safety rules but provides no penalties for their failure to comply. Employers are liable under OSHA for all acts of employees except when the employer did not and could not, with the exercise of reasonable diligence, know of the presence of the violation.

15.6 ENFORCEMENT OF OSHA

Though OSHA requires that each employer in interstate commerce provide his employees with employment and a place of employment that are free from recognized hazards, the matter of duty and responsibility under the act becomes very involved on a multiemployer site such as a construction project. On a workplace of this type, employees of different employers are subjected to common hazards that exist at the site, such hazards often being created by different employers. Who is responsible for safety violations under these conditions is not entirely clear, and the law in this area is still emerging in the form of court decisions.

The rule for multicontractor sites has been that a prime contractor or a subcontractor is in violation of the law if its employees are exposed to a hazard that is in violation of OSHA standards. Who created the hazard is not a consideration. In addition, both general contractors and subcontractors are responsible for violative conditions which they created and to which employees of other contractors have been exposed. General contractors are held responsible for safety violations of subcontractors which they could have reasonably been expected to prevent or abate by reasons of their supervisory capacity. Subcontractors on multiemployer sites are responsible for violative conditions that they have created and to which employees of other contractors have been exposed, even though the employees of the subcontractor who created the violative conditions were not exposed to such conditions.

Under present conditions, it appears that any employer on a multiemployer site is liable for the hazards it creates or controls, regardless of whether its own employees are exposed. The effect is to place the legal responsibility for safety with the party who creates or who has the authority to remove the hazard. A prime contractor or a subcontractor cannot delegate or contract away its responsibility under the act. A general contractor can, however, put a clause in its subcontracts requiring the subcontractor to reimburse the general contractor for any losses sustained by reason of the subcontractor's failure to abide by safety regulations or general duty of care in conducting its activities at the worksite. To assure that safety and health requirements are met, the general contractor is well advised to see that its subcontracts require scrupulous adherence to safety and health regulations.

OSHA permits the U.S. Department of Labor to transfer enforcement of safety provisions to any state that demonstrates its ability to handle enforcement at least as effectively as the federal government. About half of the states had OSHA-approved plans at one time, but the current trend is away from state enforcement. In those states that now have enforcement powers, most jobsite inspections are conducted by state inspectors and reports are filed with a state agency rather than with the U.S. Department of Labor.

15.7 CONTRACT SAFETY REQUIREMENTS

Construction contracts routinely contain provisions requiring the prime contractor to conform to all applicable laws, ordinances, rules, and regulations that pertain to project safety. Subcontracts, in turn, extend this responsibility down to the subcontractors. Contracts with some public agencies require that the contractor conform with the requirements of the safety code of that particular agency. These standards constitute a contractual obligation with which the contractor must comply or be in breach of contract. Many state highway departments include a safety code in their construction contracts. Some federal agencies, including the U.S. Army Corps of Engineers, the Naval Facilities Engineering Command, and the U.S. Bureau of Reclamation, use construction contracts that include health and safety standards in their provisions.

Although construction projects for these agencies are not exempted from OSHA, this act provides that these agencies may continue to use their own safety codes and enforce them. Contractors on such projects are required to observe OSHA standards as well as those contractual requirements that OSHA does not cover or that are more stringent than those of OSHA.

Labor agreements may also impose contractual safety requirements on the contractor. The NLRB has ruled that safety regulations, as an essential part of the terms and conditions of employment, are mandatory subjects of bargaining whenever either party places the issue on the bargaining table.

15.8 WORK INJURY AND ILLNESS RECORDING

OSHA requires that employers keep certain records pertaining to recordable occupational injuries and illnesses. Recordable occupational injuries and illnesses are the following.

1. Occupational deaths, regardless of the time between injury and death, or the length of the illness.
2. Occupational injuries which involve one or more of the following: loss of consciousness, restriction of work or motion, transfer to another job, or medical treatment other than first aid.
3. Occupational illnesses which involve one or more of the following: loss of consciousness, restriction of work or motion, transfer to another job, or medical treatment other than first aid.

Recordable cases are classified as follows:

1. Total recordable cases, which is the sum of all recordable occupational injuries and illnesses, including deaths; lost workday cases; and nonfatal cases without lost workdays.
2. Deaths.
3. Total lost workday cases. This is the sum of cases involving days away from work and/or days of restricted activity.
4. Nonfatal cases without lost workdays. These are recordable injuries or illnesses which do not result in death or lost workdays, either days away from work or days of restricted work activity.

In addition, the employer must maintain records on number of days away from work and number of days of restricted work activity.

15.9 WORK INJURY AND ILLNESS RATES

For purposes of analyzing, summarizing, and presenting work injury and illness data, incidence rates are computed and published. These rates are expressed as the number of cases or days per 100 full-time employees or 200,000 employee hours per year. (100 employees at 2000 hours per year = 200,000 employee hours per year.)

$$\text{Incidence rate} = \frac{\text{Number of cases or days per year} \times 200,000}{\text{Total employee hours per year}}$$

An incidence rate can be computed for each category of cases or days.

Figures 15.1 and 15.2 present recent work injury information for selected major industries, including construction. These data show that the construction industry has one of the worst safety records of all the major American industries. For the past several years, with the construction industry employ-

Industry Group	Workers (×1000)	Deaths		Disabling Injuries	
		Total	Per 100,000 Workers	Total	Per 100,000 Workers
All industries	**94,800**	**13,000**	**14**	**2,200,000**	**2,321**
Trade	22,200	1,300	6	400,000	1,802
Service	22,900	1,700	7	360,000	1,572
Manufacturing	20,300	1,800	9	490,000	2,414
Government	15,400	1,700	11	310,000	2,013
Transportation and public utilities	5,100	1,500	29	179,000	3,510
Agriculture	3,500	1,900	54	190,000	5,429
Construction	**4,600**	**2,600**	**57**	**240,000**	**5,217**
Mining and quarrying	800	500	63	40,000	5,000

Adapted from data published by the National Safety Council.

Figure 15.1 Work injuries in industry groups.

Industry	Total Recordable Cases	Cases Involving Days Away From Work and Deaths	Nonfatal Cases Without Lost Workdays
All industries	**7.77**	**2.54**	**4.30**
Textile	6.49	0.92	4.93
Chemical	4.56	1.10	2.77
Motor vehicles	6.61	1.41	3.21
Electrical service	6.88	1.96	3.74
Steel	7.17	2.17	3.71
Machinery	9.37	2.43	5.68
Gas	7.91	2.69	4.36
Rubber and plastics	7.50	2.77	3.36
Furniture	12.91	2.86	9.31
Printing and publishing	6.83	2.99	3.37
Metal mining	14.80	3.00	10.43
Paper	9.89	3.05	6.22
Leather	10.26	3.48	6.23
Transit	6.17	3.81	1.37
Cement	16.43	3.86	12.41
Construction	**11.84**	**3.94**	**7.57**
Bituminous coal	12.56	4.21	8.29
Food	14.59	4.59	9.48
Air transportation	7.88	5.29	2.12
Meat products	23.02	5.90	15.88
Wood products	14.29	5.96	7.53
Trucking	22.64	11.75	10.04

Adapted from data published by the National Safety Council.

Figure 15.2   Recordable occupational injury and illness incidence rates.

ing about five percent of the total labor force, it has accounted for about 11 percent of all occupational injuries and illnesses and 20 percent of all deaths resulting from occupational accidents. Figure 15.3 presents a breakdown of incidence rates for the construction industry.

In Figures 15.2 and 15.3, "Cases Involving Days Away From Work And Deaths" do not include cases involving days of restricted work activity. By computing its own incidence rates, any construction company can compare its accident experience with such national averages.

Occupational safety and health statistics are compiled by both the National Safety Council and the Bureau of Labor Statistics of the U.S. Department of Labor. For any given year, the figures of the Bureau of Labor Statistics vary somewhat from those of the National Safety Council. The principal reason for this is that each set of data represents the experience of a different group. Incidence rates of the National Safety Council are based primarily on data furnished by its members, most of which are larger firms. The Bureau of Labor Statistics figures are based on a much broader sampling and are probably more representative of the industry as a whole.

15.10 ECONOMIC BENEFITS OF SAFETY

In addition to the humanitarian aspect, there is also a compelling economic motivation in accident prevention. Many financial benefits accrue to the contractor who conducts field operations in a safe manner and whose accident ex-

Construction Category	Total Recordable Cases	Cases Involving Days Away From Work and Deaths	Nonfatal Cases Without Lost Workdays
Construction industry	**11.84**	**3.94**	**7.57**
General building construction	12.94	2.84	9.87
Nonresidential building construction	11.97	2.95	8.98
Heavy construction, except highway	12.54	3.25	8.35
Bridge, tunnel, and elevated highway construction	23.13	3.35	19.04
Heavy construction	11.43	3.99	7.12
Highway and street construction	11.24	4.47	6.69
Public utility construction	14.26	4.53	9.64
Special trade contractors	18.98	5.21	12.70
Structural steel erection	29.75	6.38	23.28

Adapted from data published by the National Safety Council.

Figure 15.3   Recordable occupational injury and illness incidence rates in construction.

perience is low. The most immediate and obvious financial benefit is the savings realized because of accidents that do not happen. Mention has already been made of the indirect costs of accidents that are not covered by insurance and that can constitute serious financial loss.

Construction workers appreciate and value job safety even though they sometimes tend to be careless in their work habits. Employees who feel that their employers are genuinely concerned about safety and who see tangible evidence of this concern are more likely to be loyal and cooperative workers. The incorporation of safety measures as an integral part of operations, instead of setting them apart to be handled as a necessary evil, is a key factor in obtaining high morale and employee loyalty. Safety is one of the potent forces that makes workers proud of the company they work for, proud of the manner in which they perform their jobs, and proud of their record in preventing accidents. The fruits of good worker morale are higher production and better workmanship, two economic benefits of no trifling proportion.

Another important financial benefit to the contractor that results from fewer accidents is the reduced cost of insurance. As discussed in Chapter 8, the premiums of certain types of insurance are adjusted up or down for the individual contracting firm in accordance with its loss experience. Workmen's compensation insurance is of this type, with variations of more than 50 percent from the manual rates being possible. The following instance illustrates the magnitude of savings in the cost of this type of insurance that can result from an effective job safety program. Assume a building contractor does an annual volume of $10 million worth of work a year. Considering a typical amount of subcontracting and the cost of materials, this general contractor's annual payroll will be on the order of magnitude of $2.5 million. If its present workmen's compensation rate averages about 8 percent, the annual premium cost will be about $200,000. Now assume an effective accident prevention program results in an experience-modification rate reduction for this contractor of 30 percent below the workmen's compensation insurance rates formerly paid. Annual savings on the order of $60,000 are thereby realized on the cost of this one insurance coverage alone. A direct result of this is the improved competitive situation of the contractor. Lower insurance premiums mean lower bids.

Safety is also an important public relations tool, there being few other activities with such great potential for building goodwill. Good public relations have important financial and business implications for the contractor. A serious job accident can adversely affect a contractor's reputation, and the attendant unfavorable publicity can undo years of favorable public relations.

15.11 THE ROLE OF MANAGEMENT IN SAFETY

Top management bears the ultimate responsibility for the company's accident record, and the impetus for improved safety performance must emanate from this level. The attitude of management toward safe work practices will be reflected by the supervisors and workers. A mood of top-level interest and concern keeps everybody safety conscious. Conversely, if the top executives are not genuinely interested in preventing accidents and injuries, no one else is likely to be concerned either. If employee cooperation and participation are to be obtained, the accident control program must start with the demonstrated interest and backing of company management. Additionally, adequate funds must be provided to assure the successful operation of the safety program.

Every level of company management, therefore, must reflect a concern for safety and set a good example of compliance with safety regulations. Management interest must be vocal, visible, and vigorous. The logical beginning is a written company safety policy, announced by management and implemented by rules that are enforced. This policy should be widely publicized so that every employee becomes familiar with it, especially with the aspects that pertain directly to him. A carefully worded safety policy, signed by the company president, that is personally and forcefully brought to the attention of every employee will emphasize management's desire and determination to reduce accidents. In addition, a written policy is very useful in the enforcement of safety rules by supervisors. It is important that company management place the administration and enforcement of the safety policy squarely in the mainstream of company operations. Safety must be included as a company objective along with productivity and quality performance.

Top administrators personally should give timely credit and commendation for good safety performance whenever and wherever it occurs. The attendance of executives at employee safety meetings will impress workers with management's sincere desire to eliminate job accidents. A recommended procedure is that company accidents be broken down by projects so top management can see where safety problems exist. Field supervisors should be evaluated for promotions and salary increases in terms of accident records as well as production and costs. Making promotions and salary decisions on the basis of safety as well as company profits can be an effective way to reduce job accidents.

15.12 THE COMPANY SAFETY PROGRAM

A company safety program is just as much a part of a contractor's business as estimating and cost accounting. Fundamentally, the company safety plan must be one of identifying specific job hazards and educating the employees to conduct their work in a way that will minimize the risk of injury. The cause of an accident is an unsafe act or an unsafe condition. The objective of the company accident prevention program is to eliminate both of these from the jobs. Unsafe conditions result from either inadequate safety planning or no planning at all and are a consequence of the way the job is planned or the way it is conducted. In most cases unsafe conditions can be corrected by changing job procedures. Unsafe acts result from personal carelessness or lack of safety training. Such acts can be prevented by safety education for the individual worker and enforcement of safety regulations.

Despite OSHA and its inspections and penalties, the injury and illness incidence rates in the construction industry have not decreased appreciably in the last several years. At best, the current trend could be called static. Because insurance company studies disclose that an unsafe act by a worker is present in about 85 percent of all construction accidents, the company safety plan should certainly direct itself to this point and have a highly personal focus. Merely checking the field work for unsafe conditions will not suffice.

Training programs to assist supervisors with the safety planning of their projects and to educate workers in how to properly perform their tasks are essential to any company plan. The U.S. Department of Labor, insurance companies, the Associated General Contractors of America, the American National Standards Institute, the National Safety Council, and others provide various kinds of aid and assistance with regard to safety training and instruction. Organized labor plays an active role in health and safety training and sponsors educational programs in this regard. First aid training for selected personnel, sponsored by local offices of the American National Red Cross, can be a valuable part of the company safety training program.

Rules, safety devices, and mechanical safeguards are important to the prevention of job accidents, but the mental attitude of the workers is even more significant. Until an awareness of their individual responsibility and a desire to ensure their personal safety are established in the minds of all employees, efforts to reduce accidents will not be fully effective. Accident prevention in construction is largely a human relations problem and is achieved primarily through education, persuasion, and eternal vigilance. People cause accidents, and only people can prevent them. The company plan must emphasize the personal approach to job safety.

Because of differences in organization, type of activity, and scope of operations, each contractor must develop an accident prevention plan that fits its own particular situation. Job hazards differ considerably among housing, building, highway, heavy, utility, and industrial construction, a fact that must be reflected by the detailed workings of the program. Overall responsibility for the company's safety program must be placed with an individual who is capable, energetic, qualified, and interested in safety. This person may be a company executive or a staff assistant. In any event, he must have the authority to carry out his functions effectively. He is made responsible for project safety planning, safety training, distribution and use of safety equipment, maintenance of first aid facilities on the projects, job inspection, investigation of accidents, writing and filing of accident reports, and associated duties.

Company safety meetings of supervisory personnel held at regular intervals can be very effective. A formal program consisting of a guest safety speaker or a film may be presented. Information concerning how the company ranked in state and national safety contests can be discussed. Accompanying this could be a discussion of recent company lost-time accidents with information on how they could have been prevented. Regular safety inspections of each project and a detailed investigation of all lost-time accidents are important aspects of a company safety plan.

Safety contests between company projects may also be conducted. At a dinner meeting, project superintendents with the best safety records are recognized and presented with cash awards substantial enough to be appreciated. The participation of rank-and-file workers can also be encouraged through the awarding of cash prizes for their safety suggestions and slogans.

15.13 THE PROJECT SAFETY PLAN

Accident prevention aimed at the avoidance of specific happenings must be planned into each construction project. The constantly changing nature of a project under construction does not allow the detection and elimination of hazards purely on an experience basis. The contractor must establish in advance of each project the particular hazards that the proposed methods, procedures, and equipment will create and then devise an accident prevention plan to combat them. Analysis of accident experience is a valuable first step.

Having established the ground rules of job safety, the commencement of field operations must be accompanied by the implementation of the plan. The following steps are suggested for the conduct of the safety program during construction activities.

1. Assign prime responsibility for the project safety plan and its enforcement to the top field supervisor—the project manager or the job superintendent. Under him, each craft foreman is made responsible for safety measures as applied to his group. It is the foreman who is with his men all the time, and it is the foreman who must watch for unsafe practices or conditions and who must promote safety by instruction, precept, and example. On large projects there may be a safety engineer who coordinates the overall plan and devotes his time and energies exclusively to matters of safety, first aid, sanitation, fire prevention, and other such activities.

2. Make suitable and adequate first aid facilities readily available. These facilities may range from a well supplied first aid kit on small projects to a nurse-staffed infirmary on very large ones. On the usual job with no professional medical assistance available on the site, first aid training for supervisors and foremen is desirable. Locations of the nearest hospital facilities and ambulance service telephone numbers should be prominently posted in the job office. First aid kits should be dustproof, easily available, and checked at frequent intervals for any replenishments needed.

Every person on the job should be informed as to where first aid facilities are available.

3. Acquaint all employees with the company's safety policy and stress that strict conformance with safety regulations is a condition of employment. Special safety instruction must be given for particularly hazardous work. Every employee should be instructed to report immediately any injury, however trivial, to his foreman and to obtain suitable first aid treatment.

4. Insist on the wearing and proper use of personal protective clothing and equipment, with no exceptions made.

5. Conduct periodic "tool-box" safety talks and demonstrations on the project for all work crews. Merely prodding workers "to be careful" is not likely to accomplish much. Information as to the proper use of tools, handling of materials, building of scaffolds, and operation of equipment can be topics of successful meetings. The material must be specific, practical, and pertinent to current operations. Suggestions for improved safety should be solicited from the workers.

6. Utilize safety posters, safety instruction cards, and warning signs. Prominent display of the project accident record, as well as of notices that remind workers of specific project safety requirements, can be very effective.

7. Periodic meetings of the superintendent, craft foremen, and other supervisors are essential to review job safety and to make necessary revisions to the program. Investigate all lost-time accidents and devise corrective measures to prevent their recurrence.

8. Provide adequate, suitable, and easily accessible fire-fighting equipment and materials. Because welding and flame cutting are among the most frequent causes of construction fires, special regulations must apply to these activities. Specific areas should be provided for the storage of flammable, combustible, and explosive materials.

9. Establish a program of periodic job safety inspections. The inspection team must include the company safety specialist and the top field supervisor. Notes of safety violations and job hazards should be made and immediate action taken to correct them.

10. Insist on good project housekeeping. Designated storage areas for materials, tools, and supplies should be maintained and used. Rubbish and waste material should be removed promptly from the area of operations.

11. See that regular equipment maintenance includes safety inspection. This maintenance should include inspection of accident hazards such as frayed cables, bad tires, slipping clutches, and electrical grounds. Inspection and maintenance must not be limited to mechanical equipment but should be extended to scaffolding, towers, ladders, and other nonoperating items.

12. Seek and obtain the full cooperation of all subcontractors on the project. All of the measures previously described must include subcontractor personnel as well.

As a general guide, the most frequently encountered safety violations on construction projects have been reported to be the following.

1. Improperly shored or sloped excavations.
2. Ungrounded or unguarded portable electric tools.

3. Improperly maintained and substandard scaffolding.
4. Absence of or inadequate guard rails at floor, wall, and roof openings.
5. Absence of or inadequate barricades for the site perimeter.
6. Improper and unsecured ladders.
7. Improper handling and storing of flammable liquids and compressed gas cylinders.
8. Poor general housekeeping.
9. Lack of qualified first aid attendant at site.
10. Failure to wear personal protective equipment.
11. Failure to backfill holes and trenches promptly.
12. Unsafe crane operation.

15.14 THE FIELD SUPERVISOR

Top management has the major responsibility for establishing safety policies, procedures, and safe working conditions. However, most of what is planned and established must reach the worker on the job by way of the field supervisor. The craft foremen are the real key to the success of any project safety plan. To be effective, any campaign for the prevention of accidents in construction must be communicated to the individual worker in a clear, practical, and understandable form. Although the executives of the company may prescribe safe practices, the foreman, who has the authority to direct the workers in the field and is in daily contact with them, plays a dominant role in implementing the company safety policy.

Construction workers mirror their supervisors' attitude toward safety. For this reason, the wholehearted cooperation of the superintendent and foremen is indispensable to the success of any safety program. The best way for supervisors to sell accident prevention is to practice what they preach. A worker is much more likely to follow a supervisor's example than that same supervisor's instructions. If supervisors break a safety rule, they not only reduce the importance of the rule but also lose some of the confidence of their workers. If the field supervisors clearly believe in safety and reflect the fact by word and deed, the people under them will be much more cognizant of the advantages of safe work practices and the costly results of any alternative. It is largely up to the supervisors to find and control the potential hazards on the job. They must teach by doing and be able to demonstrate the safe way to do any particular job. Job instructions to workers should include not only what is to be done but also how it is to be done. Accident potential decreases when the workman is given complete procedural instructions in advance.

Safety regulations must be enforced. The worker must be taught safe practices and be required to follow them. Safety discipline is a delicate area, but it is very important to an accident prevention program. If a rule has been established, the supervisor must always enforce it. An established safety regulation that is not enforced will not be obeyed. If a rule is enforced only some of the time, a worker who is reprimanded will feel that he is being "picked on" and singled out for unfair treatment. Those who violate the rule and are not reprimanded will begin to believe the regulation is not very important. The supervisor must, of course, exercise care to keep personal feelings out of safety discipline.

The objective of safety discipline is to improve the safety performance of the crew. Workers who are convinced that a safety procedure is designed to protect their welfare will support its enforcement. Similarly, workers are more likely to accept a reprimand if they believe that it is for their own protection. A spirit of group responsibility for safety and a sense of safety competition among crews help to establish a self-disciplining attitude about safety violations.

15.15 ACCIDENT RECORDS

The keeping of accident records is an important part of a company's safety and health program. These records serve to pinpoint the locations and underlying causes of job injuries and illnesses, information which is vital to the planning of more effective accident prevention programs. They also provide information on the efficacy of the overall safety effort and how the company compares with other construction firms. Accurate, complete, and detailed records can be invaluable in defending against charges of safety law violations or claims for damages. Accident information can be valuable in quite another way. It can be used to arouse the competitive spirit of workmen and supervisors on the various projects to establish a safety record that compares favorably with the experience of other projects or with their own past record.

Accident recording and reporting might be thought of as starting with the first aid log that is maintained on each project. Every job injury or illness is made a matter of record, regardless of how inconsequential it may appear to be. A daily record book is maintained in which entry is made of the date, name of the employee affected, nature of the injury or illness, first aid treatment given on the site, and any further information deemed desirable. This is followed by a first report of injury which is required by workmen's compensation laws in most states. It is prepared for every incident that requires off-site medical treatment regardless of whether time is lost from work or not. Another report that is used by many construction companies is prepared by the appropriate foreman for each recordable injury or illness and is directed toward analyzing the accident and determining how it could have been prevented.

OSHA requires that occupational injury and illness records be kept for all employees. These records include (1) a log and summary of occupational injuries and illnesses (OSHA Form 200), (2) a supplementary record of occupational injuries and illnesses which contains detailed information on individual accidents (OSHA Form 101), and (3) an annual summary of occupational injuries and illnesses.The log must record all cases resulting in medical treatment, loss of consciousness, restriction of work or motion, or transfer to another less taxing job. Any fatality or accident that hospitalizes five or more employees must be reported to the OSHA area director within 48 hours. OSHA records provide management with valuable information concerning the efficacy of the company safety program.

15.16 PROTECTION OF THE PUBLIC

A very important aspect of a company's accident prevention program is the safety of the general public. People are innately curious and are capable of many thoughtless actions in their attempts to observe construction operations. The contracting firm must reconcile itself to the fact that the public will know what is going on; how it finds out is up to the contractor. If an attempt is made to shut people out completely, they will feel compelled to climb over the fences, follow trucks through the gates, or do some other equally human but hazardous act. Verbal admonitions or warning signs seem to do little good, and positive action must be taken to protect the public against its own unthinking actions. The best scheme is for the contractor to allow the public to view the proceedings from controlled vantage points. The contractor who provides means for the public to see the work and simultaneously be protected from its hazards is wise from the standpoint of both safety and public relations.

The problem of protecting the public becomes even more difficult during weekends and at other times when job operations are not in progress. Children, in particular, seem to find construction projects irresistible, and insurance company records are filled with cases involving the deaths or injuries of youngsters playing on projects during off-hours. On projects located in areas where children are likely to be playing, the job safety plan must make specific provision for this additional hazard.

15.17 THE COST OF A SAFETY PROGRAM

A company safety program does, of course, cost money. The fact is, however, that safety is just as necessary for the conduct of a construction business as is estimating or purchasing. There is, however, an important distinction between safety costs and other items of company expense that must be recognized. The distinction is that the spending of one dollar for safety can save the contractor two dollars. Although this ratio is only figurative, it has been well demonstrated that the costs of safety programs are more than compensated for by savings on accidents that do not happen.

Accident reports make it evident that most job accidents can be prevented at only moderate if any extra cost. The additional expense involved in building a proper scaffold, shoring an excavation, grounding an electric drill, or otherwise doing the work in a safe and sane manner is insignificant compared to the costs of accident or injury. The contractor cannot look on its safety program as an extra source of expense. Rather, because an effective accident prevention program is necessary to achieve fast-moving, smoothly functioning jobs, any costs entailed should be considered merely as normal business expenses associated with efficient operation.

APPENDIX A

STANDARD FORM OF AGREEMENT BETWEEN OWNER AND ARCHITECT, AIA DOCUMENT B141*

THE AMERICAN INSTITUTE OF ARCHITECTS

AIA Document B141

Standard Form of Agreement Between Owner and Architect

1977 EDITION

*THIS DOCUMENT HAS IMPORTANT LEGAL CONSEQUENCES; CONSULTATION WITH
AN ATTORNEY IS ENCOURAGED WITH RESPECT TO ITS COMPLETION OR MODIFICATION*

AGREEMENT

made as of the   day of   in the year of Nineteen
Hundred and

BETWEEN the Owner:

and the Architect:

For the following Project:
(Include detailed description of Project location and scope.)

The Owner and the Architect agree as set forth below.

AIA DOCUMENT B141 • OWNER-ARCHITECT AGREEMENT • THIRTEENTH EDITION • JULY 1977 • AIA® • © 1977
THE AMERICAN INSTITUTE OF ARCHITECTS, 1735 NEW YORK AVENUE, N.W., WASHINGTON, D.C. 20006
B141-1977 1

TERMS AND CONDITIONS OF AGREEMENT BETWEEN OWNER AND ARCHITECT

ARTICLE 1

ARCHITECT'S SERVICES AND RESPONSIBILITIES

BASIC SERVICES

The Architect's Basic Services consist of the five phases described in Paragraphs 1.1 through 1.5 and include normal structural, mechanical and electrical engineering services and any other services included in Article 15 as part of Basic Services.

1.1 SCHEMATIC DESIGN PHASE

1.1.1 The Architect shall review the program furnished by the Owner to ascertain the requirements of the Project and shall review the understanding of such requirements with the Owner.

1.1.2 The Architect shall provide a preliminary evaluation of the program and the Project budget requirements, each in terms of the other, subject to the limitations set forth in Subparagraph 3.2.1.

1.1.3 The Architect shall review with the Owner alternative approaches to design and construction of the Project.

1.1.4 Based on the mutually agreed upon program and Project budget requirements, the Architect shall prepare, for approval by the Owner, Schematic Design Documents consisting of drawings and other documents illustrating the scale and relationship of Project components.

1.1.5 The Architect shall submit to the Owner a Statement of Probable Construction Cost based on current area, volume or other unit costs.

1.2 DESIGN DEVELOPMENT PHASE

1.2.1 Based on the approved Schematic Design Documents and any adjustments authorized by the Owner in the program or Project budget, the Architect shall prepare, for approval by the Owner, Design Development Documents consisting of drawings and other documents to fix and describe the size and character of the entire Project as to architectural, structural, mechanical and electrical systems, materials and such other elements as may be appropriate.

1.2.2 The Architect shall submit to the Owner a further Statement of Probable Construction Cost.

1.3 CONSTRUCTION DOCUMENTS PHASE

1.3.1 Based on the approved Design Development Documents and any further adjustments in the scope or quality of the Project or in the Project budget authorized by the Owner, the Architect shall prepare, for approval by the Owner, Construction Documents consisting of Drawings and Specifications setting forth in detail the requirements for the construction of the Project.

1.3.2 The Architect shall assist the Owner in the preparation of the necessary bidding information, bidding forms, the Conditions of the Contract, and the form of Agreement between the Owner and the Contractor.

1.3.3 The Architect shall advise the Owner of any adjust-

ments to previous Statements of Probable Construction Cost indicated by changes in requirements or general market conditions.

1.3.4 The Architect shall assist the Owner in connection with the Owner's responsibility for filing documents required for the approval of governmental authorities having jurisdiction over the Project.

1.4 BIDDING OR NEGOTIATION PHASE

1.4.1 The Architect, following the Owner's approval of the Construction Documents and of the latest Statement of Probable Construction Cost, shall assist the Owner in obtaining bids or negotiated proposals, and assist in awarding and preparing contracts for construction.

1.5 CONSTRUCTION PHASE—ADMINISTRATION OF THE CONSTRUCTION CONTRACT

1.5.1 The Construction Phase will commence with the award of the Contract for Construction and, together with the Architect's obligation to provide Basic Services under this Agreement, will terminate when final payment to the Contractor is due, or in the absence of a final Certificate for Payment or of such due date, sixty days after the Date of Substantial Completion of the Work, whichever occurs first.

1.5.2 Unless otherwise provided in this Agreement and incorporated in the Contract Documents, the Architect shall provide administration of the Contract for Construction as set forth below and in the edition of AIA Document A201, General Conditions of the Contract for Construction, current as of the date of this Agreement.

1.5.3 The Architect shall be a representative of the Owner during the Construction Phase, and shall advise and consult with the Owner. Instructions to the Contractor shall be forwarded through the Architect. The Architect shall have authority to act on behalf of the Owner only to the extent provided in the Contract Documents unless otherwise modified by written instrument in accordance with Subparagraph 1.5.16.

1.5.4 The Architect shall visit the site at intervals appropriate to the stage of construction or as otherwise agreed by the Architect in writing to become generally familiar with the progress and quality of the Work and to determine in general if the Work is proceeding in accordance with the Contract Documents. However, the Architect shall not be required to make exhaustive or continuous on-site inspections to check the quality or quantity of the Work. On the basis of such on-site observations as an architect, the Architect shall keep the Owner informed of the progress and quality of the Work, and shall endeavor to guard the Owner against defects and deficiencies in the Work of the Contractor.

1.5.5 The Architect shall not have control or charge of and shall not be responsible for construction means, methods, techniques, sequences or procedures, or for safety precautions and programs in connection with the Work, for the acts or omissions of the Contractor, Subcontractors or any other persons performing any of the Work, or for the failure of any of them to carry out the Work in accordance with the Contract Documents.

1.5.6 The Architect shall at all times have access to the Work wherever it is in preparation or progress.

1.5.7 The Architect shall determine the amounts owing to the Contractor based on observations at the site and on evaluations of the Contractor's Applications for Payment, and shall issue Certificates for Payment in such amounts, as provided in the Contract Documents.

1.5.8 The issuance of a Certificate for Payment shall constitute a representation by the Architect to the Owner, based on the Architect's observations at the site as provided in Subparagraph 1.5.4 and on the data comprising the Contractor's Application for Payment, that the Work has progressed to the point indicated; that, to the best of the Architect's knowledge, information and belief, the quality of the Work is in accordance with the Contract Documents (subject to an evaluation of the Work for conformance with the Contract Documents upon Substantial Completion, to the results of any subsequent tests required by or performed under the Contract Documents, to minor deviations from the Contract Documents correctable prior to completion, and to any specific qualifications stated in the Certificate for Payment); and that the Contractor is entitled to payment in the amount certified. However, the issuance of a Certificate for Payment shall not be a representation that the Architect has made any examination to ascertain how and for what purpose the Contractor has used the moneys paid on account of the Contract Sum.

1.5.9 The Architect shall be the interpreter of the requirements of the Contract Documents and the judge of the performance thereunder by both the Owner and Contractor. The Architect shall render interpretations necessary for the proper execution or progress of the Work with reasonable promptness on written request of either the Owner or the Contractor, and shall render written decisions, within a reasonable time, on all claims, disputes and other matters in question between the Owner and the Contractor relating to the execution or progress of the Work or the interpretation of the Contract Documents.

1.5.10 Interpretations and decisions of the Architect shall be consistent with the intent of and reasonably inferable from the Contract Documents and shall be in written or graphic form. In the capacity of interpreter and judge, the Architect shall endeavor to secure faithful performance by both the Owner and the Contractor, shall not show partiality to either, and shall not be liable for the result of any interpretation or decision rendered in good faith in such capacity.

1.5.11 The Architect's decisions in matters relating to artistic effect shall be final if consistent with the intent of the Contract Documents. The Architect's decisions on any other claims, disputes or other matters, including those in question between the Owner and the Contractor, shall be subject to arbitration as provided in this Agreement and in the Contract Documents.

1.5.12 The Architect shall have authority to reject Work which does not conform to the Contract Documents. Whenever, in the Architect's reasonable opinion, it is

necessary or advisable for the implementation of the intent of the Contract Documents, the Architect will have authority to require special inspection or testing of the Work in accordance with the provisions of the Contract Documents, whether or not such Work be then fabricated, installed or completed.

1.5.13 The Architect shall review and approve or take other appropriate action upon the Contractor's submittals such as Shop Drawings, Product Data and Samples, but only for conformance with the design concept of the Work and with the information given in the Contract Documents. Such action shall be taken with reasonable promptness so as to cause no delay. The Architect's approval of a specific item shall not indicate approval of an assembly of which the item is a component.

1.5.14 The Architect shall prepare Change Orders for the Owner's approval and execution in accordance with the Contract Documents, and shall have authority to order minor changes in the Work not involving an adjustment in the Contract Sum or an extension of the Contract Time which are not inconsistent with the intent of the Contract Documents.

1.5.15 The Architect shall conduct inspections to determine the Dates of Substantial Completion and final completion, shall receive and forward to the Owner for the Owner's review written warranties and related documents required by the Contract Documents and assembled by the Contractor, and shall issue a final Certificate for Payment.

1.5.16 The extent of the duties, responsibilities and limitations of authority of the Architect as the Owner's representative during construction shall not be modified or extended without written consent of the Owner, the Contractor and the Architect.

1.6 PROJECT REPRESENTATION BEYOND BASIC SERVICES

1.6.1 If the Owner and Architect agree that more extensive representation at the site than is described in Paragraph 1.5 shall be provided, the Architect shall provide one or more Project Representatives to assist the Architect in carrying out such responsibilities at the site.

1.6.2 Such Project Representatives shall be selected, employed and directed by the Architect, and the Architect shall be compensated therefor as mutually agreed between the Owner and the Architect as set forth in an exhibit appended to this Agreement, which shall describe the duties, responsibilities and limitations of authority of such Project Representatives.

1.6.3 Through the observations by such Project Representatives, the Architect shall endeavor to provide further protection for the Owner against defects and deficiencies in the Work, but the furnishing of such project representation shall not modify the rights, responsibilities or obligations of the Architect as described in Paragraph 1.5.

1.7 ADDITIONAL SERVICES

The following Services are not included in Basic Services unless so identified in Article 15. They shall be provided if authorized or confirmed in writing by the Owner, and they shall be paid for by the Owner as provided in this Agreement, in addition to the compensation for Basic Services.

AIA DOCUMENT B141 • OWNER-ARCHITECT AGREEMENT • THIRTEENTH EDITION • JULY 1977 • AIA® • © 1977
THE AMERICAN INSTITUTE OF ARCHITECTS, 1735 NEW YORK AVENUE, N.W., WASHINGTON, D.C. 20006

1.7.1 Providing analyses of the Owner's needs, and programming the requirements of the Project.

1.7.2 Providing financial feasibility or other special studies.

1.7.3 Providing planning surveys, site evaluations, environmental studies or comparative studies of prospective sites, and preparing special surveys, studies and submissions required for approvals of governmental authorities or others having jurisdiction over the Project.

1.7.4 Providing services relative to future facilities, systems and equipment which are not intended to be constructed during the Construction Phase.

1.7.5 Providing services to investigate existing conditions or facilities or to make measured drawings thereof, or to verify the accuracy of drawings or other information furnished by the Owner.

1.7.6 Preparing documents of alternate, separate or sequential bids or providing extra services in connection with bidding, negotiation or construction prior to the completion of the Construction Documents Phase, when requested by the Owner.

1.7.7 Providing coordination of Work performed by separate contractors or by the Owner's own forces.

1.7.8 Providing services in connection with the work of a construction manager or separate consultants retained by the Owner.

1.7.9 Providing Detailed Estimates of Construction Cost, analyses of owning and operating costs, or detailed quantity surveys or inventories of material, equipment and labor.

1.7.10 Providing interior design and other similar services required for or in connection with the selection, procurement or installation of furniture, furnishings and related equipment.

1.7.11 Providing services for planning tenant or rental spaces.

1.7.12 Making revisions in Drawings, Specifications or other documents when such revisions are inconsistent with written approvals or instructions previously given, are required by the enactment or revision of codes, laws or regulations subsequent to the preparation of such documents or are due to other causes not solely within the control of the Architect.

1.7.13 Preparing Drawings, Specifications and supporting data and providing other services in connection with Change Orders to the extent that the adjustment in the Basic Compensation resulting from the adjusted Construction Cost is not commensurate with the services required of the Architect, provided such Change Orders are required by causes not solely within the control of the Architect.

1.7.14 Making investigations, surveys, valuations, inventories or detailed appraisals of existing facilities, and services required in connection with construction performed by the Owner.

1.7.15 Providing consultation concerning replacement of any Work damaged by fire or other cause during con-

struction, and furnishing services as may be required in connection with the replacement of such Work.

1.7.16 Providing services made necessary by the default of the Contractor, or by major defects or deficiencies in the Work of the Contractor, or by failure of performance of either the Owner or Contractor under the Contract for Construction.

1.7.17 Preparing a set of reproducible record drawings showing significant changes in the Work made during construction based on marked-up prints, drawings and other data furnished by the Contractor to the Architect.

1.7.18 Providing extensive assistance in the utilization of any equipment or system such as initial start-up or testing, adjusting and balancing, preparation of operation and maintenance manuals, training personnel for operation and maintenance, and consultation during operation.

1.7.19 Providing services after issuance to the Owner of the final Certificate for Payment, or in the absence of a final Certificate for Payment, more than sixty days after the Date of Substantial Completion of the Work.

1.7.20 Preparing to serve or serving as an expert witness in connection with any public hearing, arbitration proceeding or legal proceeding.

1.7.21 Providing services of consultants for other than the normal architectural, structural, mechanical and electrical engineering services for the Project.

1.7.22 Providing any other services not otherwise included in this Agreement or not customarily furnished in accordance with generally accepted architectural practice.

1.8   TIME

1.8.1 The Architect shall perform Basic and Additional Services as expeditiously as is consistent with professional skill and care and the orderly progress of the Work. Upon request of the Owner, the Architect shall submit for the Owner's approval a schedule for the performance of the Architect's services which shall be adjusted as required as the Project proceeds, and shall include allowances for periods of time required for the Owner's review and approval of submissions and for approvals of authorities having jurisdiction over the Project. This schedule, when approved by the Owner, shall not, except for reasonable cause, be exceeded by the Architect.

ARTICLE 2

THE OWNER'S RESPONSIBILITIES

2.1 The Owner shall provide full information regarding requirements for the Project including a program, which shall set forth the Owner's design objectives, constraints and criteria, including space requirements and relationships, flexibility and expandability, special equipment and systems and site requirements.

2.2 If the Owner provides a budget for the Project it shall include contingencies for bidding, changes in the Work during construction, and other costs which are the responsibility of the Owner, including those described in this Article 2 and in Subparagraph 3.1.2. The Owner shall, at the request of the Architect, provide a statement of funds available for the Project, and their source.

AIA DOCUMENT B141 • OWNER-ARCHITECT AGREEMENT • THIRTEENTH EDITION • JULY 1977 • AIA® • © 1977
THE AMERICAN INSTITUTE OF ARCHITECTS, 1735 NEW YORK AVENUE, N.W., WASHINGTON, D.C. 20006   **B141-1977   5**

2.3 The Owner shall designate, when necessary, a representative authorized to act in the Owner's behalf with respect to the Project. The Owner or such authorized representative shall examine the documents submitted by the Architect and shall render decisions pertaining thereto promptly, to avoid unreasonable delay in the progress of the Architect's services.

2.4 The Owner shall furnish a legal description and a certified land survey of the site, giving, as applicable, grades and lines of streets, alleys, pavements and adjoining property; rights-of-way, restrictions, easements, encroachments, zoning, deed restrictions, boundaries and contours of the site; locations, dimensions and complete data pertaining to existing buildings, other improvements and trees; and full information concerning available service and utility lines both public and private, above and below grade, including inverts and depths.

2.5 The Owner shall furnish the services of soil engineers or other consultants when such services are deemed necessary by the Architect. Such services shall include test borings, test pits, soil bearing values, percolation tests, air and water pollution tests, ground corrosion and resistivity tests, including necessary operations for determining subsoil, air and water conditions, with reports and appropriate professional recommendations.

2.6 The Owner shall furnish structural, mechanical, chemical and other laboratory tests, inspections and reports as required by law or the Contract Documents.

2.7 The Owner shall furnish all legal, accounting and insurance counseling services as may be necessary at any time for the Project, including such auditing services as the Owner may require to verify the Contractor's Applications for Payment or to ascertain how or for what purposes the Contractor uses the moneys paid by or on behalf of the Owner.

2.8 The services, information, surveys and reports required by Paragraphs 2.4 through 2.7 inclusive shall be furnished at the Owner's expense, and the Architect shall be entitled to rely upon the accuracy and completeness thereof.

2.9 If the Owner observes or otherwise becomes aware of any fault or defect in the Project or nonconformance with the Contract Documents, prompt written notice thereof shall be given by the Owner to the Architect.

2.10 The Owner shall furnish required information and services and shall render approvals and decisions as expeditiously as necessary for the orderly progress of the Architect's services and of the Work.

ARTICLE 3

CONSTRUCTION COST

3.1 DEFINITION

3.1.1 The Construction Cost shall be the total cost or estimated cost to the Owner of all elements of the Project designed or specified by the Architect.

3.1.2 The Construction Cost shall include at current market rates, including a reasonable allowance for overhead and profit, the cost of labor and materials furnished by the Owner and any equipment which has been de-

signed, specified, selected or specially provided for by the Architect.

3.1.3 Construction Cost does not include the compensation of the Architect and the Architect's consultants, the cost of the land, rights-of-way, or other costs which are the responsibility of the Owner as provided in Article 2.

3.2 RESPONSIBILITY FOR CONSTRUCTION COST

3.2.1 Evaluations of the Owner's Project budget, Statements of Probable Construction Cost and Detailed Estimates of Construction Cost, if any, prepared by the Architect, represent the Architect's best judgment as a design professional familiar with the construction industry. It is recognized, however, that neither the Architect nor the Owner has control over the cost of labor, materials or equipment, over the Contractor's methods of determining bid prices, or over competitive bidding, market or negotiating conditions. Accordingly, the Architect cannot and does not warrant or represent that bids or negotiated prices will not vary from the Project budget proposed, established or approved by the Owner, if any, or from any Statement of Probable Construction Cost or other cost estimate or evaluation prepared by the Architect.

3.2.2 No fixed limit of Construction Cost shall be established as a condition of this Agreement by the furnishing, proposal or establishment of a Project budget under Subparagraph 1.1.2 or Paragraph 2.2 or otherwise, unless such fixed limit has been agreed upon in writing and signed by the parties hereto. If such a fixed limit has been established, the Architect shall be permitted to include contingencies for design, bidding and price escalation, to determine what materials, equipment, component systems and types of construction are to be included in the Contract Documents, to make reasonable adjustments in the scope of the Project and to include in the Contract Documents alternate bids to adjust the Construction Cost to the fixed limit. Any such fixed limit shall be increased in the amount of any increase in the Contract Sum occurring after execution of the Contract for Construction.

3.2.3 If the Bidding or Negotiation Phase has not commenced within three months after the Architect submits the Construction Documents to the Owner, any Project budget or fixed limit of Construction Cost shall be adjusted to reflect any change in the general level of prices in the construction industry between the date of submission of the Construction Documents to the Owner and the date on which proposals are sought.

3.2.4 If a Project budget or fixed limit of Construction Cost (adjusted as provided in Subparagraph 3.2.3) is exceeded by the lowest bona fide bid or negotiated proposal, the Owner shall (1) give written approval of an increase in such fixed limit, (2) authorize rebidding or renegotiating of the Project within a reasonable time, (3) if the Project is abandoned, terminate in accordance with Paragraph 10.2, or (4) cooperate in revising the Project scope and quality as required to reduce the Construction Cost. In the case of (4), provided a fixed limit of Construction Cost has been established as a condition of this Agreement, the Architect, without additional charge, shall modify the Drawings and Specifications as necessary to comply

AIA DOCUMENT B141 • OWNER-ARCHITECT AGREEMENT • THIRTEENTH EDITION • JULY 1977 • AIA® • © 1977
THE AMERICAN INSTITUTE OF ARCHITECTS, 1735 NEW YORK AVENUE, N.W., WASHINGTON, D.C. 20006

with the fixed limit. The providing of such service shall be the limit of the Architect's responsibility arising from the establishment of such fixed limit, and having done so, the Architect shall be entitled to compensation for all services performed, in accordance with this Agreement, whether or not the Construction Phase is commenced.

ARTICLE 4

DIRECT PERSONNEL EXPENSE

4.1 Direct Personnel Expense is defined as the direct salaries of all the Architect's personnel engaged on the Project, and the portion of the cost of their mandatory and customary contributions and benefits related thereto, such as employment taxes and other statutory employee benefits, insurance, sick leave, holidays, vacations, pensions and similar contributions and benefits.

ARTICLE 5

REIMBURSABLE EXPENSES

5.1 Reimbursable Expenses are in addition to the Compensation for Basic and Additional Services and include actual expenditures made by the Architect and the Architect's employees and consultants in the interest of the Project for the expenses listed in the following Subparagraphs:

5.1.1 Expense of transportation in connection with the Project; living expenses in connection with out-of-town travel; long distance communications; and fees paid for securing approval of authorities having jurisdiction over the Project.

5.1.2 Expense of reproductions, postage and handling of Drawings, Specifications and other documents, excluding reproductions for the office use of the Architect and the Architect's consultants.

5.1.3 Expense of data processing and photographic production techniques when used in connection with Additional Services.

5.1.4 If authorized in advance by the Owner, expense of overtime work requiring higher than regular rates.

5.1.5 Expense of renderings, models and mock-ups requested by the Owner.

5.1.6 Expense of any additional insurance coverage or limits, including professional liability insurance, requested by the Owner in excess of that normally carried by the Architect and the Architect's consultants.

ARTICLE 6

PAYMENTS TO THE ARCHITECT

6.1 PAYMENTS ON ACCOUNT OF BASIC SERVICES

6.1.1 An initial payment as set forth in Paragraph 14.1 is the minimum payment under this Agreement.

6.1.2 Subsequent payments for Basic Services shall be made monthly and shall be in proportion to services performed within each Phase of services, on the basis set forth in Article 14.

6.1.3 If and to the extent that the Contract Time initially established in the Contract for Construction is exceeded

or extended through no fault of the Architect, compensation tor any Basic Services required for such extended period of Administration of the Construction Contract shall be computed as set forth in Paragraph 14.4 for Additional Services.

6.1.4 When compensation is based on a percentage of Construction Cost, and any portions of the Project are deleted or otherwise not constructed, compensation for such portions of the Project shall be payable to the extent services are performed on such portions, in accordance with the schedule set forth in Subparagraph 14.2.2, based on (1) the lowest bona fide bid or negotiated proposal or, (2) if no such bid or proposal is received, the most recent Statement of Probable Construction Cost or Detailed Estimate of Construction Cost for such portions of the Project.

**6.2 PAYMENTS ON ACCOUNT OF
ADDITIONAL SERVICES**

6.2.1 Payments on account of the Architect's Additional Services as defined in Paragraph 1.7 and for Reimbursable Expenses as defined in Article 5 shall be made monthly upon presentation of the Architect's statement of services rendered or expenses incurred.

6.3 PAYMENTS WITHHELD

6.3.1 No deductions shall be made from the Architect's compensation on account of penalty, liquidated damages or other sums withheld from payments to contractors, or on account of the cost of changes in the Work other than those for which the Architect is held legally liable.

6.4 PROJECT SUSPENSION OR TERMINATION

6.4.1 If the Project is suspended or abandoned in whole or in part for more than three months, the Architect shall be compensated for all services performed prior to receipt of written notice from the Owner of such suspension or abandonment, together with Reimbursable Expenses then due and all Termination Expenses as defined in Paragraph 10.4. If the Project is resumed after being suspended for more than three months, the Architect's compensation shall be equitably adjusted.

ARTICLE 7

ARCHITECT'S ACCOUNTING RECORDS

7.1 Records of Reimbursable Expenses and expenses pertaining to Additional Services and services performed on the basis of a Multiple of Direct Personnel Expense shall be kept on the basis of generally accepted accounting principles and shall be available to the Owner or the Owner's authorized representative at mutually convenient times.

ARTICLE 8

OWNERSHIP AND USE OF DOCUMENTS

8.1 Drawings and Specifications as instruments of service are and shall remain the property of the Architect whether the Project for which they are made is executed or not. The Owner shall be permitted to retain copies, including reproducible copies, of Drawings and Specifications for information and reference in connection with the Owner's use and occupancy of the Project. The Drawings and Specifications shall not be used by the Owner on other projects, for additions to this Project, or for completion of this Project by others provided the Architect is not in default under this Agreement, except by agreement in writing and with appropriate compensation to the Architect.

8.2   Submission or distribution to meet official regulatory requirements or for other purposes in connection with the Project is not to be construed as publication in derogation of the Architect's rights.

ARTICLE 9

ARBITRATION

9.1   All claims, disputes and other matters in question between the parties to this Agreement, arising out of or relating to this Agreement or the breach thereof, shall be decided by arbitration in accordance with the Construction Industry Arbitration Rules of the American Arbitration Association then obtaining unless the parties mutually agree otherwise. No arbitration, arising out of or relating to this Agreement, shall include, by consolidation, joinder or in any manner, any additional person not a party to this Agreement except by written consent containing a specific reference to this Agreement and signed by the Architect, the Owner, and any other person sought to be joined. Any consent to arbitration involving an additional person or persons shall not constitute consent to arbitration of any dispute not described therein or with any person not named or described therein. This Agreement to arbitrate and any agreement to arbitrate with an additional person or persons duly consented to by the parties to this Agreement shall be specifically enforceable under the prevailing arbitration law.

9.2   Notice of the demand for arbitration shall be filed in writing with the other party to this Agreement and with the American Arbitration Association. The demand shall be made within a reasonable time after the claim, dispute or other matter in question has arisen. In no event shall the demand for arbitration be made after the date when institution of legal or equitable proceedings based on such claim, dispute or other matter in question would be barred by the applicable statute of limitations.

9.3   The award rendered by the arbitrators shall be final, and judgment may be entered upon it in accordance with applicable law in any court having jurisdiction thereof.

ARTICLE 10

TERMINATION OF AGREEMENT

10.1   This Agreement may be terminated by either party upon seven days' written notice should the other party fail substantially to perform in accordance with its terms through no fault of the party initiating the termination.

10.2   This Agreement may be terminated by the Owner upon at least seven days' written notice to the Architect in the event that the Project is permanently abandoned.

10.3   In the event of termination not the fault of the Architect, the Architect shall be compensated for all services performed to termination date, together with Reimbursable Expenses then due and all Termination Expenses as defined in Paragraph 10.4.

10.4   Termination Expenses include expenses directly attributable to termination for which the Architect is not otherwise compensated, plus an amount computed as a percentage of the total Basic and Additional Compensation earned to the time of termination, as follows:

> **.1**   20 percent if termination occurs during the Schematic Design Phase; or
>
> **.2**   10 percent if termination occurs during the Design Development Phase; or
>
> **.3**   5 percent if termination occurs during any subsequent phase.

ARTICLE 11

MISCELLANEOUS PROVISIONS

11.1   Unless otherwise specified, this Agreement shall be governed by the law of the principal place of business of the Architect.

11.2   Terms in this Agreement shall have the same meaning as those in AIA Document A201, General Conditions of the Contract for Construction, current as of the date of this Agreement.

11.3   As between the parties to this Agreement: as to all acts or failures to act by either party to this Agreement, any applicable statute of limitations shall commence to run and any alleged cause of action shall be deemed to have accrued in any and all events not later than the relevant Date of Substantial Completion of the Work, and as to any acts or failures to act occurring after the relevant Date of Substantial Completion, not later than the date of issuance of the final Certificate for Payment.

11.4   The Owner and the Architect waive all rights against each other and against the contractors, consultants, agents and employees of the other for damages covered by any property insurance during construction as set forth in the edition of AIA Document A201, General Conditions, current as of the date of this Agreement. The Owner and the Architect each shall require appropriate similar waivers from their contractors, consultants and agents.

ARTICLE 12

SUCCESSORS AND ASSIGNS

12.1   The Owner and the Architect, respectively, bind themselves, their partners, successors, assigns and legal representatives to the other party to this Agreement and to the partners, successors, assigns and legal representatives of such other party with respect to all covenants of this Agreement. Neither the Owner nor the Architect shall assign, sublet or transfer any interest in this Agreement without the written consent of the other.

ARTICLE 13

EXTENT OF AGREEMENT

13.1   This Agreement represents the entire and integrated agreement between the Owner and the Architect and supersedes all prior negotiations, representations or agreements, either written or oral. This Agreement may be amended only by written instrument signed by both Owner and Architect.

8   B141-1977   AIA DOCUMENT B141 • OWNER-ARCHITECT AGREEMENT • THIRTEENTH EDITION • JULY 1977 • AIA® • © 1977
THE AMERICAN ~~INSTITUTE~~ OF ARCHITECTS, 1735 NEW YORK AVENUE, N.W., WASHINGTON, D.C. 20006

ARTICLE 14

BASIS OF COMPENSATION

The Owner shall compensate the Architect for the Scope of Services provided, in accordance with Article 6, Payments to the Architect, and the other Terms and Conditions of this Agreement, as follows:

14.1   AN INITIAL PAYMENT of   dollars ($

shall be made upon execution of this Agreement and credited to the Owner's account as follows:

14.2   BASIC COMPENSATION

14.2.1   FOR BASIC SERVICES, as described in Paragraphs 1.1 through 1.5, and any other services included in Article 15 as part of Basic Services, Basic Compensation shall be computed as follows:

(Here insert basis of compensation, including fixed amounts, multiples or percentages, and identify Phases to which particular methods of compensation apply, if necessary.)

14.2.2   Where compensation is based on a Stipulated Sum or Percentage of Construction Cost, payments for Basic Services shall be made as provided in Subparagraph 6.1.2, so that Basic Compensation for each Phase shall equal the following percentages of the total Basic Compensation payable:

(Include any additional Phases as appropriate.)

Schematic Design Phase:	percent (%)
Design Development Phase:	percent (%)
Construction Documents Phase:	percent (%)
Bidding or Negotiation Phase:	percent (%)
Construction Phase:	percent (%)

14.3   FOR PROJECT REPRESENTATION BEYOND BASIC SERVICES, as described in Paragraph 1.6, Compensation shall be computed separately in accordance with Subparagraph 1.6.2.

14.4 COMPENSATION FOR ADDITIONAL SERVICES

14.4.1 FOR ADDITIONAL SERVICES OF THE ARCHITECT, as described in Paragraph 1.7, and any other services included in Article 15 as part of Additional Services, but excluding Additional Services of consultants, Compensation shall be computed as follows:

(Here insert basis of compensation, including rates and/or multiples of Direct Personnel Expense for Principals and employees, and identify Principals and classify employees, if required. Identify specific services to which particular methods of compensation apply, if necessary.)

14.4.2 FOR ADDITIONAL SERVICES OF CONSULTANTS, including additional structural, mechanical and electrical engineering services and those provided under Subparagraph 1.7.21 or identified in Article 15 as part of Additional Services, a multiple of ( ) times the amounts billed to the Architect for such services.

(Identify specific types of consultants in Article 15, if required.)

14.5 FOR REIMBURSABLE EXPENSES, as described in Article 5, and any other items included in Article 15 as Reimbursable Expenses, a multiple of ( ) times the amounts expended by the Architect, the Architect's employees and consultants in the interest of the Project.

14.6 Payments due the Architect and unpaid under this Agreement shall bear interest from the date payment is due at the rate entered below, or in the absence thereof, at the legal rate prevailing at the principal place of business of the Architect.

(Here insert any rate of interest agreed upon.)

(Usury laws and requirements under the Federal Truth in Lending Act, similar state and local consumer credit laws and other regulations at the Owner's and Architect's principal places of business, the location of the Project and elsewhere may affect the validity of this provision. Specific legal advice should be obtained with respect to deletion, modification, or other requirements such as written disclosures or waivers.)

14.7 The Owner and the Architect agree in accordance with the Terms and Conditions of this Agreement that:

14.7.1 IF THE SCOPE of the Project or of the Architect's Services is changed materially, the amounts of compensation shall be equitably adjusted.

14.7.2 IF THE SERVICES covered by this Agreement have not been completed within

( ) months of the date hereof, through no fault of the Architect, the amounts of compensation, rates and multiples set forth herein shall be equitably adjusted.

AIA DOCUMENT B141 • OWNER-ARCHITECT AGREEMENT • THIRTEENTH EDITION • JULY 1977 • AIA® • © 1977
THE AMERICAN INSTITUTE OF ARCHITECTS, 1735 NEW YORK AVENUE, N.W., WASHINGTON, D.C. 20006

ARTICLE 15
OTHER CONDITIONS OR SERVICES

AIA DOCUMENT B141 • OWNER-ARCHITECT AGREEMENT • THIRTEENTH EDITION • JULY 1977 • AIA® • © 1977
THE AMERICAN INSTITUTE OF ARCHITECTS, 1735 NEW YORK AVENUE, N.W., WASHINGTON, D.C. 20006

B141-1977   11

This Agreement entered into as of the day and year first written above.

OWNER   ARCHITECT

_____   _____

_____   _____

_____   _____

BY_____   BY_____

AIA DOCUMENT B141 • OWNER-ARCHITECT AGREEMENT • THIRTEENTH EDITION • JULY 1977 • AIA® • © 1977
THE AMERICAN INSTITUTE OF ARCHITECTS, 1735 NEW YORK AVENUE, N.W., WASHINGTON, D.C. 20006

APPENDIX B

CONSTRUCTION SPECIFICATIONS

MASTERFORMAT, BROADSCOPE SECTION TITLES*

DIVISION 0 – BIDDING AND CONTRACT REQUIREMENTS

00010	PRE-BID INFORMATION
00100	INSTRUCTIONS TO BIDDERS
00200	INFORMATION AVAILABLE TO BIDDERS
00300	BID/TENDER FORMS
00400	SUPPLEMENTS TO BID/TENDER FORMS
00500	AGREEMENT FORMS
00600	BONDS AND CERTIFICATES
00700	GENERAL CONDITIONS OF THE CONTRACT
00800	SUPPLEMENTARY CONDITIONS
00950	DRAWINGS INDEX
00900	ADDENDA AND MODIFICATIONS

SPECIFICATIONS—DIVISIONS 1-16

DIVISION 1 – GENERAL REQUIRMENTS

01010	SUMMARY OF WORK
01020	ALLOWANCES
01030	SPECIAL PROJECT PROCEDURES
01040	COORDINATION
01050	FIELD ENGINEERING
01060	REGULATORY REQUIRMENTS
01070	ABBREVIATIONS AND SYMBOLS
01080	IDENTIFICATION SYSTEMS
01100	ALTERNATES/ALTERNATIVES
01150	MEASUREMENT AND PAYMENT
01200	PROJECT MEETINGS
01300	SUBMITTALS
01400	QUALITY CONTROL
01500	CONSTRUCTION FACILITIES AND TEMPORARY CONTROLS
01600	MATERIAL AND EQUIPMENT
01650	STARTING OF SYSTEMS
01660	TESTING, ADJUSTING, AND BALANCING OF SYSTEMS
01700	CONTRACT CLOSEOUT
01800	MAINTENANCE MATERIALS

DIVISION 2 – SITEWORK

02010	SUBSURFACE INVESTIGATION
02050	DEMOLITION
02100	SITE PREPARATION
02150	UNDERPINNING
02200	EARTHWORK
02300	TUNNELLING
02350	PILES, CAISSONS AND COFFERDAMS
02400	DRAINAGE
02440	SITE IMPROVEMENTS
02480	LANDSCAPING
02500	PAVING AND SURFACING
02580	BRIDGES
02590	PONDS AND RESERVOIRS
02600	PIPED UTILITY MATERIALS AND METHODS
02700	PIPED UTILITIES
02800	POWER AND COMMUNICATION UTILITIES
02850	RAILROAD WORK
02880	MARINE WORK

DIVISION 3 – CONCRETE

03010	CONCRETE MATERIALS
03050	CONCRETING PROCEDURES
03100	CONCRETE FORMWORK
03150	FORMS
03180	FORM TIES AND ACCESSORIES
03200	CONCRETE REINFORCEMENT
03250	CONCRETE ACCESSORIES
03300	CAST-IN-PLACE CONCRETE
03350	SPECIAL CONCRETE FINISHES
03360	SPECIALLY PLACED CONCRETE
03370	CONCRETE CURING
03400	PRECAST CONCRETE
03500	CEMENTITIOUS DECKS
03600	GROUT
03700	CONCRETE RESTORATION AND CLEANING

DIVISION 4 – MASONRY

04050	MASONRY PROCEDURES
04100	MORTAR
04150	MASONRY ACCESSORIES
04200	UNIT MASONRY
04400	STONE
04500	MASONRY RESTORATION AND CLEANING
04550	REFRACTORIES
04600	CORROSION RESISTANT MASONRY

DIVISION 5 – METALS

05010	METAL MATERIALS AND METHODS
05050	METAL FASTENING
05100	STRUCTURAL METAL FRAMING
05200	METAL JOISTS
05300	METAL DECKING
05400	COLD-FORMED METAL FRAMING
05500	METAL FABRICATIONS
05700	ORNAMENTAL METAL
05800	EXPANSION CONTROL
05900	METAL FINISHES

DIVISION 6 – WOOD AND PLASTICS

06050	FASTENERS AND SUPPORTS
06100	ROUGH CARPENTRY
06130	HEAVY TIMBER CONSTRUCTION
06150	WOOD-METAL SYSTEMS
06170	PREFABRICATED STRUCTURAL WOOD
06200	FINISH CARPENTRY
06300	WOOD TREATMENT
06400	ARCHITECTURAL WOODWORK
06500	PREFABRICATED STRUCTURAL PLASTICS
06600	PLASTIC FABRICATIONS

DIVISION 7 – THERMAL AND MOISTURE PROTECTION

07100	WATERPROOFING
07150	DAMPPROOFING
07200	INSULATION
07250	FIREPROOFING
07300	SHINGLES AND ROOFING TILES
07400	PREFORMED ROOFING AND SIDING
07500	MEMBRANE ROOFING
07570	TRAFFIC TOPPING
07600	FLASHING AND SHEET METAL
07800	ROOF ACCESSORIES
07900	SEALANTS

DIVISION 8 – DOORS AND WINDOWS

08100	METAL DOORS AND FRAMES
08200	WOOD AND PLASTIC DOORS
08250	DOOR OPENING ASSEMBLIES
08300	SPECIAL DOORS
08400	ENTRANCES AND STOREFRONTS
08500	METAL WINDOWS
08600	WOOD AND PLASTIC WINDOWS
08650	SPECIAL WINDOWS
08700	HARDWARE
08800	GLAZING
08900	GLAZED CURTAIN WALLS

DIVISION 9 – FINISHES

09100	METAL SUPPORT SYSTEMS
09200	LATH AND PLASTER
09230	AGGREGATE COATINGS
09250	GYPSUM WALLBOARD
09300	TILE
09400	TERRAZZO
09500	ACOUSTICAL TREATMENT
09550	WOOD FLOORING
09600	STONE AND BRICK FLOORING
09650	RESILIENT FLOORING
09680	CARPETING
09700	SPECIAL FLOORING
09760	FLOOR TREATMENT
09800	SPECIAL COATINGS
09900	PAINTING
09950	WALL COVERING

*Reproduced by permission of the copyright owner, The Construction Specifications Institute, Inc., Washington, D.C. 20036.

DIVISION 10 – SPECIALITIES

10100	CHALKBOARDS AND TACKBOARDS
10150	COMPARTMENTS AND CUBICLES
10200	LOUVERS AND VENTS
10240	GRILLES AND SCREENS
10250	SERVICE WALL SYSTEMS
10260	WALL AND CORNER GUARDS
10270	ACCESS FLOORING
10280	SPECIALTY MODULES
10290	PEST CONTROL
10300	FIREPLACES AND STOVES
10340	PREFABRICATED STEEPLES, SPIRES, AND CUPOLAS
10350	FLAGPOLES
10400	IDENTIFYING DEVICES
10450	PEDESTRIAN CONTROL DEVICES
10500	LOCKERS
10520	FIRE EXTINGUISHERS, CABINETS, AND ACCESSORIES
10530	PROTECTIVE COVERS
10550	POSTAL SPECIALTIES
10600	PARTITIONS
10650	SCALES
10670	STORAGE SHELVING
10700	EXTERIOR SUN CONTROL DEVICES
10750	TELEPHONE ENCLOSURES
10800	TOILET AND BATH ACCESSORIES
10900	WARDROBE SPECIALTIES

DIVISION 11 – EQUIPMENT

11010	MAINTENANCE EQUIPMENT
11020	SECURITY AND VAULT EQUIPMENT
11030	CHECKROOM EQUIPMENT
11040	ECCLESIASTICAL EQUIPMENT
11050	LIBRARY EQUIPMENT
11060	THEATER AND STAGE EQUIPMENT
11070	MUSICAL EQUIPMENT
11080	REGISTRATION EQUIPMENT
11100	MERCANTILE EQUIPMENT
11110	COMMERCIAL LAUNDRY AND DRY CLEANING EQUIPMENT
11120	VENDING EQUIPMENT
11130	AUDIO-VISUAL EQUIPMENT
11140	SERVICE STATION EQUIPMENT
11150	PARKING EQUIPMENT
11160	LOADING DOCK EQUIPMENT
11170	WASTE HANDLING EQUIPMENT
11190	DETENTION EQUIPMENT
11200	WATER SUPPLY AND TREATMENT EQUIPMENT
11300	FLUID WASTE DISPOSAL AND TREATMENT EQUIPMENT
11400	FOOD SERVICE EQUIPMENT
11450	RESIDENTIAL EQUIPMENT
11460	UNIT KITCHENS
11470	DARKROOM EQUIPMENT
11480	ATHLETIC, RECREATIONAL, AND THERAPEUTIC EQUIPMENT
11500	INDUSTRIAL AND PROCESS EQUIPMENT
11600	LABORATORY EQUIPMENT
11650	PLANETARIUM AND OBSERVATORY EQUIPMENT
11700	MEDICAL EQUIPMENT
11780	MORTUARY EQUIPMENT
11800	TELECOMMUNICATION EQUIPMENT
11850	NAVIGATION EQUIPMENT

DIVISION 12 – FURNISHINGS

12100	ARTWORK
12300	MANUFACTURED CABINETS AND CASEWORK
12500	WINDOW TREATMENT
12550	FABRICS
12600	FURNITURE AND ACCESSORIES
12670	RUGS AND MATS
12700	MULTIPLE SEATING
12800	INTERIOR PLANTS AND PLANTINGS

DIVISION 13 – SPECIAL CONSTRUCTION

13010	AIR SUPPORTED STRUCTURES
13020	INTEGRATED ASSEMBLIES
13030	AUDIOMETRIC ROOMS
13040	CLEAN ROOMS
13050	HYPERBARIC ROOMS
13060	INSULATED ROOMS
13070	INTEGRATED CEILINGS
13080	SOUND, VIBRATION, AND SEISMIC CONTROL
13090	RADIATION PROTECTION
13100	NUCLEAR REACTORS
13110	OBSERVATORIES
13120	PRE-ENGINEERED STRUCTURES
13130	SPECIAL PURPOSE ROOMS AND BUILDINGS
13140	VAULTS
13150	POOLS
13160	ICE RINKS
13170	KENNELS AND ANIMAL SHELTERS
13200	SEISMOGRAPHIC INSTRUMENTATION
13210	STRESS RECORDING INSTRUMENTATION
13220	SOLAR AND WIND INSTRUMENTATION
13410	LIQUID AND GAS STORAGE TANKS
13510	RESTORATION OF UNDERGROUND PIPELINES
13520	FILTER UNDERDRAINS AND MEDIA
13530	DIGESTION TANK COVERS AND APPURTENANCES
13540	OXYGENATION SYSTEMS
13550	THERMAL SLUDGE CONDITIONING SYSTEMS
13560	SITE CONSTRUCTED INCINERATORS
13600	UTILITY CONTROL SYSTEMS
13700	INDUSTRIAL AND PROCESS CONTROL SYSTEMS
13800	OIL AND GAS REFINING INSTALLATIONS AND CONTROL SYSTEMS
13900	TRANSPORTATION INSTRUMENTATION
13940	BUILDING AUTOMATION SYSTEMS
13970	FIRE SUPPRESSION AND SUPERVISORY SYSTEMS
13980	SOLAR ENERGY SYSTEMS
13990	WIND ENERGY SYSTEMS

DIVISION 14 – CONVEYING SYSTEMS

14100	DUMBWAITERS
14200	ELEVATORS
14300	HOISTS AND CRANES
14400	LIFTS
14500	MATERIAL HANDLING SYSTEMS
14600	TURNTABLES
14700	MOVING STAIRS AND WALKS
14800	POWERED SCAFFOLDING
14900	TRANSPORTATION SYSTEMS

DIVISION 15 – MECHANICAL

15050	BASIC MATERIALS AND METHODS
15200	NOISE, VIBRATION, AND SEISMIC CONTROL
15250	INSULATION
15300	SPECIAL PIPING SYSTEMS
15400	PLUMBING SYSTEMS
15450	PLUMBING FIXTURES AND TRIM
15500	FIRE PROTECTION
15600	POWER OR HEAT GENERATION
15650	REFRIGERATION
15700	LIQUID HEAT TRANSFER
15800	AIR DISTRIBUTION
15900	CONTROLS AND INSTRUMENTATION

DIVISION 16 – ELECTRICAL

16050	BASIC MATERIALS AND METHODS
16200	POWER GENERATION
16300	POWER TRANSMISSION
16400	SERVICE AND DISTRIBUTION
16500	LIGHTING
16600	SPECIAL SYSTEMS
16700	COMMUNICATIONS
16850	HEATING AND COOLING
16900	CONTROLS AND INSTRUMENTATION

TYPICAL SPECIFICATION OUTLINE
FOR ENGINEERING CONSTRUCTION

GENERAL PROVISIONS
1. Definitions and terms
2. Proposal requirements and conditions
3. Award and execution of contract
4. Scope of work
5. Control of the work
6. Control of material
7. Legal relations and responsibility
8. Prosecution and progress
9. Measurement and payment

Division 1. OBSTRUCTIONS

Division 2. CLEARING AND GRUBBING

Division 3. EARTHWORK
1. General
2. Roadway excavation
3. Excavation for structures
4. Ditch and channel excavation
5. Embankment
6. Selected rock slope protection.
7. Shoulders
8. Surplus and borrow excavation
9. Overhaul

Division 4. EROSION CONTROL AND
PREPARATORY LANDSCAPING

Division 5. SUBGRADE
1. General requirements
2. Class "A" subgrade
3. Class "B" subgrade
4. Class "C" subgrade
5. Class "D" subgrade
6. Class "E" subgrade
7. Class "F" subgrade

Division 6. WATERING

Division 7. FINISH ROADWAY

Division 8. CEMENT-TREATED SUBGRADE

Division 9. ROAD-MIXED CEMENT
TREATED BASE

Division 10. PLANT-MIXED CEMENT
TREATED BASE

Division 11. UNTREATED ROCK SURFACING

Division 12. CRUSHER RUN BASE

Division 13. PENETRATION TREATMENT

Division 14. SEAL COATS

Division 15. BITUMINOUS SURFACE
TREATMENT

Division 16. ARMOR COAT

Division 17. NON-SKID SURFACE
TREATMENT

Division 18. ROAD-MIXED SURFACING

Division 19. PLANT-MIXED SURFACING
1. Description
2. Materials
3. Drying, proportioning and mixing
   materials.
4. Subgrade
5. Prime coat
6. Spreading and compacting equipment
7. Spreading and compacting mixture
8. Miscellaneous details
9. Payment

Division 20. SIDE FORMS

Division 21. BITUMINOUS MACADAM
SURFACE

Division 22. ASPHALT CONCRETE
PAVEMENT

Division 23. PORTLAND CEMENT
CONCRETE PAVEMENT

Division 24. TIMBER STRUCTURES
1. General requirements
2. Structural timber

3. Construction
4. Payment

Division 25. CONCRETE STRUCTURES

Division 26. STEEL STRUCTURES
1. General requirements
2. Materials and testing
3. Workmanship
4. Payment

Division 27. PILING
1. General requirements
2. Timber piles
3. Precast concrete piles
4. Cast-in-place concrete piles
5. Steel piles
6. Payment

Division 28. TREATMENT OF TIMBER AND PILES

Division 29. WATERPROOFING

Division 30. PAINTING

Division 31. RUBBLE MASONRY

Division 32. RIPRAP

Division 33. CONCRETE SLOPE PAVING AND APRONS

Division 34. CONCRETE CURBS AND GUTTERS

Division 35. PORTLAND CEMENT CONCRETE SIDEWALKS

Division 36. RIGHT-OF-WAY MONUMENTS AND FREEWAY ACCESS OPENING MARKERS

Division 37. CONCRETE BARRIER POSTS

Division 38. GUARD RAILING
1. New construction
2. Salvaging and reconstructing guard railing

Division 39. PIPE HANDRAIL

Division 40. CULVERT MARKERS, CLEARANCE MARKERS, AND GUIDEPOSTS
1. New installation
2. Salvaging and resetting

Division 41. FENCES
1. New construction
2. Salvaging and reconstruction

Division 42. REINFORCED CONCRETE PIPE CULVERTS AND SIPHONS
1. General requirements
2. Manufacture
3. Installation
4. Headwalls
5. Payment

Division 43. CORRUGATED METAL PIPE CULVERTS AND SIPHONS

Division 44. FIELD-ASSEMBLED PLATE CULVERTS

Division 45. NON-REINFORCED CONCRETE PIPE LINES
1. General requirements
2. Installation
3. Headwalls
4. Payment

Division 46. SEWER PIPE LINES

Division 47. UNDERDRAINS

Division 48. SPILLWAY ASSEMBLIES AND DOWN DRAINS

Division 49. SALVAGING AND RELAYING EXISTING DRAINAGE FACILITIES

Division 50. REINFORCEMENT

Division 51. PORTLAND CEMENT CONCRETE

Division 52. ASPHALTIC PAINT BINDER

Division 53. PAINT
1. Materials
2. Standard paint for steel work
3. Standard paint for reinforcing bars, drift bolts, and iron hand rails
4. Standard paint for timber
5. Black paint for culvert markers
6. Wood preservative
7. Reflective paints

Division 54. ASPHALTS

Division 55. LIQUID ASPHALTS

Division 56. ASPHALTIC EMULSIONS

Division 57. EXPANSION JOINT FILLER

APPENDIX C

GENERAL CONDITIONS OF THE CONTRACT FOR CONSTRUCTION, AIA DOCUMENT A201*

THE AMERICAN INSTITUTE OF ARCHITECTS

AIA Document A201

General Conditions of the Contract for Construction

*THIS DOCUMENT HAS IMPORTANT LEGAL CONSEQUENCES; CONSULTATION
WITH AN ATTORNEY IS ENCOURAGED WITH RESPECT TO ITS MODIFICATION*

1976 EDITION
TABLE OF ARTICLES

This document has been approved and endorsed by The Associated General Contractors of America.

AIA DOCUMENT A201 • GENERAL CONDITIONS OF THE CONTRACT FOR CONSTRUCTION • THIRTEENTH EDITION • AUGUST 1976
AIA® • © 1976 • THE AMERICAN INSTITUTE OF ARCHITECTS, 1735 NEW YORK AVENUE, N.W., WASHINGTON, D.C. 20006 **A201-1976 1**

*AIA copyrighted material has been reproduced with permission of The American Institute of Architects under application number 80063. Further reproduction is prohibited.

INDEX

AIA DOCUMENT A201 • GENERAL CONDITIONS OF THE CONTRACT FOR CONSTRUCTION • THIRTEENTH EDITION • AUGUST 1976
AIA® • © 1976 • THE AMERICAN INSTITUTE OF ARCHITECTS, 1735 NEW YORK AVENUE, N.W., WASHINGTON, D.C. 20006   **A201-1976** **3**

AIA DOCUMENT A201 • GENERAL CONDITIONS OF THE CONTRACT FOR CONSTRUCTION • THIRTEENTH EDITION • AUGUST 1976
AIA® • © 1976 • THE AMERICAN INSTITUTE OF ARCHITECTS, 1735 NEW YORK AVENUE, N.W., WASHINGTON, D.C. 20006

GENERAL CONDITIONS OF THE CONTRACT FOR CONSTRUCTION

ARTICLE 1

CONTRACT DOCUMENTS

1.1   DEFINITIONS

1.1.1   THE CONTRACT DOCUMENTS

The Contract Documents consist of the Owner-Contractor Agreement, the Conditions of the Contract (General, Supplementary and other Conditions), the Drawings, the Specifications, and all Addenda issued prior to and all Modifications issued after execution of the Contract. A Modification is (1) a written amendment to the Contract signed by both parties, (2) a Change Order, (3) a written interpretation issued by the Architect pursuant to Subparagraph 2.2.8, or (4) a written order for a minor change in the Work issued by the Architect pursuant to Paragraph 12.4. The Contract Documents do not include Bidding Documents such as the Advertisement or Invitation to Bid, the Instructions to Bidders, sample forms, the Contractor's Bid or portions of Addenda relating to any of these, or any other documents, unless specifically enumerated in the Owner-Contractor Agreement.

1.1.2   THE CONTRACT

The Contract Documents form the Contract for Construction. This Contract represents the entire and integrated agreement between the parties hereto and supersedes all prior negotiations; representations, or agreements, either written or oral. The Contract may be amended or modified only by a Modification as defined in Subparagraph 1.1.1. The Contract Documents shall not be construed to create any contractual relationship of any kind between the Architect and the Contractor, but the Architect shall be entitled to performance of obligations intended for his benefit, and to enforcement thereof. Nothing contained in the Contract Documents shall create any contractual relationship between the Owner or the Architect and any Subcontractor or Sub-subcontractor.

1.1.3   THE WORK

The Work comprises the completed construction required by the Contract Documents and includes all labor necessary to produce such construction, and all materials and equipment incorporated or to be incorporated in such construction.

1.1.4   THE PROJECT

The Project is the total construction of which the Work performed under the Contract Documents may be the whole or a part.

1.2   EXECUTION, CORRELATION AND INTENT

1.2.1 The Contract Documents shall be signed in not less than triplicate by the Owner and Contractor. If either the Owner or the Contractor or both do not sign the Conditions of the Contract, Drawings, Specifications, or any of the other Contract Documents, the Architect shall identify such Documents.

1.2.2 By executing the Contract, the Contractor represents that he has visited the site, familiarized himself with the local conditions under which the Work is to be performed, and correlated his observations with the requirements of the Contract Documents.

1.2.3 The intent of the Contract Documents is to include all items necessary for the proper execution and completion of the Work. The Contract Documents are complementary, and what is required by any one shall be as binding as if required by all. Work not covered in the Contract Documents will not be required unless it is consistent therewith and is reasonably inferable therefrom as being necessary to produce the intended results. Words and abbreviations which have well-known technical or trade meanings are used in the Contract Documents in accordance with such recognized meanings.

1.2.4 The organization of the Specifications into divisions, sections and articles, and the arrangement of Drawings shall not control the Contractor in dividing the Work among Subcontractors or in establishing the extent of Work to be performed by any trade.

1.3   OWNERSHIP AND USE OF DOCUMENTS

1.3.1 All Drawings, Specifications and copies thereof furnished by the Architect are and shall remain his property. They are to be used only with respect to this Project and are not to be used on any other project. With the exception of one contract set for each party to the Contract, such documents are to be returned or suitably accounted for to the Architect on request at the completion of the Work. Submission or distribution to meet official regulatory requirements or for other purposes in connection with the Project is not to be construed as publication in derogation of the Architect's common law copyright or other reserved rights.

ARTICLE 2

ARCHITECT

2.1   DEFINITION

2.1.1 The Architect is the person lawfully licensed to practice architecture, or an entity lawfully practicing architecture identified as such in the Owner-Contractor Agreement, and is referred to throughout the Contract Documents as if singular in number and masculine in gender. The term Architect means the Architect or his authorized representative.

2.2   ADMINISTRATION OF THE CONTRACT

2.2.1 The Architect will provide administration of the Contract as hereinafter described.

2.2.2 The Architect will be the Owner's representative during construction and until final payment is due. The Architect will advise and consult with the Owner. The Owner's instructions to the Contractor shall be forwarded through the Architect. The Architect will have authority to act on behalf of the Owner only to the extent provided in the Contract Documents, unless otherwise modified by written instrument in accordance with Subparagraph 2.2.18.

2.2.3 The Architect will visit the site at intervals appropriate to the stage of construction to familiarize himself generally with the progress and quality of the Work and to determine in general if the Work is proceeding in accordance with the Contract Documents. However, the Architect will not be required to make exhaustive or continuous on-site inspections to check the quality or quantity of the Work. On the basis of his on-site observations as an architect, he will keep the Owner informed of the progress of the Work, and will endeavor to guard the Owner against defects and deficiencies in the Work of the Contractor.

2.2.4 The Architect will not be responsible for and will not have control or charge of construction means, methods, techniques, sequences or procedures, or for safety precautions and programs in connection with the Work, and he will not be responsible for the Contractor's failure to carry out the Work in accordance with the Contract Documents. The Architect will not be responsible for or have control or charge over the acts or omissions of the Contractor, Subcontractors, or any of their agents or employees, or any other persons performing any of the Work.

2.2.5 The Architect shall at all times have access to the Work wherever it is in preparation and progress. The Contractor shall provide facilities for such access so the Architect may perform his functions under the Contract Documents.

2.2.6 Based on the Architect's observations and an evaluation of the Contractor's Applications for Payment, the Architect will determine the amounts owing to the Contractor and will issue Certificates for Payment in such amounts, as provided in Paragraph 9.4.

2.2.7 The Architect will be the interpreter of the requirements of the Contract Documents and the judge of the performance thereunder by both the Owner and Contractor.

2.2.8 The Architect will render interpretations necessary for the proper execution or progress of the Work, with reasonable promptness and in accordance with any time limit agreed upon. Either party to the Contract may make written request to the Architect for such interpretations.

2.2.9 Claims, disputes and other matters in question between the Contractor and the Owner relating to the execution or progress of the Work or the interpretation of the Contract Documents shall be referred initially to the Architect for decision which he will render in writing within a reasonable time.

2.2.10 All interpretations and decisions of the Architect shall be consistent with the intent of and reasonably inferable from the Contract Documents and will be in writing or in the form of drawings. In his capacity as interpreter and judge, he will endeavor to secure faithful performance by both the Owner and the Contractor, will not

show partiality to either, and will not be liable for the result of any interpretation or decision rendered in good faith in such capacity.

2.2.11 The Architect's decisions in matters relating to artistic effect will be final if consistent with the intent of the Contract Documents.

2.2.12 Any claim, dispute or other matter in question between the Contractor and the Owner referred to the Architect, except those relating to artistic effect as provided in Subparagraph 2.2.11 and except those which have been waived by the making or acceptance of final payment as provided in Subparagraphs 9.9.4 and 9.9.5, shall be subject to arbitration upon the written demand of either party. However, no demand for arbitration of any such claim, dispute or other matter may be made until the earlier of (1) the date on which the Architect has rendered a written decision, or (2) the tenth day after the parties have presented their evidence to the Architect or have been given a reasonable opportunity to do so, if the Architect has not rendered his written decision by that date. When such a written decision of the Architect states (1) that the decision is final but subject to appeal, and (2) that any demand for arbitration of a claim, dispute or other matter covered by such decision must be made within thirty days after the date on which the party making the demand receives the written decision, failure to demand arbitration within said thirty days' period will result in the Architect's decision becoming final and binding upon the Owner and the Contractor. If the Architect renders a decision after arbitration proceedings have been initiated, such decision may be entered as evidence but will not supersede any arbitration proceedings unless the decision is acceptable to all parties concerned.

2.2.13 The Architect will have authority to reject Work which does not conform to the Contract Documents. Whenever, in his opinion, he considers it necessary or advisable for the implementation of the intent of the Contract Documents, he will have authority to require special inspection or testing of the Work in accordance with Subparagraph 7.7.2 whether or not such Work be then fabricated, installed or completed. However, neither the Architect's authority to act under this Subparagraph 2.2.13, nor any decision made by him in good faith either to exercise or not to exercise such authority, shall give rise to any duty or responsibility of the Architect to the Contractor, any Subcontractor, any of their agents or employees, or any other person performing any of the Work.

2.2.14 The Architect will review and approve or take other appropriate action upon Contractor's submittals such as Shop Drawings, Product Data and Samples, but only for conformance with the design concept of the Work and with the information given. in the Contract Documents. Such action shall be taken with reasonable promptness so as to cause no delay. The Architect's approval of a specific item shall not indicate approval of an assembly of which the item is a component.

2.2.15 The Architect will prepare Change Orders in accordance with Article 12, and will have authority to order minor changes in the Work as provided in Subparagraph 12.4.1.

2.2.16 The Architect will conduct inspections to determine the dates of Substantial Completion and final completion, will receive and forward to the Owner for the Owner's review written warranties and related documents required by the Contract and assembled by the Contractor, and will issue a, final Certificate for Payment upon compliance with the requirements of Paragraph 9.9.

2.2.17 If the Owner and Architect agree, the Architect will provide one or more Project Representatives to assist the Architect in carrying out his responsibilities at the site. The duties, responsibilities and limitations of authority of any such Project Representative shall be as set forth in an exhibit to be incorporated in the Contract Documents.

2.2.18 The duties, responsibilities and limitations of authority of the Architect as the Owner's representative during construction as set forth in the Contract Documents will not be modified or extended without written consent of the Owner, the Contractor and the Architect.

2.2.19 In case of the termination of the employment of the Architect, the Owner shall appoint an architect against whom the Contractor makes no reasonable objection whose status under the Contract Documents shall be that of the former architect. Any dispute in connection with such appointment shall be subject to arbitration.

ARTICLE 3

OWNER

3.1   DEFINITION

3.1.1 The Owner is the person or entity identified as such in the Owner-Contractor Agreement and is referred to throughout the Contract Documents as if singular in number and masculine in gender. The term Owner means the Owner or his authorized representative.

**3.2   INFORMATION AND SERVICES REQUIRED
          OF THE OWNER**

3.2.1 The Owner shall, at the request of the Contractor, at the time of execution of the Owner-Contractor Agreement, furnish to the Contractor reasonable evidence that he has made financial arrangements to fulfill his obligations under the Contract. Unless such reasonable evidence is furnished, the Contractor is not required to execute the Owner-Contractor Agreement or to commence the Work.

3.2.2 The Owner shall furnish all surveys describing the physical characteristics, legal limitations and utility locations for the site of the Project, and a legal description of the site.

3.2.3 Except as provided in Subparagraph 4.7.1, the Owner shall secure and pay for necessary approvals, easements, assessments and charges required for the construction, use or occupancy of permanent structures or for permanent changes in existing facilities.

3.2.4 Information or services under the Owner's control shall be furnished by the Owner with reasonable promptness to avoid delay in the orderly progress of the Work.

3.2.5 Unless otherwise provided in the Contract Documents, the Contractor will be furnished, free of charge, all copies of Drawings and Specifications reasonably necessary for the execution of the Work.

3.2.6 The Owner shall forward all instructions to the Contractor through the Architect.

3.2.7 The foregoing are in addition to other duties and responsibilities of the Owner enumerated herein and especially those in respect to Work by Owner or by Separate Contractors, Payments and Completion, and Insurance in Articles 6, 9 and 11 respectively.

3.3   OWNER'S RIGHT TO STOP THE WORK

3.3.1 If the Contractor fails to correct defective Work as required by Paragraph 13.2 or persistently fails to carry out the Work in accordance with the Contract Documents, the Owner, by a written order signed personally or by an agent specifically so empowered by the Owner in writing, may order the Contractor to stop the Work, or.any portion thereof, until the cause for such order has been eliminated; however, this right of the Owner to stop the Work shall not give rise to any duty on the part of the Owner to exercise this right for the benefit of the Contractor or any other person or entity, except to the extent required by Subparagraph 6.1.3.

3.4   OWNER'S RIGHT TO CARRY OUT THE WORK

3.4.1 If the Contractor defaults or neglects to carry out the Work in accordance with the Contract Documents and fails within seven days after receipt of written notice from the Owner to commence and continue correction of such default or neglect with diligence and promptness, the Owner may, after seven days following receipt by the Contractor of an additional written notice and without prejudice to any other remedy he may have, make good such deficiencies. In such case an appropriate Change Order shall be issued deducting from the payments then or thereafter due the Contractor the cost of correcting such deficiencies, including compensation for the Architect's additional services made necessary by such default, neglect or failure. Such action by the Owner and the amount charged to the Contractor are both subject to the prior approval of the Architect. If the payments then or thereafter due the Contractor are not sufficient to cover such amount, the Contractor shall pay the difference to the Owner.

ARTICLE 4

CONTRACTOR

4.1   DEFINITION

4.1.1 The Contractor is the person or entity identified as such in the Owner-Contractor Agreement and is referred to throughout the Contract Documents as if singular in number and masculine in gender. The term Contractor means the Contractor or his authorized representative.

4.2   REVIEW OF CONTRACT DOCUMENTS

4.2.1 The Contractor shall carefully study and compare the Contract Documents and shall at once·report to the Architect any error, inconsistency or omission he may discover. The Contractor shall not be liable to the Owner or the Architect for any damage resulting from any such errors, inconsistencies or omissions in the Contract Documents. The Contractor shall perform no portion of the Work at any time without Contract Documents or, where required, approved Shop Drawings, Product Data or Samples for such portion of the Work.

4.3   SUPERVISION AND CONSTRUCTION PROCEDURES

4.3.1 The Contractor shall supervise and direct the Work, using his best skill and attention. He shall be solely responsible for all construction means, methods, techniques, sequences and procedures and for coordinating all portions of the Work under the Contract.

4.3.2 The Contractor shall be responsible to the Owner for the acts and omissions of his employees, Subcontractors and their agents and employees, and other persons performing any of the Work under a contract with the Contractor.

4.3.3 The Contractor shall not be relieved from his obligations to perform the Work in accordance with the Contract Documents either by the activities or duties of the Architect in his administration of the Contract, or by inspections, tests or approvals required or performed under Paragraph 7.7 by persons other than the Contractor.

4.4   LABOR AND MATERIALS

4.4.1 Unless otherwise provided in the Contract Documents, the Contractor shall provide and pay for all labor, materials, equipment, tools, construction equipment and machinery, water, heat, utilities, transportation, and other facilities and services necessary for the proper execution and completion of the Work, whether temporary or permanent and whether or not incorporated or to be incorporated in the Work.

4.4.2 The Contractor shall at all times enforce strict discipline and good order among his employees and shall not employ on the Work any unfit person or anyone not skilled in the task assigned to him.

4.5   WARRANTY

4.5.1 The Contractor warrants to the Owner and the Architect that all materials and equipment furnished under this Contract will be new unless otherwise specified, and that all Work will be of good quality, free from faults and defects and in conformance with the Contract Documents. All Work not conforming to these requirements, including substitutions not properly approved and authorized, may be considered defective. If required by the Architect, the Contractor shall furnish satisfactory evidence as to the kind and quality of materials and equipment. This warranty is not limited by the provisions of Paragraph 13.2.

4.6   TAXES

4.6.1 The Contractor shall pay all sales, consumer, use and other similar taxes for the Work or portions thereof provided by the Contractor which are legally enacted at the time bids are received, whether or not yet effective.

4.7   PERMITS, FEES AND NOTICES

4.7.1 Unless otherwise provided in the Contract Documents, the Contractor shall secure and pay for the building permit and for all other permits and governmental fees, licenses and inspections necessary for the proper execution and completion of the Work which are customarily secured after execution of the Contract and which are legally required at the time the bids are received.

4.7.2 The Contractor shall give all notices and comply with all laws, ordinances, rules, regulations and lawful orders of any public authority bearing on the performance of the Work.

4.7.3 It is not the responsibility of the Contractor to make certain that the Contract Documents are in accordance with applicable laws, statutes, building codes and regulations. If the Contractor observes that any of the Contract Documents are at variance therewith in any respect, he shall promptly notify the Architect in writing, and any necessary changes shall be accomplished by appropriate Modification.

4.7.4 If the Contractor performs any Work knowing it to be contrary to such laws, ordinances, rules and regulations, and without such notice to the Architect, he shall assume full responsibility therefor and shall bear all costs attributable thereto.

4.8   ALLOWANCES

4.8.1 The Contractor shall include in the Contract Sum all allowances stated in the Contract Documents. Items covered by these allowances shall be supplied for such amounts and by such persons as the Owner may direct, but the Contractor will not be required to employ persons against whom he makes a reasonable objection.

4.8.2 Unless otherwise provided in the Contract Documents:

.1 these allowances shall cover the cost to the Contractor, less any applicable trade discount, of the materials and equipment required by the allowance delivered at the site, and all applicable taxes;

.2 the Contractor's costs for unloading and handling on the site, labor, installation costs, overhead, profit and other expenses contemplated for the original allowance shall be included in the Contract Sum and not in the allowance;

.3 whenever the cost is more than or less than the allowance, the Contract Sum shall be adjusted accordingly by Change Order, the amount of which will recognize changes, if any, in handling costs on the site, labor, installation costs, overhead, profit and other expenses.

4.9   SUPERINTENDENT

4.9.1 The Contractor shall employ a competent superintendent and necessary assistants who shall be in attendance at the Project site during the progress of the Work. The superintendent shall represent the Contractor and all communications given to the superintendent shall be as binding as if given to the Contractor. Important communications shall be confirmed in writing. Other communications shall be so confirmed on written request in each case.

4.10   PROGRESS SCHEDULE

4.10.1 The Contractor, immediately after being awarded the Contract, shall prepare and submit for the Owner's and Architect's information an estimated progress schedule for the Work. The progress schedule shall be related to the entire Project to the extent required by the Contract Documents, and shall provide for expeditious and practicable execution of the Work.

4.11 DOCUMENTS AND SAMPLES AT THE SITE

4.11.1 The Contractor shall maintain at the site for the Owner one record copy of all Drawings, Specifications, Addenda, Change Orders and other Modifications, in good order and marked currently to record all changes made during construction, and approved Shop Drawings, Product Data and Samples. These shall be available to the Architect and shall be delivered to him for the Owner upon completion of the Work.

4.12 SHOP DRAWINGS, PRODUCT DATA AND SAMPLES

4.12.1 Shop Drawings are drawings, diagrams, schedules and other data specially prepared for the Work by the Contractor or any Subcontractor, manufacturer, supplier or distributor to illustrate some portion of the Work.

4.12.2 Product Data are illustrations, standard schedules, performance charts, instructions, brochures, diagrams and other information furnished by the Contractor to illustrate a material, product or system for some portion of the Work.

4.12.3 Samples are physical examples which illustrate materials, equipment or workmanship and establish standards by which the Work will be judged.

4.12.4 The Contractor shall review, approve and submit, with reasonable promptness and in such sequence as to cause no delay in the Work or in the work of the Owner or any separate contractor, all Shop Drawings, Product Data and Samples required by the Contract Documents.

4.12.5 By approving and submitting Shop Drawings, Product Data and Samples, the Contractor represents that he has determined and verified all materials, field measurements, and field construction criteria related thereto, or will do so, and that he has checked and coordinated the information contained within such submittals with the requirements of the Work and of the Contract Documents.

4.12.6 The Contractor shall not be relieved of responsibility for any deviation from the requirements of the Contract Documents by the Architect's approval of Shop Drawings, Product Data or Samples under Subparagraph 2.2.14 unless the Contractor has specifically informed the Architect in writing of such deviation at the time of submission and the Architect has given written approval to the specific deviation. The Contractor shall not be relieved from responsibility for errors or omissions in the Shop Drawings, Product Data or Samples by the Architect's approval thereof.

4.12.7 The Contractor shall direct specific attention, in writing or on resubmitted Shop Drawings, Product Data or Samples, to revisions other than those requested by the Architect on previous submittals.

4.12.8 No portion of the Work requiring submission of a Shop Drawing, Product Data or Sample shall be commenced until the submittal has been approved by the Architect as provided in Subparagraph 2.2.14. All such

portions of the Work shall be in accordance with approved submittals.

4.13 USE OF SITE

4.13.1 The Contractor shall confine operations at the site to areas permitted by law, ordinances, permits and the Contract Documents and shall not unreasonably encumber the site with any materials or equipment.

4.14 CUTTING AND PATCHING OF WORK

4.14.1 The Contractor shall be responsible for all cutting, fitting or patching that may be required to complete the Work or to make its several parts fit together properly.

4.14.2 The Contractor shall not damage or endanger any portion of the Work or the work of the Owner or any separate contractors by cutting, patching or otherwise altering any work, or by excavation. The Contractor shall not cut or otherwise alter the work of the Owner or any separate contractor except with the written consent of the Owner and of such separate contractor. The Contractor shall not unreasonably withhold from the Owner or any separate contractor his consent to cutting or otherwise altering the Work.

4.15 CLEANING UP

4.15.1 The Contractor at all times shall keep the premises free from accumulation of waste materials or rubbish caused by his operations. At the completion of the Work he shall remove all his waste materials and rubbish from and about the Project as well as all his tools, construction equipment, machinery and surplus materials.

4.15.2 If the Contractor fails to clean up at the completion of the Work, the Owner may do so as provided in Paragraph 3.4 and the cost thereof shall be charged to the Contractor.

4.16 COMMUNICATIONS

4.16.1 The Contractor shall forward all communications to the Owner through the Architect.

4.17 ROYALTIES AND PATENTS

4.17.1 The Contractor shall pay all royalties and license fees. He shall defend all suits or claims for infringement of any patent rights and shall save the Owner harmless from loss on account thereof, except that the Owner shall be responsible for all such loss when a particular design, process or the product of a particular manufacturer or manufacturers is specified, but if the Contractor has reason to believe that the design, process or product specified is an infringement of a patent, he shall be responsible for such loss unless he promptly gives such information to the Architect.

4.18 INDEMNIFICATION

4.18.1 To the fullest extent permitted by law, the Contractor shall indemnify and hold harmless the Owner and the Architect and their agents and employees from and against all claims, damages, losses and expenses, including but not limited to attorneys' fees, arising out of or resulting from the performance of the Work, provided that any such claim, damage, loss or expense (1) is attributable to bodily injury, sickness, disease or death, or to injury to or destruction of tangible property (other than the Work itself) including the loss of use resulting therefrom, and (2) is caused in whole or in part by any negligent act or omission of the Contractor, any Subcontractor, anyone directly or indirectly employed by any of them or anyone for whose acts any of them may be liable, regardless of whether or not it is caused in part by a party indemnified hereunder. Such obligation shall not be construed to negate, abridge, or otherwise reduce any other right or obligation of indemnity which would otherwise exist as to any party or person described in this Paragraph 4.18.

4.18.2 In any and all claims against the Owner or the Architect or any of their agents or employees by any employee of the Contractor, any Subcontractor, anyone directly or indirectly employed by any of them or anyone for whose acts any of them may be liable, the indemnification obligation under this Paragraph 4.18 shall not be limited in any way by any limitation on the amount or type of damages, compensation or benefits payable by or for the Contractor or any Subcontractor under workers' or workmen's compensation acts, disability benefit acts or other employee benefit acts.

4.18.3 The obligations of the Contractor under this Paragraph 4.18 shall not extend to the liability of the Architect, his agents or employees, arising out of (1) the preparation or approval of maps, drawings, opinions, reports, surveys, change orders, designs or specifications, or (2) the giving of or the failure to give directions or instructions by the Architect, his agents or employees provided such giving or failure to give is the primary cause of the injury or damage.

ARTICLE 5

SUBCONTRACTORS

5.1   DEFINITION

5.1.1 A Subcontractor is a person or entity who has a direct contract with the Contractor to perform any of the Work at the site. The term Subcontractor is referred to throughout the Contract Documents as if singular in number and masculine in gender and means a Subcontractor or his authorized representative. The term Subcontractor does not include any separate contractor or his subcontractors.

5.1.2 A Sub-subcontractor is a person or entity who has a direct or indirect contract with a Subcontractor to perform any of the Work at the site. The term Sub-subcontractor is referred to throughout the Contract Documents as if singular in number and masculine in gender and means a Sub-subcontractor or an authorized representative thereof.

5.2   AWARD OF SUBCONTRACTS AND OTHER CONTRACTS FOR PORTIONS OF THE WORK

5.2.1 Unless otherwise required by the Contract Documents or the Bidding Documents, the Contractor, as soon as practicable after the award of the Contract, shall furnish to the Owner and the Architect in writing the names of the persons or entities (including those who are to furnish materials or equipment fabricated to a special design) proposed for each of the principal portions of the Work. The Architect will promptly reply to the Contractor in writing stating whether or not the Owner or the Architect, after due investigation, has reasonable objection to any

such proposed person or entity. Failure of the Owner or Architect to reply promptly shall constitute notice of no reasonable objection.

5.2.2 The Contractor shall not contract with any such proposed person or entity to whom the Owner or the Architect has made reasonable objection under the provisions of Subparagraph 5.2.1. The Contractor shall not be required to contract with anyone to whom he has a reasonable objection.

5.2.3 If the Owner or the Architect has reasonable objection to any such proposed person or entity, the Contractor shall submit a substitute to whom the Owner or the Architect has no reasonable objection, and the Contract Sum shall be increased or decreased by the difference in cost occasioned by such substitution and an appropriate Change Order shall be issued; however, no increase in the Contract Sum shall be allowed for any such substitution unless the Contractor has acted promptly and responsively in submitting names as required by Subparagraph 5.2.1.

5.2.4 The Contractor shall make no substitution for any Subcontractor, person or entity previously selected if the Owner or Architect makes reasonable objection to such substitution.

5.3   SUBCONTRACTUAL RELATIONS

5.3.1 By an appropriate agreement, written where legally required for validity, the Contractor shall require each Subcontractor, to the extent of the Work to be performed by the Subcontractor, to be bound to the Contractor by the terms of the Contract Documents, and to assume toward the Contractor all the obligations and responsibilities which the Contractor, by these Documents, assumes toward the Owner and the Architect. Said agreement shall preserve and protect the rights of the Owner and the Architect under the Contract Documents with respect to the Work to be performed by the Subcontractor so that the subcontracting thereof will not prejudice such rights, and shall allow to the Subcontractor, unless specifically provided otherwise in the Contractor-Subcontractor agreement, the benefit of all rights, remedies and redress against the Contractor that the Contractor, by these Documents, has against the Owner. Where appropriate, the Contractor shall require each Subcontractor to enter into similar agreements with his Sub-subcontractors. The Contractor shall make available to each proposed Subcontractor, prior to the execution of the Subcontract, copies of the Contract Documents to which the Subcontractor will be bound by this Paragraph 5.3, and identify to the Subcontractor any terms and conditions of the proposed Subcontract which may be at variance with the Contract Documents. Each Subcontractor shall similarly make copies of such Documents available to his Sub-subcontractors.

ARTICLE 6

WORK BY OWNER OR BY SEPARATE CONTRACTORS

6.1   OWNER'S RIGHT TO PERFORM WORK AND TO AWARD SEPARATE CONTRACTS

6.1.1 The Owner reserves the right to perform work related to the Project with his own forces, and to award separate contracts in connection with other portions of the Project or other work on the site under these or similar Conditions of the Contract. If the Contractor claims that delay or additional cost is involved because of such action by the Owner, he shall make such claim as provided elsewhere in the Contract Documents.

6.1.2 When separate contracts are awarded for different portions of the Project or other work on the site, the term Contractor in the Contract Documents in each case shall mean the Contractor who executes each separate Owner-Contractor Agreement.

6.1.3 The Owner will provide for the coordination of the work of his own forces and of each separate contractor with the Work of the Contractor, who shall cooperate therewith as provided in Paragraph 6.2.

6.2   MUTUAL RESPONSIBILITY

6.2.1 The Contractor shall afford the Owner and separate contractors reasonable opportunity for the introduction and storage of their materials and equipment and the execution of their work, and shall connect and coordinate his Work with theirs as required by the Contract Documents.

6.2.2 If any part of the Contractor's Work depends for proper execution or results upon the work of the Owner or any separate contractor, the Contractor shall, prior to proceeding with the Work, promptly report to the Architect any apparent discrepancies or defects in such other work that render it unsuitable for such proper execution and results. Failure of the Contractor so to report shall constitute an acceptance of the Owner's or separate contractors' work as fit and proper to receive his Work, except as to defects which may subsequently become apparent in such work by others.

6.2.3 Any costs caused by defective or ill-timed work shall be borne by the party responsible therefor.

6.2.4 Should the Contractor wrongfully cause damage to the work or property of the Owner, or to other work on the site, the Contractor shall promptly remedy such damage as provided in Subparagraph 10.2.5.

6.2.5 Should the Contractor wrongfully cause damage to the work or property of any separate contractor, the Contractor shall upon due notice promptly attempt to settle with such other contractor by agreement, or otherwise to resolve the dispute. If such separate contractor sues or initiates an arbitration proceeding against the Owner on account of any damage alleged to have been caused by the Contractor, the Owner shall notify the Contractor who shall defend such proceedings at the Owner's expense, and if any judgment or award against the Owner arises therefrom the Contractor shall pay or satisfy it and shall reimburse the Owner for all attorneys' fees and court or arbitration costs which the Owner has incurred.

6.3   OWNER'S RIGHT TO CLEAN UP

6.3.1 If a dispute arises between the Contractor and separate contractors as to their responsibility for cleaning up as required by Paragraph 4.15, the Owner may clean up

and charge the cost thereof to the contractors responsible therefor as the Architect shall determine to be just.

ARTICLE 7

MISCELLANEOUS PROVISIONS

7.1   GOVERNING LAW

7.1.1 The Contract shall be governed by the law of the place where the Project is located.

7.2   SUCCESSORS AND ASSIGNS

7.2.1 The Owner and the Contractor each binds himself, his partners, successors, assigns and legal representatives to the other party hereto and to the partners, successors, assigns and legal representatives of such other party in respect to all covenants, agreements and obligations contained in the Contract Documents. Neither party to the Contract shall assign the Contract or sublet it as a whole without the written consent of the other, nor shall the Contractor assign any moneys due or to become due to him hereunder, without the previous written consent of the Owner.

7.3   WRITTEN NOTICE

7.3.1 Written notice shall be deemed to have been duly served if delivered in person to the individual or member of the firm or entity or to an officer of the corporation for whom it was intended, or if delivered at or sent by registered or certified mail to the last business address known to him who gives the notice.

7.4   CLAIMS FOR DAMAGES

7.4.1 Should either party to the Contract suffer injury or damage to person or property because of any act or omission of the other party or of any of his employees, agents or others for whose acts he is legally liable, claim shall be made in writing to such other party within a reasonable time after the first observance of such injury or damage.

7.5   PERFORMANCE BOND AND LABOR AND
**   MATERIAL PAYMENT BOND**

7.5.1 The Owner shall have the right to require the Contractor to furnish bonds covering the faithful performance of the Contract and the payment of all obligations arising thereunder if and as required in the Bidding Documents or in the Contract Documents.

7.6   RIGHTS AND REMEDIES

7.6.1 The duties and obligations imposed by the Contract Documents and the rights and remedies available thereunder shall be in addition to and not a limitation of any duties, obligations, rights and remedies otherwise imposed or available by law.

7.6.2 No action or failure to act by the Owner, Architect or Contractor shall constitute a waiver of any right or duty afforded any of them under the Contract, nor shall any such action or failure to act constitute an approval of or acquiescence in any breach thereunder, except as may be specifically agreed in writing.

AIA DOCUMENT A201 • GENERAL CONDITIONS OF THE CONTRACT FOR CONSTRUCTION • THIRTEENTH EDITION • AUGUST 1976
AIA® • © 1976 • THE AMERICAN INSTITUTE OF ARCHITECTS, 1735 NEW YORK AVENUE, N.W., WASHINGTON, D.C. 20006   **A201-1976   11**

7.7   TESTS

7.7.1 If the Contract Documents, laws, ordinances, rules, regulations or orders of any public authority having jurisdiction require any portion of the Work to be inspected, tested or approved, the Contractor shall give the Architect timely notice of its readiness so the Architect may observe such inspection, testing or approval. The Contractor shall bear all costs of such inspections, tests or approvals conducted by public authorities. Unless otherwise provided, the Owner shall bear all costs of other inspections, tests or approvals.

7.7.2 If the Architect determines that any Work requires special inspection, testing, or approval which Subparagraph 7.7.1 does not include, he will, upon written authorization from the Owner, instruct the Contractor to order such special inspection, testing or approval, and the Contractor shall give notice as provided in Subparagraph 7.7.1. If such special inspection or testing reveals a failure of the Work to comply with the requirements of the Contract Documents, the Contractor shall bear all costs thereof, including compensation for the Architect's additional services made necessary by such failure; otherwise the Owner shall bear such costs, and an appropriate Change Order shall be issued.

7.7.3 Required certificates of inspection, testing or approval shall be secured by the Contractor and promptly delivered by him to the Architect.

7.7.4 If the Architect is to observe the inspections, tests or approvals required by the Contract Documents, he will do so promptly and, where practicable, at the source of supply.

7.8   INTEREST

7.8.1 Payments due and unpaid under the Contract Documents shall bear interest from the date payment is due at such rate as the parties may agree upon in writing or, in the absence thereof, at the legal rate prevailing at the place of the Project.

7.9   ARBITRATION

7.9.1 All claims, disputes and other matters in question between the Contractor and the Owner arising out of, or relating to, the Contract Documents or the breach thereof, except as provided in Subparagraph 2.2.11 with respect to the Architect's decisions on matters relating to artistic effect, and except for claims which have been waived by the making or acceptance of final payment as provided by Subparagraphs 9.9.4 and 9.9.5, shall be decided by arbitration in accordance with the Construction Industry Arbitration Rules of the American Arbitration Association then obtaining unless the parties mutually agree otherwise. No arbitration arising out of or relating to the Contract Documents shall include, by consolidation, joinder or in any other manner, the Architect, his employees or consultants except by written consent containing a specific reference to the Owner-Contractor Agreement and signed by the Architect, the Owner, the Contractor and any other person sought to be joined. No arbitration shall include by consolidation, joinder or in any other manner, parties other than the Owner, the Contractor and any other persons substantially involved in a common question of fact or law, whose presence is required if complete relief is to be accorded in the arbitration. No person other than the Owner or Contractor shall be included as an original third party or additional third party to an arbitration whose interest or responsibility is insubstantial. Any consent to arbitration involving an additional person or persons shall not constitute consent to arbitration of any dispute not described therein or with any person not named or described therein. The foregoing agreement to arbitrate and any other agreement to arbitrate with an additional person or persons duly consented to by the parties to the Owner-Contractor Agreement shall be specifically enforceable under the prevailing arbitration law. The award rendered by the arbitrators shall be final, and judgment may be entered upon it in accordance with applicable law in any court having jurisdiction thereof.

7.9.2 Notice of the demand for arbitration shall be filed in writing with the other party to the Owner-Contractor Agreement and with the American Arbitration Association, and a copy shall be filed with the Architect. The demand for arbitration shall be made within the time limits specified in Subparagraph 2.2.12 where applicable, and in all other cases within a reasonable time after the claim, dispute or other matter in question has arisen, and in no event shall it be made after the date when institution of legal or equitable proceedings based on such claim, dispute or other matter in question would be barred by the applicable statute of limitations.

7.9.3 Unless otherwise agreed in writing, the Contractor shall carry on the Work and maintain its progress during any arbitration proceedings, and the Owner shall continue to make payments to the Contractor in accordance with the Contract Documents.

ARTICLE 8

TIME

8.1   DEFINITIONS

8.1.1 Unless otherwise provided, the Contract Time is the period of time allotted in the Contract Documents for Substantial Completion of the Work as defined in Subparagraph 8.1.3, including authorized adjustments thereto.

8.1.2 The date of commencement of the Work is the date established in a notice to proceed. If there is no notice to proceed, it shall be the date of the Owner-Contractor Agreement or such other date as may be established therein.

8.1.3 The Date of Substantial Completion of the Work or designated portion thereof is the Date certified by the Architect when construction is sufficiently complete, in accordance with the Contract Documents, so the Owner can occupy or utilize the Work or designated portion thereof for the use for which it is intended.

8.1.4 The term day as used in the Contract Documents shall mean calendar day unless otherwise specifically designated.

8.2   PROGRESS AND COMPLETION

8.2.1 All time limits stated in the Contract Documents are of the essence of the Contract.

8.2.2 The Contractor shall begin the Work on the date of commencement as defined in Subparagraph 8.1.2. He shall carry the Work forward expeditiously with adequate forces and shall achieve Substantial Completion within the Contract Time.

8.3 DELAYS AND EXTENSIONS OF TIME

8.3.1 If the Contractor is delayed at any time in the progress of the Work by any act or neglect of the Owner or the Architect, or by any employee of either, or by any separate contractor employed by the Owner, or by changes ordered in the Work, or by labor disputes, fire, unusual delay in transportation, adverse weather conditions not reasonably anticipatable, unavoidable casualties, or any causes beyond the Contractor's control, or by delay authorized by the Owner pending arbitration, or by any other cause which the Architect determines may justify the delay, then the Contract Time shall be extended by Change Order for such reasonable time as the Architect may determine.

8.3.2 Any claim for extension of time shall be made in writing to the Architect not more than twenty days after the commencement of the delay; otherwise it shall be waived. In the case of a continuing delay only one claim is necessary. The Contractor shall provide an estimate of the probable effect of such delay on the progress of the Work.

8.3.3 If no agreement is made stating the dates upon which interpretations as provided in Subparagraph 2.2.8 shall be furnished, then no claim for delay shall be allowed on account of failure to furnish such interpretations until fifteen days after written request is made for them, and not then unless such claim is reasonable.

8.3.4 This Paragraph 8.3 does not exclude the recovery of damages for delay by either party under other provisions of the Contract Documents.

ARTICLE 9

PAYMENTS AND COMPLETION

9.1 CONTRACT SUM

9.1.1 The Contract Sum is stated in the Owner-Contractor Agreement and, including authorized adjustments thereto, is the total amount payable by the Owner to the Contractor for the performance of the Work under the Contract Documents.

9.2 SCHEDULE OF VALUES

9.2.1 Before the first Application for Payment, the Contractor shall submit to the Architect a schedule of values allocated to the various portions of the Work, prepared in such form and supported by such data to substantiate its accuracy as the Architect may require. This schedule, unless objected to by the Architect, shall be used only as a basis for the Contractor's Applications for Payment.

9.3 APPLICATIONS FOR PAYMENT

9.3.1 At least ten days before the date for each progress payment established in the Owner-Contractor Agreement, the Contractor shall submit to the Architect an itemized Application for Payment, notarized if required, supported by such data substantiating the Contractor's right to payment as the Owner or the Architect may require, and reflecting retainage, if any, as provided elsewhere in the Contract Documents.

9.3.2 Unless otherwise provided in the Contract Documents, payments will be made on account of materials or equipment not incorporated in the Work but delivered and suitably stored at the site and, if approved in advance by the Owner, payments may similarly be made for materials or equipment suitably stored at some other location agreed upon in writing. Payments for materials or equipment stored on or off the site shall be conditioned upon submission by the Contractor of bills of sale or such other procedures satisfactory to the Owner to establish the Owner's title to such materials or equipment or otherwise protect the Owner's interest, including applicable insurance and transportation to the site for those materials and equipment stored off the site.

9.3.3 The Contractor warrants that title to all Work, materials and equipment covered by an Application for Payment will pass to the Owner either by incorporation in the construction or upon the receipt of payment by the Contractor, whichever occurs first, free and clear of all liens, claims, security interests or encumbrances, hereinafter referred to in this Article 9 as "liens"; and that no Work, materials or equipment covered by an Application for Payment will have been acquired by the Contractor, or by any other person performing Work at the site or furnishing materials and equipment for the Project, subject to an agreement under which an interest therein or an encumbrance thereon is retained by the seller or otherwise imposed by the Contractor or such other person.

9.4 CERTIFICATES FOR PAYMENT

9.4.1 The Architect will, within seven days after the receipt of the Contractor's Application for Payment, either issue a Certificate for Payment to the Owner, with a copy to the Contractor, for such amount as the Architect determines is properly due, or notify the Contractor in writing his reasons for withholding a Certificate as provided in Subparagraph 9.6.1.

9.4.2 The issuance of a Certificate for Payment will constitute a representation by the Architect to the Owner, based on his observations at the site as provided in Subparagraph 2.2.3 and the data comprising the Application for Payment, that the Work has progressed to the point indicated; that, to the best of his knowledge, information and belief, the quality of the Work is in accordance with the Contract Documents (subject to an evaluation of the Work for conformance with the Contract Documents upon Substantial Completion, to the results of any subsequent tests required by or performed under the Contract Documents, to minor deviations from the Contract Documents correctable prior to completion, and to any specific qualifications stated in his Certificate); and that the Contractor is entitled to payment in the amount certified. However, by issuing a Certificate for Payment, the Architect shall not thereby be deemed to represent that he has made exhaustive or continuous on-site inspections to check the quality or quantity of the Work or that he has reviewed the construction means, methods, techniques, sequences or procedures, or that he has made any examination to ascertain how or for what purpose the Contractor has used the moneys previously paid on account of the Contract Sum.

9.5 PROGRESS PAYMENTS

9.5.1 After the Architect has issued a Certificate for Payment, the Owner shall make payment in the manner and within the time provided in the Contract Documents.

9.5.2 The Contractor shall promptly pay each Subcontractor, upon receipt of payment from the Owner, out of the amount paid to the Contractor on account of such Subcontractor's Work, the amount to which said Subcontractor is entitled, reflecting the percentage actually retained, if any, from payments to the Contractor on account of such Subcontractor's Work. The Contractor shall, by an appropriate agreement with each Subcontractor, require each Subcontractor to make payments to his Subsubcontractors in similar manner.

9.5.3 The Architect may, on request and at his discretion, furnish to any Subcontractor, if practicable, information regarding the percentages of completion or the amounts applied for by the Contractor and the action taken thereon by the Architect on account of Work done by such Subcontractor.

9.5.4 Neither the Owner nor the Architect shall have any obligation to pay or to see to the payment of any moneys to any Subcontractor except as may otherwise be required by law.

9.5.5 No Certificate for a progress payment, nor any progress payment, nor any partial or entire use or occupancy of the Project by the Owner, shall constitute an acceptance of any Work not in accordance with the Contract Documents.

9.6 PAYMENTS WITHHELD

9.6.1 The Architect may decline to certify payment and may withhold his Certificate in whole or in part, to the extent necessary reasonably to protect the Owner, if in his opinion he is unable to make representations to the Owner as provided in Subparagraph 9.4.2. If the Architect is unable to make representations to the Owner as provided in Subparagraph 9.4.2 and to certify payment in the amount of the Application, he will notify the Contractor as provided in Subparagraph 9.4.1. If the Contractor and the Architect cannot agree on a revised amount, the Architect will promptly issue a Certificate for Payment for the amount for which he is able to make such representations to the Owner. The Architect may also decline to certify payment or, because of subsequently discovered evidence or subsequent observations, he may nullify the whole or any part of any Certificate for Payment previously issued, to such extent as may be necessary in his opinion to protect the Owner from loss because of:

.1 defective Work not remedied,

.2 third party claims filed or reasonable evidence indicating probable filing of such claims,

.3 failure of the Contractor to make payments properly to Subcontractors or for labor, materials or equipment,

.4 reasonable evidence that the Work cannot be completed for the unpaid balance of the Contract Sum,

.5 damage to the Owner or another contractor,

.6 reasonable evidence that the Work will not be completed within the Contract Time, or

.7 persistent failure to carry out the Work in accordance with the Contract Documents.

9.6.2 When the above grounds in Subparagraph 9.6.1 are removed, payment shall be made for amounts withheld because of them.

9.7 FAILURE OF PAYMENT

9.7.1 If the Architect does not issue a Certificate for Payment, through no fault of the Contractor, within seven days after receipt of the Contractor's Application for Payment, or if the Owner does not pay the Contractor within seven days after the date established in the Contract Documents any amount certified by the Architect or awarded by arbitration, then the Contractor may, upon seven additional days' written notice to the Owner and the Architect, stop the Work until payment of the amount owing has been received. The Contract Sum shall be increased by the amount of the Contractor's reasonable costs of shut-down, delay and start-up, which shall be effected by appropriate Change Order in accordance with Paragraph 12.3.

9.8 SUBSTANTIAL COMPLETION

9.8.1 When the Contractor considers that the Work, or a designated portion thereof which is acceptable to the Owner, is substantially complete as defined in Subparagraph 8.1.3, the Contractor shall prepare for submission to the Architect a list of items to be completed or corrected. The failure to include any items on such list does not alter the responsibility of the Contractor to complete all Work in accordance with the Contract Documents. When the Architect on the basis of an inspection determines that the Work or designated portion thereof is substantially complete, he will then prepare a Certificate of Substantial Completion which shall establish the Date of Substantial Completion, shall state the responsibilities of the Owner and the Contractor for security, maintenance, heat, utilities, damage to the Work, and insurance, and shall fix the time within which the Contractor shall complete the items listed therein. Warranties required by the Contract Documents shall commence on the Date of Substantial Completion of the Work or designated portion thereof unless otherwise provided in the Certificate of Substantial Completion. The Certificate of Substantial Completion shall be submitted to the Owner and the Contractor for their written acceptance of the responsibilities assigned to them in such Certificate.

9.8.2 Upon Substantial Completion of the Work or designated portion thereof and upon application by the Contractor and certification by the Architect, the Owner shall make payment, reflecting adjustment in retainage, if any, for such Work or portion thereof, as provided in the Contract Documents.

9.9 FINAL COMPLETION AND FINAL PAYMENT

9.9.1 Upon receipt of written notice that the Work is ready for final inspection and acceptance and upon receipt of a final Application for Payment, the Architect will promptly make such inspection and, when he finds the Work acceptable under the Contract Documents and the Contract fully performed, he will promptly issue a final Certificate for Payment stating that to the best of his knowledge, information and belief, and on the basis of his observations and inspections, the Work has been completed in accordance with the terms and conditions of the Contract Documents and that the entire balance found to be due the Contractor, and noted in said final Certificate, is due and payable. The Architect's final Certificate for Payment will constitute a further representation that the conditions precedent to the Contractor's being entitled to final payment as set forth in Subparagraph 9.9.2 have been fulfilled.

9.9.2 Neither the final payment nor the remaining retained percentage shall become due until the Contractor submits to the Architect (1) an affidavit that all payrolls, bills for materials and equipment, and other indebtedness connected with the Work for which the Owner or his property might in any way be responsible, have been paid or otherwise satisfied, (2) consent of surety, if any, to final payment and (3), if required by the Owner, other data establishing payment or satisfaction of all such obligations, such as receipts, releases and waivers of liens arising out of the Contract, to the extent and in such form as may be designated by the Owner. If any Subcontractor refuses to furnish a release or waiver required by the Owner, the Contractor may furnish a bond satisfactory to the Owner to indemnify him against any such lien. If any such lien remains unsatisfied after all payments are made, the Contractor shall refund to the Owner all moneys that the latter may be compelled to pay in discharging such lien, including all costs and reasonable attorneys' fees.

9.9.3 If, after Substantial Completion of the Work, final completion thereof is materially delayed through no fault of the Contractor or by the issuance of Change Orders affecting final completion, and the Architect so confirms, the Owner shall, upon application by the Contractor and certification by the Architect, and without terminating the Contract, make payment of the balance due for that portion of the Work fully completed and accepted. If the remaining balance for Work not fully completed or corrected is less than the retainage stipulated in the Contract Documents, and if bonds have been furnished as provided in Paragraph 7.5, the written consent of the surety to the payment of the balance due for that portion of the Work fully completed and accepted shall be submitted by the Contractor to the Architect prior to certification of such payment. Such payment shall be made under the terms and conditions governing final payment, except that it shall not constitute a waiver of claims.

9.9.4 The making of final payment shall constitute a waiver of all claims by the Owner except those arising from:

.1 unsettled liens,
.2 faulty or defective Work appearing after Substantial Completion,
.3 failure of the Work to comply with the requirements of the Contract Documents, or
.4 terms of any special warranties required by the Contract Documents.

9.9.5 The acceptance of final payment shall constitute a waiver of all claims by the Contractor except those previously made in writing and identified by the Contractor as unsettled at the time of the final Application for Payment.

ARTICLE 10

PROTECTION OF PERSONS AND PROPERTY

10.1 SAFETY PRECAUTIONS AND PROGRAMS

10.1.1 The Contractor shall be responsible for initiating, maintaining and supervising all safety precautions and programs in connection with the Work.

10.2 SAFETY OF PERSONS AND PROPERTY

10.2.1 The Contractor shall take all reasonable precautions for the safety of, and shall provide all reasonable protection to prevent damage, injury or loss to:

.1 all employees on the Work and all other persons who may be affected thereby;
.2 all the Work and all materials and equipment to be incorporated therein, whether in storage on or off the site, under the care, custody or control of the Contractor or any of his Subcontractors or Sub-subcontractors; and
.3 other property at the site or adjacent thereto, including trees, shrubs, lawns, walks, pavements, roadways, structures and utilities not designated for removal, relocation or replacement in the course of construction.

10.2.2 The Contractor shall give all notices and comply with all applicable laws, ordinances, rules, regulations and lawful orders of any public authority bearing on the safety of persons or property or their protection from damage, injury or loss.

10.2.3 The Contractor shall erect and maintain, as required by existing conditions and progress of the Work, all reasonable safeguards for safety and protection, including posting danger signs and other warnings against hazards, promulgating safety regulations and notifying owners and users of adjacent utilities.

10.2.4 When the use or storage of explosives or other hazardous materials or equipment is necessary for the execution of the Work, the Contractor shall exercise the utmost care and shall carry on such activities under the supervision of properly qualified personnel.

10.2.5 The Contractor shall promptly remedy all damage or loss (other than damage or loss insured under Paragraph 11.3) to any property referred to in Clauses 10.2.1.2 and 10.2.1.3 caused in whole or in part by the Contractor, any Sub-contractor, any Sub-subcontractor, or anyone directly or indirectly employed by any of them, or by anyone for whose acts any of them may be liable and for which the Contractor is responsible under Clauses 10.2.1.2 and 10.2.1.3, except damage or loss attributable to the acts or omissions of the Owner or Architect or anyone directly or indirectly employed by either of them, or by anyone for whose acts either of them may be liable, and not attributable to the fault or negligence of the Contractor. The foregoing obligations of the Contractor are in addition to his obligations under Paragraph 4.18.

10.2.6 The Contractor shall designate a responsible member of his organization at the site whose duty shall be the prevention of accidents. This person shall be the Contractor's superintendent unless otherwise designated by the Contractor in writing to the Owner and the Architect.

10.2.7 The Contractor shall not load or permit any part of the Work to be loaded so as to endanger its safety.

10.3   EMERGENCIES

10.3.1 In any emergency affecting the safety of persons or property, the Contractor shall act, at his discretion, to prevent threatened damage, injury or loss. Any additional compensation or extension of time claimed by the Contractor on account of emergency work shall be determined as provided in Article 12 for Changes in the Work.

ARTICLE 11

INSURANCE

11.1   CONTRACTOR'S LIABILITY INSURANCE

11.1.1 The Contractor shall purchase and maintain such insurance as will protect him from claims set forth below which may arise out of or result from the Contractor's operations under the Contract, whether such operations be by himself or by any Subcontractor or by anyone directly or indirectly employed by any of them, or by anyone for whose acts any of them may be liable:

   .1 claims under workers' or workmen's compensation, disability benefit and other similar employee benefit acts;

   .2 claims for damages because of bodily injury, occupational sickness or disease, or death of his employees;

   .3 claims for damages because of bodily injury, sickness or disease, or death of any person other than his employees;

   .4 claims for damages insured by usual personal injury liability coverage which are sustained (1) by any person as a result of an offense directly or indirectly related to the employment of such person by the Contractor, or (2) by any other person;

   .5 claims for damages, other than to the Work itself, because of injury to or destruction of tangible property, including loss of use resulting therefrom; and

   .6 claims for damages because of bodily injury or death of any person or property damage arising out of the ownership, maintenance or use of any motor vehicle.

11.1.2 The insurance required by Subparagraph 11.1.1 shall be written for not less than any limits of liability specified in the Contract Documents, or required by law, whichever is greater.

11.1.3 The insurance required by Subparagraph 11.1.1 shall include contractual liability insurance applicable to the Contractor's obligations under Paragraph 4.18.

11.1.4 Certificates of Insurance acceptable to the Owner shall be filed with the Owner prior to commencement of the Work. These Certificates shall contain a provision that coverages afforded under the policies will not be cancelled until at least thirty days' prior written notice has been given to the Owner.

11.2   OWNER'S LIABILITY INSURANCE

11.2.1 The Owner shall be responsible for purchasing and maintaining his own liability insurance and, at his option, may purchase and maintain such insurance as will protect him against claims which may arise from operations under the Contract.

11.3   PROPERTY INSURANCE

11.3.1 Unless otherwise provided, the Owner shall purchase and maintain property insurance upon the entire Work at the site to the full insurable value thereof. This insurance shall include the interests of the Owner, the Contractor, Subcontractors and Sub-subcontractors in the Work and shall insure against the perils of fire and extended coverage and shall include "all risk" insurance for physical loss or damage including, without duplication of coverage, theft, vandalism and malicious mischief. If the Owner does not intend to purchase such insurance for the full insurable value of the entire Work, he shall inform the Contractor in writing prior to commencement of the Work. The Contractor may then effect insurance which will protect the interests of himself, his Subcontractors and the Sub-subcontractors in the Work, and by appropriate Change Order the cost thereof shall be charged to the Owner. If the Contractor is damaged by failure of the Owner to purchase or maintain such insurance and to so notify the Contractor, then the Owner shall bear all reasonable costs properly attributable thereto. If not covered under the all risk insurance or otherwise provided in the Contract Documents, the Contractor shall effect and maintain similar property insurance on portions of the Work stored off the site or in transit when such portions of the Work are to be included in an Application for Payment under Subparagraph 9.3.2.

11.3.2 The Owner shall purchase and maintain such boiler and machinery insurance as may be required by the Contract Documents or by law. This insurance shall include the interests of the Owner, the Contractor, Subcontractors and Sub-subcontractors in the Work.

11.3.3 Any loss insured under Subparagraph 11.3.1 is to be adjusted with the Owner and made payable to the Owner as trustee for the insureds, as their interests may appear, subject to the requirements of any applicable mortgagee clause and of Subparagraph 11.3.8. The Contractor shall pay each Subcontractor a just share of any insurance moneys received by the Contractor, and by appropriate agreement, written where legally required for validity, shall require each Subcontractor to make payments to his Sub-subcontractors in similar manner.

11.3.4 The Owner shall file a copy of all policies with the Contractor before an exposure to loss may occur.

11.3.5 If the Contractor requests in writing that insurance for risks other than those described in Subparagraphs 11.3.1 and 11.3.2 or other special hazards be included in the property insurance policy, the Owner shall, if possible, include such insurance, and the cost thereof shall be charged to the Contractor by appropriate Change Order.

11.3.6 The Owner and Contractor waive all rights against (1) each other and the Subcontractors, Sub-subcontractors, agents and employees each of the other, and (2) the Architect and separate contractors, if any, and their subcontractors, sub-subcontractors, agents and employees, for damages caused by fire or other perils to the extent covered by insurance obtained pursuant to this Paragraph 11.3 or any other property insurance applicable to the Work, except such rights as they may have to the proceeds of such insurance held by the Owner as trustee. The foregoing waiver afforded the Architect, his agents and employees shall not extend to the liability imposed by Subparagraph 4.18.3. The Owner or the Contractor, as appropriate, shall require of the Architect, separate contractors, Subcontractors and Sub-subcontractors by appropriate agreements, written where legally required for validity, similar waivers each in favor of all other parties enumerated in this Subparagraph 11.3.6.

11.3.7 If required in writing by any party in interest, the Owner as trustee shall, upon the occurrence of an insured loss, give bond for the proper performance of his duties. He shall deposit in a separate account any money so received, and he shall distribute it in accordance with such agreement as the parties in interest may reach, or in accordance with an award by arbitration in which case the procedure shall be as provided in Paragraph 7.9. If after such loss no other special agreement is made, replacement of damaged work shall be covered by an appropriate Change Order.

11.3.8 The Owner as trustee shall have power to adjust and settle any loss with the insurers unless one of the parties in interest shall object in writing within five days after the occurrence of loss to the Owner's exercise of this power, and if such objection be made, arbitrators shall be chosen as provided in Paragraph 7.9. The Owner as trustee shall, in that case, make settlement with the insurers in accordance with the directions of such arbitrators. If distribution of the insurance proceeds by arbitration is required, the arbitrators will direct such distribution.

11.3.9 If the Owner finds it necessary to occupy or use a portion or portions of the Work prior to Substantial Completion thereof, such occupancy or use shall not commence prior to a time mutually agreed to by the Owner and Contractor and to which the insurance company or companies providing the property insurance have consented by endorsement to the policy or policies. This insurance shall not be cancelled or lapsed on account of such partial occupancy or use. Consent of the Contractor and of the insurance company or companies to such occupancy or use shall not be unreasonably withheld.

11.4   LOSS OF USE INSURANCE

11.4.1 The Owner, at his option, may purchase and maintain such insurance as will insure him against loss of use of his property due to fire or other hazards, however caused. The Owner waives all rights of action against the Contractor for loss of use of his property, including consequential losses due to fire or other hazards however caused, to the extent covered by insurance under this Paragraph 11.4.

ARTICLE 12

CHANGES IN THE WORK

12.1   CHANGE ORDERS

12.1.1 A Change Order is a written order to the Contractor signed by the Owner and the Architect, issued after execution of the Contract, authorizing a change in the Work or an adjustment in the Contract Sum or the Contract Time. The Contract Sum and the Contract Time may be changed only by Change Order. A Change Order signed by the Contractor indicates his agreement therewith, including the adjustment in the Contract Sum or the Contract Time.

12.1.2 The Owner, without invalidating the Contract, may order changes in the Work within the general scope of the Contract consisting of additions, deletions or other revisions, the Contract Sum and the Contract Time being adjusted accordingly. All such changes in the Work shall be authorized by Change Order, and shall be performed under the applicable conditions of the Contract Documents.

12.1.3 The cost or credit to the Owner resulting from a change in the Work shall be determined in one or more of the following ways:

.1 by mutual acceptance of a lump sum properly itemized and supported by sufficient substantiating data to permit evaluation;

.2 by unit prices stated in the Contract Documents or subsequently agreed upon;

.3 by cost to be determined in a manner agreed upon by the parties and a mutually acceptable fixed or percentage fee; or

.4 by the method provided in Subparagraph 12.1.4.

12.1.4 If none of the methods set forth in Clauses 12.1.3.1, 12.1.3.2 or 12.1.3.3 is agreed upon, the Contractor, provided he receives a written order signed by the Owner, shall promptly proceed with the Work involved. The cost of such Work shall then be determined by the Architect on the basis of the reasonable expenditures and savings of those performing the Work attributable to the change, including, in the case of an increase in the Contract Sum, a reasonable allowance for overhead and profit. In such case, and also under Clauses 12.1.3.3 and 12.1.3.4 above, the Contractor shall keep and present, in such form as the Architect may prescribe, an itemized accounting together with appropriate supporting data for inclusion in a Change Order. Unless otherwise provided in the Contract Documents, cost shall be limited to the following: cost of materials, including sales tax and cost of delivery; cost of labor, including social security, old age and unemployment insurance, and fringe benefits required by agreement or custom; workers' or workmen's compensation insurance; bond premiums; rental value of equipment and machinery; and the additional costs of supervision and field office personnel directly attributable to the change. Pending final determination of cost to the Owner, payments on account shall be made on the Architect's Certificate for Payment. The amount of credit to be allowed by the Contractor to the Owner for any deletion or change which results in a net decrease in the Contract Sum will be the amount of the actual net cost as confirmed by the Architect. When both additions and credits covering related Work or substitutions are involved in any one change, the allowance for overhead and profit shall be figured on the basis of the net increase, if any, with respect to that change.

12.1.5 If unit prices are stated in the Contract Documents or subsequently agreed upon, and if the quantities originally contemplated are so changed in a proposed Change Order that application of the agreed unit prices to the quantities of Work proposed will cause substantial inequity to the Owner or the Contractor, the applicable unit prices shall be equitably adjusted.

12.2   CONCEALED CONDITIONS

12.2.1 Should concealed conditions encountered in the performance of the Work below the surface of the ground or should concealed or unknown conditions in an existing structure be at variance with the conditions indicated by the Contract Documents, or should unknown physical conditions below the surface of the ground or should concealed or unknown conditions in an existing structure of an unusual nature, differing materially from those ordinarily encountered and generally recognized as inherent in work of the character provided for in this Contract, be encountered, the Contract Sum shall be equitably adjusted by Change Order upon claim by either party made within twenty days after the first observance of the conditions.

12.3   CLAIMS FOR ADDITIONAL COST

12.3.1 If the Contractor wishes to make a claim for an increase in the Contract Sum, he shall give the Architect written notice thereof within twenty days after the occurrence of the event giving rise to such claim. This notice shall be given by the Contractor before proceeding to execute the Work, except in an emergency endangering life or property in which case the Contractor shall proceed in accordance with Paragraph 10.3. No such claim shall be valid unless so made. If the Owner and the Contractor cannot agree on the amount of the adjustment in the Contract Sum, it shall be determined by the Architect. Any change in the Contract Sum resulting from such claim shall be authorized by Change Order.

12.3.2 If the Contractor claims that additional cost is involved because of, but not limited to, (1) any written interpretation pursuant to Subparagraph 2.2.8, (2) any order by the Owner to stop the Work pursuant to Paragraph 3.3 where the Contractor was not at fault, (3) any written order for a minor change in the Work issued pursuant to Paragraph 12.4, or (4) failure of payment by the Owner pursuant to Paragraph 9.7, the Contractor shall make such claim as provided in Subparagraph 12.3.1.

12.4   MINOR CHANGES IN THE WORK

12.4.1 The Architect will have authority to order minor changes in the Work not involving an adjustment in the Contract Sum or an extension of the Contract Time and not inconsistent with the intent of the Contract Documents. Such changes shall be effected by written order, and shall be binding on the Owner and the Contractor.

The Contractor shall carry out such written orders promptly.

ARTICLE 13

UNCOVERING AND CORRECTION OF WORK

13.1   UNCOVERING OF WORK

13.1.1 If any portion of the Work should be covered contrary to the request of the Architect or to requirements specifically expressed in the Contract Documents, it must, if required in writing by the Architect, be uncovered for his observation and shall be replaced at the Contractor's expense.

13.1.2 If any other portion of the Work has been covered which the Architect has not specifically requested to observe prior to being covered, the Architect may request to see such Work and it shall be uncovered by the Contractor. If such Work be found in accordance with the Contract Documents, the cost of uncovering and replacement shall, by appropriate Change Order, be charged to the Owner. If such Work be found not in accordance with the Contract Documents, the Contractor shall pay such costs unless it be found that this condition was caused by the Owner or a separate contractor as provided in Article 6, in which event the Owner shall be responsible for the payment of such costs.

13.2   CORRECTION OF WORK

13.2.1 The Contractor shall promptly correct all Work rejected by the Architect as defective or as failing to conform to the Contract Documents whether observed before or after Substantial Completion and whether or not fabricated, installed or completed. The Contractor shall bear all costs of correcting such rejected Work, including compensation for the Architect's additional services made necessary thereby.

13.2.2 If, within one year after the Date of Substantial Completion of the Work or designated portion thereof or within one year after acceptance by the Owner of designated equipment or within such longer period of time as may be prescribed by law or by the terms of any applicable special warranty required by the Contract Documents, any of the Work is found to be defective or not in accordance with the Contract Documents, the Contractor shall correct it promptly after receipt of a written notice from the Owner to do so unless the Owner has previously given the Contractor a written acceptance of such condition. This obligation shall survive termination of the Contract. The Owner shall give such notice promptly after discovery of the condition.

13.2.3 The Contractor shall remove from the site all portions of the Work which are defective or non-conforming and which have not been corrected under Subparagraphs 4.5.1, 13.2.1 and 13.2.2, unless removal is waived by the Owner.

13.2.4 If the Contractor fails to correct defective or non-conforming Work as provided in Subparagraphs 4.5.1, 13.2.1 and 13.2.2, the Owner may correct it in accordance with Paragraph 3.4.

AIA DOCUMENT A201 • GENERAL CONDITIONS OF THE CONTRACT FOR CONSTRUCTION • THIRTEENTH EDITION • AUGUST 1976
18   A201-1976   AIA® • © 1976 • THE AMERICAN INSTITUTE OF ARCHITECTS, 1735 NEW YORK AVENUE, N.W., WASHINGTON, D.C. 20006

13.2.5 If the Contractor does not proceed with the correction of such defective or non-conforming Work within a reasonable time fixed by written notice from the Architect, the Owner may remove it and may store the materials or equipment at the expense of the Contractor. If the Contractor does not pay the cost of such removal and storage within ten days thereafter, the Owner may upon ten additional days' written notice sell such Work at auction or at private sale and shall account for the net proceeds thereof, after deducting all the costs that should have been borne by the Contractor, including compensation for the Architect's additional services made necessary thereby. If such proceeds of sale do not cover all costs which the Contractor should have borne, the difference shall be charged to the Contractor and an appropriate Change Order shall be issued. If the payments then or thereafter due the Contractor are not sufficient to cover such amount, the Contractor shall pay the difference to the Owner.

13.2.6 The Contractor shall bear the cost of making good all work of the Owner or separate contractors destroyed or damaged by such correction or removal.

13.2.7 Nothing contained in this Paragraph 13.2 shall be construed to establish a period of limitation with respect to any other obligation which the Contractor might have under the Contract Documents, including Paragraph 4.5 hereof. The establishment of the time period of one year after the Date of Substantial Completion or such longer period of time as may be prescribed by law or by the terms of any warranty required by the Contract Documents relates only to the specific obligation of the Contractor to correct the Work, and has no relationship to the time within which his obligation to comply with the Contract Documents may be sought to be enforced, nor to the time within which proceedings may be commenced to establish the Contractor's liability with respect to his obligations other than specifically to correct the Work.

13.3 ACCEPTANCE OF DEFECTIVE OR NON-CONFORMING WORK

13.3.1 If the Owner prefers to accept defective or non-conforming Work, he may do so instead of requiring its removal and correction, in which case a Change Order will be issued to reflect a reduction in the Contract Sum where appropriate and equitable. Such adjustment shall be effected whether or not final payment has been made.

ARTICLE 14

TERMINATION OF THE CONTRACT

14.1 TERMINATION BY THE CONTRACTOR

14.1.1 If the Work is stopped for a period of thirty days under an order of any court or other public authority having jurisdiction, or as a result of an act of government, such as a declaration of a national emergency making materials unavailable, through no act or fault of the Contractor or a Subcontractor or their agents or employees or any other persons performing any of the Work under a contract with the Contractor, or if the Work should be stopped for a period of thirty days by the Contractor because the Architect has not issued a Certificate for Payment as provided in Paragraph 9.7 or because the Owner has not made payment thereon as provided in Paragraph 9.7, then the Contractor may, upon seven additional days' written notice to the Owner and the Architect, terminate the Contract and recover from the Owner payment for all Work executed and for any proven loss sustained upon any materials, equipment, tools, construction equipment and machinery, including reasonable profit and damages.

14.2 TERMINATION BY THE OWNER

14.2.1 If the Contractor is adjudged a bankrupt, or if he makes a general assignment for the benefit of his creditors, or if a receiver is appointed on account of his insolvency, or if he persistently or repeatedly refuses or fails, except in cases for which extension of time is provided, to supply enough properly skilled workmen or proper materials, or if he fails to make prompt payment to Subcontractors or for materials or labor, or persistently disregards laws, ordinances, rules, regulations or orders of any public authority having jurisdiction, or otherwise is guilty of a substantial violation of a provision of the Contract Documents, then the Owner, upon certification by the Architect that sufficient cause exists to justify such action, may, without prejudice to any right or remedy and after giving the Contractor and his surety, if any, seven days' written notice, terminate the employment of the Contractor and take possession of the site and of all materials, equipment, tools, construction equipment and machinery thereon owned by the Contractor and may finish the Work by whatever method he may deem expedient. In such case the Contractor shall not be entitled to receive any further payment until the Work is finished.

14.2.2 If the unpaid balance of the Contract Sum exceeds the costs of finishing the Work, including compensation for the Architect's additional services made necessary thereby, such excess shall be paid to the Contractor. If such costs exceed the unpaid balance, the Contractor shall pay the difference to the Owner. The amount to be paid to the Contractor or to the Owner, as the case may be, shall be certified by the Architect, upon application, in the manner provided in Paragraph 9.4, and this obligation for payment shall survive the termination of the Contract.

APPENDIX D

SUPPLEMENTARY CONDITIONS

1. LOCATION OF THE PROJECT. The general location of the work covered in this specification is Portland, Ohio.

2. SCOPE OF THE WORK. The work to be performed under this Contract consists of furnishing all plant, materials, equipment, supplies, labor, and transportation, including fuel, power, and water, and performing all work in strict accordance with specifications, schedules, and drawings, all of which are made a part hereof, and including such detail drawings as may be furnished by the Architect-Engineer from time to time during the construction.

3. EXAMINATION OF SITE. Bidders should visit the site of the building, compare the drawings and specifications with any work in place, and inform themselves of all conditions, including other work, if any, being performed. Failure to visit the site will in no way relieve the successful bidder from the necessity of furnishing any materials or performing any work that may be required to complete the work in accordance with drawings and specifications.

4. LAYING OUT WORK. The Contractor shall, immediately upon entering the project site for the purpose of beginning work, locate all Owner-provided reference points and take such action as is necessary to prevent their destruction, lay out his own work and be responsible for all lines, elevations, and measurements of buildings, grading, paving, utilities, and other work executed by him under the Contract.

5. COMMENCEMENT, PROSECUTION, AND COMPLETION.

a. The Contractor will be required to commence work under this Contract within ten (10) calendar days after the date of receipt by him of Notice to Proceed, to prosecute said work with faithfulness and energy and to complete the entire work, ready for use, within 380 calendar days after receipt of Notice to Proceed. The time stated for completion shall include final cleanup of the premises.

b. It is mutually agreed that the time for the commencement and completion of the work will materially affect the progress of other work and that the Owner will suffer financial damages in an amount not now possible to ascertain if this work is not completed on schedule, and in view of these facts, it is agreed that the Owner will withhold from the Contractor, as liquidated damages and not as a penalty, the sum of $100.00 per day for each calendar day that the work remains uncompleted beyond the date specified for the completion of the work.

c. If completion of the work to be performed under the terms of this Contract is delayed by reasons of delay in the performance of any work to be performed by the Owner, or other contractors, and which is essential to the work performed under this Contract, such delay shall not constitute a basis for any claim against the Owner, but the time of performance will be extended for a period equal to such delay or as otherwise mutually agreed upon.

6. WATCHMAN. The Contractor shall employ a responsible watchman to guard the site and premises at all times except during regular working hours, from the beginning of work until acceptance by the Owner.

7. OWNER-FURNISHED MATERIALS AND EQUIPMENT. With the following exception there will be no Owner-furnished materials and/or equipment.

a. Hardware consisting of removable cylinders for locks will be Owner-furnished and installed by the Contractor as specified herein.

8. TAXES. Except as may be otherwise provided in this Contract, the contract price is to include all applicable federal, state, and local taxes, but does not include any tax from which the Contractor is exempt. Upon request of the Contractor, the Owner shall furnish a tax exemption certificate or similar evidence of exemption with respect to any such tax not included in the contract price pursuant to this provision.

9. RATES OF WAGES.

a. There shall be paid each laborer or mechanic of the Contractor or subcontractor engaged in work on the project under this Contract in the trade or occupation listed below, not less than the hourly wage rate opposite the same, regardless of any contractual relationship which may be alleged to exist between the Contractor or any subcontractor and such laborers and mechanics.

Classification	Wage Rates per Hour
Asbestos workers	$15.30
Bricklayers	15.68
Carpenters	14.70
Cement masons	14.25
Electricians	15.75
Glaziers	13.60
Ironworkers	14.25
Lathers	14.40
Linoleum layers	13.12
Marble setters	15.70
Mosaic and terrazzo workers	14.70
Painters	13.30
Plasterers	14.85
Plumbers	15.90
Roofers	13.35
Sheet metal workers	15.10
Steamfitters	15.85
Stonemasons	15.68
Tile setters	14.70
Waterproofers	13.35
LABORERS	
Air and power tool operator	10.80
Cement mason tender	10.80
Power buggy operator	10.95
Sandblaster, potman, nozzleman	10.95
Mason tender	10.95
Pipe layer, nonmetallic, sewer and drainage	11.20
Pumpcrete nozzle placementman	10.58
Carpenter tender	10.35
Unskilled and common laborer	10.35
Concrete buggy operator	10.35
Concrete puddler	10.95
Vibrator operator	10.95

POWER EQUIPMENT OPERATORS

Air compressor, power plant, pump operator	12.70
Gunite and pumpcrete machine	13.20
Concrete mixers over 1 yd³	13.20
Concrete batching plants	13.20
Cranes, hysters, side boom tractors	12.90
Winch truck	12.90
Hoists:	
One drum	13.50
Two or more drums	14.05
Guy and stiff leg derricks	14.05
Loaders:	
Front end	12.10
Elevating belt, fork lift	12.70
Pile driver	13.20
Sheeps foot rollers	12.15
Rubber tired rollers	12.15
Shovel, backhoe, clamshell, dragline, under ¾ yd³	12.70
Shovel, backhoe, clamshell, dragline, over ¾ yd³	13.15
Loaders, bulldozers, patrol, scraper	13.35
Trenching machine	12.70

TRUCK DRIVERS

Dumpster	11.95
Dump trucks:	
Batch and under 8 yd³	11.55
8 and over yd³	11.80
Lowboy, heavy equipment	12.35
Lowboy, light equipment	11.85
Flat bed truck, under ½ ton	11.55
Pickup truck	11.50
Transit mix	12.00
Tank truck, no trailer	11.70
Tank truck, with trailer	11.80
Swamper or riding helper	10.35

Welders: receive rate prescribed for craft performing operation to which welding is incidental.

b. The foregoing specified wage rates are minimum rates only, and the Owner will not consider any claims for additional compensation made by the Contractor because of payment by the Contractor of any wage rate in excess of the applicable rate contained in this Contract. All disputes in regard to the payment of wages in excess of those specified in this Contract shall be adjusted by the Contractor.

10. TEMPORARY FACILITIES. The Contractor shall furnish materials and labor to build all temporary buildings on the project site for use during the construction of the project. All such buildings and/or utilities shall remain the property of the Contractor and shall be removed by him, at his expense, upon completion of the work under this Contract.

a. Sanitary Facilities. The Contractor shall provide and maintain ample toilet accommodations for all workmen employed on the project under this Contract. The latrines shall be weather tight, fly-proof, and shall conform to the standards established by the Owner. Toilets shall be flush-type water closets connected to the sewer, and/or chemical type. Toilet facilities shall be maintained in sanitary condition as approved by the Architect-Engineer at all times during the work on this project.

b. Temporary Enclosures. The Contractor shall provide protection against entry, rain, wind, frost, and heat, at all times, and shall maintain all materials, apparatus, equipment, and fixtures free from damage and injury.

c. Temporary Heat. The Contractor shall provide temporary heat at all times when weather conditions are such that good construction will be hampered and delayed. Uniform temperature, not lower than 50°F. shall be maintained at all times, including Saturdays, Sundays, and holidays, in that portion of the building where plastering, ceramic tile work, resilient flooring, and painting are being performed. When the permanent heating system is used to provide temporary heat requirements, all costs of operation shall be at the expense of the Contractor.

d. Water Supply. Water supply for all trades for construction uses and domestic consumption shall be provided and paid for by the Contractor. Temporary lines and connections shall be removed in a manner satisfactory to the Architect-Engineer before final acceptance of work under this Contract.

e. Electricity. Electric current required for power and light for all trades, for construction uses, and temporary lines, lamps, and equipment as required, shall be provided, connected, and maintained by the Contractor at his expense and shall be removed in like manner at the completion of construction work.

f. Telephone. The Contractor shall provide and pay for such telephone service as he may require.

11. SHOP DRAWINGS. The Contractor shall submit to the Architect-Engineer for approval five copies of all shop drawings as may be required. These drawings shall be complete and shall contain all required detailed information. If approved by the Architect-Engineer, each copy of the drawings will be identified as having received such approval by being so stamped and dated. The Contractor shall make any corrections required by the Architect-Engineer. Two sets of all shop drawings will be retained by the Architect-Engineer and three sets will be returned to the Contractor. The approval of the drawings by the Architect-Engineer shall not be construed as a complete check but will indicate only that the general method of construction and detailing is satisfactory. Approval of such drawings will not relieve the Contractor of the responsibility of any error which may exist as the Contractor shall be responsible for the dimensions and design of adequate connections, details, and satisfactory construction of all work.

12. PLAN OF OPERATIONS. The Contractor shall coordinate his work with the operations and work of the Owner and other contractors, as directed by the Architect-Engineer.

13. REPORTING OF COST INFORMATION. The Owner will need cost information regarding certain items of property which will be furnished or installed under this Contract. Accordingly, at the written request of the Architect-Engineer, the Contractor shall furnish to the Owner cost information pertaining to such items of property furnished or installed under this Contract as the Architect-Engineer may designate, in such form and in such detail as may be required.

14. FIELD OFFICE. The Contractor shall provide adequate facilities for inspection of the project, including office space of not less than 100 square feet of floor space, properly heated, ventilated, and lighted, on the site of the project, for the exclusive use of the Architect-Engineer's representative. The Contractor shall construct a built-in work table approximately 3'6" wide by 6'0" long in the field office and plan racks as directed by the Architect-Engineer.

15. PAYMENT. Partial payments under the Contract shall be made at the request of the Contractor once each month, based upon partial estimates to be furnished by the Contractor and approved by the Architect-Engineer. In making such partial payments, there shall be retained 10 percent of the estimated amounts until final completion and acceptance of all work covered by the Contract; provided, however, that the Architect-Engineer at any time after 50 percent of the work has been completed, if he finds that satisfactory progress is being made, with written consent of surety, shall recommend that the remaining partial payments be paid in full. Payments for work, under subcontracts of the Contractor, shall be subject to the above conditions applying to the Contract after the work under a subcontract has been 50 percent completed. In preparing estimates for partial payments, the material delivered on the site and preparatory work done may be taken into consideration.

16. CONSTRUCTION SCHEDULE. Immediately after execution and delivery of the Contract, and before the first partial payment is made, the Contractor shall deliver to the Architect-Engineer an estimated construction progress schedule showing the proposed dates of commencement and completion of each of the various subdivisions of the work required under the Contract, and the anticipated amount of each monthly payment that will become due the Contractor in accordance with the progress schedule. The Contractor shall also furnish on forms to be supplied by the Owner (a) a detailed estimate giving a complete breakdown of the contract price and (b) periodical itemized estimates of work done for the purpose of making partial payments thereon.

17. DRAWINGS FURNISHED. Subparagraph 3.2.5, of General Conditions, shall be changed to read as follows: "The Architect-Engineer shall furnish free of charge 25 sets of working Drawings and 25 sets of Specifications. The Contractor shall pay the cost of reproduction for all other copies of Drawings and Specifications furnished to him."

18. PERMITS. Subparagraph 4.7.1, of General Conditions, shall be modified as follows: "The Owner shall procure the required building permit at no cost to the Contractor."

19. INSURANCE.

    a. Subparagraph 11.2.1, of General Conditions, shall be modified as follows: "The Contractor shall procure, maintain, and pay for Owner's contingent liability insurance. The amount of Owner's contingent liability insurance shall be $100,000 for one person and $300,000 for one accident. A certificate of insurance shall be filed with the Architect-Engineer."

    b. Subparagraph 11.3.1, of General Conditions, shall be modified as follows: "The Contractor shall procure, maintain, and pay for fire insurance in the amount of 100 percent of the insurable value of the Project and, in addition, shall procure, maintain, and pay for insurance to protect the Owner from damage by hail, tornado, and hurricane upon the entire work in the amount of 100 percent of the insurable value thereof. Certificates of insurance shall be filed with the Architect-Engineer."

20. CLEANING UP. Subparagraph 4.15.1, of General Conditions, shall be modified as follows: "In addition to removal of rubbish and leaving the work broom-clean, the Contractor shall remove stains, spots, marks, and dirt from decorated work; clean hardware; remove paint spots and smears from all surfaces; clean light and plumbing fixtures; and wash all concrete, tile, and terrazzo floors."

APPENDIX E

BID FORM
UNIT-PRICE CONTRACT

STANDARD FORM 21 DECEMBER 1965 EDITION GENERAL SERVICES ADMINISTRATION FED. PROC. REG. (41 CFR) 1-16.401	**BID FORM** **(CONSTRUCTION CONTRACT)**	REFERENCE Serial No. Eng. 33-9-81

Read the Instructions to Bidders (Standard Form 22) *This form to be submitted in*	DATE OF INVITATION 9 July 19--

NAME AND LOCATION OF PROJECT	NAME OF BIDDER *(Type or print)*
Taxiways and Aprons Holloman Air Force Base New Mexico	The Excello Company, Inc. Albuquerque, New Mexico

9 August 19--
_____
(Date)

TO: District Engineer
     Albuquerque District
     Corps of Engineers
     Albuquerque, New Mexico 87101

In compliance with the above-dated invitation for bids, the undersigned hereby proposes to perform all work for   construction of Taxiways and Aprons
                         Holloman Air Force Base
                         New Mexico

in strict accordance with the General Provisions (Standard Form 23–A), Labor Standards Provisions Applicable to Contracts in Excess of $2,000 (Standard Form 19–A), specifications, schedules, drawings, and conditions, for the following amount(s)

Unit prices as set forth in the attached

Bidding Schedule.

21-108   *(Continue on other side)*

435

The undersigned agrees that, upon written acceptance of this bid, mailed or otherwise furnished within calendar days (60 calendar days unless a different period be inserted by the bidder) after the date of opening of bids, he will within 10 calendar days (unless a longer period is allowed) after receipt of the prescribed forms, execute Standard Form 23, Construction Contract, and give performance and payment bonds on Government standard forms with good and sufficient surety.

The undersigned agrees, if awarded the contract, to commence the work within 10 calendar days after the date of receipt of notice to proceed, and to complete the work within 420 calendar days after the date of receipt of notice to proceed.

RECEIPT OF AMENDMENTS: *The undersigned acknowledges receipt of the following amendments of the invitation for bids, drawings, and/or specifications, etc. (Give number and date of each):*

None

The representations and certifications on the accompanying STANDARD FORM 19-B are made a part of this bid.

ENCLOSED IS BID GUARANTEE, CONSISTING OF	IN THE AMOUNT OF
Bid Bond	20 percent

NAME OF BIDDER (*Type or print*)	FULL NAME OF ALL PARTNERS (*Type or print*)
The Excello Company, Inc.	

BUSINESS ADDRESS (*Type or print*) (*Include "ZIP Code"*)
2000 Random Road N.W.
Albuquerque, New Mexico 87107

BY (*Signature in ink. Type or print name under signature*)

V. J. Excello

TITLE (*Type or print*)

President

DIRECTIONS FOR SUBMITTING BIDS: *Envelopes containing bids, guarantee, etc., must be sealed, marked, and addressed as follows:*

In the upper left corner, enter: "Bid under Serial No. Eng. 33-9-81 to be opened 9 August 19--."

Address to: District Engineer
Albuquerque District
Corps of Engineers
Albuquerque,
New Mexico

CAUTION—Bids should not be qualified by exceptions to the bidding conditions.

☼ U.S. GOVERNMENT PRINTING OFFICE : 1966 O—205-147—(15-F)

BIDDING SCHEDULE

Serial No. Eng. 33-9-81

Bid Item No.	Description	Estimated Quantity	Unit	Unit Price	Estimated Amount
1	Clearing	Job	Lump Sum	$13,433.00	$ 13,433
2	Demolition	Job	Lump Sum	10,845.00	10,845
3	Excavation	127,000	Cu. yd.	1.63	207,010
4	Base course	79,500	Ton	12.11	962,745
5	Concrete pavement, 9 in.	90,000	Sq. yd.	18.34	1,650,600
6	Concrete pavement, 11 in.	115,400	Sq. yd.	22.46	2,591,884
7	Asphalt concrete surface	150	Ton	25.08	3,762
8	Concrete pipe, 12 in.	1,000	Lin. ft.	21.60	21,600
9	Concrete pipe, 36 in.	300	Lin. ft.	37.23	11,169
10	Inlet	2	Each	444.50	889
11	Fiber duct, 4-way	600	Lin. ft.	25.89	15,534
12	Fiber duct, 8-way	1,200	Lin. ft.	31.51	37,812
13	Electrical manhole	6	Each	1,284.00	7,704
14	Underground cable	34,000	Lin. ft.	6.22	211,480
15	Taxiway lights	120	Each	315.00	37,800
16	Apron lights	70	Each	311.00	21,770
17	Taxiway marking	Job	Lump Sum	7,378.00	7,378
18	Fence	26,000	Lin. ft.	5.28	137,280
	TOTAL ESTIMATED AMOUNT				$5,950,695

NOTES:

   1.   All extensions of the unit prices shown will be subject to verification by the Government. In case of variation between the unit price and the extension, the unit price will be considered to be the bid.

   2.   If a bid or modification to a bid, based on unit prices, is submitted and provides for a lump-sum adjustment to the total estimated cost, the application of the lump-sum adjustment to each unit price, including lump-sum units, in the bid schedule, must be stated, or if it is not stated, the bidder agrees that the lump-sum adjustment shall be applied on a pro rata basis to every unit price in the bid schedule.

APPENDIX F

INSTRUCTIONS TO BIDDERS

1. CONTRACT DOCUMENTS. The Notice to Bidders, the Instructions to Bidders, the General Conditions, the Supplementary Conditions, the Drawings and Specifications, the Contractor's Proposal Form, and the Agreement as finally negotiated compose the Contract Documents.

Copies of these documents can be obtained from the office of Jones and Smith, Architect-Engineers, 142 Welsh St., Portland, Ohio, upon deposit of $100. 00 for each set thereof, said deposit being refundable upon return of the documents in good order within 10 days after the bidding date.

2. PRINTED FORM FOR PROPOSAL. All proposals must be made upon the Contractor's Proposal Form attached hereto and should give the amounts bid for the work, both in words and in figures, and must be signed and acknowledged by the Contractor. In order to ensure consideration, the Proposal should be enclosed in a sealed envelope marked "Proposal for Municipal Airport Terminal Building to be opened at 2:30 P.M. (E.S.T.), May 19, 19—," showing the return address of the sender and addressed to John Doe, City Manager, Portland, Ohio.

If the proposal is made by a partnership, it shall contain the names of each partner and shall be signed in the firm name, followed by the signature of the person authorized to sign. If the proposal is made by a corporation, it shall be signed by the name of the corporation, followed by the written signature of the officer signing, and the printed or typewritten designation of the office he holds in the corporation, together with the corporation seal. All blank spaces in the proposal form shall be properly filled in.

3. ALTERNATES. Each bidder shall submit with his proposal, on forms provided, alternate proposals stating the differences in price (additions or deductions) from the base bid for substituting, omitting, or changing the materials or construction from that shown on the drawings and as specified in a manner as described in the Division "Alternate Proposals" of these specifications.

The difference in price shall include all omissions, additions, and adjustments of all trades as may be necessary because of each change, substitution, or omission as described.

4. PAYMENT OF EMPLOYEES. For work done in the State of Ohio the payment of employees of the Contractor and any and all subcontractors shall comply with the current minimum wage scale as published by the Labor Commission of the State of Ohio, a copy of which is made a part of the Supplementary Conditions.

The Contractor and each of his subcontractors shall pay each of their employees engaged in work on the project under this contract in full, less deductions made mandatory by law, and not less often than once each week. All forms required by local authorities, the State of Ohio, and the United States Government, shall be properly submitted.

5. TELEGRAPHIC MODIFICATION. Any bidder may modify his bid by telegraphic communication at any time prior to the scheduled closing time for receipt of bids provided such telegraphic communication is received by the City Manager prior to said

closing time, and provided further that the City Manager is satisfied that a written confirmation of such telegraphic modification over the signature of the bidder was mailed prior to said closing time. If such written confirmation is not received within two (2) days from said closing time, no consideration will be given to the said telegraphic modification.

6. DELIVERY OF PROPOSALS. It is the bidder's responsibility to deliver his proposal at the proper time to the proper place. The mere fact that a proposal was dispatched will not be considered. The bidder must have the proposal actually delivered. Any proposal received after the scheduled closing time will be returned unopened to the bidder.

7. OPENING OF PROPOSALS. At 2:30 P.M. (E.S.T.), May 19, 19—, in the Office of the City Manager, City Hall, Portland, Ohio, each and every proposal (except those which have been withdrawn in accordance with Item 10, "Withdrawal of Proposals,") received prior to the scheduled closing time for receipt of proposals, will be publicly opened and read aloud, irrespective of any irregularities or informalities in such proposals.

8. ALTERATIONS IN PROPOSAL. Except as otherwise provided herein, proposals which are incomplete or which are conditioned in any way or which contain erasures not authenticated as provided herein or items not called for in the proposal or which may have been altered or are not in conformity with the law, may be rejected as informal.

The proposal form invites bids on definite plans and specifications. Only the amounts and information asked on the proposal form furnished will be considered as the bid. Each bidder shall bid upon the work exactly as specified and as provided in the proposal.

9. ERASURES. The proposal submitted must not contain erasures, interlineations, or other corrections unless each such correction is suitably authenticated by affixing in the margin immediately opposite the correction the surname or surnames of the person or persons signing the bid.

10. WITHDRAWAL OF PROPOSALS. At any time prior to the scheduled closing time for receipt of proposals, any bidder may withdraw his proposal, either personally or by telegraphic or written request. If withdrawal is made personally, proper receipt shall be given therefor.

After the scheduled closing time for the receipt of proposals and before award of contract, no bidder will be permitted to withdraw his proposal unless said award is delayed for a period exceeding thirty (30) days. Negligence on the part of the bidder in preparing his bid confers no rights for the withdrawal of the proposal after it has been opened.

11. DETERMINATION OF LOW BID. In making award of contract, the owner reserves the right to take into consideration the plant facilities of the bidders and the bidder's ability to complete the contract within the time specified in the proposal. The owner also reserves the right to evaluate factors that in his opinion would affect the final total cost.

12. REJECTION OF PROPOSALS. The owner reserves the right to reject any or all proposals. Without limiting the generality of the foregoing, any proposal which is incomplete, obscure, or irregular may be rejected; any proposal which omits a bid for any one or more items in the price sheet may be rejected; any proposal in which unit prices are omitted or in which unit prices are obviously unbalanced may be rejected; any proposal accompanied by an insufficient or irregular certified check, cashier's check, or bid bond may be rejected.

13. PROPOSAL AND PERFORMANCE GUARANTEES. A certified check, cashier's check, or bid bond for an amount equal to least five per cent (5%) of the total amount bid shall accompany each proposal as evidence of good faith and as a guarantee that if awarded the Contract, the bidder will execute the Contract and give bond as required. The successful bidder's check or bid bond will be retained until he has entered into a satisfactory contract and furnished required contract bonds. The owner reserves the right to hold the certified checks, cashier's checks, or bid bonds of the three lowest bidders, until the successful bidder has entered into a contract and furnished the required contract bonds.

14. ACCEPTANCE OF PROPOSALS. Within thirty (30) days after receipt of the proposals the owner will act upon them. The acceptance of a proposal will be a Notice of Acceptance in writing signed by a duly authorized representative of the owner and no other act of the owner shall constitute the acceptance of a proposal. The acceptance of a proposal shall bind the successful bidder to execute the Contract. The rights and obligations provided for in the Contract shall become effective and binding upon the parties only upon its formal execution.

15. TIME FOR EXECUTING CONTRACT AND PROVIDING CONTRACT BOND. Any contractor whose proposal shall be accepted will be required to execute the Contract and furnish contract bonds as required within ten (10) days after notice that the Contract has been awarded to him. Failure or neglect to do so shall constitute a breach of the agreement effected by the acceptance of the proposal.

16. PRICES. In the event of a discrepancy between the prices quoted in words and those quoted in figures in the proposal, the words shall control. The prices are to include the furnishing of all materials, plant, equipment, tools, and all other facilities, and the performance of all labor and services necessary or proper for the completion of the work except as may be otherwise expressly provided in the Contract Documents.

17. EXAMINATION OF DRAWINGS. Bidders shall thoroughly examine and be familiar with the drawings and specifications. The failure or omission of any bidder to receive or examine any form, instrument, addendum, or other document shall in no way relieve any bidder from any obligation with respect to his proposal or to the Contract. The submission of a bid shall be taken as prima facie evidence of compliance with this Section.

18. INTERPRETATIONS. No oral interpretations will be made to any bidder as to the meaning of the drawings and specifications or other contract documents. Every request for such an interpretation shall be made in writing and addressed and forwarded to the owner's authorized representative (Architect-Engineer) five (5) or more days before the date fixed for opening of proposals. Every interpretation made to a bidder will be in the form of an addendum to the Contract Documents which, if issued, will be sent as promptly as is practicable to all persons to whom the drawings and specifications have been issued. All such addenda shall become part of the Contract Documents.

19. POSTPONEMENT OF DATE FOR OPENING PROPOSALS. The owner reserves the right to postpone the date of presentation and opening of proposals and will give telegraphic notice of any such postponement to each interested party.

APPENDIX G

A STRATEGY OF BIDDING

G.1 MARKUP

When a contractor is bidding on a new job, one of the final actions taken is adding the markup to the estimated cost of the construction. This markup is customarily determined as a percentage of the cost. Although markup can include certain cost items as well as an allowance for profit, the term, as used in this appendix, is construed to be profit only. It is assumed that all items of job expense, including contingency and office overhead, are included in the contractor's estimate of cost.

Customarily, the contractor will select a markup that hopefully will enable him to bid the largest amount possible and still be the low bidder. By minimizing the difference between his bid and the second lowest, he will realize the maximum profit possible if he actually becomes the successful bidder. How the contractor goes about selecting his markup figure to accomplish this objective is normally a very inexact process and is based more on his subjective judgment than on any rational procedure. Using his past experience as a guide, he attempts to evaluate several bidding variables such as the competition, the type of work, the geographical area, the architect-engineer, and the terms of the contract documents. Although these variables cannot be precisely defined, they indicate in a general way whether high or low markups are in order.

This is not to say that the selection of a markup figure in such an intuitive fashion is altogether wrong or that more objective criteria can be devised which will suit all bidding conditions. Nevertheless, under the right circumstances, there are ways in which markup percentages can be selected that could increase the contractor's profits over the long term. One such circumstance may exist when a contractor's bidding is largely confined to a specific type of work in the same general geographical area.

The customary basis for most construction bidding is to bid each job essentially as an independent entity, attempting to maximize the potential profit on a particular project with little systematic attention being given to how it may relate to the panorama of past and future projects. Where a history of bidding experience has been built up, competitive patterns in the past biddings can be used as a guide for future ones. There are a number of ways in which this can be done, these procedures being known as "bidding strategies." It is the purpose of this appendix to discuss one such method. Although this discussion is oriented to lump-sum projects, the procedures are adaptable to unit-price biddings.

G.2 EXPECTED PROFIT

When a contractor bids a lump-sum job, he can anticipate a "potential profit" of $(b - c)$, where b is the amount of his bid and c is the actual cost of the work. The bid b is the sum of the estimated cost of the work and the markup, which is a variable. However, the contractor does not get every job on which he bids. Obviously, his chances of being the successful bidder are related to the amount of his bid, b. He can make his bid so high that his chances of being the low bidder would be zero. Here is a case in which the potential profit is large but the chances of actually realizing it are zero. To use slightly different terms, the "probability" p of being the low bidder is said to be zero when there is no chance of this occurrence. On the other hand, the contractor could bid so low that his winning would be a certainty (probability, $p = 1$), but the potential profit would be very small or even negative (loss). In between, of course, there are many possible bids with the probability of success varying inversely with the size of the bid; that is, the higher the bid the smaller the probability of success. Clearly, there is some optimum bid between the two extremes.

To determine this optimum bid, the concept of expectation or "expected profit" is useful. Expected profit is defined as $p(b - c)$, where p is the probability of the contractor's being the low bidder when his submitted bid is equal to b. As previously explained, p can vary between 0 (no chance) and 1 (certainty), with higher values indicating higher probability or better chances of winning. If the probability, p, is equal to 0.4, this indicates that the contractor's chances of being the low bidder are 4 out of 10.

To illustrate by a simple example the idea of expected profit, let us consider a project whose actual cost, c, will be \$50,000. Suppose that a contractor is able to determine that a bid of \$56,000 has a probability of 0.3 of being low and that a bid of \$53,000 has a probability of 0.8. The objective is to determine which of the two bids is better with respect to expected profit.

> When $b = \$56,000$:
> Expected profit $= p(b-c) = 0.3(\$56,000 - \$50,000) = \$1,800$
> When $b = \$53,000$:
> Expected profit $= p(b-c) = 0.8(\$53,000 - \$50,000) = \$2,400$

The bid of \$53,000 is the better of the two because it yields the larger expected profit. The reasoning behind maximizing expected profit is based upon long-term considerations. If the \$56,000 bid were used, the potential profit would be \$6,000. However, there are only 3 chances out of 10 that our contractor will get the job if he uses this bid. If it were possible to bid this same job 10 times and our contractor submitted a bid of \$56,000 each time, he would expect to be the successful bidder 3 times and would realize a total actual profit of $3(\$6000) = \$18,000$ from the series of biddings. This is an average profit of \$1800 per bidding. On the other hand, the \$53,000 bid would be successful 8 times out of the 10 and would produce a total actual profit of \$24,000 or an average profit of \$2400 per bid.

As can now be seen, expected profit represents the *average return per bid* if the biddings were to be repeated a large number of times. Biddings for the same project are, of course, not repeated, but the concept of increasing a contractor's actual profits by maximizing the expected profit for each bid submitted is still valid when utilized over a span of many biddings. It amounts to what might be described as "playing the odds." Our bidding strategy is based on the concept of maximizing expected profit.

G.3 COST OF CONSTRUCTION

The cost, c, as used herein is the actual cost of the work, a quantity that cannot possibly be known until the construction has been completed. Unfortunately, c is an important element in our bidding strategy, and bidding strategies are designed for use when the job is just being bid and long before c is known. A common assumption is that the actual cost, c, will be equal to the contractor's estimate of cost used in his bid.

Whether this is a valid assumption depends on the contractor's past record of bidding accuracy. It is an easy matter for a contractor to check the precision with which his estimating staff has been able to estimate the costs of completed projects. To illustrate the procedure, suppose that a contractor wishes to study his bidding performance over the past five years, a period during which he has completed 60 competitively bid contracts. For each of these contracts, he computes the value, V, which is the actual construction expense divided by his originally estimated cost. After determining these values, he groups them together into equal intervals and obtains the data shown in Figure G.1. This figure shows the number of occurrences of V in each 0.05 interval.

These data are then plotted to obtain the histogram shown in Figure G.2, and a frequency polygon is drawn (dashed lines) utilizing the midpoints of the horizontal bars of the histogram. The frequency polygon can then be smoothed by sketching in by eye a smooth curve that closely approximates the linear segments of the polygon. If the smoothed frequency polygon of past biddings resembles a normal distribution of small variance with its mode at or very near a value, V, of 1.0, such as that shown by Curve A in Figure G.3, the overall bidding accuracy of the contractor has been consistently good. Although curve B has its mode at 1.0, the large dispersion of the values indicates loose estimating with a serious tendency to both under- and overestimate. A biased distribution such as curve C is indicative of systematic estimating errors that result in a pronounced tendency to consistently either under- or overestimate a majority of the jobs bid.

There are statistical ways to make allowance for the bias and variability of past cost estimates. However, it is believed that the bidding records of most successful contractors will generally resemble curve A. In this case, the assumption that the actual cost, c, will be equal to the cost as estimated is statistically acceptable and the bidding strategy will be developed on this basis. Notice that this does not say that the costs of one contractor will necessarily equal those of another competing contractor.

G.4 MAXIMIZING EXPECTED PROFIT

To illustrate how a bid is selected to maximize expected profit, consider the bidding of a project whose estimated cost is $200,000. Assume for the moment that the probability of success for each of several different bids is known and is as shown in the second column of Figure G.4. How these probability values are obtained will be discussed subsequently. For the figures shown in Figure G.4, a bid of $210,000, or a markup of 5 percent, yields the maximum expected profit and is the best bid.

$V = \dfrac{\text{Actual project cost}}{\text{Estimated project cost}}$	Number of Projects
$V < 0.85$	0
$0.85 \leqslant V < 0.90$	3
$0.90 \leqslant V < 0.95$	12
$0.95 \leqslant V < 1.00$	19
$1.00 \leqslant V < 1.05$	16
$1.05 \leqslant V < 1.10$	8
$1.10 \leqslant V < 1.15$	2
$1.15 \leqslant V$	0
Total	60

Figure G.1 Bidding accuracy distribution.

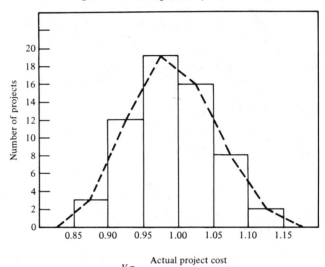

$$V = \frac{\text{Actual project cost}}{\text{Estimated project cost}}$$

Figure G.2 Histogram and frequency polygon.

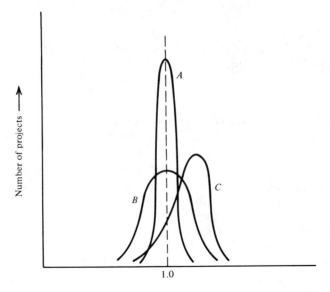

$$V = \frac{\text{Actual project cost}}{\text{Estimated project cost}}$$

Figure G.3 Bidding accuracy distribution.

446

Bid, b	Probability, p	Expected Profit, $p(b - c)$
$195,000	1.00	1.00($195,000 — $200,000) = —$5,000 (loss)
200,000	0.97	0.97($200,000 — $200,000) = 0
205,000	0.81	0.81($205,000 — $200,000) = 4,050
210,000	0.50	0.50($210,000 — $200,000) = 5,000
215,000	0.30	0.30($215,000 — $200,000) = 4,500
220,000	0.18	0.18($220,000 — $200,000) = 3,600
225,000	0.05	0.05($225,000 — $200,000) = 1,250
230,000	0.00	0.00($230,000 — $200,000) = 0

Figure G.4   Maximizing expected profit.

G.5 CASE 1—SINGLE KNOWN COMPETITOR

Before our bidding strategy can be developed further, we must determine how to find the values of probability of bidding success. As we shall see, these values are derived from an analysis of the historical bidding experience of our contractor and his competitors. For simplicity of explanation, let us first confine our discussion to bidding against a single, known competitor called Competitor A. Identification as a "known" competitor implies that our contractor has reason to believe that this competitor will be bidding the project and that our contractor has had past bidding experience with him.

Because bids are customarily opened and read aloud to all attendees at a bid opening, the past bids of competitors are known. In addition, area trade magazines publish the results of most biddings. Suppose that our contractor, using these past bidding records, compiles the information shown in Figure G.5 concerning Competitor A, against whom he has bid 62 times during recent years. It is suggested that only comparatively recent biddings be included, say within the past five or six years. The reason is that the bidding behavior of most contractors changes with time, reflecting econom-

$R = \dfrac{b_A}{c}$	Number of Times
$R < 0.98$	0
$0.98 \leq R < 1.00$	1
$1.00 \leq R < 1.02$	3
$1.02 \leq R < 1.04$	5
$1.04 \leq R < 1.06$	13
$1.06 \leq R < 1.08$	18
$1.08 \leq R < 1.10$	14
$1.10 \leq R < 1.12$	5
$1.12 \leq R < 1.14$	2
$1.14 \leq R < 1.16$	1
$1.16 \leq R$	0
	Total = $\overline{62}$

Figure G.5   Bidding history of Competitor A.

ic conditions, changes of company policy, and other temporal factors. Consequently, it is judged that recent bidding data reflect more accurately the probable temper of competitors at present biddings. In Figure G.5, $R = b_A/c$ (b_A divided by c), where b_A is Competitor A's bid and c is our contractor's estimate of construction cost.

The data in Figure G.5 now enable us to compute the probabilities, p_A, shown in Figure G.6, that various bids by our contractor will be lower than the bid submitted by Competitor A. To illustrate, if our contractor bids a ratio of b/c of 0.98 (2 percent less than his estimated cost), the probability that he will underbid Competitor A is 1.00 because at no time in the past has Competitor A ever bid this low. If our contractor bids the job at cost($b/c = 1.00$), there is 1 chance in 62 that Competitor A will bid lower. Another way of stating the same thing is to say that our contractor has 61 chances out of 62, or a probability of $^{61}/_{62} = 0.98$, of being the low bidder. Figure G.5 shows that, in the past, Competitor A has bid a value of b_A/c less than 1.02 four times. If our contractor submits a bid-to-cost ratio of 1.02 (markup = 2 percent), the probability is $^{58}/_{62}$ or 0.94 that he will be the low bidder. Continuing this process wll yield the values of p_A shown in Figure G.6.

The values of b/c in Figure G.6 are the ratios of our contractor's bid to his estimated cost. The expected profit has been computed in terms of c, the estimated cost. When $b/c = 1.08$ (a markup of 8 percent), then $b = 1.08c$, the probability of success is 0.36, and the expected profit is $0.029c$. The values of expected profit in the last column of Figure G.6 indicate that a bid of $1.06c$, or a markup of 6 percent, is optimum when Competitor A is the only other bidder. If our contractor's estimated cost, c, is

$\dfrac{b}{c}$	p_A	Expected Profit, $p_A(b - c)$
0.98	$\dfrac{62}{62} = 1.00$	$1.00(0.98c - c) = -0.020c$
1.00	$\dfrac{61}{62} = 0.98$	$0.98(1.00c - c) = 0$
1.02	$\dfrac{58}{62} = 0.94$	$0.94(1.02c - c) = 0.019c$
1.04	$\dfrac{53}{62} = 0.85$	$0.85(1.04c - c) = 0.034c$
1.06	$\dfrac{40}{62} = 0.65$	$0.65(1.06c - c) = 0.039c$
1.08	$\dfrac{22}{62} = 0.36$	$0.36(1.08c - c) = 0.029c$
1.10	$\dfrac{8}{62} = 0.13$	$0.13(1.10c - c) = 0.013c$
1.12	$\dfrac{3}{62} = 0.05$	$0.05(1.12c - c) = 0.006c$
1.14	$\dfrac{1}{62} = 0.02$	$0.02(1.14c - c) = 0.003c$
1.16	$\dfrac{0}{62} = 0.00$	$0.00(1.16c - c) = 0$

Figure G.6   Maximizing expected profit when bidding against Competitor A.

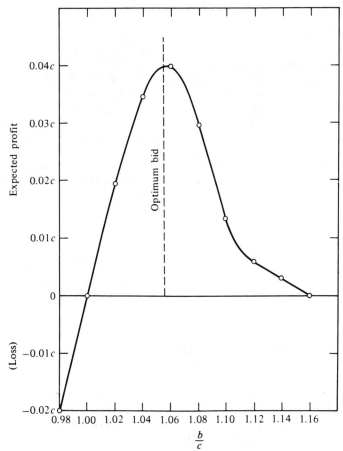

Figure G.7   Plot of data from Figure G.6.

$200,000, his bid would be $212,000. Some additional accuracy in determining the optimum bid can be achieved by plotting the results of Figure G.6, as shown in Figure G.7. This figure shows that the optimum markup would be closer to 5.5 than to 6 percent. Whether this kind of accuracy is significant is problematical.

It is interesting to note that the optimum markup percentage is independent of estimated cost and can be determined in advance of bidding. The result is that the optimum markup will be the same for a small project as for a large one. This statement may well run contrary to actual bidding practice, and the answer is to analyze Competitor A's bidding record separately for small jobs and for large ones. For example, probability values could be obtained for all projects less than $500,000 and for all projects over this amount. It should be obvious that probability values must be updated each time our contractor bids another job against Competitor A.

It may be well at this point to emphasize the fact that any bidding strategy attempts to predict future bidding behavior on the basis of past performance. The entire strategy is based on the supposition that competitors will continue to follow the same general bidding patterns as they have in the past. What changes in a competitor's future bidding habits might result from the use of a bidding strategy by our contractor is an open question.

G.6 CASE 2—MULTIPLE KNOWN COMPETITORS

Now, let us assume that our contractor is going to bid against two known bidders, Competitors A and B. Our contractor must first, by analyzing the past bidding record of Competitor B, determine the same kind of cumulative probability distribution for this competitor as he obtained for Competitor A in the second column of Figure G.6. Assume that the results of his analysis are shown in Figure G.8.

Figure G.8 discloses that if our contractor uses a markup of 8 percent, there is a probability of 0.36 of beating Competitor A, a probability of 0.52 of beating Competitor B, and a probability of 0.19 of beating both of them. For any given value of b/c the probability of beating both Competitors A and B, p_{AB}, is found as the product of the values of p_A and p_B. This applies because the probability of the simultaneous occurrence of a number of independent events is the product of their separate probabilities.

Figure G.9 reveals that the optimum markup for our contractor to use when bidding against both Competitors A and B is 6 percent. The extension of the process to include any number of known competitors should be obvious. The inclusion of more competitors invariably reduces the value of the optimum markup. In addition, more competitors reduce the values of the expected profit such that to have enough significant figures the computations must be carried out to more decimal places.

G.7 CASE 3—THE AVERAGE COMPETITOR

The preceding discussion has been based on the supposition that the competing contractors are all known; that is, our contractor knows who his competitors will be and that he possesses past bidding information about each of them. Consideration is now given to the case in which some or all of the bidders are not known. It is in this regard that the concept of "average bidder" is useful.

An average bidder is a hypothetical competitor whose bidding behavior is a statistical composite of the behaviors of all of our contractor's competitors. The collective bidding pattern of his competitors can be obtained by combining all of the

$\dfrac{b}{c}$	p_A	p_B	p_{AB}
0.98	1.00	1.00	1.00
1.00	0.98	0.99	0.97
1.02	0.94	0.96	0.90
1.04	0.85	0.90	0.76
1.06	0.65	0.84	0.55
1.08	0.36	0.52	0.19
1.10	0.13	0.31	0.04
1.12	0.05	0.14	0.01
1.14	0.02	0.03	0.00
1.16	0.00	0.00	0.00

Figure G.8   Probability values of known Competitors A and B.

$\dfrac{b}{c}$	p_{AB}	Expected Profit, $p_{AB}(b-c)$
0.98	1.00	$1.00(0.98c - c) = -0.020c$
1.00	0.97	$0.97(1.00c - c) = \quad 0$
1.02	0.90	$0.90(1.02c - c) = \quad 0.018c$
1.04	0.76	$0.76(1.04c - c) = \quad 0.030c$
1.06	0.55	$0.55(1.06c - c) = \quad 0.033c$
1.08	0.19	$0.19(1.08c - c) = \quad 0.015c$
1.10	0.04	$0.04(1.10c - c) = \quad 0.004c$
1.12	0.01	$0.01(1.12c - c) = \quad 0.001c$
1.14	0.00	$0.00(1.14c - c) = \quad 0$
1.16	0.00	$0.00(1.16c - c) = \quad 0$

Figure G.9 Maximizing expected profit when bidding against Competitors A and B.

$\dfrac{b}{c}$	p_{av}	Expected Profit, $p_{av}(b-c)$
0.98	1.00	$-0.020c$
1.00	0.98	0
1.02	0.95	$0.019c$
1.04	0.89	$0.036c$
1.06	0.72	$0.043c$
1.08	0.51	$0.041c$
1.10	0.30	$0.030c$
1.12	0.12	$0.014c$
1.14	0.05	$0.007c$
1.16	0.00	0

Figure G.10 Maximizing expected profit when bidding against a single average competitor.

$\dfrac{b}{c}$	p_{3av}	Expected Profit, $p_{3av}(b-c)$
0.98	$(1.00)^3 = 1.00$	$-0.020c$
1.00	$(0.98)^3 = 0.94$	0
1.02	$(0.95)^3 = 0.86$	$0.017c$
1.04	$(0.89)^3 = 0.70$	$0.028c$
1.06	$(0.72)^3 = 0.37$	$0.022c$
1.08	$(0.51)^3 = 0.13$	$0.010c$
1.10	$(0.30)^3 = 0.03$	$0.003c$
1.12	$(0.12)^3 = 0.00$	0
1.14	$(0.05)^3 = 0.00$	0
1.16	$(0.00)^3 = 0.00$	0

Figure G.11 Maximizing expected profit when bidding against three average competitors.

competitors into one probability distribution. The procedure here is exactly the same as that followed for Competitor A except that all competitors are included with no differentiation made among them.

For illustrative purposes, suppose that the result of this all-inclusive analysis is as shown in the second column of Figure G.10, where p_{av} is the probability that our contractor will submit a lower bid than any single, unspecified competitor. If only one unknown competitor is anticipated, an optimum markup of 6 percent is indicated, although the use of 7 or 8 percent might be equally in order, because the difference in expected profit is small and the methods used are not this precise.

In the event that our contractor expects three unknown competitors, the same procedure is followed as with multiple known competitors except that all competitors are the same. That is, each is an average competitor. To illustrate, Figure G.10 shows that a probability of 0.51 is associated with a markup of 8 percent and a single, unknown competitor. If there were three such competitors, the probability would be $(0.51)(0.51)(0.51) = (0.51)^3 = 0.13$ for a markup of 8 percent. In other words, if our contractor uses a markup of 8 percent, there is a probability of 0.13 that he will underbid all three unknown competitors. Figure G.11 discloses that the optimum markup when three unknown competitors are anticipated is between 4 and 5 percent. This example illustrates the variation of optimum markup with the number of competitors. For this reason, accuracy in estimating the number of competing firms is important.

G.8 THE USE OF BIDDING STRATEGIES

Bidding strategies of one kind or another have been in use since the concept of competitive bidding began. Every bidding involves a complex game of trying to outguess and out-smart the opposition. The application of bidding procedures based on a systematic analysis of past bidding experience, however, is not common in the construction industry. Nevertheless, available evidence indicates that such methods are now catching on, although their practitioners are understandably reticent to divulge procedures or results.

The probabilistic procedure discussed in this appendix is basic and is easily applied. Refinements of the procedure are possible, such as computing probabilities for different categories of projects. For example, the past projects could be subdivided into different ranges of construction cost and into different types of construction. The reason for this refinement is that different markups are commonly applied to jobs, depending upon their size and type. Another modification could be to give greater weight to recent bidding information than to that from older projects.

Several other forms of bidding models have been devised, many of which involve relatively sophisticated statistical procedures. Whether these more refined methods have any real advantage over the simple and straightforward procedure discussed in this appendix is not yet clear.

APPENDIX H

PROPOSAL FOR LUMP-SUM CONTRACT

Portland, Ohio

(Place)

May 19, 19—

(Date)

PROPOSAL of The Blank Construction Company, Inc., a corporation organized and existing under the laws of the State of Ohio, a partnership consisting of_____

an individual doing business as_____

TO: The City of Portland, Ohio
PROJECT: Municipal Airport Terminal Building
    For the City of Portland, Ohio

Gentlemen:

  The Undersigned, in compliance with your Invitation for Bids for the General Construction of the above-described project, having examined the drawings and specifications with related contract documents carefully, with all addenda thereto, and the site of the work, and being familiar with all of the conditions surrounding the construction of the proposed project, hereby proposes to furnish all plant, labor, equipment, appliances, supplies, and materials and to perform all work for the construction of the project as required by and in strict accordance with the contract documents, specifications, schedules, and drawings with all addenda issued by the Architect-Engineer, at the prices stated below.

  The Undersigned hereby acknowledges receipt of the following Addenda:
Addendum No. 1 dated April 28, 19—.
Addendum No. 2 dated May 6, 19—.

BASE PROPOSAL: For all work described in the detailed specifications and shown on the contract drawings for the building, I (or We) agree to perform all the work for the sum of
Three million, six hundred sixty-eight thousand, eight hundred sixty-one and no/100 ($3,668,861.00) dollars. (Amount shall be shown in both written form and figures. In case of discrepancy between the written amount and the figures, the written amount will govern.)

  The above-stated compensation covers all expenses incurred in performing the work, including premium for contract bonds, required under the contract documents, of which this proposal is a part.

ALTERNATE NO. 1: QUARRY TILE IN PLACE OF TERRAZZO: If the substitutions speci-fied under this alternate are made, you may (~~deduct from~~) (add to) the base proposal the sum of Nine thousand, seven hundred fourteen and no/100 ($9,714.00) dollars.

ALTERNATE NO. 2: CHANGE STRUCTURAL GLAZED TILE TO BRICK IN CONCOURSE: If the substitutions specified under this alternate are made, you may (deduct from) (~~add to~~) the base proposal the sum of Four thousand, two hundred eighty and no/100 ($4,280.00) dollars.

ALTERNATE NO. 3: OMIT KITCHEN EQUIPMENT: If the substitutions specified under this alternate are made, you may (deduct from) (~~add to~~) the base proposal the sum of Sixty-six thousand, seven hundred twenty-three and no/100 ($66,723.00) dollars.

BID SECURITY: Attached cashier's check (certified check) (Bid Bond) payable without condition, in the sum of 5% of maximum possible bid amount ($_____) dollars (equal to 5% of the largest possible combination) is to become the property of the City of Portland, Ohio, in the event the Contract and contract bonds are not executed with-in the time set forth hereinafter, as liquidated damages for the delay and additional work caused thereby.

CONTRACT SECURITY: The Undersigned hereby agrees, if awarded the contract, to fur-nish the contract bonds, as specified, with the Hartford Accident and Indemnity Company Surety Company of Hartford, Connecticut.

Upon receipt of notice of the acceptance of this bid, the Undersigned hereby agrees that he will execute and deliver the formal written Contract in the form prescribed, in accordance with the bid as accepted and that he will give contract bonds, all within ten days after the prescribed forms are presented to him for signature.

If awarded the Contract, the Undersigned proposes to commence work within 10 calendar days after receipt of notice to proceed and to fully complete all of the work under his Contract, ready for occupancy, within 380 calendar days thereafter.

Respectfully submitted,

The Blank Construction Company, Inc.

By K.O. Acme

Vice-President

(Title)

1938 Cranbrook Lane

Portland, Ohio

(Business Address)
216-344-5507

(Telephone Number)

SEAL:
(If bid is by a Corporation)

APPENDIX I

STANDARD FORM OF AGREEMENT BETWEEN OWNER AND CONTRACTOR, STIPULATED SUM, AIA DOCUMENT A101*

THE AMERICAN INSTITUTE OF ARCHITECTS

AIA Document A101

Standard Form of Agreement Between Owner and Contractor

where the basis of payment is a

STIPULATED SUM

1977 EDITION

THIS DOCUMENT HAS IMPORTANT LEGAL CONSEQUENCES; CONSULTATION WITH AN ATTORNEY IS ENCOURAGED WITH RESPECT TO ITS COMPLETION OR MODIFICATION

Use only with the 1976 Edition of AIA Document A201, General Conditions of the Contract for Construction.

This document has been approved and endorsed by The Associated General Contractors of America.

AGREEMENT

made as of the   **Sixteenth**   day of   **June**   in the year of Nineteen Hundred and

BETWEEN the Owner: The City of Portland, Ohio

and the Contractor: The Blank Construction Company, Inc.
Portland, Ohio

The Project: A Municipal Airport Terminal Building for the
City of Portland, Ohio

The Architect: Jones and Smith, Architect-Engineers
Portland, Ohio

The Owner and the Contractor agree as set forth below.

AIA DOCUMENT A101 • OWNER-CONTRACTOR AGREEMENT • ELEVENTH EDITION • JUNE 1977 • AIA®
©1977 • THE AMERICAN INSTITUTE OF ARCHITECTS, 1735 NEW YORK AVE., N.W., WASHINGTON, D. C. 20006   **A101-1977   1**

ARTICLE 1

THE CONTRACT DOCUMENTS

The Contract Documents consist of this Agreement, the Conditions of the Contract (General, Supplementary and other Conditions), the Drawings, the Specifications, all Addenda issued prior to and all Modifications issued after execution of this Agreement. These form the Contract, and all are as fully a part of the Contract as if attached to this Agreement or repeated herein. An enumeration of the Contract Documents appears in Article 7.

ARTICLE 2

THE WORK

The Contractor shall perform all the Work required by the Contract Documents for
(Here insert the caption descriptive of the Work as used on other Contract Documents.)

Municipal Airport Terminal Building
Portland, Ohio

ARTICLE 3

TIME OF COMMENCEMENT AND SUBSTANTIAL COMPLETION

The Work to be performed under this Contract shall be commenced   within ten days after receipt of Notice to Proceed
and, subject to authorized adjustments, Substantial Completion shall be achieved not later than 380 calendar days after receipt of Notice to Proceed.
(Here insert any special provisions for liquidated damages relating to failure to complete on time.)

It is mutually agreed that the Owner shall withhold from the Contractor, as liquidated damages and not as a penalty, the sum of one hundred dollars ($100.00) per day for each calendar day that the work remains uncompleted beyond this date.

AIA DOCUMENT A101 • OWNER-CONTRACTOR AGREEMENT • ELEVENTH EDITION • JUNE 1977 • AIA®
©1977 • THE AMERICAN INSTITUTE OF ARCHITECTS, 1735 NEW YORK AVE., N.W., WASHINGTON, D. C. 20006   A101-1977   2

ARTICLE 4

CONTRACT SUM

The Owner shall pay the Contractor in current funds for the performance of the Work, subject to additions and deductions by Change Order as provided in the Contract Documents, the Contract Sum of

The lump-sum amount of $3,602,138.00

The Contract Sum is determined as follows:
(State here the base bid or other lump sum amount, accepted alternates, and unit prices, as applicable.)

```
Base Proposal                                         $3,668,861.00
   Alternate No. 1:  Add $9,714.00 (not accepted)
   Alternate No. 2:  Deduct $4,280.00 (not accepted)
   Alternate No. 3:  Deduct $66,723.00 (accepted)
Total Alternates Accepted, deduct                         66,723.00
Total Contract Amount                                 $3,602,138.00
```

ARTICLE 5

PROGRESS PAYMENTS

Based upon Applications for Payment submitted to the Architect by the Contractor and Certificates for Payment issued by the Architect, the Owner shall make progress payments on account of the Contract Sum to the Contractor as provided in the Contract Documents for the period ending the   **last**   day of the month as follows:

Not later than   **ten**   days following the end of the period covered by the Application for Payment **ninety**   percent (90 %) of the portion of the Contract Sum properly allocable to labor, materials and equipment incorporated in the Work and   **ninety**   percent (90 %) of the portion of the Contract Sum properly allocable to materials and equipment suitably stored at the site or at some other location agreed upon in writing, for the period covered by the Application for Payment, less the aggregate of previous payments made by the Owner; and upon Substantial Completion of the entire Work, a sum sufficient to increase the total payments to **ninety-five**   percent (95 %) of the Contract Sum, less such amounts as the Architect shall determine for all incomplete Work and unsettled claims as provided in the Contract Documents.

(If not covered elsewhere in the Contract Documents, here insert any provision for limiting or reducing the amount retained after the Work reaches a certain stage of completion.)

It is provided, however, that the Architect-Engineer at any time after fifty (50) percent of the work has been completed, if he finds that satisfactory progress is being made, with written consent of Surety, shall recommend that the remaining progress payments be paid in full.

Payments due and unpaid under the Contract Documents shall bear interest from the date payment is due at the rate entered below, or in the absence thereof, at the legal rate prevailing at the place of the Project.
(Here insert any rate of interest agreed upon.)

(Usury laws and requirements under the Federal Truth in Lending Act, similar state and local consumer credit laws and other regulations at the Owner's and Contractor's principal places of business, the location of the Project and elsewhere may affect the validity of this provision. Specific legal advice should be obtained with respect to deletion, modification, or other requirements such as written disclosures or waivers.)

AIA DOCUMENT A101 • OWNER-CONTRACTOR AGREEMENT • ELEVENTH EDITION • JUNE 1977 • AIA®
©1977 • THE AMERICAN INSTITUTE OF ARCHITECTS, 1735 NEW YORK AVE., N.W., WASHINGTON, D. C. 20006   **A101-1977 3**

ARTICLE 6

FINAL PAYMENT

Final payment, constituting the entire unpaid balance of the Contract Sum, shall be paid by the Owner to the Contractor when the Work has been completed, the Contract fully performed, and a final Certificate for Payment has been issued by the Architect.

ARTICLE 7

MISCELLANEOUS PROVISIONS

7.1 Terms used in this Agreement which are defined in the Conditions of the Contract shall have the meanings designated in those Conditions.

7.2 The Contract Documents, which constitute the entire agreement between the Owner and the Contractor, are listed in Article 1 and, except for Modifications issued after execution of this Agreement, are enumerated as follows:

(List below the Agreement, the Conditions of the Contract (General, Supplementary, and other Conditions), the Drawings, the Specifications, and any Addenda and accepted alternates, showing page or sheet numbers in all cases and dates where applicable.)

All Drawings and Specifications entitled "A Municipal Airport Terminal Building for The City of Portland, Ohio", and dated March 6, 19--.

```
Drawings:    Sheets A-1 through A-44
                    S-1 through S-18
                    P-1 through P-6
                    M-1 through M-8
                    E-1 through E-11

Specifications:   General Conditions
                  Supplementary Conditions
                  Technical Sections 1 through 16

Addendum No. 1, dated April 28, 19--
Addendum No. 2, dated May 6, 19--
Alternate No. 3, Omit Kitchen Equipment
Agreement, dated June 16, 19--
```

This Agreement entered into as of the day and year first written above.

OWNER　　　　　　　　　　　　　　　　　CONTRACTOR

_____　　_____

_____　　_____

_____　　_____

AIA DOCUMENT A101 • OWNER-CONTRACTOR AGREEMENT • ELEVENTH EDITION • JUNE 1977 • AIA®
©1977 • THE AMERICAN INSTITUTE OF ARCHITECTS, 1735 NEW YORK AVE., N.W., WASHINGTON, D. C. 20006　　**A101-1977　4**

APPENDIX J

STANDARD FORM OF AGREEMENT BETWEEN OWNER AND CONTRACTOR, COST OF THE WORK PLUS A FEE, AIA DOCUMENT A111*

THE AMERICAN INSTITUTE OF ARCHITECTS

AIA Document A111

Standard Form of Agreement Between Owner and Contractor

where the basis of payment is the

COST OF THE WORK PLUS A FEE

1978 EDITION

THIS DOCUMENT HAS IMPORTANT LEGAL CONSEQUENCES; CONSULTATION WITH AN ATTORNEY IS ENCOURAGED WITH RESPECT TO ITS COMPLETION OR MODIFICATION

Use only with the 1976 Edition of AIA Document A201, General Conditions of the Contract for Construction.

This document has been approved and endorsed by The Associated General Contactors of America

---

AGREEMENT

made as of the   day of   in the year of Nineteen
Hundred and

BETWEEN the Owner:

and the Contractor:

the Project:

the Architect:

The Owner and the Contractor agree as set forth below.

---

*AIA copyrighted material has been reproduced with permission of The American Institute of Architects under application number 80063. Further reproduction is prohibited.

ARTICLE 1

THE CONTRACT DOCUMENTS

1.1 The Contract Documents consist of this Agreement, the Conditions of the Contract (General, Supplementary and other Conditions), the Drawings, the Specifications, all Addenda issued prior to and all Modifications issued after execution of this Agreement. These form the Contract, and all are as fully a part of the Contract as if attached to this Agreement or repeated herein. An enumeration of the Contract Documents appears in Article 16. If anything in the Contract Documents is inconsistent with this Agreement, the Agreement shall govern.

ARTICLE 2

THE WORK

2.1 The Contractor shall perform all the Work required by the Contract Documents for

(Here insert the caption descriptive of the Work as used on other Contract Documents.)

ARTICLE 3

THE CONTRACTOR'S DUTIES AND STATUS

3.1 The Contractor accepts the relationship of trust and confidence established between him and the Owner by this Agreement. He covenants with the Owner to furnish his best skill and judgment and to cooperate with the Architect in furthering the interests of the Owner. He agrees to furnish efficient business administration and superintendence and to use his best efforts to furnish at all times an adequate supply of workmen and materials, and to perform the Work in the best way and in the most expeditious and economical manner consistent with the interests of the Owner.

ARTICLE 4

TIME OF COMMENCEMENT AND SUBSTANTIAL COMPLETION

4.1 The Work to be performed under this Contract shall be commenced

and, subject to authorized adjustments,

Substantial Completion shall be achieved not later than

(Here insert any special provisions for liquidated damages relating to failure to complete on time.)

ARTICLE 5

COST OF THE WORK AND GUARANTEED MAXIMUM COST

5.1   The Owner agrees to reimburse the Contractor for the Cost of the Work as defined in Article 8. Such reimbursement shall be in addition to the Contractor's Fee stipulated in Article 6.

5.2   The maximum cost to the Owner, including the Cost of the Work and the Contractor's Fee, is guaranteed not to exceed the sum of   dollars ($); such Guaranteed Maximum Cost shall be increased or decreased for Changes in the Work as provided in Article 7.

(Here insert any provision for distribution of any savings. Delete Paragraph 5.2 if there is no Guaranteed Maximum Cost.)

ARTICLE 6

CONTRACTOR'S FEE

6.1   In consideration of the performance of the Contract, the Owner agrees to pay the Contractor in current funds as compensation for his services a Contractor's Fee as follows:

6.2   For Changes in the Work, the Contractor's Fee shall be adjusted as follows:

6.3   The Contractor shall be paid   percent (%) of the proportional amount of his Fee with each progress payment, and the balance of his Fee shall be paid at the time of final payment.

ARTICLE 7

CHANGES IN THE WORK

7.1 The Owner may make Changes in the Work as provided in the Contract Documents. The Contractor shall be reimbursed for Changes in the Work on the basis of Cost of the Work as defined in Article 8.

7.2 The Contractor's Fee for Changes in the Work shall be as set forth in Paragraph 6.2, or in the absence of specific provisions therein, shall be adjusted by negotiation on the basis of the Fee established for the original Work.

ARTICLE 8

COSTS TO BE REIMBURSED

8.1 The term Cost of the Work shall mean costs necessarily incurred in the proper performance of the Work and paid by the Contractor. Such costs shall be at rates not higher than the standard paid in the locality of the Work except with prior consent of the Owner, and shall include the items set forth below in this Article 8.

8.1.1 Wages paid for labor in the direct employ of the Contractor in the performance of the Work under applicable collective bargaining agreements, or under a salary or wage schedule agreed upon by the Owner and Contractor, and including such welfare or other benefits, if any, as may be payable with respect thereto.

8.1.2 Salaries of Contractor's personnel when stationed at the field office, in whatever capacity employed. Personnel engaged, at shops or on the road, in expediting the production or transportation of materials or equipment, shall be considered as stationed at the field office and their salaries paid for that portion of their time spent on this Work.

8.1.3 Cost of contributions, assessments or taxes incurred during the performance of the Work for such items as unemployment compensation and social security, insofar as such cost is based on wages, salaries, or other remuneration paid to employees of the Contractor and included in the Cost of the Work under Subparagraphs 8.1.1 and 8.1.2.

8.1.4 The portion of reasonable travel and subsistence expenses of the Contractor or of his officers or employees incurred while traveling in discharge of duties connected with the Work.

8.1.5 Cost of all materials, supplies and equipment incorporated in the Work, including costs of transportation thereof.

8.1.6 Payments made by the Contractor to Subcontractors for Work performed pursuant to Subcontracts under this Agreement.

8.1.7 Cost, including transportation and maintenance, of all materials, supplies, equipment, temporary facilities and hand tools not owned by the workers, which are consumed in the performance of the Work, and cost less salvage value on such items used but not consumed which remain the property of the Contractor.

8.1.8 Rental charges of all necessary machinery and equipment, exclusive of hand tools, used at the site of the Work, whether rented from the Contractor or others, including installation, minor repairs and replacements, dismantling, removal, transportation and delivery costs thereof, at rental charges consistent with those prevailing in the area.

8.1.9 Cost of premiums for all bonds and insurance which the Contractor is required by the Contract Documents to purchase and maintain.

8.1.10 Sales, use or similar taxes related to the Work and for which the Contractor is liable imposed by any governmental authority.

8.1.11 Permit fees, royalties, damages for infringement of patents and costs of defending suits therefor, and deposits lost for causes other than the Contractor's negligence.

8.1.12 Losses and expenses, not compensated by insurance or otherwise, sustained by the Contractor in connection with the Work, provided they have resulted from causes other than the fault or neglect of the Contractor. Such losses shall include settlements made with the written consent and approval of the Owner. No such losses and expenses shall be included in the Cost of the Work for the purpose of determining the Contractor's Fee. If, however, such loss requires reconstruction and the Contractor is placed in charge thereof, he shall be paid for his services a Fee proportionate to that stated in Paragraph 6.1.

8.1.13 Minor expenses such as telegrams, long distance telephone calls, telephone service at the site, expressage, and similar petty cash items in connection with the Work.

8.1.14 Cost of removal of all debris.

8.1.15 Costs incurred due to an emergency affecting the safety of persons and property.

8.1.16 Other costs incurred in the performance of the Work if and to the extent approved in advance in writing by the Owner.

(Here insert modifications or limitations to any of the above Subparagraphs, such as equipment rental charges and small tool charges applicable to the Work.)

ARTICLE 9

COSTS NOT TO BE REIMBURSED

9.1 The term Cost of the Work shall not include any of the items set forth below in this Article 9.

9.1.1 Salaries or other compensation of the Contractor's personnel at the Contractor's principal office and branch offices.

9.1.2 Expenses of the Contractor's principal and branch offices other than the field office.

9.1.3 Any part of the Contractor's capital expenses, including interest on the Contractor's capital employed for the Work.

9.1.4 Except as specifically provided for in Subparagraph 8.1.8 or in modifications thereto, rental costs of machinery and equipment.

9.1.5 Overhead or general expenses of any kind, except as may be expressly included in Article 8.

9.1.6 Costs due to the negligence of the Contractor, any Subcontractor, anyone directly or indirectly employed by any of them, or for whose acts any of them may be liable, including but not limited to the correction of defective or nonconforming Work, disposal of materials and equipment wrongly supplied, or making good any damage to property.

9.1.7 The cost of any item not specifically and expressly included in the items described in Article 8.

9.1.8 Costs in excess of the Guaranteed Maximum Cost, if any, as set forth in Article 5 and adjusted pursuant to Article 7.

ARTICLE 10

DISCOUNTS, REBATES AND REFUNDS

10.1   All cash discounts shall accrue to the Contractor unless the Owner deposits funds with the Contractor with which to make payments, in which case the cash discounts shall accrue to the Owner. All trade discounts, rebates and refunds, and all returns from sale of surplus materials and equipment shall accrue to the Owner, and the Contractor shall make provisions so that they can be secured.

(Here insert any provisions relating to deposits by the Owner to permit the Contractor to obtain cash discounts.)

ARTICLE 11

SUBCONTRACTS AND OTHER AGREEMENTS

11.1   All portions of the Work that the Contractor's organization does not perform shall be performed under Subcontracts or by other appropriate agreement with the Contractor. The Contractor shall request bids from Subcontractors and shall deliver such bids to the Architect. The Owner will then determine, with the advice of the Contractor and subject to the reasonable objection of the Architect, which bids will be accepted.

11.2   All Subcontracts shall conform to the requirements of the Contract Documents. Subcontracts awarded on the basis of the cost of such work plus a fee shall also be subject to the provisions of this Agreement insofar as applicable.

ARTICLE 12

ACCOUNTING RECORDS

12.1   The Contractor shall check all materials, equipment and labor entering into the Work and shall keep such full and detailed accounts as may be necessary for proper financial management under this Agreement, and the system shall be satisfactory to the Owner. The Owner shall be afforded access to all the Contractor's records, books, correspondence, instructions, drawings, receipts, vouchers, memoranda and similar data relating to this Contract, and the Contractor shall preserve all such records for a period of three years, or for such longer period as may be required by law, after the final payment.

ARTICLE 13

APPLICATIONS FOR PAYMENT

13.1   The Contractor shall, at least ten days before each payment falls due, deliver to the Architect an itemized statement, notarized if required, showing in complete detail all moneys paid out or costs incurred by him on account of the Cost of the Work during the previous month for which he is to be reimbursed under Article 5 and the amount of the Contractor's Fee due as provided in Article 6, together with payrolls for all labor and such other data supporting the Contractor's right to payment for Subcontracts or materials as the Owner or the Architect may require.

ARTICLE 14

PAYMENTS TO THE CONTRACTOR

14.1   The Architect will review the Contractor's Applications for Payment and will promptly take appropriate action thereon as provided in the Contract Documents. Such amount as he may recommend for payment shall be payable by the Owner not later than the   day of the month.

14.1.1   In taking action on the Contractor's Applications for Payment, the Architect shall be entitled to rely on the accuracy and completeness of the information furnished by the Contractor and shall not be deemed to represent that he has made audits of the supporting data, exhaustive or continuous on-site inspections or that he has made any examination to ascertain how or for what purposes the Contractor has used the moneys previously paid on account of the Contract.

14.2   Final payment, constituting the entire unpaid balance of the Cost of the Work and of the Contractor's Fee, shall be paid by the Owner to the Contractor   days after Substantial Completion of the Work unless otherwise stipulated in the Certificate of Substantial Completion, provided the Work has been completed, the Contract fully performed, and final payment has been recommended by the Architect.

14.3   Payments due and unpaid under the Contract Documents shall bear interest from the date payment is due at the rate entered below, or in the absence thereof, at the legal rate prevailing at the place of the Project.

(Here insert any rate of interest agreed upon.)

(Usury laws and requirements under the Federal Truth in Lending Act, similar state and local consumer credit laws and other regulations at the Owner's and Contractor's principal places of business, the location of the Project and elsewhere may affect the validity of this provision. Specific legal advice should be obtained with respect to deletion, modification, or other requirements such as written disclosures or waivers.)

ARTICLE 15

TERMINATION OF CONTRACT

15.1   The Contract may be terminated by the Contractor as provided in the Contract Documents.

15.2   If the Owner terminates the Contract as provided in the Contract Documents, he shall reimburse the Contractor for any unpaid Cost of the Work due him under Article 5, plus (1) the unpaid balance of the Fee computed upon the Cost of the Work to the date of termination at the rate of the percentage named in Article 6, or (2) if the Contractor's Fee be stated as a fixed sum, such an amount as will increase the payments on account of his Fee to a sum which bears the same ratio to the said fixed sum as the Cost of the Work at the time of termination bears to the adjusted Guaranteed Maximum Cost, if any, otherwise to a reasonable estimated Cost of the Work when completed. The Owner shall also pay to the Contractor fair compensation, either by purchase or rental at the election of the Owner, for any equipment retained. In case of such termination of the Contract the Owner shall further assume and become liable for obligations, commitments and unsettled claims that the Contractor has previously undertaken or incurred in good faith in connection with said Work. The Contractor shall, as a condition of receiving the payments referred to in this Article 15, execute and deliver all such papers and take all such steps, including the legal assignment of his contractual rights, as the Owner may require for the purpose of fully vesting in himself the rights and benefits of the Contractor under such obligations or commitments.

AIA DOCUMENT A111 • COST-PLUS OWNER-CONTRACTOR AGREEMENT • NINTH EDITION • APRIL 1978 • AIA®
© 1978 • THE AMERICAN INSTITUTE OF ARCHITECTS, 1735 NEW YORK AVE., N.W., WASHINGTON, D.C. 20006   **A111-1978**   **7**

ARTICLE 16

MISCELLANEOUS PROVISIONS

16.1 Terms used in this Agreement which are defined in the Contract Documents shall have the meanings designated in those Contract Documents.

16.2 The Contract Documents, which constitute the entire agreement between the Owner and the Contractor, are listed in Article 1 and, except for Modifications issued after execution of this Agreement, are enumerated as follows:

(List below the Agreement, the Conditions of the Contract, [General, Supplementary, and other Conditions], the Drawings, the Specifications, and any Addenda and accepted alternates, showing page or sheet numbers in all cases and dates where applicable.)

This Agreement entered into as of the day and year first written above.

OWNER   CONTRACTOR

_____   _____

_____   _____

_____   _____

AIA DOCUMENT A111 • COST-PLUS OWNER-CONTRACTOR AGREEMENT • NINTH EDITION • APRIL 1978 • AIA®
© 1978 • THE AMERICAN INSTITUTE OF ARCHITECTS, 1735 NEW YORK AVE., N.W., WASHINGTON, D.C. 20006   **A111-1978 8**

APPENDIX K

CONSTRUCTION INDUSTRY ARBITRATION RULES*

CONSTRUCTION INDUSTRY ARBITRATION RULES

Effective November 15, 1979

AMERICAN CONSULTING ENGINEERS COUNCIL

AMERICAN INSTITUTE OF ARCHITECTS

AMERICAN SOCIETY OF CIVIL ENGINEERS

AMERICAN SOCIETY OF LANDSCAPE ARCHITECTS

AMERICAN SUBCONTRACTORS ASSOCIATION

ASSOCIATED GENERAL CONTRACTORS

ASSOCIATED SPECIALTY CONTRACTORS, INC.

CONSTRUCTION SPECIFICATIONS INSTITUTE

NATIONAL SOCIETY OF PROFESSIONAL ENGINEERS

NATIONAL UTILITY CONTRACTORS ASSOCIATION, INC.

ADMINISTERED BY

AMERICAN ARBITRATION ASSOCIATION

140 West 51st Street   New York, N.Y. 10020

*This Document is reproduced here with the permission of the American Arbitration Association.

467

AMERICAN ARBITRATION ASSOCIATION

Arbitration is the voluntary submission of a dispute to a disinterested person or persons for final determination. And to achieve orderly, economical and expeditious arbitration, in accordance with federal and state laws, the American Arbitration Association is available to administer arbitration cases under various specialized rules.

The American Arbitration Association maintains throughout the United States a National Panel of Arbitrators consisting of experts in all trades and professions. By arranging for arbitration under the Construction Industry Arbitration Rules, parties may obtain the services of arbitrators who are familiar with the construction industry.

The American Arbitration Association shall establish and maintain as members of its National Panel of Arbitrators individuals competent to hear and determine disputes administered under the Construction Industry Arbitration Rules. The Association shall consider for appointment to the Construction Industry Panel persons recommended by the National Construction Industry Arbitration Committee as qualified to serve by virtue of their experience in the construction field.

The Association does not act as arbitrator. Its function is to administer arbitrations in accordance with the agreement of the parties and to maintain Panels from which arbitrators may be chosen by parties. Once designated, the arbitrator decides the issues and an award is final and binding.

When an agreement to arbitrate is written into a construction contract, it may expedite peaceful settlement without the necessity of going to arbitration at all. Thus, the arbitration clause is a form of insurance against loss of good will.

3

TABLE OF CONTENTS

4

CONSTRUCTION INDUSTRY
ARBITRATION RULES

Section 1.   AGREEMENT OF PARTIES — The parties shall be deemed to have made these Rules a part of their arbitration agreement whenever they have provided for arbitration under the Construction Industry Arbitration Rules. These Rules and any amendment thereof shall apply in the form obtaining at the time the arbitration is initiated.

Section 2.   NAME OF TRIBUNAL — Any Tribunal constituted by the parties for the settlement of their dispute under these Rules shall be called the Construction Industry Arbitration Tribunal, hereinafter called the Tribunal.

Section 3.   ADMINISTRATOR — When parties agree to arbitrate under these Rules, or when they provide for arbitration by the American Arbitration Association, hereinafter called AAA, and an arbitration is initiated hereunder, they thereby constitute AAA the administrator of the arbitration. The authority and duties of the administrator are prescribed in the agreement of the parties and in these Rules.

Section 4.   DELEGATION OF DUTIES — The duties of the AAA under these Rules may be carried out through Tribunal Administrators, or such other officers or committees as the AAA may direct.

Section 5.   NATIONAL PANEL OF ARBITRATORS — In cooperation with the Construction Industry Arbitration Committee, the AAA shall establish and maintain a National Panel of Construction Arbitrators, hereinafter called the Panel, and shall appoint an arbitrator or arbitrators therefrom as hereinafter provided. A neutral arbitrator selected by mutual choice of both parties or their appointees, or appointed by the AAA, is hereinafter called the arbitrator, whereas an arbitrator selected unilaterally by one party is hereinafter called the party-appointed arbitrator. The term arbitrator may hereinafter be used to refer to one arbitrator or to a Tribunal of multiple arbitrators.

Section 6.   OFFICE OF TRIBUNAL — The general office of a Tribunal is the headquarters of the AAA, which may, however, assign the administration of an arbitration to any of its Regional Offices.

Section 7.   INITIATION UNDER AN ARBITRATION PROVISION IN A CONTRACT — Arbitration under an arbitration provision in a contract shall be initiated in the following manner:

The initiating party shall, within the time specified by the contract, if any, file with the other party a notice of an intention to arbitrate (Demand), which notice shall contain a statement setting forth the nature of the dispute, the amount involved, and the remedy sought; and shall file two copies of said notice with any Regional Office of the AAA, together with two copies of the arbitration provisions of the contract and the appropriate filing fee as provided in Section 48 hereunder.

5

The AAA shall give notice of such filing to the other party. A party upon whom the demand for arbitration is made may file an answering statement in duplicate with the AAA within seven days afer notice from the AAA, simultaneously sending a copy to the other party. If a monetary claim is made in the answer the appropriate administrative fee provided in the Fee Schedule shall be forwarded to the AAA with the answer. If no answer is filed within the stated time, it will be treated as a denial of the claim. Failure to file an answer shall not operate to delay the arbitration.

Section 8.   CHANGE OF CLAIM OR COUNTER-CLAIM — After filing of the claim or counterclaim, if either party desires to make any new or different claim or counterclaim, same shall be made in writing and filed with the AAA, and a copy thereof shall be mailed to the other party who shall have a period of seven days from the date of such mailing within which to file an answer with the AAA. However, after the arbitrator is appointed no new or different claim or counterclaim may be submitted without the arbitrator's consent.

Section 9.   INITIATION UNDER A SUBMIS-SION — Parties to any existing dispute may commence an arbitration under these Rules by filing at any Regional Office two (2) copies of a written agreement to arbitrate under these Rules (Submission), signed by the parties. It shall contain a statement of the matter in dispute, the amount of money involved, and the remedy sought, together with the appropriate filing fee as provided in the Fee Schedule.

Section 10.   PRE-HEARING CONFERENCE — At the request of the parties or at the discretion of the AAA a pre-hearing conference with the administrator and the parties or their counsel will be scheduled in appropriate cases to arrange for an exchange of information and the stipulation of uncontested facts so as to expedite the arbitration proceedings.

Section 11.   FIXING OF LOCALE — The parties may mutually agree on the locale where the arbitration is to be held. If any party requests that the hearing be held in a specific locale and the other party files no objection thereto within seven days after notice of the request is mailed to such party, the locale shall be the one requested. If a party objects to the locale requested by the other party, the AAA shall have power to determine the locale and its decision shall be final and binding.

Section 12.   QUALIFICATIONS OF ARBITRA-TOR — Any arbitrator appointed pursuant to Section 13 or Section 15 shall be neutral, subject to disqualification for the reasons specified in Section 19. If the agreement of the parties names an arbitrator or specifies any other method of appointing an arbitrator, or if the parties specifically agree in writing, such arbitrator shall not be subject to disqualification for said reasons.

Section 13.   APPOINTMENT FROM PANEL — If the parties have not appointed an arbitrator and have not provided any other method of appointment,

6

the arbitrator shall be appointed in the following manner: Immediately after the filing of the Demand or Submission, the AAA shall submit simultaneously to each party to the dispute an identical list of names of persons chosen from the Panel. Each party to the dispute shall have seven days from the mailing date in which to cross off any names to which it objects, number the remaining names to indicate the order of preference, and return the list to the AAA. If a party does not return the list within the time specified, all persons named therein shall be deemed acceptable. From among the persons who have been approved on both lists, and in accordance with the designated order of mutual preference, the AAA shall invite the acceptance of an arbitrator to serve. If the parties fail to agree upon any of the persons named, or if acceptable arbitrators are unable to act, or if for any other reason the appointment cannot be made from the submitted lists, the AAA shall have the power to make the appointment from other members of the Panel without the submission of any additional lists.

Section 14. DIRECT APPOINTMENT BY PARTIES — If the agreement of the parties names an arbitrator or specifies a method of appointing an arbitrator, that designation or method shall be followed. The notice of appointment, with name and address of such arbitrator, shall be filed with the AAA by the appointing party. Upon the request of any such appointing party, the AAA shall submit a list of members of the Panel from which the party may make the appointment.

If the agreement specifies a period of time within which an arbitrator shall be appointed, and any party fails to make such appointment within that period, the AAA shall make the appointment.

If no period of time is specified in the agreement, the AAA shall notify the parties to make the appointment, and if within seven days after mailing of such notice such arbitrator has not been so appointed, the AAA shall make the appointment.

Section 15. APPOINTMENT OF ARBITRATOR BY PARTY-APPOINTED ARBITRATORS — If the parties have appointed their party-appointed arbitrators or if either or both of them have been appointed as provided in Section 14, and have authorized such arbitrator to appoint an arbitrator within a specified time and no appointment is made within such time or any agreed extension thereof, the AAA shall appoint an arbitrator who shall act as Chairperson.

If no period of time is specified for appointment of the third arbitrator and the party-appointed arbitrators do not make the appointment within seven days from the date of the appointment of the last party-appointed arbitrator, the AAA shall appoint the arbitrator who shall act as Chairperson.

If the parties have agreed that their party-appointed arbitrators shall appoint the arbitrator from the Panel, the AAA shall furnish to the party-appointed arbitrators, in the manner prescribed in Section 13, a list selected from the Panel, and the appointment of the arbitrator shall be made as prescribed in such Section.

7

Section 16. NATIONALITY OF ARBITRATOR IN INTERNATIONAL ARBITRATION — If one of the parties is a national or resident of a country other than the United States, the arbitrator shall, upon the request of either party, be appointed from among the nationals of a country other than that of any of the parties.

Section 17. NUMBER OF ARBITRATORS — If the arbitration agreement does not specify or the parties are unable to agree as to the number of arbitrators, the dispute shall be heard and determined by three arbitrators, unless the AAA, in its discretion, directs that a single arbitrator or a greater number of arbitrators be appointed.

Section 18. NOTICE TO ARBITRATOR OF APPOINTMENT — Notice of the appointment of the arbitrator, whether mutually appointed by the parties or appointed by the AAA, shall be mailed to the arbitrator by the AAA, together with a copy of these Rules, and the signed acceptance of the arbitrator shall be filed prior to the opening of the first hearing.

Section 19. DISCLOSURE AND CHALLENGE PROCEDURE — A person appointed as neutral arbitrator shall disclose to the AAA any circumstances likely to affect his or her impartiality, including any bias or any financial or personal interest in the result of the arbitration or any past or present relationship with the parties or their counsel. Upon receipt of such information from such arbitrator or other source, the AAA shall communicate such information to the parties, and, if it deems it appropriate to do so, to the arbitrator and others. Thereafter, the AAA shall determine whether the arbitrator should be disqualified and shall inform the parties of its decision, which shall be conclusive.

Section 20. VACANCIES — If any arbitrator should resign, die, withdraw, refuse, be disqualified or be unable to perform the duties of office, the AAA shall, on proof satisfactory to it, declare the office vacant. Vacancies shall be filled in accordance with the applicable provision of these Rules. In the event of a vacancy in a panel of neutral arbitrators, the remaining arbitrator or arbitrators may continue with the hearing and determination of the controversy, unless the parties agree otherwise.

Section 21. TIME AND PLACE — The arbitrator shall fix the time and place for each hearing. The AAA shall mail to each party notice thereof at least five days in advance, unless the parties by mutual agreement waive such notice or modify the terms thereof.

Section 22 REPRESENTATION BY COUNSEL — Any party may be represented by counsel. A party intending to be so represented shall notify the other party and the AAA of the name and address of counsel at least three days prior to the date set for the hearing at which counsel is first to appear. When an arbitration is initiated by counsel, or where an attorney replies for the other party, such notice is deemed to have been given.

Section 23. STENOGRAPHIC RECORD — The AAA shall make the necessary arrangements for the taking of a stenographic record whenever such record is requested by a party. The requesting party

8

or parties shall pay the cost of such record as provided in Section 50.

Section 24. INTERPRETER — The AAA shall make the necessary arrangements for the services of an interpreter upon the request of one or both parties, who shall assume the cost of such services.

Section 25. ATTENDANCE AT HEARINGS — Persons having a direct interest in the arbitration are entitled to attend hearings. The arbitrator shall otherwise have the power to require the retirement of any witness or witnesses during the testimony of other witnesses. It shall be discretionary with the arbitrator to determine the propriety of the attendance of any other persons.

Section 26. ADJOURNMENTS — The arbitrator may adjourn the hearing, and must take such adjournment when all of the parties agree thereto.

Section 27. OATHS — Before proceeding with the first hearing or with the examination of the file, each arbitrator may take an oath of office, and if required by law, shall do so. The arbitrator may require witnesses to testify under oath administered by any duly qualified person or, if required by law or demanded by either party, shall do so.

Section 28. MAJORITY DECISION — Whenever there is more than one arbitrator, all decisions of the arbitrators must be by at least a majority. The award must also be made by at least a majority unless the concurrence of all is expressly required by the arbitration agreement or by law.

Section 29. ORDER OF PROCEEDINGS — A hearing shall be opened by the filing of the oath of the arbitrator, where required, and by the recording of the place, time and date of the hearing, the presence of the arbitrator and parties, and counsel, if any, and by the receipt by the arbitrator of the statement of the claim and answer, if any.

The arbitrator may, at the beginning of the hearing, ask for statements clarifying the issues involved.

The complaining party shall then present its claims, proofs and witnesses, who shall submit to questions or other examination. The defending party shall then present its defenses, proofs and witnesses, who shall submit to questions or other examination. The arbitrator may vary this procedure but shall afford full and equal opportunity to the parties for the presentation of any material or relevant proofs.

Exhibits, when offered by either party, may be received in evidence by the arbitrator.

The names and addresses of all witnesses and exhibits in order received shall be made a part of the record.

Section 30. ARBITRATION IN THE ABSENCE OF A PARTY — Unless the law provides to the contrary, the arbitration may proceed in the absence of any party, who, after due notice, fails to be present or fails to obtain an adjournment. An award shall not be made solely on the default of a party. The arbitrator shall require the party who is present to submit such evidence as deemed necessary for the making of an award.

9

Section 31. EVIDENCE — The parties may offer such evidence as they desire and shall produce such additional evidence as the arbitrator may deem necessary to an understanding and determination of the dispute. An arbitrator authorized by law to subpoena witnesses or documents may do so upon the request of any party, or independently. The arbitrator shall be the judge of the admissibility of the evidence offered and conformity to legal rules of evidence shall not be necessary. All evidence shall be taken in the presence of all of the arbitrators and all of the parties, except where any of the parties is absent in default or has waived his or her right to be present.

Section 32. EVIDENCE BY AFFIDAVIT AND FILING OF DOCUMENTS — The arbitrator may receive and consider the evidence of witnesses by affidavit, giving it such weight as seems appropriate after consideration of any objections made to its admission.

All documents not filed with the arbitrator at the hearing, but arranged for at the hearing or subsequently by agreement of the parties, shall be filed with the AAA for transmission to the arbitrator. All parties shall be afforded opportunity to examine such documents.

Section 33. INSPECTION OR INVESTIGATION — An arbitrator finding it necessary to make an inspection or investigation in connection with the arbitration shall direct the AAA to so advise the parties. The arbitrator shall set the time and the AAA shall notify the parties thereof. Any party who so desires may be present at such inspection or investigation. In the event that one or both parties are not present at the inspection or investigation, the arbitrator shall make a verbal or written report to the parties and afford them an opportunity to comment.

Section 34. CONSERVATION OF PROPERTY — The arbitrator may issue such orders as may be deemed necessary to safeguard the property which is the subject matter of the arbitration without prejudice to the rights of the parties or to the final determination of the dispute.

Section 35. CLOSING OF HEARINGS — The arbitrator shall specifically inquire of the parties whether they have any further proofs to offer or witnesses to be heard. Upon receiving negative replies, the arbitrator shall declare the hearings closed and a minute thereof shall be recorded. If briefs are to be filed, the hearings shall be declared closed as of the final date set by the arbitrator for the receipt of briefs. If documents are to be filed as provided for in Section 32 and the date set for their receipt is later than that set for the receipt of briefs, the later date shall be the date of closing the hearing. The time limit within which the arbitrator is required to make an award shall commence to run, in the absence of other agreements by the parties, upon the closing of the hearings.

Section 36. REOPENING OF HEARINGS — The hearings may be reopened by the arbitrator at will, or upon application of a party at any time be-

10

fore the award is made. If the reopening of the hearing would prevent the making of the award within the specific time agreed upon by the parties in the contract out of which the controversy has arisen, the matter may not be reopened, unless the parties agree upon the extension of such time limit. When no specific date is fixed in the contract, the arbitrator may reopen the hearings, and the arbitrator shall have thirty days from the closing of the reopened hearings within which to make an award.

Section 37. WAIVER OF ORAL HEARINGS — The parties may provide, by written agreement, for the waiver of oral hearings. If the parties are unable to agree as to the procedure, the AAA shall specify a fair and equitable procedure.

Section 38. WAIVER OF RULES — Any party who proceeds with the arbitration after knowledge that any provision or requirement of these Rules has not been complied with and who fails to state an objection thereto in writing, shall be deemed to have waived the right to object.

Section 39. EXTENSIONS OF TIME — The parties may modify any period of time by mutual agreement. The AAA for good cause may extend any period of time established by these Rules, except the time for making the award. The AAA shall notify the parties of any such extension of time and its reason therefor.

Section 40. COMMUNICATION WITH ARBITRATOR AND SERVING OF NOTICES — There shall be no communication between the parties and an arbitrator other than at oral hearings. Any other oral or written communications from the parties to the arbitrator shall be directed to the AAA for transmittal to the arbitrator.

Each party to an agreement which provides for arbitration under these Rules shall be deemed to have consented that any papers, notices or process necessary or proper for the initiation or continuation of an arbitration under these Rules and for any court action in connection therewith or for the entry of judgment on any award made thereunder may be served upon such party by mail addressed to such party or its attorney at the last known address or by personal service, within or without the state wherein the arbitration is to be held (whether such party be within or without the United States of America), provided that reasonable opportunity to be heard with regard thereto has been granted such party.

Section 41. TIME OF AWARD — The award shall be made promptly by the arbitrator and, unless otherwise agreed by the parties, or specified by law, not later than thirty days from the date of closing the hearings, or if oral hearings have been waived, from the date of transmitting the final statements and proofs to the arbitrator.

Section 42. FORM OF AWARD — The award shall be in writing and shall be signed either by the sole arbitrator or by at least a majority if there be

11

more than one. It shall be executed in the manner required by law.

Section 43. SCOPE OF AWARD -- The arbitrator may grant any remedy or relief which is just and equitable and within the terms of the agreement of the parties. The arbitrator, in the award, shall assess arbitration fees and expenses as provided in Sections 48 and 50 equally or in favor of any party and, in the event any administrative fees or expenses are due the AAA, in favor of the AAA.

Section 44. AWARD UPON SETTLEMENT -- If the parties settle their dispute during the course of the arbitration, the arbitrator, upon their request, may set forth the terms of the agreed settlement in an award.

Section 45. DELIVERY OF AWARD TO PARTIES — Parties shall accept as legal delivery of the award the placing of the award or a true copy thereof in the mail by the AAA, addressed to such party or its attorney at the last known address or by personal service, within or without the state wherein the arbitration is to be held (whether such party be within or without the United States of America), provided that reasonable opportunity to be heard with regard thereto has been granted such party.

Section 46. RELEASE OF DOCUMENTS FOR JUDICIAL PROCEEDINGS — The AAA shall, upon the written request of a party, furnish to such party, at its expense, certified facsimiles of any papers in the AAA's possession that may be required in judicial proceedings relating to the arbitration.

Section 47. APPLICATIONS TO COURT — No judicial proceedings by a party relating to the subject matter of the arbitration shall be deemed a waiver of the party's right to arbitrate.

The AAA is not a necessary party in judicial proceedings relating to the arbitration.

Parties to these Rules shall be deemed to have consented that judgment upon the award rendered by the arbitrator(s) may be entered in any Federal or State Court having jurisdiction thereof.

Section 48. ADMINISTRATIVE FEES — As a nonprofit organization, the AAA shall prescribe an administrative fee schedule and a refund schedule to compensate it for the cost of providing administrative services. The schedule in effect at the time of filing or the time of refund shall be applicable.

The administrative fees shall be advanced by the initiating party or parties in accordance with the administrative fee schedule, subject to final apportionment by the arbitrator in the award.

When a matter is withdrawn or settled, the refund shall be made in accordance with the refund schedule.

The AAA, in the event of extreme hardship on the part of any party, may defer or reduce the administrative fee.

Section 49. FEE WHEN ORAL HEARINGS ARE WAIVED — Where all oral hearings are waived under Section 37 the Administrative Fee Schedule shall apply.

12

Section 50. EXPENSES — The expenses of witnesses for either side shall be paid by the party producing such witnesses.

The cost of the stenographic record, if any is made, and all transcripts thereof, shall be prorated equally between the parties ordering copies unless they shall otherwise agree and shall be paid for by the responsible parties directly to the reporting agency.

All other expenses of the arbitration, including required traveling and other expenses of the arbitrator and of AAA representatives, and the expenses of any witness or the cost of any proofs produced at the direct request of the arbitrator, shall be borne equally by the parties, unless they agree otherwise, or unless the arbitrator in the award assesses such expenses or any part thereof against any specified party or parties.

Section 51. ARBITRATOR'S FEE — Unless the parties agree to terms of compensation, members of the National Panel of Construction Arbitrators will serve without compensation for the first two days of service.

Thereafter, compensation shall be based upon the amount of service involved and the number of hearings. An appropriate daily rate and other arrangements will be discussed by the administrator with the parties and the arbitrator(s). If the parties fail to agree to the terms of compensation, an appropriate rate shall be established by the AAA, and communicated in writing to the parties.

Any arrangement for the compensation of an arbitrator shall be made through the AAA and not directly by the arbitrator with the parties. The terms of compensation of neutral arbitrators on a Tribunal shall be identical.

Section 52. DEPOSITS — The AAA may require the parties to deposit in advance such sums of money as it deems necessary to defray the expense of the arbitration, including the arbitrator's fee if any, and shall render an accounting to the parties and return any unexpended balance.

Section 53. INTERPRETATION AND APPLICATION OF RULES — The arbitrator shall interpret and apply these Rules insofar as they relate to the arbitrator's powers and duties. When there is more than one arbitrator and a difference arises among them concerning the meaning or application of any such Rules, it shall be decided by a majority vote. If that is unobtainable, either an arbitrator or a party may refer the question to the AAA for final decision. All other Rules shall be interpreted and applied by the AAA.

13

ADMINISTRATIVE FEE SCHEDULE

A filing fee of $150 will be paid at the time the case is initiated.

The balance of the administrative fee of the AAA is based upon the amount of each claim and counterclaim as disclosed when the claim and counterclaim are filed, and is due and payable prior to the notice of appointment of the neutral arbitrator.

In those claims and counterclaims which are not for a monetary amount, an appropriate administrative fee will be determined by the AAA, payable prior to such notice of appointment.

Amount of Claim or Counterclaim	Fee for Claim or Counterclaim
Up to $10,000	3% (minimum $150)
$ 10,000 to $ 25,000	$ 300, plus 2% of excess over $10,000
$ 25,000 to $100,000	$ 600, plus 1% of excess over $25,000
$100,000 to $200,000	$1,350, plus ½% of excess over $100,000
$200,000 to $5,000,000	$1,850, plus ¼% of excess over $200,000

Where the claim or counterclaim exceeds $5 million, an appropriate fee will be determined by the AAA. If there are more than two parties represented in the arbitration, an additional 10% of the administrative fee will be due for each additional represented party.

OTHER SERVICE CHARGES

$50.00 payable by a party causing an adjournment of any scheduled hearing;

$100.00 payable by a party causing a second or additional adjournment of any scheduled hearing;

$25.00 payable by each party for each second and subsequent hearing which is either clerked by the AAA or held in a hearing room provided by the AAA.

REFUND SCHEDULE

If the AAA is notified that a case has been settled or withdrawn before it mails a notice of appointment to a neutral arbitrator, all of the fee in excess of $150 will be refunded.

If the AAA is notified that a case is settled or withdrawn thereafter but at least 48 hours before the date and time set for the first hearing, one-half of the fee in excess of $150 will be refunded.

14

APPENDIX L

STANDARD SUBCONTRACT AGREEMENT FOR BUILDING CONSTRUCTION*

THE ASSOCIATED GENERAL CONTRACTORS

STANDARD SUBCONTRACT AGREEMENT FOR BUILDING CONSTRUCTION

This Document has important legal and insurance consequences; consultation with an attorney and insurance consultants and carriers is encouraged with respect to its completion or modification.

THIS AGREEMENT made at

this   day of   , 19   , by

and between

hereinafter referred to as the Contractor, and

hereinafter referred to as the Subcontractor, to perform part of the Work on the following Project:

PROJECT:

OWNER:

ARCHITECT:

AGC DOCUMENT NO. 5   STANDARD SUBCONTRACT AGREEMENT FOR BUILDING CONSTRUCTION   APRIL 1980
©1980 ASSOCIATED GENERAL CONTRACTORS OF AMERICA

ARTICLE 1

Scope of Work

1.1   The Contractor employs the Subcontractor as an independent contractor, to perform the following part of the Work which the Contractor has contracted with the Owner to provide on the Project:

The Subcontractor agrees to perform such part of the Work (hereinafter called "Subcontractor's Work") under the general direction of the Contractor and subject to the final approval of the Architect/Engineer or other specified representative of the Owner, in accordance with the Contract Documents. Subcontractor will furnish all of the labor and materials, along with competent supervision, shop drawings and samples, tools, equipment, scaffolding, and permits which are necessary for such performance.

1.2   The Contract Documents are:

The Subcontrator binds himself to the Contractor for the performance of Subcontractor's Work in the same manner as the Contractor is bound to the Owner for such performance under Contractor's contract with the Owner. The pertinent parts of such contract will be made available upon Subcontractor's request.

1.3   Should any question arise with respect to the interpretation of the drawings and specifications, such questions shall be submitted to the Architect/Engineer and his decision shall be final and binding. If there is no Architect/Engineer for this Project, the Contractor's decision shall be followed by the Subcontractor.

ARTICLE 2

Payments

2.1   The Contractor agrees to pay to the Subcontractor for the satisfactory completion of Subcontractor's Work the sum of _____ ($ _____) in monthly payments of _____ percent of the work performed in any preceding month, in accordance with estimates prepared by the Subcontractor and approved by the Contractor and _____
Payments made on account of materials not incorporated in the work, but delivered and suitably stored at the site, or at some other location agreed upon in writing, shall be in accordance with the terms and conditions of the Contract Documents. Subcontractor will provide monthly completed lien waivers and supplier affidavit forms, in a form satisfactory to the Owner and Contractor. Payment of the approved portion of the Subcontractor's monthly estimate shall be conditioned upon receipt by the Contractor of his payment from the Owner. Approval and payment of Subcontractor's monthly estimate is specifically agreed not to constitute or imply acceptance by the Contractor or Owner of any portion of the Subcontractor's Work.

2.2   In the event the Subcontractor does not submit to the Contractor such monthly estimates by _____ then the Contractor may at his option include in his monthly estimate to the Owner for Work performed during the preceding month such amount as he may deem proper for the Work of the Subcontractor for the preceding month and the Subcontractor agrees to accept such approved portion thereof in lieu of monthly payment based upon the Subcontractor's estimate.

2.3   In the event it appears to the Contractor that the labor, material and other bills incurred in the performance of Subcontractor's Work are not being currently paid, the Contractor may take such steps as he deems necessary to insure that the money paid with any progress payment will be utilized to pay such bills.

2.4   Final payment shall be paid to the Subcontractor upon approval by the Owner, Architect and the Contractor of the Subcontractor's Work and, upon payment having been received by the Contractor for all of Subcontractor's Work and satisfactory evidence having been received by the Contractor that all labor, including customary fringe benefits and payments due under collective bargaining agreements, and all subcontractors and materialmen have been paid to date and are waiving their lien rights upon the final payment of a specific balance due.

2.5   The Contractor may deduct from any amounts due or to become due to the Subcontractor any sum or sums owing by the Subcontractor to the Contractor; and in the event of any breach by the Subcontractor of any provision or obligation of this Subcontract, or in the event of the assertion by other parties of any claim or lien against the Owner, the Contractor, Contractor's Surety, or the premises upon which the Work was performed, which claim or lien arises out of the Subcontractor's performance of this Agreement, the Contractor shall have the right, but is not required, to retain out of any payments due or to become due to the Subcontractor an amount sufficient to completely protect the Contractor from any and all loss, damage or expense therefrom, until the claim or lien has been adjusted by the Subcontractor to the satisfaction of the Contractor. This paragraph shall be applicable even though the Subcontractor has posted a full payment and performance bond.

ARTICLE 3

Prosecution of the Work

3.1   Time is of the essence for both parties, and they mutually agree to see to the performance of their Work and the Work of their subcontractors so that the entire project may be completed in accordance with the Contract Documents. The Subcontractor shall provide the Contractor with scheduling information and Subcontractor's proposed schedule for the Subcontractor's Work. The Contractor shall then prepare the Schedule of the Work and, as may be necessary, revise such schedule as the Work progresses. Subcontractor acknowledges that revisions may be made in such schedule and agrees to make no claim for acceleration or delay by reason of such revisions so long as such revisions are of the type normally experienced in Work of this scope and complexity.

3.2   The Subcontractor shall prosecute Subcontractor's Work in a prompt and diligent manner in accordance with the Schedule of Work without hindering the Work of the Contractor or any other subcontractor. If work of others is damaged by Subcontractor, the Subcontractor will cause such damage to be corrected to the satisfaction of and without cost to the Contractor and Owner. In the event Subcontractor fails to maintain his part of the Schedule of the Work, he shall, without additional compensation, work such overtime as the Contractor may direct until Subcontractor's Work is in accordance with such schedule.

3.3   The Subcontractor shall be responsible for and will prepare for performance of Subcontractor's Work, including without limitation thereto, the submission of shop drawings, samples, tests, field dimensions, determination of labor requirements and ordering of materials as required to meet the Schedule of Work. Subcontractor shall notify Contractor when portions of his Work are ready for inspection.

3.4   The Subcontractor will furnish periodic progress reports of the Subcontractor's Work as mutually agreed including the progress of materials or equipment to be provided under this Agreement that may be in the course of preparation or manufacture.

3.5   The Subcontractor shall cooperate with the Contractor and subcontractors whose work may interfere with the Subcontractor's Work and participate in the preparation of coordinated drawings and work schedules in areas of congestion, specifically noting and advising the Contractor of any interference by other contractors or subcontractors.

3.6   The Subcontractor shall keep the building and premises reasonably clean of debris resulting from the performance of Subcontractor's Work. If the Subcontractor fails to comply with this paragraph within 48 hours after receipt of notice of noncompliance from the Contractor, the Contractor may perform such necessary clean-up and deduct the cost from any amounts due to the Subcontractor.

3.7   The Subcontractor shall give adequate notices pertaining to the Work of the Subcontractor to proper authorities and secure and pay for all necessary licenses and permits to carry on Subcontractor's Work, the furnishing of which is required by the Contract Documents.

3.8 The Subcontractor shall comply with all Federal, State and local laws, Social Security Laws and Unemployment Compensation Laws, Workers' Compensation Laws and Safety Laws insofar as applicable to the performance of this Agreement. He shall pay all taxes applicable to the performance of Subcontractor's Work. He shall also maintain his own safety program for compliance with such laws.

3.9 The Subcontractor will not assign this subcontract nor subcontract the whole or any part of the Work to be performed hereunder without the prior written consent of the Contractor, with the exception of those subcontractors listed by the Subcontractor and furnished to the Contractor at the time this Agreement is executed.

ARTICLE 4

Changes in the Work

4.1 The Contractor and Subcontractor agree that the Contractor may add to or deduct from the amount of Work covered by this Agreement, and any changes so made in the amount of Work involved, or any other parts of this Agreement, shall be by a written amendment hereto setting forth in detail the changes involved and the value thereof which shall be mutually agreed upon between the Contractor and Subcontractor. The Subcontractor agrees to proceed with the Work as changed when so ordered in writing by the Contractor so as not to delay the progress of the Work, and pending any determination of the value thereof unless Contractor first requests a proposal of cost before the change is effected. If the Contractor requests a proposal of cost for a change, the Subcontractor shall promptly comply with such request.

4.2 Subcontractor shall be entitled to receive no extra compensation for extra Work or materials or changes of any kind regardless of whether the same was ordered by the Contractor or any of his representatives unless a Change Order therefor has been issued in writing by the Contractor. If extra work was ordered by the Contractor and the Subcontractor performed same but did not receive a written order therefor, the Subcontractor shall be deemed to have waived any claim for extra compensation therefor, regardless of any written or verbal protests or claims by the Subcontractor. The Subcontractor shall be responsible for any costs incurred by the Contractor for changes of any kind made by the Subcontractor that increase the cost of the work for either the Contractor or other subcontractors when the Subcontractor proceeds with such changes without a written order therefor.

4.3 The Subcontractor agrees that no claim for additional services rendered or materials furnished by the Subcontractor to the Contractor shall be valid unless notice is given to the Contractor prior to the furnishing of the services or material or unless written notice of the claim therefor is given by the Subcontractor to the Contractor not later than the last day of the calendar month following that in which the claim originated, with the amount of the claim to be given in writing by the Subcontractor as soon as practicable.

4.4 The Subcontractor will make all claims for extra compensation and for extension of time to the Contractor promptly in accordance with this Article and consistent with the Contract Documents.

4.5 Notwithstanding any other provision, if the Work for which the Subcontractor claims extra compensation is determined by the Owner or Architect not to entitle the Contractor to a Change Order or extra compensation, then the Contractor shall not be liable to the Subcontractor for any extra compensation for such Work, unless Contractor agreed in writing to such extra compensation.

ARTICLE 5

Insurance and Indemnity

5.1 Prior to starting Work the Subcontractor shall procure and maintain in force, Workers' Compensation Insurance, Employers Liability Insurance, Comprehensive General Liability Insurance with contractual coverage and Automobile Liability Insurance and such other insurance, to the extent required by the Contract Documents for the Subcontractor's Work.

5.2 The Subcontractor's Comprehensive General and Automobile Liability Insurance, as required by Paragraph 5.1 shall be written for not less than limits of liability as follows:

AGC DOCUMENT NO. 5 • STANDARD SUBCONTRACT AGREEMENT FOR BUILDING CONSTRUCTION • APRIL 1980

a. Comprehensive General Liability

    1. Bodily Injury $ _____ Each Occurrence
                                                       (Completed Operations)

                                 $ _____ Aggregate

    2. Property Damage $ _____ Each Occurrence

                                 $ _____ Aggregate

b. Comprehensive Automobile Liability

    1. Bodily Injury $ _____ Each Person

                                 $ _____ Each Occurrence

    2. Property Damage $ _____ Each Occurence

5.3 Comprehensive General Liability Insurance may be arranged under a single policy for the full limits required or by a combination of underlying policies with the balance provided by an Excess or Umbrella Liability policy.

5.4 The foregoing policies shall contain a provision that coverages afforded under the policies will not be cancelled or not renewed until at least thirty (30) days' prior written notice has been given to the Contractor. Certificates of Insurance acceptable to the Contractor shall be filed with the Contractor prior to the commencement of Work.

5.5 The Contractor and Subcontractor waive all rights against each other and against the Owner, the Architect/Engineer, separate contractors, and all other subcontractors for damages caused by fire or other perils to the extent covered by Builder's Risk or any other property insurance, except such rights as they may have to the proceeds of such insurance.

5.6 To the fullest extent permitted by law, the Subcontractor agrees to indemnify and hold harmless the Contractor, the Owner, the Architect/Engineer and all of their agents and employees from and against all claims, damages, losses and expenses, including but not limited to attorney's fees, arising out of or resulting from the performance, or failure in performance, of the Subcontractor's Work under this Subcontract, provided that any such claim, damage, loss or expense (1) is attributable to bodily injury, sickness, disease, or death, or to injury to or destruction of tangible property (other than the Work itself) including the loss of use resulting therefrom, and (2) is caused in whole or in part by any negligent act or omission of the Subcontractor or anyone directly or indirectly employed by him or anyone for whose acts he may be liable regardless of whether it is caused in part by a party indemnified hereunder. Such obligations shall not be construed to negate, abridge, or otherwise reduce any other right or obligation of indemnity which would otherwise exist as to any party or person described in this Paragraph 5.6.

5.6.1 In any and all claims against the Contractor or any of his agents or employees by any employee of the Subcontractor, anyone directly or indirectly employed by him or anyone for whose acts he may be liable, the indemnification obligation under this Paragraph 5.6 shall not be limited in any way by any limitation on the amount or type of damages, compensation or benefits payable by or for the Subcontractor under Workers' Compensation acts, disability benefit acts or other employee benefit acts.

5.6.2 The obligations of the Subcontractor under this Paragraph 5.6 shall not extend to the liability of the Architect/Engineer, his agents or employees, arising out of (a) the preparation or approval of maps, drawings, opinions, reports, surveys, Change Orders, designs or specifications, or (b) the giving of or failure to give directions or instructions by the Architect/Engineer, his agents or employees, providing such giving or failure to give is the primary cause of the injury or damage.

ARTICLE 6

Performance Bond and Labor and Material Payment Bond

A Performance Bond and a Labor and Material Payment Bond in a form satisfactory to the Contractor shall be furnished in the full amount of this Agreement, if required by the Contractor. This obligation shall continue throughout the agreement and may be required at any time during the performance of Subcontractor's Work by a change under Article 4.

ARTICLE 7

Warranty

The Subcontractor agrees to promptly make good without cost to the Owner or Contractor any and all defects due to faulty workmanship and/or materials which may appear within the guarantee or warranty period so established in the Contract Documents; and if no such period be stipulated in the Contract Documents, then such guarantee shall be for a period of one year from date of completion and acceptance of the project by the Owner. The Subcontractor further agrees to execute any special guarantees as provided by the terms of the Contract Documents, prior to final payment.

ARTICLE 8

Contractors' Obligations

8.1   The Contractor agrees to be bound to the Subcontractor by all the obligations that the Owner assumes to the Contractor under the Contract Documents and by all provisions thereof affording remedies and redress to the Contractor from the Owner insofar as applicable to this Agreement.

8.2   Upon request, the Contractor will give the Subcontractor written authorization to obtain direct from the Architect/Engineer or Owner's authorized agent, evidence of amount and percentages of completion certified on his account.

8.3   The Contractor shall not issue or give any instruction, order or directions directly to employees or workmen of the Subcontractor other than to the persons designated as the authorized representative(s) of the Subcontractor.

8.4   The Contractor shall make no demand for liquidated damages in any sum in excess of the amount specifically named in this Agreement or the Contract Documents. Liquidated damages shall not be assessed for delays not caused by the Subcontractor. Liquidated damages, when assessed, shall not exceed the Subcontractor's proportionate share of the responsibility for such delay. This provision does not preclude any claim the Contractor may have for direct damages under law.

8.5   The Subcontractor will furnish those temporary facilities and services required by the Subcontractor except for those to be provided by the Contractor set forth in the Attachment A to this Agreement. Adequate storage areas, if available, will be allocated by the Contractor for the Subcontractor's materials and equipment during the course of the Work.

8.6   The Contractor agrees that no claim for services rendered or materials furnished by the Contractor to the Subcontractor shall be valid unless notice is given to the Subcontractor prior to furnishing of the services or material or unless written notice of the claim therefor is given by the Contractor to the Subcontractor not later than the last day of the calendar month following that in which the claim originated, with the amount of the claim to be given in writing by the Contractor as soon as practicable.

ARTICLE 9

Termination

9.1   Should the Subcontractor fail at any time to supply a sufficient number of properly skilled workmen or sufficient materials and equipment of the proper quality, or fail in any respect to prosecute the Work with promptness and diligence, or fail to promptly correct defective Work or fail in the performance of any of the agreements herein contained, the Contractor may, at his option, provide such labor, materials and equipment and to deduct the cost thereof, together with all loss or damage occasioned thereby, from any money then due or thereafter to become due to the Subcontractor under this Agreement.

9.2   If the Subcontractor at any time shall refuse or neglect to supply sufficient properly skilled workmen, or materials or equipment of the proper quality and quantity, or fail in any respect to prosecute Subcontractor's Work with promptness and diligence, or cause by any action or omission the stoppage or interference with the work of the Contractor or other subcontractors, or fail in the performance of any of the covenants herein contained, or be unable to meet his debts as they mature, the Contractor may at his option at any time after serving written notice of such default with direction to cure in a specific period, but not less than two (2) working days, and the Subcontractor's failure to cure the default, terminate the Subcontractor's employment by delivering written notice of termination to the Subcontractor. Thereafter, the Contractor may take possession of the plant and work, materials, tools, appliances and equipment of the Subcontractor at the building site, and through himself or others provide labor, equipment and materials to prosecute Subcontractor's Work on such terms and conditions as shall be deemed necessary, and shall deduct the cost thereof, including without restriction thereto all charges, expenses, losses, costs, damages, and attorney's fees, incurred as a result of the Subcontractor's failure to perform, from any money then due or thereafter to become due to the Subcontractor under this Agreement.

9.3   If the Contrator so terminates the employment of the Subcontractor, the Subcontractor shall not be entitled to any further payments under this agreement until Subcontractor's Work has been completed and accepted by the Owner, and payment has been received by the Contractor from the Owner with respect thereto. In the event that the unpaid balance due exceeds the Contractor's cost of completion, the difference shall be paid to the Subcontractor; but if such expense exceeds the balance due, the Subcontractor agrees promptly to pay the difference to the Contractor.

ARTICLE 10

Claims

10.1   All claims, disputes and other matters in question arising out of, or relating to, this Subcontract or the breach thereof shall be decided by Arbitration in accordance with the Construction Industry Arbitration Rules of the American Arbitration Association then obtaining unless the parties mutually agree otherwise. This agreement to arbitrate shall be specifically enforceable under the prevailing arbitration law. The award rendered by the arbitrators shall be final, and judgment may be entered upon in accordance with applicable law in any court having jurisdiction thereof.

10.2   In the event the Contractor and Owner or others arbitrate matters relating to this Subcontract, it shall be the responsibility of the Subcontractor to prepare and present the Contractor's case, to the extent the proceedings are related to this Subcontract.

10.3   Should the Contractor enter into arbitration with the Owner or others regarding matters relating to this Agreement, the Subcontractor shall be bound by the result of the arbitration to the same degree as the Contractor.

10.4   The Subcontractor shall carry on Subcontractor's Work and maintain his progress during any arbitration proceedings.

ARTICLE 11

Prevailing Law

This Agreement shall be governed by the law in effect in _____

IN WITNESS WHEREOF the parties hereto have executed this Agreement under seal, the day and year first above written.

_____
Subcontractor

ATTEST:

_____   By_____
                                                               (Title)

_____
Contractor

ATTEST:

_____   By_____
                                                               (Title)

APPENDIX M

GENERAL LEDGER ACCOUNTS

ASSETS

10.		Petty Cash
11.		Bank Deposits
	.1	General Bank Account
	.2	Payroll Bank Account
	.3	Project Bank Accounts
	.4	
12.		Accounts Receivable
	.1	
	.2	Parent, Associated, or Affiliated Companies
	.3	Notes Receivable
	.4	Employees' Accounts
	.5	Sundry Debtors
	.6	
13.		Deferred Receivables

All construction contracts are charged to this account, being diminished by progress payments as received. This account is offset by Account 48.0, Deferred Income.

14.　　Property, Plant, and Equipment

Property and General Plant

.100　Real Estate and Improvements
.200　Leasehold Improvements
.300　Shops and Yards

Mobile Equipment

.400　Motor Vehicles
.500　Tractors
　　.01　Repairs, parts, and labor
　　.05　Outside service
　　.12　Tire replacement
　　.15　Tire repair
　　.20　Fuel
　　.25　Oil, lubricants, filters
　　.30　Licenses, permits
　　.35　Depreciation
　　.40　Insurance
　　.45　Taxes
.510　Power Shovels
.520　Bottom Dumps
.525

Stationary Equipment

.530 Concrete Mixing Plant
.540 Concrete Pavers
.550 Air Compressors
.560

Small Power Tools and Portable Equipment

.600 Welders
.610 Concrete Power Buggies
.620 Electric Drills
.630

Marine Equipment

.700

Miscellaneous Construction Equipment

.800 Scaffolding
.810 Concrete Forms
.820 Wheelbarrows
.830

Office and Engineering Equipment

.900 Office Equipment
.910 Office Furniture
.920 Engineering Instruments
.930

15. Reserve for Depreciation
16. Amortization for Leasehold
17. Inventory of Materials and Supplies
.1 Lumber
.2 Hand Shovels
.3 Spare Parts
.4

These accounts show the values of all expendable materials and supplies. Charges against these accounts are made by authenticated requisitions showing project where used.

18. Returnable Deposits
.1 Plan Deposits
.2 Utilities
.3
19. Prepaid Expenses
.1 Insurance
.2 Bonds
.3

LIABILITIES
40. Accounts Payable
41. Subcontracts Payable
42. Notes Payable
43. Interest Payable
44. Contracts Payable
45. Taxes Payable
.1 Old-Age, Survivors, and Disability Insurance (withheld from employees' pay)

.2   Federal Income Taxes (withheld from employees' pay)

.3   State Income Taxes (withheld from employees' pay)

.4

46.   Accrued Expenses

  .1   Wages and Salaries

  .2   Old-Age, Survivors, and Disability Insurance (employer's portion)

  .3   Federal Unemployment Tax

  .4   State Unemployment Tax

  .51   Payroll Insurance (public liability and property damage)

  .52   Payroll Insurance (workmen's compensation)

  .6   Interest

  .7

47.   Payrolls Payable

48.   Deferred Income

49.   Advances by Clients

NET WORTH

50.   Capital Stock

51.   Earned Surplus

52.   Paid-in Surplus

53.

INCOME

70.   Income Accounts

  .101   Project Income

  .102

  .2   Cash Discount Earned

  .3   Profit or Loss from Sale of Capital Assets

  .4   Equipment Rental Income

  .5   Interest Income

  .6   Other Income

EXPENSE

80.   Project Expense (Expenses directly chargeable to the projects. See Figure 12.1.)

  .100   Project Work Accounts

  .700   Project Overhead Accounts

           These are control accounts for the detail project cost accounts that are maintained in the detail cost ledgers.

81.   Office Expense

  .10   Officer Salaries

  .11   Insurance on Property and Equipment

  .20   Donations

  .21   Utilities

  .22   Telephone and Telegraph

  .23   Postage

  .30   Repairs and Maintenance

82.   Yard and Warehouse Expense (not assignable to a particular project)

  .10   Yard Salaries

  .11   Yard Supplies

83.   Estimating Department Expense Accounts

  .10   Estimating Salaries

  .11   Estimating Supplies

.12 Estimating Travel
84. Engineering Department Expense Accounts
.10
85. Cost of Equipment Ownership
.1 Depreciation
.2 Interest
.3 Taxes and Licenses
.4 Insurance
.5 Storage
86. Loss on Bad Debts
87. Interest
90. Expense on Office Employees
.1 Workmen's Compensation Insurance
.2 Old-Age, Survivors, and Disability Insurance
.3 Employees' Insurance
.4 Other Insurance
.5 Federal and State Unemployment Taxes
.6
91. Taxes and Licenses
.1 Sales Taxes
.2 Compensating Taxes
.3 State Income Taxes
.4 Federal Income Taxes

INDEX